Computer Animation

Computer Animation

Algorithms and Techniques

Third Edition

Rick Parent

Ohio State University

ELSEVIER

AMSTERDAM • BOSTON • HEIDELBERG • LONDON
NEW YORK • OXFORD • PARIS • SAN DIEGO
SAN FRANCISCO • SINGAPORE • SYDNEY • TOKYO
Morgan Kaufmann is an imprint of Elsevier

Acquiring Editor: Steven Elliot
Development Editor: Robyn Day
Project Manager: Paul Gottehrer
Designer: Joanne Blank

Morgan Kaufmann is an imprint of Elsevier
225 Wyman Street, Waltham, MA 02451, USA

Library of Congress Cataloging-in-Publication Data
Application submitted

British Library Cataloguing-in-Publication Data
A catalogue record for this book is available from the British Library.

ISBN: 978-0-12-415842-9

Printed and bound by CPI Group (UK) Ltd, Croydon, CR0 4YY

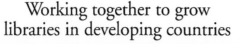

Working together to grow
libraries in developing countries

www.elsevier.com | www.bookaid.org | www.sabre.org

ELSEVIER   BOOK AID International   Sabre Foundation

For information on all MK publications visit our website at http://store.elsevier.com

To Kim and John, for teaching me to keep things in perspective.
And to my wife, Arlene, for her attention-to-detail approach to life, especially
when juxtaposed to my 'big picture' way of doing things.

Contents

Preface

Overview

This book surveys computer algorithms and programming techniques for specifying and generating motion for graphical objects, that is, *computer animation*. It is primarily concerned with three-dimensional (3D) computer animation. The main audience is advanced undergraduate or beginning graduate students in Computer Science. Computer graphics programmers who want to learn the basics of computer animation programming and artists who use software packages to generate computer animation (*digital animators*) who want to better understand the underlying computational issues of animation software will also benefit from this book.

It should come as no surprise to anyone reading this book that activity in Computer Animation has exploded in recent years - as a research area, as an academic field of study, as a career, and even as a hobby. Feature length films are now often stored digitally and incorporate digital special effects (often referred to as *computer generated imagery* and abbreviated *CGI*). As listed by the Internet Movie Database (imdb.com) [1] as of March 2012, all of the top 10 U.S. films (All-Time U.S. Box Office) depend on extensive use of CGI. Computer animated films have become top box office attractions - according to the same movie database, 2 of the top 10 feature length films are computer animations (*Shrek 2* and *Toy Story 3*) with a third having a significant computer animation component (*Avatar*). Recent *Technical Achievement* and *Scientific and Engineering* awards from the Motion Picture Academy of Arts and Sciences have been for digital image technology including render queue management, facial motion retargeting, tools to review digital effects, and efficient rendering of volumetric effects, just to name a few [2]. And, of course, the computer game industry has exploded. The Entertainment Software Association estimate that, in 2010, consumers spent $25.1 billion on video games, hardware and accessories [3].

Computer animation is more accessible that ever. Desktop, high-quality, computer animation is now possible because of sophisticated off-the-shelf animation software, cheap CPU cycles, and cheap storage coupled with digital video recording. Many technical programs and computer science departments now offer courses in computer animation and the proliferating artistic programs train digital artists in the use of off-the-shelf animation software. There are now major technical conferences and journals that archive developments in computer animation and video game algorithms and techniques.

This book addresses practical issues, provides accessible techniques, and offers straightforward implementations. Purely theoretical discussions have been avoided except to point out avenues of current and future research. In some cases, programming examples are complete working code segments—in C, which can be copied, compiled, and run to produce basic examples of the algorithms discussed; other programming examples are C-like pseudocode that can be translated into working code. C was chosen because it forms the common basis for languages such as C++ and Java, and it lends itself to illustrating the step-by-step nature of algorithms. The Appendixes cover basic material that the reader may find useful as a refresher as well as specific algorithms for use in implementations.

This text is not intended for animators using off-the-shelf animation software (except to the extent that it might help in understanding the underlying computations required for a particular technique). It does not attempt to cover the theory of computer animation, address the aesthetics of computer animation, or discuss the artistic issues involved in designing animations. It does not detail the production issues in the actual commercial enterprise of producing a finished piece of animation. And, finally, it does not address the issue of *computer-assisted animation,* which, for our purposes, is taken to mean the computerization of conventional hand-drawn techniques; for the most part, that area has its own set of separate issues [4] [5]. The book does concentrate on full 3D computer animation and identifies the useful algorithms and techniques that animators and programmers can use to move objects in interesting ways. While 3D techniques are the emphasis, 2D is not completely ignored.

The fundamental objective of computer animation programming is to select techniques and design tools that are expressive enough for animators to specify what they intend, yet at the same time are powerful enough to relieve animators from specifying any details they are not interested in. Obviously, no one tool is going to be right for every animator, for every animation, or even for every scene in a single animation. The appropriateness of a particular animation tool depends on the effect desired and the control required by the animator. An artistic piece of animation will usually require tools different from those required by an animation that simulates reality or educates a patient. In this spirit, alternative approaches are presented whenever possible.

Organization of the Book

This book presents background information in the first couple of chapters. Techniques that directly specify motion (*kinematic* - not based on underlying forces) are presented in the next 4 chapters followed by 2 chapters that cover force-based (*dynamics*) animation. Character animation is then covered in 3 chapters. The last chapter covers special geometric models. Appendices provide extensive support material. More detail about the chapters is given below.

Chapter 1 discusses general issues related to animation, including motion perception, the heritage of conventional animation paying particular attention to its technological innovations, overviews of animation production and computer animation production, and a snapshot of the ever-evolving history of computer animation. These provide a broad perspective of the art and craft that is animation.

Chapter 2 presents background material and reviews the basics of computer graphics necessary for animation. It reviews computational issues in computer graphics to ensure a solid background in the techniques that are important in understanding the remainder of the book. This includes a review of the rendering pipeline and a discussion of the ordering of transformations to reduce round-off errors that can creep into a series of calculations as one builds on another. A detailed section on quaternion representation of orientation is presented in this chapter as well. If the reader is well versed in computer graphics, this chapter may be skimmed to pick up relevant terminology or skipped altogether.

Chapters 3 and 4 cover interpolation. Chapter 3 presents the fundamentals. It introduces time-space curves, arc-length parameterization of a curve, and speed control along a curve. Interpolation of orientation with an emphasis on using quaternions is then covered. Various ways to work with paths are then presented. Chapter 4 presents animation techniques based on interpolation including key frame interpolation, animation languages, shape deformation, and shape interpolation including morphing.

Chapters 5 and 6 are primarily concerned with kinematic control of articulated figures. Chapter 5 is concerned with kinematics of linked appendages. It covers both forward and inverse kinematics. Chapter 6 covers the basics of motion capture (*mocap*). First, the basic technology is reviewed. Then the chapter discusses how the images are processed to reconstruct articulated figure kinematics, including some techniques to modify the resultant mocap data.

Chapters 7 and 8 cover animation that is more concerned with simulating real-world (e.g. physics-based) processes. Chapter 7 covers physics-based animation as well as mass-spring-damper systems, particle systems, rigid body dynamics, and enforcing constraints. It has an additional section on ways to model cloth. Chapter 8 covers the modeling and animation of fluids. It first covers models that handle specific macro-features of fluids and then covers computational fluid dynamics (CFD) as it relates to computer animation.

Chapters 9 through 11cover animation concerned with people and other critters. Chapter 9 covers human figure animation: modeling, reaching, walking, clothing, and hair. Chapter 10 covers facial animation: facial modeling, expressions, and lip-sync animation. Chapter 11 covers behavioral animation including flocking, predator–prey models, intelligent behavior and crowd behavior.

Finally, Chapter 12 covers a few special models that are useful to animation: implicit surfaces, L-systems, and subdivision surfaces.

Appendix A presents rendering issues often involved in producing images for computer animation: double buffering, compositing, computing motion blur, drop shadows, and billboarding. It assumes a general knowledge of the use of frame buffers, how a *z*-buffer display algorithm works, and aliasing.

Appendix B is a collection of relevant material from a variety of disciplines. It contains a survey of interpolation and approximation techniques, vector algebra and matrices, quaternion conversion code, the first principles of physics, several useful numeric techniques, optimization, and attributes of film, video, and image formats, and a few other topics.

The Web page associated with the book, containing images, code, and figures can be found at *textbooks.elsevier.com/9780125320009*.

Acknowledgments

Many people contributed in various ways to this book. First and foremost, I'd like to thank my wife, Arlene, who both tolerated my preoccupation with this project and greatly increased the readability of the book.

In general, I owe much to the students I have had the pleasure of knowing and/or working with over the years and whose collective interest in, knowledge of, and enthusiasm for the field has fueled my own. Their contribution cannot be overstated. These include Doug Roble, John Chadwick, Dave Haumann, Dave Ebert, Matt Lewis, Karan Singh, Steve May, James Hahn, Ferdi Scheepers, Dave Miller, Beth Hofer, Madhavi Muppala, Domin Lee, Kevin Rogers, Brent Watkins, Brad Winemiller, Meg Geroch, Lawson Wade, Arun Somasundaram, Scott King, and Scott (Slim) Whitman (apologies to anyone left out).

I would also like to thank the readers who, over the years, have given me feedback (both good and bad) concerning the book. In particular, I would like to note the critical contributions of Dr. Philip Schlup of Colorado State University, Dr. Brian Wyvill of the University of Calgary, and Dr. Przemyslaw Kiciak of the University of Warsaw.

And finally, I would like to acknowledge the support of the Department of Computer Science and Engineering at Ohio State (Xiaodong Zhang, Chair) as well as the support of the Advanced Computing Center for Art and Design (Maria Palazzi, Director). I would also like to thank Morgan Kaufmann's reviewers and multiple editors for seeing this project through.

References

[1] All Time Grossing Movies. In: The Internet Movie Database (IMDB). IMDb.com, Inc; 2012. Web. 26 March 2012. http://www.imdb.com/boxoffice/alltimegross.

[2] Scientific and Technical Awards. In: Academy of Motion Picture Arts and Sciences. 2012. Web. 26 March 2012. http://www.oscars.org/awards/scitech/index.html.

[3] Industry facts. In: The Entertainment Software Association (ESA). 2012. Web. 26 March 2012. http://www.theesa.com/facts/index.asp.

[4] Catmull E. The Problems of Computer-Assisted Animation. In: Computer Graphics. Proceedings of SIGGRAPH 78, vol. 12(3). August Atlanta, Ga.; 1978. p. 348–53.

[5] Levoy M. A Color Animation System Based on the Multiplane Technique. In: George J, editor. Computer Graphics. Proceedings of SIGGRAPH 77, vol 11(2). San Jose, Calif.; July 1977. p. 65–71.

About the Author

Rick Parent is a Professor in the Computer Science and Engineering Department of Ohio State University (OSU). As a graduate student, Rick worked at the Computer Graphics Research Group (CGRG) at OSU under the direction of Charles Csuri. In 1977, he received his Ph.D. from the Computer and Information Science (CIS) Department, majoring in Artificial Intelligence. For the next three years, he worked at CGRG first as a Research Associate, and then as Associate Director. In 1980 he co-founded and was President of The Computer Animation Company. In 1985, he joined the faculty of the CIS Department (now the Department of Computer -Science and Engineering) at Ohio State. Rick's research interests include various aspects of computer animation with special focus on animation of the human figure. Currently, he is working on facial animation and on using model-based techniques to track human figures in video.

Introduction

Computer animation, for many people, is synonymous with big-screen events such as *Star Wars*, *Toy Story*, and *Avatar*. But not all, or arguably even most, computer animation is done in Hollywood. It is not unusual for Saturday morning cartoons to be entirely computer generated. Computer games take advantage of state-of-the-art computer graphics techniques and have become a major motivating force driving research in computer animation. Real-time performance-driven computer animation has appeared at SIGGRAPH[1] and on *Sesame Street*. Desktop computer animation is now possible at a reasonable cost. Computer animation on the Web is routine. Digital simulators for training pilots, SWAT teams, and nuclear reactor operators are commonplace. The distinguishing characteristics of these various venues are the cost, the image quality desired, and the amount and type of interaction allowed. This book does not address the issues concerned with a particular venue, but it does present algorithms and techniques used to do computer animation in all of them.

Computer animation, as used here, refers to any computer-based computation used in producing images intended to create the perception of motion. The emphasis in this book is on algorithms and techniques that process three-dimensional graphical data. In general, any value that can be changed can be animated. An object's position and orientation are obvious candidates for animation, but all of the following can be animated as well: the object's shape, its shading parameters, its texture coordinates, the light source parameters, and the camera parameters.

This book is organized as follows. To lay a firm foundation for the rest of the book, Chapter 2 surveys the technical background of computer graphics relevant to computer animation. This includes the fundamental geometric transformations and associated representations of graphical data. It can be skipped by those well versed in the mathematics of the computer graphics display pipeline. Chapters 3–11 cover various computer animation algorithms and techniques: Chapters 3–5 deal with directly specifying motion (kinematics), Chapter 6 covers digitizing motion (motion capture), Chapters 7 and 8 consider physically based animation (dynamics), and Chapters 9–11 concentrate on (mostly human) figure animation. Finally, Chapter 12 surveys some modeling techniques that have been used in computer animation. The appendices provide ancillary material. Appendix A covers rendering issues that are relevant for animation, and Appendix B provides detail of the mathematics used in the text.

In considering computer animation techniques, there are basically three general approaches to motion control. The first is *artistic animation* in which the animator has the prime responsibility for crafting the motion. The foundation of artistic animation is interpolation. Various animation

---

[1]SIGGRAPH is the Association for Computing Machinery's (ACM) special interest group on computer graphics. The ACM is the main professional group for computer scientists.

techniques based on interpolation are concentrated in the early chapters (Chapters 3–5). The second is *data-driven animation* in which live motion is digitized and then mapped onto graphical objects. The primary technology for data-driven animation is referred to as *motion capture* and is the topic of Chapter 6. The third is *procedural animation*, in which there is a computational model that is used to control the motion. Usually, this is in the form of setting initial conditions for some type of physical or behavioral simulation. Procedural animation is concentrated in the later chapters (Chapters 7–11).

To set the context for computer animation, it is important to understand its heritage, its history, and certain relevant concepts. The rest of this chapter discusses motion perception, the technical evolution of animation, animation production, and notable works in computer animation. It provides a grounding in computer animation as a field of endeavor.

1.1 Motion perception

A picture can quickly convey a large amount of information because the human visual system is a sophisticated information processor. It follows, then, that moving images have the potential to convey even more information in a short time. Indeed, the human visual system has evolved to provide for survival in an ever-changing world; it is designed to notice and interpret movement.

It is widely recognized that a series of images, when displayed in rapid succession, are perceived by an observer as a single moving image. This is possible because the eye–brain complex has the ability, under sufficient viewing conditions and within certain playback rates, to create a sensation of continuous imagery from such a sequence of still images. A commonly held view is that this experience is due to *persistence of vision*, whereby the eye retains a visual imprint of an image for a brief instant once the stimulus is removed. It is argued that these imprints, called *positive afterimages* of the individual stills, fill in the gaps between the images to produce the perception of a continuously changing image. Peter Roget (of *Roget's Thesaurus* fame) presented the idea of impressions of light being retained on the retina in 1824 [35]. But persistence of vision is not the same as perception of motion. Rotating a white light source fast enough will create the impression of a stationary white ring. Although this effect can be attributed to persistence of vision, the result is static. The sequential illumination of a group of lights typical of a movie theater marquee produces the illusion of a lighted object circling the signage. Motion is perceived, yet persistence of vision does not appear to be involved because no individual images are present. Recently, the causality of the (physiological) persistence of vision mechanism has been called into question and the perception of motion has been attributed to a (psychological) mechanism known as the *phi phenomenon* (as is the case in the movie marquee example given above). A related phenomenon, for example the apparent motion of a disk traveling between two flickering disks, is referred to as *beta movement* [1] [2] [13] [39].

Whatever the underlying mechanism is, the result is that in both film and video, a sequence of images can be displayed at rates fast enough to fool the eye into interpreting it as continuous imagery. When the perception of continuous imagery fails to be created, the display is said to *flicker*. In this case, the animation appears as a rapid sequence of still images to the eye–brain complex. Depending on conditions such as room lighting and viewing distance, the rate at which individual images must be played back in order to maintain the perception of continuous imagery varies. This rate is referred to as the *critical flicker frequency* [8].

While perception of motion addresses the lower limits for establishing the perception of continuous imagery, there are also upper limits to what the eye can perceive. The receptors in the eye continually sample light in the environment. The limitations on motion perception are determined, in part, by the reaction time of those sensors and by other mechanical limitations such as blinking and tracking. If an object moves too quickly with respect to the viewer, then the receptors in the eye will not be able to respond fast enough for the brain to distinguish sharply defined, individual detail; *motion blur* results [11]. In a sequence of still images, motion blur is produced by a combination of the object's speed and the time interval over which the scene is sampled. In a still camera, a fast-moving object will not blur if the shutter speed is fast enough relative to the object's speed. In computer graphics, motion blur will never result if the scene is sampled at a precise instant in time; to compute motion blur, the scene needs to be sampled over an interval of time or manipulated to appear as though it were [21] [32]. (See Appendix A.3 for a discussion of motion blur calculations.) If motion blur is not calculated, then images of a fast-moving object can appear disjointed, similar to viewing live action under the effects of a strobe light. This effect is often referred to as *strobing*. In hand-drawn animation, fast-moving objects are typically stretched in the direction of travel so that the object's images in adjacent frames overlap [49], reducing the strobing effect.

As reflected in the previous discussion, there are actually two rates of concern. One is the *playback* or *refresh* rate—the number of images per second displayed in the viewing process. The other is the *sampling* or *update* rate—the number of different images that occur per second. The playback rate is the rate related to flicker. The sampling rate determines how jerky the motion appears. For example, a television signal conforming to the National Television Standards Committee (NTSC) format displays images at a rate of roughly 30 frames per second (fps),[2] but because it is *interlaced*,[3] *fields* are played at 60 frames per second to prevent flicker under normal viewing conditions [34]. In some programs (e.g., some Saturday morning cartoons) there may be only six different images per second, with each image repeatedly displayed five times. Often, lip-sync animation is drawn *on twos* (every other frame) because drawing it *on ones* (animating it in every frame) makes it appear too hectic. Film is typically shown in movie theatres at playback rates of 24 fps (in the United States) but, to reduce the flicker, each frame is actually displayed twice (*double-shuttered*) to obtain an effective refresh rate of 48 fps. On the other hand, because an NTSC television signal is interlaced, smoother motion can be produced by sampling the scene every 60th of a second even though the complete frames are only played back at 30 fps [8]. Computer displays are typically progressive scan (*noninterlaced*) devices with refresh rates above 70 fps [34]. See Appendix B.10 for some details concerning various film and video formats.

The display and perception of animation using a sequence of still images imposes certain requirements on how those images are computed and played effectively. Understanding the operation, limits, and trade-offs of the human visual system are essential when making intelligent decisions about designing any type of visual and auditory content, including computer animation.

---

[2]More accurately, the format for broadcast television system, established by the NTSC, specifies a frame rate of 29.97 fps [29].

[3]An *interlaced* display is one in which a frame is divided into two *fields*. Each field consists of odd or even numbered scanlines. The odd and even fields are displayed in alternate scans of the display device [8].

1.2 The heritage of animation

In the most general sense, *animate*[4] means "give life to" and includes live-action puppetry such as that found on *Sesame Street* and the use of electromechanical devices to move puppets, such as *animatronics*. History is replete with attempts to bring objects to life. This history is a combination of myth, deception, entertainment, science, and medicine. Many of the references to animation are in the form of stories about conjuring a life force into some humanoid form: from Pygmalion to Prometheus to Wagner's homunculus in Goethe's *Faust* to Shelley's Dr. Frankenstein. Some of the history is about trying to create mechanical devices that mimic certain human activity: from Jacque Vaucanson's mechanical flute player, drummer, and defecating duck in the 1730s to Wolfgang von Kempelen's chess player in 1769 to Pierre Jaquet-Droz's writing automaton of 1774 to the electromechanical humanoid robots (*animatronics*) popular today. The early mechanisms from the 1700s and 1800s were set in the milieu of scientific debate over the mechanistic nature of the human body (e.g., *L'Homme Machine*, translated as *Man a Machine*, was written by Julien Offray de La Mettrie in 1747 and was quite controversial). This activity in humanoid mechanical devices was propelled by a confluence of talents contributed by magicians, clock makers, philosophers, scientists, artists, anatomists, glove makers, and surgeons (see Gaby Wood's book for an entertaining survey on the quest for mechanical life [50]). Here, however, the focus is on devices that use a sequence of individual still images to create the effect of a single moving image, because these devices have a closer relationship to hand-drawn animation.

1.2.1 Early devices

Persistence of vision and the ability to interpret a series of stills as a moving image were actively investigated in the 1800s [5], well before the film camera was invented. The recognition and subsequent investigation of this effect led to a variety of devices intended as parlor toys [23] [38]. Perhaps the simplest of these early devices is the *thaumatrope*, a flat disk with images drawn on both sides with two strings connected opposite each other on the rim of the disk (see Figure 1.1). The disk could be quickly flipped back and forth by twirling the strings. When flipped rapidly enough, the two images appear to be superimposed. The classic example uses the image of a bird on one side and the image of a birdcage on the other; the rotating disk visually places the bird inside the birdcage. An equally primitive technique is the *flip book*, a tablet of paper with an individual drawing on each page. When the pages are flipped rapidly, the viewer has the impression of motion.

   One of the most well known early animation devices is the *zoetrope*, or wheel of life. The zoetrope has a short fat cylinder that rotates on its axis of symmetry. Around the inside of the cylinder is a sequence of drawings, each one slightly different from the ones next to it. The cylinder has long vertical slits cut into its side between each adjacent pair of images so that when it is spun on its axis each slit allows the eye to see the image on the opposite wall of the cylinder (see Figure 1.2). The sequence of slits passing in front of the eye as the cylinder is spun on its axis presents a sequence of images to the eye, creating the illusion of motion.

---

[4]A more restricted definition of *animation*, also found in the literature, requires the use of a sequence of stills to create the visual impression of motion. The restricted definition does not admit techniques such as animatronics or shadow puppets under the rubric animation.

FIGURE 1.1

A thaumatrope.

FIGURE 1.2

A zoetrope.

Related gizmos that use a rotating mechanism to present a sequence of stills to the viewer are the *phenakistoscope* and the *praxinoscope*. The phenakistoscope also uses a series of rotating slots to present a sequence of images to the viewer by positioning two disks rotating in unison on an axis; one disk has slits, and the other contains images facing the slits. One sights along the axis of rotation so the slits pass in front of the eye, which can thus view a sequence of images from the other disk.

The praxinoscope uses a cylindrical arrangement of rotating mirrors inside a large cylinder of images facing the mirrors. The mirrors are angled so that, as the cylinders rotate in unison, each image is successively reflected to the observer.

Just before the turn of the century, the moving image began making its way on stage. The *magic lantern* (an image projector powered by candle or lamp) and shadow puppets became popular theater entertainment [3]. On the educational front, Etienne-Jules Marey [27] and Eadweard Muybridge [30] [31] investigated the motions of humans and animals. To show image sequences during his lectures, Muybridge invented the *zoopraxinoscope*, a projection device also based on rotating slotted disks. Then, in 1891, the seed of a revolution was planted: Thomas Edison invented the motion picture viewer (the *kinetoscope*), giving birth to a new industry [38].

1.2.2 The early days of "conventional" animation

Animation in America exploded in the twentieth century in the form of filming hand-drawn, two-dimensional images (referred to here also as *conventional* or *traditional* animation). Studying the early days of conventional animation is interesting in itself [26] [38] [44] [45], but the purpose of this overview is to provide an appreciation of the technological advances that drove the progress of animation during the early years.

Following Edison's kinetoscope, there were several rapid developments in film technology. One of the most notable developments was the motion picture projector by the Lumiere brothers, Auguste and Louis, in France. They are credited with the first commercial, public screening of film in Paris on December 28, 1895. They called their device the *cinematograph*. It is a camera that could both project and develop film. They used it to film everyday events including a train coming into a train station; this footage, when shown to the audience, sent everyone scrambling for cover. It was also used for aerial photography (years before the airplane took to the skies).

The earliest use of a camera to make lifeless things appear to move occurred in 1896 by Georges Méliès. Méliès used simple camera tricks such as multiple exposures and stop-motion techniques to make objects appear, disappear, and change shape [18] [47]. His best known trick film is *A Trip to the Moon* (1902). One of the earliest pioneers in film animation was J. Stuart Blackton, an American who animated "smoke" in a scene in 1900, ushering in the field of *visual effects*. Blackton is credited with creating the first animated cartoon, *Humorous Phases of Funny Faces* (1906), by drawing and erasing on a chalkboard between takes. Emile Cohl, a Frenchman, produced several vignettes including *Fantasmagorie* (1908), which is considered to be the first fully animated film ever made. The American Winsor McCay is the first celebrated animator, best known for his works *Little Nemo* (1911) and *Gertie the Dinosaur* (1914). McCay is considered by many to have produced the first popular animations [26].

Like many of the early animators, McCay was an accomplished newspaper cartoonist. He redrew each complete image on rice paper mounted on cardboard and then filmed them individually. He was also the first to experiment with color in animation. Much of his early work was incorporated into vaudeville acts in which he "interacted" with an animated character on a screen. Similarly, early cartoons often incorporated live action with animated characters. To appreciate the impact of such a popular entertainment format, keep in mind the relative naïveté of audiences at the time; they had no idea how film worked, much less what hand-drawn animation was. As Arthur C. Clarke stated about sufficiently advanced technology, it must have been indistinguishable from magic.

The first major technical developments in the animation process can be traced to the efforts of John Bray, one of the first to recognize that patenting aspects of the animation process would result in a competitive advantage [26]. Starting in 1910, his work laid the foundation for conventional animation as it exists today. Earl Hurd, who joined forces with Bray in 1914, patented the use of translucent *cels*[5] in the compositing of multiple layers of drawings into a final image and also patented gray scale drawings as opposed to black and white. Later developments by Bray and others enhanced the overlay idea to include a peg system for registration and the drawing of the background on long sheets of paper so that *panning* (translating the background relative to the camera, perpendicular to the view direction) could be performed more easily. Out of Bray's studio came Max Fleischer (Betty Boop), Paul Terry (Terrytoons), George Stallings (Tom and Jerry), and Walter Lantz (Woody Woodpecker). In 1915, Fleischer patented *rotoscoping* (drawing images on cells by tracing over previously recorded live action). Several years later, in 1920, Bray experimented with color in the short *The Debut of Thomas Cat*.

While the technology was advancing, animation as an art form was still struggling. The first animated character with an identifiable personality was Felix the Cat, drawn by Otto Messmer of Pat Sullivan's studio [26]. Felix was the most popular and most financially successful cartoon of the mid-1920s. In the late 1920s, however, new forces had to be reckoned with: sound and Walt Disney.

1.2.3 Disney

Walt Disney was, of course, the overpowering force in the history of conventional animation. Not only did his studio contribute several technical innovations, but Disney, more than anyone else, advanced animation as an art form [45]. Disney's innovations in animation technology included the use of a storyboard to review the story and pencil sketches to review motion. In addition, he pioneered the use of sound and color in animation (although he was not the first to use color). Disney also studied live-action sequences to create more realistic motion in his films. When he used sound for the first time in *Steamboat Willie* (1928), he gained an advantage over his competitors.

One of the most significant technical innovations of the Disney studio was the development of the multiplane camera [26] [44]. The multiplane camera consists of a camera mounted above multiple planes, each of which holds an animation cell. Each of the planes can move in six directions (right, left, up, down, in, out), and the camera can move closer and farther away (see Figure 1.3).

Multiplane camera animation is more powerful than one might think. By moving the camera closer to the planes while the planes are used to move foreground images out to the sides, a more effective zoom can be performed. Moving multiple planes at different rates can produce the *parallax effect*, which is the visual effect of closer objects apparently moving faster across the field of view than objects farther away, as an observer's view pans across an environment. This is very effective in creating the illusion of depth and an enhanced sensation of three dimensions. Keeping the camera lens open during movement can produce several additional effects: figures can be extruded into shapes of higher dimension, depth cues can be incorporated into an image by blurring the figures on more distant cels, and a blurred image of a moving object can be produced.

---

[5]Cel is short for *celluloid*, which was the original material used in making the translucent layers. Currently, cels are made from acetate.

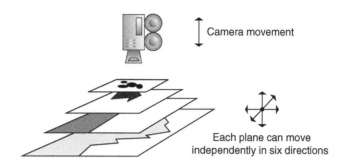

Camera movement

Each plane can move
independently in six directions

FIGURE 1.3

Directional range of the multiplane camera, inside of which the image is optically composited.

With regard to the art of animation, Disney perfected the ability to impart unique, endearing personalities in his characters, such as those exemplified in Mickey Mouse, Pluto, Goofy, the Three Little Pigs,
and the Seven Dwarfs [44] [45]. He promoted the idea that the mind of the character was the driving
force of the action and that a key to believable animated motion was the analysis of real-life motion.
He also developed mood pieces, for example, *Skeleton Dance* (1929) and *Fantasia* (1940).

1.2.4 Contributions of others

The 1930s saw the proliferation of animation studios, among them Fleischer, Iwerks, Van Beuren, Universal Pictures, Paramount, MGM, and Warner Brothers. The technological advances that are of concern here were mostly complete by this period. The differences between, and contributions of, the
various studios have to do more with the artistic aspects of animation than with the technology involved
in producing animation [26]. Many of the notable animators in these studios were graduates of Disney's
or Bray's studio. Among the most recognizable names are Ub Iwerks, George Stallings, Max Fleischer,
Bill Nolan, Chuck Jones, Paul Terry, and Walter Lantz.

1.2.5 Other media for animation

The rich heritage of hand-drawn animation in the United States makes it natural to consider it the pre-
cursor to computer animation, which also has strong roots in the United States. However, computer
animation has a close relationship to other animation techniques as well.

A good comparison can be made between computer animation and some of the stop-motion techniques, such as clay and puppet animation. Typically, in three-dimensional computer animation, one of
the first steps is the object modeling process. The models are then manipulated to create the three-
dimensional scenes that are rendered to produce the images of the animation. In much the same
way, clay and puppet stop-motion animation use three-dimensional figures that are built and then animated in separate, well-defined stages [23]. Once the physical three-dimensional figures are created,
they are used to lay out a three-dimensional environment. A camera is positioned to view the environment and record an image. One or more of the figures are manipulated, and the camera may be repositioned. The camera records another image of the scene. The figures are manipulated again, another
image is taken of the scene, and the process is repeated to produce the animated sequence.

Willis O'Brien of *King Kong* fame is generally considered the dean of this type of stop-motion animation. His understudy, who went on to create an impressive body of work in his own right, was Ray Harryhausen (*Mighty Joe Young*, *Jason and the Argonauts*, and many more). More recent impressive examples of three-dimensional stop-motion animation are Nick Park's *Wallace and Gromit* series and *Chicken Run* and Tim Burton's projects such as *The Nightmare Before Christmas*, *James and the Giant Peach*, *Corpse Bride*, and *Alice in Wonderland*.

Because of computer animation's close association with video technology, it has also been associated with video art, which depends largely on the analog manipulation of the video signal in producing effects such as colorization and warping [12]. Because creating video art is inherently a two-dimensional process, the relationship is viewed mainly in the context of computer animation post-production techniques. Even this connection has faded because the popularity of recording computer animation by digital means has eliminated most analog processing.

1.3 Animation production

Although producing a final animated film is not the subject of this book, the production process merits some discussion in order to establish the context in which an animator works. It is useful for technical animators to have some familiarity with how a piece of animation is broken into parts and how a finished piece is produced. Much of this is taken directly from conventional animation and is directly applicable to any type of animation.

A piece of animation is usually discussed using a four-level hierarchy, although the specific naming convention for the levels may vary.[6] Here, the overall animation—the entire project—is referred to as the *production*. Typically, productions are broken into major parts referred to as *sequences*. A sequence is a major episode and is usually identified by an associated staging area; a production usually consists of one to a dozen sequences. A sequence is broken down into one or more *shots*; each shot is the recording of the action from a single point of view. A shot is broken down into the individual *frames* of film. A frame is a single recorded image. This results in the hierarchy shown in Figure 1.4.

Several steps are required to successfully plan and carry out the production of a piece of animation [23] [44]. Animation is a trial-and-error process that involves feedback from one step to previous steps

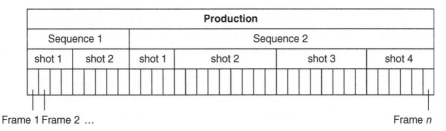

FIGURE 1.4

Sample hierarchy of a simple animation production.

---

[6]Live-action film tends to use a five-level hierarchy: film, sequence, scene, shot, and frame [9]. Here, the terminology, which is often used in feature-length computer animation, is presented.

and usually demands several iterations through multiple steps at various times. Even so, the production of animation typically follows a standard pattern. First, a *preliminary story* is decided on, including a *script*. A *storyboard* is developed that lays out the action scenes by sketching representative frames. The frames are often accompanied by text that sketches out the action taking place. This is used to present, review, and critique the action as well as to examine character development.

A model sheet is developed that consists of a number of drawings for each figure in various poses and is used to ensure that each figure's appearance is consistent as it is repeatedly drawn during the animation process. The exposure sheet records information for each frame such as sound track cues, camera moves, and compositing elements. The *route sheet* records the statistics and responsibility for each scene.

An *animatic*, or *story reel*, may be produced in which the storyboard frames are recorded, each for as long as the sequence it represents, thus creating a rough review of the timing. Often, a *scratch track*, or *rough sound track*, is built at the same time the storyboard is being developed and is included in the animatic. Once the storyboard has been decided on (see Figure 1.5), the *detailed story* is worked out to identify the actions in more detail. *Key frames* (also known as *extremes*) are then identified and produced by master animators to aid in confirmation of timing, character development, and image quality. Associate and assistant animators are responsible for producing the frames between the keys; this is called *in-betweening*. *Test shots*, short sequences rendered in full color, are used to test the rendering and motions. To completely check the motion, a *pencil test* may be shot, which is a full-motion rendering of an extended sequence using low-quality images such as pencil sketches. Problems identified in the test shots and pencil tests may require reworking of the key frames, detailed story, or even the storyboard.

Inking refers to the process of transferring the penciled frames to cels. *Opaquing*, also called *painting*, is the application of color to these cels.

1.3.1 Principles of animation

To study various techniques and algorithms used in computer animation, it is useful to first understand their relationship to the animation principles used in hand-drawn animation. In an article by Lasseter [22], the principles of animation, articulated by some of the original Disney animators [45], are related to techniques commonly used in computer animation. The principles are squash and stretch, timing, secondary action, slow in and slow out, arcs, follow through and overlapping action, exaggeration, appeal, anticipation, staging, solid drawing, and straight ahead and pose to pose. Lasseter is a conventionally trained animator who worked at Disney before going to Pixar. At Pixar, he was responsible for many celebrated computer animations including *Tin Toy* that, in 1989, was the first computer animation to win an Academy Award. Whereas Lasseter discusses each principle in terms of how it might be implemented using computer animation techniques, the principles are organized here according to the type of motion quality they contribute to in a significant way. Because several principles relate to multiple qualities, some principles appear under more than one heading.

Simulating physics

Squash and stretch, *timing*, *secondary action*, *slow in and slow out*, and *arcs* establish the physical basis of objects in the scene. A given object possesses some degree of rigidity and should appear to have some amount of mass. This is reflected in the distortion (squash and stretch) of its shape during an

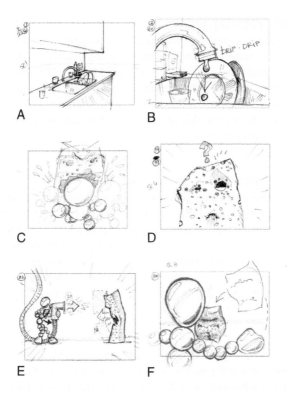

FIGURE 1.5

Sample panels from a storyboard. (a) Establishing shot: overview of the background introduces the place where the action takes place. (b) Extreme close-up: the action is shown in detail helping the viewer to get familiar with the main characters and objects. (c) Low angle: the camera position and the action happen in the same camera direction increasing the dramatic feeling of the scene. (d) POV (point of view): shows the viewer what the character would be seeing, which helps to better understand the reaction to the action. (e) Wide shot: shows the whole action making the viewer understand the motivation, the action, and the consequences all at once. (f) OTS (over the shoulder): the camera looks to one character, or action, from just behind and over the shoulder of another character to get the viewer involved with the action. (Images courtesy of Beth Albright and Iuri Lioi.)

action, especially a collision. The animation must support these notions consistently for a given object throughout the animation. Timing has to do with how actions are spaced according to the weight, size, and personality of an object or character and, in part, with the physics of movement as well as the artistic aspects of the animation. Secondary action supports the main action, possibly supplying physically based reactions to an action that just occurred. Slow in and slow out and arcs are concerned with how things move through space. Objects slow in and slow out of poses. When speaking of the actions involved, objects are said to "ease in" and "ease out." Such speed variations model inertia, friction, and viscosity. Objects, because of the physical laws of nature such as gravity, usually move not in straight lines but rather in arcs.

Designing aesthetically pleasing actions

Appeal, *solid drawing*, and *follow through/overlapping action* are principles that address the aesthetic design of an action or action sequence. To keep the audience's attention, the animator needs to make it enjoyable to watch (appeal). In addition, actions should flow into one another (follow through/ overlapping action) to make the entire shot appear to continually evolve instead of looking like disjointed movements. Solid drawing refers to making the character look pliable and not stiff or wooden. Squash and stretch can also be used in this regard. Secondary actions and timing considerations also play a role in designing pleasing motion.

Effectively presenting action

Often the animator needs to employ *exaggeration* so a motion cannot be missed or so it makes a point (Tex Avery is well known for this type of conventional animation). *Anticipation* and *staging* concern how an action is presented to the audience. Anticipation dictates that an upcoming action is set up so that the audience knows it (or something) is coming. Staging expands on this notion of presenting an action so that it is not missed by the audience. *Timing* is also involved in effective presentation to the extent that an action has to be given the appropriate duration for the intended effect to reach the audience. *Secondary action* can also be used to create an effective presentation of an action.

Production technique

Straight ahead versus *pose to pose* concerns how a motion is created. Straight ahead refers to progressing from a starting point and developing the motion continually along the way. Physically based animation could be considered a form of straight ahead processing. Pose to pose, the typical approach in conventional animation, refers to identifying key frames and then interpolating the intermediate frames, an approach the computer is particularly good at.

1.3.2 Principles of filmmaking

Basic principles of filmmaking are worth reviewing in order to get a sense of how effective imagery is constructed. Several of the basic principles are listed in the following sections, although more complete references should be consulted by the interested reader (e.g., [28]). Some of the following principals are guidelines that should be followed when composing a single image; others are options of how to present the action.

Three-point lighting

There is a standard set of three lights that are used to illuminate the central figure in a scene. These are the *key light*, the *fill light*, and the *rim light*. The key light is often positioned up and to the side of the camera, pointing directly at the central figure. This focuses the observer's attention on what is important. The rim light is positioned behind the central figure and serves to highlight the rim, thus outlining the figure and making the figure stand out from the background. The fill light is a flood light typically positioned below the camera that fills the figure with a soft light bringing out other details in the figure's appearance. See Figure 1.6 (Color Plate 1) for an example.

FIGURE 1.6

Three-point lighting example: (a) Key light: A single spot light is placed at 45 degrees from the top-right of the frame. This light has the highest intensity in the setup and is responsible for the cast shadows of the objects. (b) Fill light: A blue fill light from the front and right of the object is used to illuminate the dark areas created by the key light. This light is less intense and does not cast shadows or highlights on the objects. (c) Rim light: Multiple lights are placed opposite the direction of the key light. They highlight the edges, which are otherwise in shadow. These highlights help separate the objects from the background as well as from other overlapping objects in the scene. (d) All lights: This is the final lighting set up—a combination of key, fill, and rim lights. The scene is rendered with ray tracing, generating reflections on selective surfaces. (Images courtesy of Sucheta Bhatawadekar, ACCAD.)

180 rule

When filming a line of action, for example the conversation between two figures, it is common to show each figure in isolation during the action. The camera is positioned in front of a figure but a little off to the side. The 180 degree rule states that when showing the two figures, one after the other, in isolation, the camera should stay on the same side of the line of action. Thus, the camera's orientation should stay within the 180 degrees that is on one side of the line between the figures.

Rule of thirds

The rule of thirds says that the interesting places to place an object in an image are one-third along the way, either side-to-side or up-and-down or both. In particular, don't center your subject in the image and don't put your subject at the edge of the image.

Types of shots

Types of camera shots are categorized based on the distance from the camera to the subject and the angle at which the shot is taken. The distance-based shots are *extreme long range*, *long range*, *medium range* or *bust shot*, *close-up*, and *extreme close-up*. Which type of shot to use depends on the amount and location of detail that is to be shown and how much environmental context is to be included in the shot.

A *low angle shot*, meaning the camera is low shooting up at the subject, imparts a feeling of power or dominance to the subject. Conversely, a *high angle shot*, in which the camera shoots down on the subject, presents a feeling that the subject is insignificant or subordinate.

Tilt

Tilting the camera (rotating the camera about its view direction) can convey a sense of urgency, strangeness, or fear to the shot.

Framing

Framing refers to allowing enough room in the image for the action being captured. In a relatively static view, allow enough room so the subject does not fill the frame (unless there is a reason to do so). Allow enough room for motion. If your subject is walking, frame the motion so there is room in front of the subject so the subject does not appear to be walking out of the frame.

Focus the viewer's attention

Draw the viewer's attention to what's important in the image. Use color, lighting, movement, focus, etc., to direct the attention of the viewer to what you want the viewer to see. For example, the eye will naturally follow converging lines, the gaze of figures in the image, a progression from dark to light or dark to bright, and an identifiable path in the image.

1.3.3 Sound

Sound is an integral part of almost all animation, whether it's hand-drawn, computer-based, or stop-motion [23] [26]. Up through the 1920s, the early "silent films" were played in theaters with live mood music accompaniment. That changed as sound recording technology was developed for film and, later, for video.

Audio recording techniques had been around for 30 years by the time moving images were first recorded on film. It took another 30 years to develop techniques for playing a sound track in sync with recorded action. Since then, various formats have been developed for film sound tracks. Most formats record the audio on the same medium that records the images. In most of the various formats for film, for example, audio is recorded along the side of the images or between the sprocket holes in one to six tracks. Early formats used optical or magnetic analog tracks for sound, but more recent formats digitally print the sound track on the film. By recording the audio on the same stock as the film, the timing between the imagery and the audio is physically enforced by the structure of the recording technology. In some formats, a separate medium, such as a CD, is used to hold the audio. This allows more audio to be recorded, but creates a synchronization issue during playback. In the case of video, audio tracks are recorded alongside the tracks used to encode the video signal.

In the early film and video formats, audio was recorded as a low bandwidth analog signal resulting in very low-quality sound. Today's film and video technology acknowledges the importance of sound and provides multiple, high-quality digital audio tracks. Sound has four roles in a production: voice, body sounds, special effects, and background music.

In live action, voice is recorded with the action because of timing considerations while most of the other sounds are added in a post-processing phase. In animation, voices are recorded first and the animation made to sync with it. In addition, recording visuals of the voice talent during the audio recording can be used to guide the animators as they create the facial expressions and body language that accompanies the speech.

Nonspeech sounds made by the actors, such as rustling of clothes, footsteps, and objects being handled, are called body sounds. The recorded body sounds are usually replaced by synthesized sounds, called *foley*, for purposes of artistic control. These synthesized sounds must be synced with the motions of the actors. The people responsible for creating these sounds are called *foley artists*.

Special effects, such as door slams and the revving of car engines, must also be synced with the action, but with lesser precision than voice and foley sounds.

Recording background and mood music can be added after the fact and usually require no precise timing with the action. All the sounds other than voice are added after the live action or animation is recorded.

1.4 Computer animation production

Computer animation production has borrowed most of the ideas from conventional animation production, including the use of a storyboard, test shots, and pencil testing. The storyboard has translated directly to computer animation production, although it may be done on-line. It still holds the same functional place in the animation process and is an important component in planning animation. The use of key frames, and interpolating between them, has become a fundamental technique in computer animation.

While computer animation has borrowed the production approaches of conventional animation, there are significant differences between how computer animation and conventional animation create an individual frame of the animation. In computer animation, there is usually a strict distinction among creating the models; creating a layout of the models including camera positioning and lighting; specifying the motion of the models, lights, and camera; and the rendering process applied to those models. This allows for reusing models and lighting setups. In conventional animation, all of these processes happen simultaneously as each drawing is created; the only exception is the possible reuse of backgrounds, for example, with the multilayer approach.

The two main evaluation tools of conventional animation, test shots and pencil tests, have counterparts in computer animation. A speed/quality trade-off can be made in several stages of creating a frame of computer animation: model building, lighting, motion control, and rendering. By using high-quality techniques in only one or two of these stages, that aspect of the presentation can be quickly checked in a cost-effective manner. A test shot in computer animation is produced by a high-quality rendering of a highly detailed model to see a single frame, a short sequence of frames of the final product, or every nth frame of a longer sequence from the final animation. The equivalent of a pencil test can be performed by simplifying the sophistication of the models used, by using low-quality and/or low-resolution renderings, by eliminating all but the most important lights, or by using simplified motion.

Often, it is useful to have several representations of each model available at varying levels of detail. For example, placeholder cubes can be rendered to present the gross motion of rigid bodies in space and to see spatial and temporal relationships among objects. "Solids of revolution" objects (objects created by rotating a silhouette edge at certain intervals around an axis and then defining planar surfaces to fill the space between these silhouette slices) lend themselves quite well to multiple levels of detail for a given model based on the number of slices used. Texture maps and displacement maps can be disabled until the final renderings.

To simplify motion, articulated figures[7] can be kept in key poses as they navigate through an environment in order to avoid interpolation or inverse kinematics. Collision detection and response can be selectively "turned off" when not central to the effect created by the sequence. Complex effects such as smoke and water can be removed or represented by simple geometric shapes during testing.

---

[7]*Articulated figures* are models consisting of rigid segments usually connected together in a tree-like structure; the connections are revolute or prismatic joints, allowing a segment to rotate or translate relative to the segment to which it is connected.

Many aspects of the rendering can be selectively turned on or off to provide great flexibility in giving the animator clues to the finished product's quality without committing to the full computations required in the final presentation. Often, the resulting animation can be computed in real time for very effective motion testing before committing to a full anti-aliased, transparent, texture-mapped rendering. Wire frame rendering of objects can sometimes provide sufficient visual cues to be used in testing. Shadows, smooth shading, texture maps, environmental maps, specular reflection, and solid texturing are options the animator can use for a given run of the rendering program.

Even in finished pieces of commercial animation it is common practice to take computational short-cuts when they do not affect the quality of the final product. For example, the animator can select which objects can shadow which other objects in the scene. In addition to being a compositional issue, selective shadowing saves time over a more robust approach in which every object can shadow every other object. In animation, environmental mapping is commonly used instead of ray tracing; photorealistic rendering is typically avoided.

Computer animation is well suited for producing the equivalent of test shots and pencil tests. In fact, because the quality of the separate stages of computer animation can be independently controlled, it can be argued that it is even better suited for these evaluation techniques than conventional animation.

1.4.1 Computer animation production tasks

While motion control is the primary subject of this book, it is worth noting that motion control is only one aspect of the effort required to produce computer animation. The other tasks (and the other talents) that are integral to the final product should not be overlooked. As previously mentioned, producing quality animation is a trial-and-error iterative process wherein performing one task may require rethinking one or more previously completed tasks. Even so, these tasks can be laid out in an approximate chronological order according to the way they are typically encountered. The order presented here summarizes an article that describes the system used to produce Pixar's *Toy Story* [16]. See Figure 1.7.

- The *Story Department* translates the verbal into the visual. The screenplay enters the Story Department, the storyboard is developed, and the story reel leaves. These visuals then go to the Art Department.
- The *Art Department*, working from the storyboard, creates the designs and color studies for the film, including detailed model descriptions and lighting scenarios. The Art Department develops a consistent look to be used in the imagery. This look guides the Modeling, Layout, and Shading Departments.
- The *Modeling Department* creates the characters and the world in which they live. Every brick and stick to appear in the film must be handcrafted. Often, figures with jointed appendages, or other models with characteristic motion, are created as parameterized models. Parameters that control standard movements of the figure are defined. This facilitates the ability of animators to *stay on the model*, ensuring that the animation remains consistent with the concept of the model. The models are given to Layout and Shading.
- On one path between the Modeling Department and Lighting Department lies the *Shading Department*. Shading must translate the attributes of the object that relate to its visual appearance into texture maps, displacement shaders, and lighting models. Relevant attributes include the

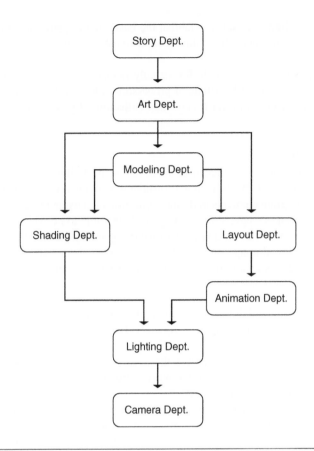

FIGURE 1.7

Computer animation production pipeline.

material the object is made of, its age, and its condition. Much of the effective appearance of an object comes not from its shape but from shading—the visual qualities of its surface.
- On the other path between Modeling and Lighting lies the *Layout Department*, followed by the Animation Department. Layout is responsible for taking the film from two dimensions to three dimensions. To ensure good flow, Layout implements proper *staging* (designing the space for the action to take place in) and *blocking* (planning out the general movements of the actors and camera). This guides the Animation Department.
- Working from audio, the story, and the blocking and staging produced by Layout, the *Animation Department* is responsible for bringing the characters to life. As mentioned above, complex figures are often parameterized by the Model Department so that a character's basic movements (e.g., smiling, taking a step) have already been defined. Animation uses these motions as well as creating the subtler gestures and movements necessary for the "actor" to effectively carry out the scene.
- The *Lighting Department* assigns to each sequence a team that is responsible for translating the Art Department's vision into digital reality. At this point the animation and camera placement

have been done. Key lights are set to establish the basic lighting environment. Subtler lighting particular to an individual shot refines this in order to establish the correct mood and bring focus to the action.

- The *Camera Department* is responsible for actually rendering the frames. During *Toy Story*, Pixar used a dedicated array of hundreds of processors called the *Render Farm*. The term render farm is now commonly used to refer to any such collection of processors for image rendering.

1.4.2 Digital editing

A revolution swept the film and video industries in the 1990s: the digital representation of images. Even if computer graphics and digital effects are not a consideration in the production process, it has become commonplace to store program elements in digital form instead of using the analog film and videotape formats. Digital representations have the advantage of being able to be copied with no image degradation. So, even if the material was originally recorded using analog means, it is often cost-effective to transcribe the images to digital image store. And, of course, once the material is in digital form, digital manipulation of the images is a natural capability to incorporate in any system.

In the old days . . .

The most useful and fundamental digital image manipulation capability is that of editing sequences of images together to create a new presentation. Originally, film sequences were edited together by physically cutting and splicing tape. This is an example of *nonlinear editing*, in which sequences can be inserted in any order at any time to assemble the final presentation. However, splicing is a time-consuming process, and making changes in the presentation or trying different alternatives can place a heavy burden on the stock material as well.

Electronic editing[8] allows the manipulation of images as electronic signals rather than using a physical process. The standard configuration uses two source videotape players, a switching box, and an output videotape recorder. More advanced configurations include a character generator (text overlays) and special effects generator (wipes, fades, etc.) on the input side, and the switching box is replaced by an editing station (see Figure 1.8). The two source tapes are searched to locate the initial desired sequence; the tape deck on which it is found is selected for recording on the output deck and the sequence is recorded. The tapes are then searched to locate the next segment, the deck is selected for input, and the segment is recorded on the output tape. This continues until the new composite sequence has been created on the output tape. The use of two source tapes allows multiple sources to be more easily integrated into the final program. Because the output is assembled in sequential order, this is referred to as *linear editing*. The linear assembly of the output is considered the main drawback of this technique. Electronic editing also has the drawback that the material is copied in the editing process, introducing some image degradation. Because the output tape is commonly used to master the tapes that are sent out to be viewed, these tapes are already third generation. Another drawback is the amount of wear on the source material as the source tapes are repeatedly played and rewound as the next desired sequence is searched for. If different output versions are required (called

---

[8]To simplify the discussion and make it more relevant to the capabilities of the personal computer, the discussion here focuses on video editing, although much of it is directly applicable to digital film editing, except that film standards require much higher resolution and therefore more expensive equipment.

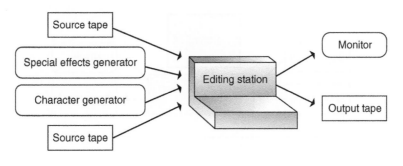

FIGURE 1.8

Linear editing system.

versioning), the source material will be subjected to even more wear and tear because the source material has to undergo more handling as it is processed for multiple purposes.

Often, to facilitate the final assemblage of the output sequence and avoid excessive wear of the original source material, copies of the source material are used in a preprocessing stage in which the final edits are determined. This is called *off-line editing*. The result of this stage is an *edit decision list* (EDL), which is a final list of the edits that need to be made to assemble the final piece. The EDL is then passed to the *on-line editing* stage, which uses the original source material to make the edits and create the finished piece. This process is referred to as *conforming*.

To keep track of edit locations, control track pulses can be incorporated onto the tape used to assemble the 30 fps NTSC video signal. Simple editing systems count the pulses, and this is called *control track editing*. However, the continual shuffling of the tape back and forth during the play and rewind of the editing process can result in the editing unit losing count of the pulses. This is something the operator must be aware of and take into account. In addition, because the edit counts are relative to the current tape location, the edit locations are lost when the editing station is turned off.

The Society of Motion Picture and Television Engineers time code is an absolute eight-digit tag on each frame in the form of HHMMSSFF, where HH is the hour, MM is the minute, SS is the second, and FF is the frame number. This tag is calculated from the beginning of the sequence. This allows an editing station to record the absolute frame number for an edit and then store the edit location in a file that can be retrieved for later use.

The process described so far is called *assemble editing*. *Insert editing* is possible if a control signal is first laid down on the output tape. Then sequences can be inserted anywhere on the tape in forming the final sequence. This provides some nonlinear editing capability, but it is still not possible to easily lengthen or shorten a sequence without repositioning other sequences on the tape to compensate for the change.

Digital on-line nonlinear editing

To incorporate a more flexible nonlinear approach, fully digital editing systems have become more accessible [17] [33] [48]. These can be systems dedicated to editing, or they can be software systems that run on standard computers. Analog tape may still be used as the source material and for the final product, but everything between is digitally represented and controlled[9] (see Figure 1.9).

---

[9]It is interesting to note that the whole process from recording to projection can now be done digitally.

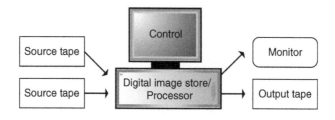

FIGURE 1.9

On-line nonlinear editing system.

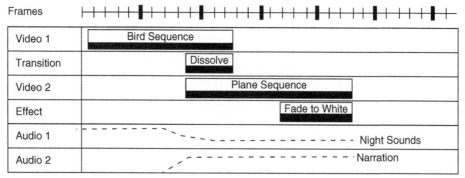

FIGURE 1.10

Simplified example of a time line used for nonlinear digital editing.

After a sequence has been digitized, an icon representing it can be dragged onto a time line provided by the editing system. Sequences can be placed relative to one another; they can be repeated, cut short, overlapped with other sequences, combined with transition effects, and mixed with other effects. A simplified example of such a time line is shown in Figure 1.10.

The positioning of the elements in the time line is conceptual only; typically the data in the digital image store is not actually copied or moved. The output sequence can be played back in real time if the disk random access and graphics display are fast enough to fetch and compile the separate tracks on the fly. In the case of overlapping sequences with transitions, the digital store must support the access of multiple tracks simultaneously so a transition can be constructed on the fly or the transition sequence needs to be precomputed (sometimes referred to as *rendering*) and explicitly stored for access during playback. When the sequence is finalized it can be assembled and stored digitally or recorded on video. Whatever the case, the flexibility of this approach, with the ability to change edits and try alternatives without generational degradation, makes nonlinear digital editing systems very powerful.

1.4.3 Digital video

As the cost of computer memory decreases and processor speeds increase, the capture, compression, storage, and playback of digital video have become more prevalent [42] [46]. This has several important ramifications. First, desktop animation has become inexpensive enough to be within the reach of the consumer. Second, in the film industry it has meant that compositing is no longer optical. Optically compositing each element in a film meant another pass of the negative through an optical film printer, which meant additional degradation of image quality. With the advent of digital compositing (see Appendix A.2), the limit on the number of composited elements is removed. Third, once films are routinely stored digitally, digital techniques can be used for wire removal and to apply special effects. These digital techniques have become the bread and butter of computer graphics in the film industry.

When one works with digital video, there are several issues that need to be addressed to determine the cost, speed, storage requirements, and overall quality of the resulting system. Compression techniques can be used to conserve space or reduce transmission time, but some compression compromises the quality of the image and the speed of compression/decompression may restrict a particular technique's suitability for a given application. During video capture, any image compression must operate in real time. Formats used for storage and playback can be encoded off-line, but the decoding must support real-time playback. Video resolution, video frame rates, and full-color imagery require that 27 MB/sec be supported for video playback.[10] An hour of uncompressed video requires just under 100 GB of storage.[11] While lossless compression (the original data can be exactly reconstructed) is possible, most video compression is lossy (not all of the original signal is recoverable) because of the favorable quality/space trade-off. There are several digital video formats used by different manufacturers of video equipment for various applications as well as video formats for streaming video and storage; these formats include D1, D2, D3, D5, miniDV, DVC, Digital8, MPEG-4, digital Betacam, H.261, and H.263. Better signal quality can be attained with the use of component instead of composite signals. Discussion of these and other issues related to digital video is beyond the scope of this book. Information on some of the more popular formats can be found in Appendix B.10.

1.4.4 Digital audio

Audio is just as important to computer animation as it is to traditional animation. Over the years, audio technology, like image technology, has gone digital. Early audio recordings used an electromechanical stylus to etch representations of the signal into wax drums or plastic platters. Later, the signal was used to modulate the magnetization of some type of ferromagnetic material on plastic tape. Digital audio has since taken over. Digital audio has the same advantages as digital imagery when it comes to duplicating and editing. Digital audio can be copied, cut and pasted, transitioned, and looped over without any degradation in signal quality—a distinct advantage over its analog counterpart. The sound capability in personal computers has dramatically improved over the years so that now high-quality sound capability is standard. As with digital imagery, there are file formats and compression standards to consider when dealing with digital audio. In addition, there is a standard for digitally controlling musical devices.

Digital musical device control

Musical instrument digital interface (MIDI) is a standard developed in 1983 to control musical instruments without being tied to any one instrument in particular. MIDI commands are keynote commands to musical devices and are intended to represent a musical performance. Mostly, the commands take the

---

[10]640 pixels/scanline $\times$ 480 scanlines/frame $\times$ 3 bytes/pixel $\times$ 30 fps = 27,630,000 bytes/sec.

[11]27,630,000 bytes/sec $\times$ 3600 sec/hour = 99,468,000,000 bytes/hour.

form of "note x on" and "note x off" where x is any of the standard musical notes. There are also control commands to set pitch, loudness, etc. Some devices can also send out MIDI commands as they are played in order to record the performance as keys are depressed and released. MIDI supports up to 16 channels and devices can be daisy-chained so that, for example, a device can be set up to respond only to a particular track.

Digital audio sampling

Sounds are pressure waves of air. In audio recording, the pressure waves are converted to some representation of the waveform. When it is recorded digitally, the wave is sampled at certain intervals—the sampling rate—with a certain number of bits per sample—the sample size—using a certain number of tracks. The sampling rate determines the highest frequency, called the Nyquist frequency, which can accurately be reconstructed. A voice signal can be sampled at a much lower rate than CD-quality music because the highest frequency of a voice signal is much lower than the highest frequency that might occur in a piece of music. The number of bits per sample determines how much distortion there is in the recorded signal. The number of tracks is how many independent signals comprise the music—one for mono, two for stereo, more for various "surround sound" recordings or for later editing. A voice signal and AM radio are sampled at approximately 10 K samples per second with 8-bits per sample using one track. CD-quality music is sampled at 44.1 K samples per second with 16-bits per sample using two tracks.

Similar to digital imagery, the digital recording can then be compressed for more efficient storage and transmission. The compression can either be lossless (the original signal can be reconstructed exactly) or lossy (some of the original signal is lost in the compression/decompression procedure— usually the very high frequencies). General data compression techniques can be used, but don't do as well on audio data as the compression schemes that use the fact that the file contains audio, referred to as being *perceptually based*. Compression of speech can use techniques that are less sensitive to preserving all of the frequency components of the sound. Common audio data compression techniques include MP3, MPEG-4 ALS, TTA, and FLAC.

1.5 A brief history of computer animation

1.5.1 Early activity (pre-1980)

The earliest computer animation of the late 1960s and early 1970s was produced by a mix of researchers in university labs and individual visionary artists [24] [25] [37]. At the time, raster displays driven by frame buffers were just being developed and digital output to television was still in the experimental stage. The displays in use were primarily storage tubes and refresh vector displays. Storage tubes retain an image indefinitely because of internal circuitry that continuously streams electrons to the display. However, because the image cannot be easily modified, storage tubes were used mainly to draw complex static models. Vector (calligraphic) displays use a display list of line- and arc-drawing instructions that an internal processor uses to repeatedly draw an image that would otherwise quickly fade on the screen. Vector displays can draw moving images by carefully changing the display list between refreshes. These displays were popular for interactive design tasks.

During this time period, static images were often recorded onto film by placing a camera in front of the display and recording an image of vectors on the screen. In this way, shaded images could be produced by opening the shutter of the film camera and essentially scan converting the elements (e.g., polygons) by drawing closely spaced horizontal vectors to fill the figure; after scan conversion was completed, the shutter was closed to terminate the image recording. The intensity of the image could be regulated by using the intensity control of the vector display or by controlling other aspects of the image recording such as by varying the density of the vectors. An image of a single color was generated by placing a colored filter in front of the camera lens. A full-color image could be produced by breaking the image into its red, green, and blue components and triple exposing the film with each exposure using the corresponding colored filter. This same approach could be used to produce animation as long as the motion camera was capable of single-frame recording. Single-frame recording required precise frame registration, usually available only in expensive film equipment. Animated sequences could be colored by triple exposing the entire film. The programmer (animator) was fortunate if both the camera and the filters could be controlled by computer.

The earliest research in computer graphics and animation occurred at the Massachusetts Institute of Technology in 1963 when Ivan Sutherland developed an interactive constraint satisfaction system on a vector refresh display [41]. The user could construct an assembly of lines by specifying constraints between the various graphical elements. If one of the graphical elements moved, the system calculated the reaction of other elements to this manipulation based on satisfying the specified constraints. By interactively manipulating one of the graphical elements, the user could produce complex motion in the rest of the assembly. Later, at the University of Utah, Sutherland helped David Evans establish the first significant research program in computer graphics and animation.

As early as the early 1960s, computer animation was produced as artistic expression. The early artistic animators in this period included Ken Knowlton, Lillian Schwartz, S. Van Der Beek, John Whitney, Sr., and A. M. Noll. Typical artistic animations consisted of animated abstract line drawings displayed on vector refresh displays. Chuck Csuri, an artist at Ohio State University, produced pieces such as *Hummingbird* (1967) that were more representational.

In the early 1970s, computer animation in university research labs became more widespread. Computer graphics, as well as computer animation, received an important impetus through government funding at the University of Utah [14]. As a result, Utah produced several groundbreaking works in animation: an animated hand and face by Ed Catmull (Hand/Face, 1972), a walking and talking human figure by Barry Wessler (*Not Just Reality*, 1973), and a talking face by Fred Parke (*Talking Face*, 1974). Although the imagery was extremely primitive by today's standards, the presentations of lip-synced facial animation and linked-appendage figure animation were impressive demonstrations well ahead of their time.

In 1972, Chuck Csuri founded the Computer Graphics Research Group (CGRG) at Ohio State with the focus of bringing computer technology to bear on creating animation [10]. Tom DeFanti produced the Graphics Symbiosis System (GRASS) in the early 1970s that scripted interactive control of animated objects on a vector display device. Later in the 1970s, CGRG produced animations using a real-time video playback system developed at North Carolina State University under the direction of John Staudhammer. Software developed at CGRG compressed frames of animation and stored them to disk. During playback, the compressed digital frames were retrieved from the disk and piped to the special-purpose hardware, which took the digital information, decompressed it on the fly, and converted it into a video signal for display on a standard television. The animation was driven by the ANIMA II language [15]. In the mid-1980s, Julian Gomez developed TWIXT [43], a track-based key-frame animation system.

In 1973, the first computer-language-based key-frame animation system, ANTICS, was developed by Alan Kitching at the Atlas Computer Laboratory under the auspices of the Royal College of Art in the United Kingdom. [19] [20]. ANTICS is a Fortran software package specifically designed for animators and graphic designers. It is a two-dimensional system that provides capabilities analogous to traditional cel animation.

In the mid-1970s, Norm Badler at the University of Pennsylvania conducted investigations into posing a human figure. He developed a constraint system to move the figure from one pose to another. He has continued this research and established the Center for Human Modeling and Simulation at the University of Pennsylvania. *Jack* is a software package developed at the center that supports the positioning and animation of anthropometrically valid human figures in a virtual world [7].

In the late 1970s, the New York Institute of Technology (NYIT) produced several computer animation systems, thanks to individuals such as Ed Catmull and Alvy Ray Smith [24]. At the end of the 1970s, NYIT embarked on an ambitious project to produce a wholly computer-generated feature film using three-dimensional computer animation, titled *The Works*. While the project was never completed, excerpts were shown at several SIGGRAPH conferences as progress was made. The excerpts demonstrated high-quality rendering, jointed figures, and interacting objects. The system used at NYIT was BBOP, a three-dimensional key-frame figure animation system [40].

In 1974, the first computer animation nominated for an Academy Award, *Hunger*, was produced by Rene Jodoin; it was directed and animated by Peter Foldes. This piece used a 2½ D system that depended heavily on object shape modification and line interpolation techniques [6]. The system was developed by Nestor Burtnyk and Marceli Wein at the National Research Council of Canada in conjunction with the National Film Board of Canada. *Hunger* was the first animated story using computer animation.

In the early 1980s Daniel Thalmann and Nadia Magnenat-Thalmann started work in computer animation at the University of Montreal [24]. Over the years, their labs have produced several impressive animations, including *Dream Flight* (N. Magnenat-Thalmann, D. Thalmann, P. Bergeron, 1982), *Tony de Peltrie* (P. Bergeron, 1985), and *Rendez-vous à Montréal* (N. Magnenat-Thalmann and D. Thalmann, 1987).

Others who advanced computer animation during this period were Ed Emshwiller at NYIT, who demonstrated moving texture maps in *Sunstone* (1979); Jim Blinn, who produced the *Voyager* flyby animations at the Jet Propulsion Laboratory (1979); Don Greenberg, who used architectural walkthroughs of the Cornell University campus (1971); and Nelson Max at the Education Development Center, who animated space-filling curves (1972).

Commercial efforts at computer animation first occurred in the late 1960s with Lee Harrison's SCANIMATE system based on analog computing technology [36]. Digital technology soon took over and the mid- to late-1970s saw the first serious hints of commercial three-dimensional digital computer animation. Tom DeFanti developed the GRASS at Ohio State University (1976), a derivative of which was used in the computer graphics sequences of the first *Star Wars* film (1977). In addition to *Star Wars*, films such as *Future World* (1976), *Alien* (1979), and *Looker*[12] (1981) began to incorporate simple computer animation as examples of advanced technology. This was an exciting time for those in the research labs wondering if computer animation would ever see the light of day. One of the earliest

---

[12]The film *Looker* is interesting as an early commentary on the potential use of digital actors in the entertainment industry.

companies to use three-dimensional computer animation was the Mathematical Application Group Inc. (MAGI), which used a ray-casting algorithm to provide scientific visualizations. MAGI also adapted its technique to produce early commercials for television.

1.5.2 **The middle years (the 1980s)**

The 1980s saw a more serious move by entrepreneurs into commercial animation. Computer hardware advanced significantly with the introduction of the VAX computer in the 1970s and the IBM PC at the beginning of the 1980s. Hardware z-buffers were produced by companies such as Raster Tech and Ikonas, Silicon Graphics was formed, and flight simulators based on digital technology were taking off because of efforts by the Evans and Sutherland Corporation. These hardware developments were making the promise of cost-effective computer animation to venture capitalists. At the same time, graphics software was getting more sophisticated: Turner Whitted introduced anti-aliased ray tracing (*The Compleat Angler*, 1980), Loren Carpenter produced a flyby of fractal terrain (*Vol Libre*, 1980), and Nelson Max produced several films about molecules as well as one of the first films animating waves (*Carla's Island*, 1981). Companies such as Alias, Wavefront, and TDI were starting to produce sophisticated software tools making advanced rendering and animation available off-the-shelf for the first time.

Animation houses specializing in three-dimensional computer animation started to appear. Television commercials, initially in the form of flying logos, provided a profitable area where companies could hone their skills. Demo reels appeared at SIGGRAPH produced by the first wave of computer graphics companies such as Information International Inc. (III, or triple-I), Digital Effects, MAGI, Robert Abel and Associates, and Real Time Design (ZGRASS).

The first four companies combined to produce the digital imagery in Disney's TRON (1982), which was a landmark movie for its (relatively) extensive use of a computer-generated environment in which graphical objects were animated. Previously, the predominant use of computer graphics in movies had been to show a monitor (or simulated projection) of something that was supposed to be a computer graphics display (*Futureworld*, 1976; *Star Wars*, 1977; *Alien*, 1979; *Looker*, 1981). Still, in *TRON*, the computer-generated imagery was not meant to simulate reality; the action takes place inside a computer, so a computer-generated look was consistent with the story line.

At the same time that computer graphics were starting to find their way into the movies it was becoming a more popular tool for generating television commercials. As a result, more computer graphics companies surfaced, including Digital Pictures, Image West, Cranston-Csuri Productions, Pacific Data Images, Lucasfilm, Marks and Marks, Digital Productions, and Omnibus Computer Graphics.

Most early use of synthetic imagery in movies was incorporated with the intent that it would appear as if computer generated. The other use of computer animation during this period was to "do animation." That is, the animations were meant not to fool the eye into thinking that what was being seen was real but rather to replace the look and feel of two-dimensional conventional animation with that of three-dimensional computer animation. Of special note are the award-winning animations produced by Lucasfilm and, later, by Pixar:

The Adventures of Andre and Wally B. (1984)—first computer animation demonstrating motion blur
Luxo Jr. (1986)—nominated for an Academy Award

Red's Dream (1987)
Tin Toy (1988)—first computer animation to win an Academy Award
Knick Knack (1989)
Geri's Game (1997)—Academy Award winner

These early animations paved the way for three-dimensional computer animation to be accepted as an art form. They were among the first fully computer-generated three-dimensional animations to be taken seriously as animations, irrespective of the technique involved. Another early piece of three-dimensional animation, which integrated computer graphics with conventional animation, was *Technological Threat* (1988, Kroyer Films). This was one of three films nominated for an Academy Award as an animated short in 1989; *Tin Toy* came out the victor.

One of the early uses of computer graphics in film was to model and animate spacecraft. Working in (virtual) outer space with spacecraft has the advantages of simple illumination models, a relatively bare environment, and relatively simple animation of rigid bodies. In addition, spacecraft are usually modeled by relatively simple geometry—as is the surrounding environment (planets)—when in flight. *The Last Starfighter* (1984, Digital Productions) used computer animation instead of building models for special effects; the computer used, the Cray X-MP, even appeared in the movie credits. The action takes place in space as well as on planets; computer graphics were used for the scenes in space, and physical models were used for the scenes on a planet. Approximately twenty minutes of computer graphics was used in the movie. While it is not hard to tell when the movie switches between graphical and physical models, this was the first time graphics were used as an extensive part of a live-action film in which the graphics were supposed to look realistic (i.e., *special effects*).

1.5.3 Animation comes of age (the mid-1980s and beyond)

As modeling, rendering, and animation became more sophisticated and the hardware became faster and inexpensive, quality computer graphics began to spread to the Internet, television commercials, computer games, and stand-alone game units. In film, computer graphics help to bring alien creatures to life. Synthetic alien creatures, while they should appear to be real, do not have to match specific audience expectations. *Young Sherlock Holmes* (1986, ILM) was the first to place a synthetic character in a live-action feature film. An articulated stained glass window comes to life and is made part of the live action. The light sources and movements of the camera in the live action had to be mimicked in the synthetic environment, and images from the live action were made to refract through the synthetic stained glass. In *The Abyss* (1989, ILM), computer graphics are used to create an alien creature that appears to be made from water. Other notable films in which synthetic alien creatures are used are *Terminator II* (1991, ILM), *Casper* (1995, ILM), *Species* (1995, Boss Film Studios), and *Men in Black* (1997, ILM).

A significant advance in the use of computer graphics for the movies came about because of the revolution in cheap digital technology, which allowed film sequences to be stored digitally. Once the film is stored digitally, it is in a form suitable for digital special effects processing, digital compositing, and the addition of synthetic elements. For example, computer graphics can be used to remove the mechanical supports of a prop or to introduce digital explosions or laser blasts. For the most part, this resides in the two-dimensional realm, thus it is not the focus of this book. However, with the advent of digital techniques for two-dimensional compositing, sequences are more routinely available in digital representations, making them amenable to a variety of digital postprocessing techniques. The first digital blue screen matte extraction was in *Willow* (1988, ILM). The first digital wire removal was in *Howard the Duck* (1986, ILM). In *True Lies* (1994, Digital Domain), digital techniques inserted atmospheric distortion to show engine heat. In *Forrest Gump* (1994, ILM), computer graphics inserted a ping-pong ball in a sequence showing an extremely fast action game, inserted a new character into old film footage, and enabled the illusion of a double amputee as played by a completely able actor. In *Babe* (1995, Rhythm & Hues), computer graphics were used to move the mouths of animals and fill in the background uncovered by the movement. In *Interview with a Vampire* (1994, Digital Domain), computer graphics were used to curl the hair of a woman during her transformation into a vampire. In this case, some of the effect was created using three-dimensional graphics, which were then integrated into the scene by two-dimensional techniques. More recently, The Matrix series (*The Matrix*, 1999; *The Matrix Reloaded*, 2003; *The Matrix Revolutions*, 2003, Groucho II Film Partnership) popularized the use of a digital visual effect used to show characters dodging bullets—slow motion was digitally enhanced to show unfilmable events such as a flying bullets.

A popular graphical technique for special effects is the use of particle systems. One of the earliest examples is in *Star Trek II: The Wrath of Khan* (1982, Lucasfilm), in which a wall of fire sweeps over the surface of a planet. Although by today's standards the wall of fire is not very convincing, it was an important step in the use of computer graphics in movies. Particle systems are also used in *Lawnmower Man* (1992, Angel Studios, Xaos), in which a character disintegrates into a swirl of small spheres. The modeling of a comet's tail in the opening sequence of the television series *Star Trek: Deep Space Nine* (1993, Paramount Television) is a more recent example of a particle system. In a much more ambitious and effective application, *Twister* (1996, ILM) uses particle systems to simulate a tornado.

More challenging is the use of computer graphics to create realistic models of creatures with which the audience is intimately familiar. *Jurassic Park* (1993, ILM) is the first example of a movie that completely integrates computer graphics characters (dinosaurs) of which the audience has fairly specific expectations. Of course, there is still some leeway here, because the audience does not have precise knowledge of how dinosaurs look. *Jumanji* (1995, ILM) takes on the ultimate task of modeling creatures for which the audience has precise expectations: various jungle animals. Most of the action is fast and blurry, so the audience does not have time to dwell on the synthetic creatures visually, but the result is very effective. To a lesser extent, *Batman Returns* (1995, PDI) does the same thing by providing "stunt doubles" of Batman in a few scenes. The scenes are quick and the stunt double is viewed from a distance, but it was the first example of a full computer graphics stunt double in a movie. More recently, the Spider Man movies (2002-present, Sony) make extensive use of synthetic stunt doubles. Use of synthetic stunt doubles in film is now commonplace.

Computer graphics show much potential for managing the complexity in crowd scenes. PDI used computer graphics to create large crowds in the Bud Bowl commercials of the mid 1980s. In feature films, some of the well-known crowd scenes occur in the wildebeest scene in *Lion King* (1994, Disney), the alien charge in *Starship Troopers* (1997, Tippet Studio), synthetic figures populating the deck of the ship in *Titanic* (1998, ILM), and various crowds in the Star Wars films (1977–2005, Lucasfilm) and The Lord of the Rings trilogy (2001–2003, New Line Cinema).

A holy grail of computer animation is to produce a synthetic human characters indistinguishable from a real person. Early examples of animations using "synthetic actors" are *Tony de Peltrie* (1985, P. Bergeron), *Rendez-vous à Montréal* (1988, D. Thalmann), *Sextone for President* (1989, Kleiser-Walziac Construction Company), and *Don't Touch Me* (1989, Kleiser-Walziac Construction

Company). However, it is obvious to viewers that these animations are computer generated. Recent advances in illumination models and texturing have produced human figures that are much more realistic and have been incorporated into otherwise live-action films.

Synthetic actors have progressed from being distantly viewed stunt doubles and passengers on a boat to assuming central roles in various movies: the dragon in *Dragonheart* (1996, Tippett Studio, ILM); the Jello-like main character in *Flubber* (1997, ILM); the aliens in *Mars Attacks* (1996, ILM); and the ghosts in *Casper* (1995, ILM). The first fully articulated humanoid synthetic actor integral to a movie was the character Jar-Jar Binks in *Star Wars: Episode I* (1999, ILM). More recently, Gollum in the *Lord of the Rings: The Return of the King* (2004, New Line Cinema[13]) and Dobby in the Harry Potter series (2001–2011, Warner Bros.) display actor-like visual qualities as well as the personality of a live actor not previously demonstrated in computer-generated characters.

However, a revolution is brewing in the realistic digital representation of human actors. At this point, we are not quite to the place where a digital human is indistinguishable from a live actor, but that time is drawing nearer. An early attempt at such realism was undertaken by *Final Fantasy: The Spirits Within* (2001, Chris Lee Productions), which incorporated full body and facial animation of human figures. The hair animation was particularly convincing; however, the facial animation was stiff and unexpressive. Several films effectively manipulated the appearance of facial features as a special effect including *The Curious Case of Benjamin Button* (2008, Paramount Pictures) and *Alice in Wonderland* (2010, Walt Disney Pictures). *Avatar* (2009, 20th Century Fox) with full body/face motion captures animated human-like creatures by mapping the performance of actors to the synthetic creatures. However, the most ambitious attempt at a synthetic actor is in the film *TRON: Legacy* (2010, Walt Disney Pictures), which animates a computer graphics version of a young Jeff Bridges. In this case, the audience (at least those of us old enough to remember what Jeff Bridges looked like when he was young) have a specific expectation of what a young Jeff Bridges should look like. The representation is effective for the most part; however, the film editing prevents the audience from closely inspecting the young Jeff Bridges and even so, there is an unreal eeriness about the figure. But this shows that synthetic actors are on their way to a theater near you. Advances in hair, clothes, and skin have paved the way, but facial animation has still not been totally conquered.

Of course, one use of computer animation is simply to "do animation"; computer graphics are used to produce animated pieces that are essentially three-dimensional cartoons that would otherwise be done by more traditional means. The animation does not attempt to fool the viewer into thinking anything is real; it is meant simply to entertain. The film *Hunger* falls into this category, as do the Lucasfilm/Pixar animations. *Toy Story* is the first full-length, fully computer-generated three-dimensional animated feature film. Other feature-length three-dimensional cartoons soon emerged, such as *Antz* (1998, PDI), *A Bug's Life* (1998, Pixar), *Toy Story 2* (1999, Pixar), *Shrek* (2001, PDI), and *Shrek 2* (2004, PDI). In 2002, *Shrek* won the first-ever Academy Award for Best Animated Feature.

Many animations of this type have been made for television. In an episode of *The Simpsons* (1995, PDI), Homer steps into a synthetic world and turns into a three-dimensional computer-generated character. There have been popular television commercials involving computer animation—too many to mention at this point. Many Saturday morning cartoons are now produced using three-dimensional

---

[13]For brevity, only the first production company itemized on the Internet Movie Database Web site (www.imdb.com) is given when more than one production company is listed.

computer animation. Because many images are generated to produce an animation, the rendering used in computer-animated weekly cartoons tends to be computationally efficient.

An example of rendering at the other extreme is *Bunny* (1999, Blue Sky Productions), which received an Academy Award for animated short. *Bunny* uses high-quality rendering in its imagery, including ray tracing and radiosity, as does *Ice Age* (2002, Blue Sky Productions). *The Incredibles* (2004, Disney/Pixar), which garnered another Academy Award for Pixar, included hair animation, sub-surface scattering for illuminating skin, cloth animation, and skin-deforming muscle models. *Polar Express* (2004, Warner Bros.) advanced the use of motion capture technology to capture full body and face motion in animating this children's story.

Computer animation is now well-established as a (some would say "the") principal medium for doing animation. Indeed, at one time, both *Jurassic Park* and *Shrek 2* were on the top ten list of all time worldwide box office grossing movies [4].

Three-dimensional computer graphics are also playing an increasing role in the production of conventional hand-drawn animation. Computer animation has been used to model three-dimensional elements in hand-drawn environments. The previously mentioned *Technological Threat* (1988) is an early animation that combined computer-animated characters with hand-drawn characters to produce an entertaining commentary on the use of technology. Three-dimensional environments were constructed for conventionally animated figures in *Beauty and the Beast* (1991, Disney) and *Tarzan* (1999, Disney); three-dimensional synthetic objects, such as the chariots, were animated in conventionally drawn environments in *Prince of Egypt* (1998, DreamWorks). Because photorealism is not the objective, the rendering in such animation is done to blend with the relatively simple rendering of hand-drawn animation.

Lastly, morphing, even though it is a two-dimensional animation technique, should be mentioned because of its use in some films and its high impact in television commercials. This is essentially a two-dimensional procedure that warps control points (or feature lines) of one image into the control points (feature lines) of another image while the images themselves are blended. In *Star Trek IV: The Voyage Home* (1986, ILM), one of the first commercial morphs occurred in the back-in-time dream sequence. In *Willow* (1988, ILM), a series of morphs changes one animal into another. This technique is also used very effectively in *Terminator II* (1991, ILM). In the early 1990s, PDI promoted the use of morphing in various productions. Michael Jackson's music video *Black and White*, in which people's faces morph into other faces, did much to popularize the technique. In a Plymouth Voyager commercial (1991, PDI) the previous year's car bodies and interiors morph into the new models, and in an Exxon commercial (1992, PDI) a car changes into a tiger. Morphing remains a useful and popular technique.

1.6 Summary

Computer graphics and computer animation have created a revolution in visual effects and animation production. Advances are still being made, and new digital techniques are finding a receptive audience. Yet there is more potential to be realized as players in the entertainment industry demand their own special look and each company tries to establish a competitive edge in the imagery it produces.

Computer animation has come a long way since the days of Ivan Sutherland and the University of Utah. Viewed in the context of the historical development of animation, the use of digital technology is indeed both a big and an important step. With the advent of low-cost computing and desktop video, animation is now within reach of more people than ever. It remains to be seen how the limits of the technology will be pushed as new and interesting ways to create moving images are explored. Of particular importance is the evolution of human figure modeling, rendering, and animation. This technology is on the verge of making a significant impact on both special effects and animated films as synthetic figures become more indistinguishable from real people. In addition to human figure animation, more sophisticated mathematical formulations for modeling cloth, water, clouds, and fire are generating more complex and interesting animations. As these and other advanced techniques are refined in the research labs, they filter down into off-the-shelf software. This software is then available to the highly talented digital artists who create the dazzling visuals that continually astound and amaze us in movie theaters and on the internet.

References

[1] Anderson J, Anderson B. The Myth of Persistence of Vision Revisited. Journal of Film and Video 1993;45(1) Spring:3–12.

[2] Anderson J, Fisher B. The Myth of Persistence of Vision. Journal of the University Film Association 1978; XXX(4) Fall:3–8.

[3] Balzer R. Optical Amusements: Magic Lanterns and Other Transforming Images; A Catalog of Popular Entertainments. Richard Balzer; 1987.

[4] Box Office Mojo . All Time Worldwide Box Office Grosses, http://www.boxofficemojo.com/alltime/world; 2006.

[5] Burns P. The Complete History of the Discovery of Cinematography, http://www.-precinemahistory.net/ introduction.htm; 2000.

[6] Burtnyk N, Wein M. Computer Animation of Free Form Images. In: Proceedings of the 2nd Annual Conference on Computer Graphics and Interactive Techniques, SIGGRAPH 75 (Bowling Green, Ohio, June 25–27, 1975). New York: ACM Press; p. 78–80.

[7] The Center for Human Modeling and Simulation. Welcome to Human Modeling and Simulation, http:// www.cis.upenn.edu~hms/home.html; June 2006.

[8] Conrac Corp . Raster Graphics Handbook. New York: Van Nostrand Reinhold; 1985.

[9] Coynik D. Film: Real to Reel. Evanston, Ill.: McDougal, Littell; 1976.

[10] Csurivison Ltd . The Digital Fine Art History of Charles Csuri, http://www.csuri.com/charles-csuri/art-history-0_0.php; June 2006.

[11] Cutting J. Perception with an Eye for Motion. Cambridge, Mass.: MIT Press; 1986.

[12] Davis D. Art and the Future: A History/Prophecy of the Collaboration Between Science, Technology, and Art. New York: Praeger Publishers; 1973.

[13] Ehrenstein WH. Basics of Seeing Motion. Arq Bras Oftalmal Sept./Oct. 2003;66(5):44–53.

[14] Geometric Design and Computation. GDC: History, http://www.cs.utah.edu/gdc/history; June 2006.

[15] Hackathorn R. Anima II: A 3-D Color Animation System. In: George J, editor. Computer Graphics (Proceedings of SIGGRAPH 77) (July 1977, San Jose, Calif.), vol. 11(2). p. 54–64.

[16] Henne M, Hickel H, Johnson E, Konishi S. The Making of *Toy Story*. In: (Proceedings of Comp Con 96) (February 25–28, 1996, San Jose, Calif.). p. 463–8.

[17] Horton M, Mumby C, Owen S, Pank B, Peters D. Quantel On-Line, Non-linear Editing, http://www.quantel. com/editingbook/index.html; 2000.

[18] Hunt M. Cinema: 100 Years of the Moving Image, http://www.angelfire.com/vt/mhunt/cinema.html; 2000.

[19] Kitching A. Computer Animation—Some New ANTICS. British Kinematography Sound and Television December 1973:372–86.

[20] Kitching A. The Antics computer animation system, www.chilton-computing.org.uk/acl/applications/graphics/p003.htm; June 2005.

[21] Korein J, Badler N. Temporal Anti-Aliasing in Computer Generated Animation. In: Computer Graphics (Proceedings of SIGGRAPH 83), (July 1983, Detroit, Mich.), vol. 17(3). p. 377–88.

[22] Lasseter J. Principles of Traditional Animation Applied to 3D Computer Animation. In: Stone MC, editor: Computer Graphics (Proceedings of SIGGRAPH 87) (July 1987, Anaheim, Calif.), vol. 21(4). p. 35–44.

[23] Laybourne K. The Animation Book: A Complete Guide to Animated Filmmaking—from Flip-Books to Sound Cartoons to 3-D Animation. New York: Three Rivers Press; 1998.

[24] Magnenat-Thalmann N, Thalmann D. Computer Animation: Theory and Practice. New York: Springer-Verlag; 1985.

[25] Magnenat-Thalmann N, Thalmann D. New Trends in Animation and Visualization. New York: John Wiley & Sons; 1991.

[26] Maltin L. Of Mice and Magic: A History of American Animated Cartoons. New York: Penguin Books; 1987.

[27] Marey E. Animal Mechanism: A Treatise on Terrestrial and Aerial Locomotion. New York: Appleton and Co; 1874.

[28] Mascelli J. The Five C's of Cinematography. Hollywood, Calif.: Cine/Grafic Publications; 1965.

[29] The Museum of Broadcast Communications. Standards, http://www.museum.tv/archives/etv/S/htmlS/standards/standards.htm.

[30] Muybridge E. Animals in Motion. New York: Dover Publications; 1957.

[31] Muybridge E. The Human Figure in Motion. New York: Dover Publications; 1955.

[32] Potmesil M, Chadkravarty I. Modeling Motion Blur in Computer Generated Images. In: Computer Graphics (Proceedings of SIGGRAPH 83) (July 1983, Detroit, Mich.), vol. 17(3). p. 389–400.

[33] Pryor B. Opus Communications, "VIDEOFAQ#1: What the Heck Is 'Non-linear' Editing Anyway?", http://www.opuskc.com/vf1.html; 2000.

34] Poynton C. Digital Video and HDTV Algorithms and Interfaces. San Francisco, Calif.: Morgan-Kaufmann; 2003.

[35] Roget P. Explanation of an Optical Deception in the Appearance of the Spokes of a Wheel Seen through Vertical Apertures. Philosophical Transactions of the Royal Society of London 1825 (presented in 1824);115:131–40.

[36] Sieg D. Welcome to the Scanimate Site—history of computer animation—early analog computers, http://scanimate.net; 2004.

[37] Siegel H. An Overview of Computer Animation & Modelling, In: Computer Animation Proceedings of the Conference Held at Computer Graphics 87. London: October 1987. p. 27–37.

[38] Solomon C. The History of Animation: Enchanted Drawings. New York: Wings Books; 1994.

[39] Steinman R, Pizlo Z, Pizlo F. Phi is not beta, and why Wertheimer's discovery launched the Gestalt revolution. Vis Res August 2000;40(17):2257–64.

[40] Stern G. Bbop: A Program for 3-Dimensional Animation. Nicograph December 1983;83:403–4.

[41] Sutherland I. SKETCHPAD: A Man-Machine Graphical Communication System. Ph.D. dissertation, MIT; 1963.

[42] Synthetic Aperture . Synthetic Aperture, http://www.synthetic-ap.com/tips/index.html; 2000.

[43] Tajchman EJ. The Incredible Electronic Machine. Videography November 1977:22–4.

[44] Thomas B. Disney's Art of Animation from Mickey Mouse to "Beauty and the Beast". New York: Hyperion; 1991.

[45] Thomas F, Johnson O. The Illusion of Life. New York: Hyperion; 1981.

[46] Tribeca Technologies, LLC . White Paper: The History of Video, http://tribecatech.com/histvide.htm; 2000.

[47] Wabash College Theater . Georges Méliès, http://www.wabash.edu/depart/theater/THAR4/Melies.htm; 2000.

[48] Whittaker R. Linear and Non-linear Editing, http://www.internetcampus.com/tvp056.htm; 2000.

[49] Williams R. The Animator's Survival Kit. New York: Faber and Faber; 2002.

[50] Wood G. Living Dolls—A Magical History of the Quest for Mechanical Life. London: Faber and Faber Limited; 2002.

Technical Background

This chapter serves as a prelude to the computer animation techniques presented in the remaining chapters. It is divided into two sections. The first serves as a quick review of the basics of the computer graphics display pipeline and discusses potential sources of error when dealing with graphical data. It is assumed that the reader has already been exposed to transformation matrices, homogeneous coordinates, and the display pipeline, including the perspective projection; this section concisely reviews these topics. The second section covers various orientation representations that are important for the discussion of orientation interpolation in Chapter 3.3.

2.1 Spaces and transformations

Much of computer graphics and computer animation involves transforming data (e.g., [2] [7]). Object data are transformed from a defining space into a world space in order to build a synthetic environment. Object data are transformed as a function of time in order to produce animation. Finally, object data are transformed in order to view the object on a screen. The workhorse transformational representation of graphics is the 4 × 4 transformation matrix, which can be used to represent combinations of three-dimensional rotations, translations, and scales as well as perspective projection.

A coordinate space can be defined by using a left- or a right-handed coordinate system (see Figure 2.1a,b). Left-handed coordinate systems have the x-, y-, and z-coordinate axes aligned as the thumb, index finger, and middle finger of the left hand are arranged when held at right angles to each other in a natural pose: extending the thumb out to the side of the hand, extending the index finger coplanar with the palm, and extending the middle finger perpendicular to the palm. The right-handed coordinate system is organized similarly with respect to the right hand. These configurations are inherently different; there is no series of pure rotations that transforms a left-handed configuration of axes into a right-handed configuration. Which configuration to use is a matter of convention. It makes no difference as long as everyone knows and understands the implications. Another arbitrary convention is the axis to use as the up vector. Some application areas assume that the y-axis is "up." Other applications assume that the z-axis is "up." As with handedness, it makes no difference as long as everyone is aware of the assumption being made. In this book, the y-axis is considered "up."

This section first reviews the transformational spaces through which object data pass as they are massaged into a form suitable for display. Then, the use of homogeneous representations of points and the 4 × 4 transformation matrix representation of three-dimensional rotations, translation, and scale are reviewed. Next come discussions of representing arbitrary position and orientation by a series

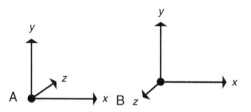

FIGURE 2.1

(a) Left-handed and (b) right-handed coordinate systems.

of matrices, representing compound transformations in a matrix, and extracting a series of basic transformations from a compound matrix. The display pipeline is then described in terms of the transformation matrices used to affect it; the discussion is focused on transforming a point in space. In the case of transforming vectors, the computation is slightly different (see Appendix B.3.2 for details). This section concludes with a discussion of error considerations, including orthonormalization of a rigid transformation matrix. Unless stated otherwise, space is assumed to be three-dimensional and right-handed.

2.1.1 The display pipeline

The *display pipeline* refers to the transformation of object data from its original defining space through a series of intermediate spaces until its final mapping onto the screen. The object data are transformed into different spaces in order to efficiently compute illumination, clip the data to the view volume, and perform the perspective transformation. This section reviews these spaces, their properties, the transformations that map data from one space to the next, and the parameters used to specify the transformations. The names used for these spaces vary from text to text, so they will be reviewed here to establish a consistent naming convention for the rest of the book. While an important process that eliminates lines and parts of lines that are not within the viewable space, clipping is not relevant to motion control and therefore is not covered.

The space in which an object is originally defined is referred to as *object space*. The data in object space are usually centered at the origin and often are created to lie within some limited standard range such as -1 to $+1$. The object, as defined by its data points (which are also referred to as its *vertices*), is transformed, usually by a series of rotations, translations, and scales, into *world space*, the space in which objects are assembled to create the environment to be viewed. Object space and world space are commonly right-handed spaces.

World space is also the space in which light sources and the observer are placed. For purposes of this discussion, *observer position* is used synonymously and interchangeably with *camera position* and *eye position*. The observer parameters include its *position* and its *orientation*. The orientation is fully specified by the *view direction* and the *up vector*. There are various ways to specify these orientation vectors. Sometimes the view direction is specified by giving a *center of interest* (COI), in which case the view direction is the vector from the observer or eye position (EYE) to the center of interest. The eye position is also known as the *look-from point*, and the COI is also known as the *look-to point*. The default orientation of "straight up" is defined as the observer's up vector being perpendicular to the view direction and in the plane defined by the view direction and the global y-axis. A rotation away from this up direction will effect a tilt of the observer's head.

In order to efficiently project the data onto a view plane, the data must be defined relative to the camera, in a camera-centric coordinate system (u, v, w); the v-axis is the observer's y-axis, or *up vector*, and the w-axis is the observer's z-axis, or *view vector*. The u-axis completes the local coordinate system of the observer. For this discussion, a left-handed coordinate system for the camera is assumed. These vectors can be computed in the right-handed world space by first taking the cross-product of the view direction vector and the y-axis, forming the u-vector, and then taking the cross-product of the u-vector and the view direction vector to form v (Eq. 2.1).

$$
\begin{aligned}
w &= COI - EYE &&\text{view direction vector} \\
u &= w \times (0, 1, 0) &&\text{cross product with y-axis} \\
v &= u \times w
\end{aligned}
\tag{2.1}
$$

After computing these vectors, they should be normalized in order to form a unit coordinate system at the eye position. A world space data point can be defined in this coordinate system by, for example, taking the dot product of the vector from the eye to the data point with each of the three coordinate system vectors.

Head-tilt information can be provided in one of two ways. It can be given by specifying an angle deviation from the straight-up direction. In this case, a head-tilt rotation matrix can be formed and incorporated in the world-to-eye-space transformation or can be applied directly to the observer's default u-vector and v-vector.

Alternatively, head-tilt information can be given by specifying an up-direction vector. The user-supplied up-direction vector is typically not required to be perpendicular to the view direction as that would require too much work on the part of the user. Instead, the vector supplied by the user, together with the view direction vector, defines the plane in which the up vector lies. The difference between the user-supplied up-direction vector and the up vector is that the up vector by definition is perpendicular to the view direction vector. The computation of the perpendicular up vector, v, is the same as that outlined in Equation 2.1, with the user-supplied up direction vector, UP, replacing the y-axis (Eq. 2.2).

$$
\begin{aligned}
w &= COI - EYE &&\text{view direction vector} \\
u &= w \times UP &&\text{cross product with user's up vector} \\
v &= u \times w
\end{aligned}
\tag{2.2}
$$

Care must be taken when using a default up vector. Defined as perpendicular to the view vector and in the plane of the view vector and global y-axis, it is undefined for straight-up and straight-down views. These situations must be dealt with as special cases or simply avoided. In addition to the undefined cases, some observer motions can result in unanticipated effects. For example, the default head-up orientation means that if the observer has a fixed center of interest and the observer's position arcs directly, or almost so, over the center of interest, then just before and just after being directly overhead, the observer's up vector will instantaneously rotate by up to 180 degrees (see Figure 2.2).

In addition to the observer's position and orientation, the *field of view* (fov) has to be specified to fully define a viewable volume of world space. This includes an *angle of view* (or the equally useful *half angle of view*), *near clipping distance*, and *far clipping distance* (sometimes the terms *hither* and *yon* are used instead of *near* and *far*). The fov information is used to set up the *perspective projection*.

The visible area of world space is formed by the observer position and orientation, angle of view, near clipping distance, and far clipping distance (Figure 2.3). The angle of view defines the angle made between the upper and lower clipping planes, symmetric around the view direction. If this angle is

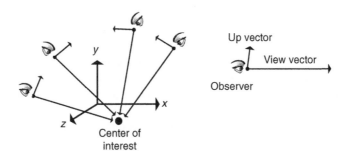

FIGURE 2.2

The up vector flips as the observer's position passes straight over the center of interest.

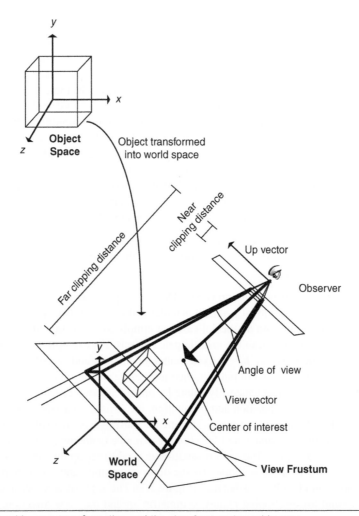

FIGURE 2.3

Object-space to world-space transformation and the view frustum in world space.

different than the angle between the left and right side clipping planes, then the two angles are identified as the *vertical angle of view* and the *horizontal angle of view*. The far clipping distance sets the distance beyond which data are not viewed. This is used, for example, to avoid processing arbitrarily complex data that is too far away to have much, if any, impact on the final image. The near clipping distance similarly sets the distance before which data are not viewed. This is primarily used to avoid division by zero in the perspective projection. These define the *view frustum*, the six-sided volume of world space containing data that need to be considered for display.

The view specification discussed above is somewhat simplified. Other view specifications use an additional vector to indicate the orientation of the projection plane, allow an arbitrary viewport to be specified on the plane of projection that is not symmetrical about the view direction to allow for off-center projections, and allow for a parallel projection. The reader should refer to standard graphics texts such as the one by Foley et al. [2] for an in-depth discussion of such view specifications.

In preparation for the perspective transformation, the data points defining the objects are usually transformed from world space to *eye space*. In *eye space*, the observer is positioned along the z-axis with the line of sight made to coincide with the z-axis. This allows the depth of a point, and therefore perspective scaling, to be dependent only on the point's z-coordinate. The exact position of the observer along the z-axis and whether the eye space coordinate system is left- or right-handed vary from text to text. For this discussion, the observer is positioned at the origin looking down the positive z-axis in left-handed space. In eye space as in world space, lines of sight emanate from the observer position and diverge as they expand into the visible view frustum, whose shape is often referred to as a *truncated pyramid*.

The *perspective transformation* transforms the objects' data points from eye space to *image space*. The perspective transformation can be considered as taking the observer back to negative infinity in z and, in doing so, makes the lines of sight parallel to each other and to the (eye space) z-axis. The pyramid-shaped view frustum becomes a rectangular solid, or cuboid, whose opposite sides are parallel. Thus, points that are farther away from the observer in eye space have their x- and y-coordinates scaled down more than points that are closer to the observer. This is sometimes referred to as *perspective foreshortening*. This is accomplished by dividing a point's x- and y-coordinates by the point's z-coordinate. Visible extents in image space are usually standardized into the -1 to $+1$ range in x and y and from 0 to 1 in z (although in some texts, visible z is mapped into the -1 to $+1$ range). Image space points are then scaled and translated (and possibly rotated) into *screen space* by mapping the visible ranges in x and y (-1 to $+1$) into ranges that coincide with the viewing area defined in the coordinate system of the window or screen; the z-coordinates can be left alone. The resulting series of spaces is shown in Figure 2.4.

Ray casting (ray tracing without generating secondary rays) differs from the above sequence of transformations in that the act of tracing rays from the observer's position out into world space implicitly accomplishes the perspective transformation. If the rays are constructed in world space based on pixel coordinates of a virtual frame buffer positioned in front of the observer, then the progression through spaces for ray casting reduces to the transformations shown in Figure 2.5. Alternatively, data can be transformed to eye space and, through a virtual frame buffer, the rays can be formed in eye space.

In any case, animation is typically produced by one or more of the following: modifying the position and orientation of objects in world space over time, modifying the shape of objects over time, modifying display attributes of objects over time, transforming the observer position and orientation in world space over time, or some combination of these transformations.

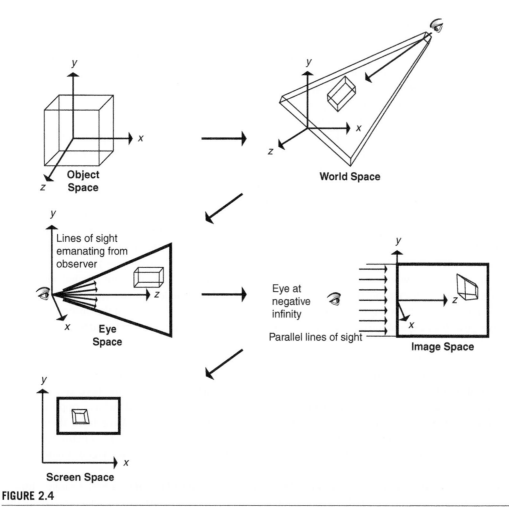

FIGURE 2.4

Display pipeline showing transformation between spaces.

2.1.2 Homogeneous coordinates and the transformation matrix

Computer graphics often uses a *homogeneous representation* of a point in space. This means that a three-dimensional point is represented by a four-element vector.[1] The coordinates of the represented point are determined by dividing the fourth component into the first three (Eq. 2.3).

$$\left(\frac{x}{w}, \frac{y}{w}, \frac{z}{w}\right) \equiv [x, y, z, w] \tag{2.3}$$

---

[1]Note the potential source of confusion in the use of the term *vector* to mean (1) a direction in space or (2) a $1 \times n$ or $n \times 1$ matrix. The context in which *vector* is used should make its meaning clear.

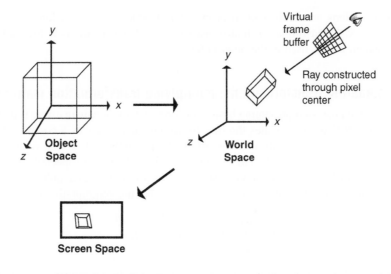

FIGURE 2.5

Transformation through spaces using ray casting.

Typically, when transforming a point in world space, the fourth component will be one. This means a point in space has a very simple homogeneous representation (Eq. 2.4).

$$(x, y, z) \equiv [x, y, z, 1] \tag{2.4}$$

The basic transformations of rotate, translate, and scale can be kept in 4×4 transformation matrices. The 4×4 matrix is the smallest matrix that can represent all of the basic transformations. Because it is a square matrix, it has the potential for having a computable inverse, which is important for texture mapping and illumination calculations. In the case of rotation, translation, and nonzero scale transformations, the matrix always has a computable inverse. It can be multiplied with other transformation matrices to produce compound transformations while still maintaining 4×4-ness. The 4×4 identity matrix has zeros everywhere except along its diagonal; the diagonal elements all equal one (Eq. 2.5).

$$\begin{bmatrix} x \\ y \\ z \\ 1 \end{bmatrix} = \begin{bmatrix} 1 & 0 & 0 & 0 \\ 0 & 1 & 0 & 0 \\ 0 & 0 & 1 & 0 \\ 0 & 0 & 0 & 1 \end{bmatrix} \begin{bmatrix} x \\ y \\ z \\ 1 \end{bmatrix} \tag{2.5}$$

Typically in the literature, a point is represented as a 4×1 column matrix (also known as a *column vector*) and is transformed by multiplying by a 4×4 matrix on the left (also known as *premultiplying* the column vector by the matrix), as shown in Equation 2.5 in the case of the identity matrix. However, some texts use a 1×4 matrix (also known as a *row vector*) to represent a point and transform it by multiplying it by a matrix on its right (the matrix *postmultiplies* the row vector). For example, postmultiplying a point by the identity transformation would appear as in Equation 2.6.

$$[x \quad y \quad z \quad 1] = [x \quad y \quad z \quad 1] \begin{bmatrix} 1 & 0 & 0 & 0 \\ 0 & 1 & 0 & 0 \\ 0 & 0 & 1 & 0 \\ 0 & 0 & 0 & 1 \end{bmatrix} \tag{2.6}$$

Because the conventions are equivalent, it is immaterial which is used as long as consistency is maintained. The 4×4 transformation matrix used in one of the notations is the transpose of the 4×4 transformation matrix used in the other notation.

2.1.3 Concatenating transformations: multiplying transformation matrices

One of the main advantages of representing basic transformations as square matrices is that they can be multiplied together, which concatenates the transformations and produces a compound transformation. This enables a series of transformations, M_i, to be premultiplied so that a single compound transformation matrix, M, can be applied to a point P (see Eq. 2.7). This is especially useful (i.e., computationally efficient) when applying the same series of transformations to a multitude of points. Note that matrix multiplication is associative $((AB)C = A(BC))$ but not commutative $(AB \neq BA)$.

$$
\begin{aligned}
P' &= M_1 \ M_2 \ M_3 \ M_4 \ M_5 \ M_6 \ P \\
M &= M_1 \ M_2 \ M_3 \ M_4 \ M_5 \ M_6 \\
P' &= MP
\end{aligned}
\tag{2.7}
$$

When using the convention of postmultiplying a point represented by a row vector by the same series of transformations used when premultiplying a column vector, the matrices will appear in reverse order in addition to being the transposition of the matrices used in the premultiplication. Equation 2.8 shows the same computation as Equation 2.7, except in Equation 2.8, a row vector is postmultiplied by the transformation matrices. The matrices in Equation 2.8 are the same as those in Equation 2.7 but are now transposed and in reverse order. The transformed point is the same in both equations, with the exception that it appears as a column vector in Equation 2.7 and as a row vector in Equation 2.8. In the remainder of this book, such equations will be in the form shown in Equation 2.7.

$$
\begin{aligned}
P'^{T} &= P^{T} \ M_6^{T} \ M_5^{T} \ M_4^{T} \ M_3^{T} \ M_2^{T} \ M_1^{T} \\
M^{T} &= M_6^{T} \ M_5^{T} \ M_4^{T} \ M_3^{T} \ M_2^{T} \ M_1^{T} \\
P'^{T} &= P^{T} \ M^{T}
\end{aligned}
\tag{2.8}
$$

2.1.4 Basic transformations

For now, only the basic transformations rotate, translate, and scale (uniform scale as well as nonuniform scale) will be considered. These transformations, and any combination of these, are *affine transformations* [4]. It should be noted that the transformation matrices are the same whether the space is left- or right-handed. The perspective transformation is discussed later. Restricting discussion to the basic transformations allows the fourth element of each point vector to be assigned the value one and the last row of the transformation matrix to be assigned the value [0 0 0 1] (Eq. 2.9).

$$
\begin{bmatrix} x' \\ y' \\ z' \\ 1 \end{bmatrix} = \begin{bmatrix} a & b & c & d \\ e & f & g & h \\ i & j & k & m \\ 0 & 0 & 0 & 1 \end{bmatrix} \begin{bmatrix} x \\ y \\ z \\ 1 \end{bmatrix}
\tag{2.9}
$$

The x, y, and z translation values of the transformation are the first three values of the fourth column (d, h, and m in Eq. 2.9). The upper left 3×3 submatrix represents rotation and scaling. Setting the upper left 3×3 submatrix to an identity transformation and specifying only translation produces Equation 2.10.

$$
\begin{bmatrix} x + t_x \\ y + t_y \\ z + t_z \\ 1 \end{bmatrix} = \begin{bmatrix} 1 & 0 & 0 & t_x \\ 0 & 1 & 0 & t_y \\ 0 & 0 & 1 & t_z \\ 0 & 0 & 0 & 1 \end{bmatrix} \begin{bmatrix} x \\ y \\ z \\ 1 \end{bmatrix}
\tag{2.10}
$$

A transformation consisting of only uniform scale is represented by the identity matrix with a scale factor, S, replacing the first three elements along the diagonal (a, f, and k in Eq. 2.9). Nonuniform scale allows for independent scale factors to be applied to the x-, y-, and z-coordinates of a point and is formed by placing S_x, S_y, and S_z along the diagonal as shown in Equation 2.11.

$$
\begin{bmatrix} S_x x \\ S_y y \\ S_z z \\ 1 \end{bmatrix} = \begin{bmatrix} S_x & 0 & 0 & 0 \\ 0 & S_y & 0 & 0 \\ 0 & 0 & S_z & 0 \\ 0 & 0 & 0 & 1 \end{bmatrix} \begin{bmatrix} x \\ y \\ z \\ 1 \end{bmatrix}
\tag{2.11}
$$

Uniform scale can also be represented by setting the lowest rightmost value to $1/S$, as in Equation 2.12. In the homogeneous representation, the coordinates of the point represented are determined by dividing the first three elements of the vector by the fourth, thus scaling up the values by the scale factor S. This technique invalidates the assumption that the only time the lowest rightmost element is not one is during perspective and therefore should be used with care or avoided altogether.

$$
\begin{bmatrix} S\,x \\ S\,y \\ S\,z \\ 1 \end{bmatrix} = \begin{bmatrix} x \\ y \\ z \\ \dfrac{1}{S} \end{bmatrix} = \begin{bmatrix} 1 & 0 & 0 & 0 \\ 0 & 1 & 0 & 0 \\ 0 & 0 & 1 & 0 \\ 0 & 0 & 0 & \dfrac{1}{S} \end{bmatrix} \begin{bmatrix} x \\ y \\ z \\ 1 \end{bmatrix}
\tag{2.12}
$$

Values to represent rotation are set in the upper left 3×3 submatrix (a, b, c, e, f, g, i, j, and k of Eq. 2.9). Rotation matrices around the x-, y-, and z-axis are shown in Equations 2.13–2.15, respectively. In a right-handed coordinate system, a positive angle of rotation produces a counterclockwise rotation as viewed from the positive end of the axis looking toward the origin (the right-hand rule). In a left-handed (right-handed) coordinate system, a positive angle of rotation produces a clockwise (counterclockwise) rotation as viewed from the positive end of an axis. This can be remembered by noting that when pointing the thumb of the left (right) hand in the direction of the positive axis, the fingers wrap clockwise (counterclockwise) around the closed hand when viewed from the end of the thumb.

$$
\begin{bmatrix} x' \\ y' \\ z' \\ 1 \end{bmatrix} = \begin{bmatrix} 1 & 0 & 0 & 0 \\ 0 & \cos\theta & -\sin\theta & 0 \\ 0 & \sin\theta & \cos\theta & 0 \\ 0 & 0 & 0 & 1 \end{bmatrix} \begin{bmatrix} x \\ y \\ z \\ 1 \end{bmatrix}
\tag{2.13}
$$

$$\begin{bmatrix} x' \\ y' \\ z' \\ 1 \end{bmatrix} = \begin{bmatrix} \cos\theta & 0 & \sin\theta & 0 \\ 0 & 1 & 0 & 0 \\ -\sin\theta & 0 & \cos\theta & 0 \\ 0 & 0 & 0 & 1 \end{bmatrix} \begin{bmatrix} x \\ y \\ z \\ 1 \end{bmatrix} \qquad (2.14)$$

$$\begin{bmatrix} x' \\ y' \\ z' \\ 1 \end{bmatrix} = \begin{bmatrix} \cos\theta & -\sin\theta & 0 & 0 \\ \sin\theta & \cos\theta & 0 & 0 \\ 0 & 0 & 1 & 0 \\ 0 & 0 & 0 & 1 \end{bmatrix} \begin{bmatrix} x \\ y \\ z \\ 1 \end{bmatrix} \qquad (2.15)$$

Combinations of rotations and translations are usually referred to as *rigid transformations* because distance is preserved and the spatial extent of the object does not change; only its position and orientation in space are changed. *Similarity transformations* also allow uniform scale in addition to rotation and translation. These transformations preserve the object's intrinsic properties[2] (e.g., dihedral angles[3]) and relative distances but not absolute distances. Nonuniform scale, however, is usually not considered a similarity transformation because object properties such as dihedral angles are changed. A *shear* transformation is a combination of rotation and nonuniform scale and creates columns (rows) that might not be orthogonal to each other. Any combination of rotations, translations, and (uniform or nonuniform) scales still retains the last row of three zeros followed by a one. Notice that any affine transformation can be represented by a multiplicative 3×3 matrix (representing rotations, scales, and shears) followed by an additive three-element vector (translation).

2.1.5 **Representing an arbitrary orientation**

Rigid transformations (consisting of only rotations and translations) are very useful for moving objects around a scene without disturbing their geometry. These rigid transformations can be represented by a (possibly compound) rotation followed by a translation. The rotation transformation represents the object's orientation relative to its definition in object space. This section considers a particular way to represent an object's orientation.

Fixed-angle representation

One way to represent an orientation is as a series of rotations around the principal axes (the *fixed-angle representation*). When illustrating the relationship between orientation and a fixed order of rotations around the principal axes, consider the problem of determining the transformations that would produce a given geometric configuration. For example, consider that an aircraft is originally defined at the origin of a right-handed coordinate system with its nose pointed down the z-axis and its up vector in the positive y-axis direction (i.e., its object space representation). Now, imagine that the objective is to position the aircraft in world space so that its center is at $(20, -10, 35)$, its nose is oriented toward the point $(23, -14, 40)$, and its up vector is pointed in the general direction of the y-axis (or, mathematically, so that its up vector lies in the plane defined by the aircraft's center, the point the plane is oriented toward, and the global y-axis) (see Figure 2.6).

---

[2]An object's intrinsic properties are those that are measured irrespective of an external coordinate system.
[3]The dihedral angle is the interior angle between adjacent polygons measured at the common edge.

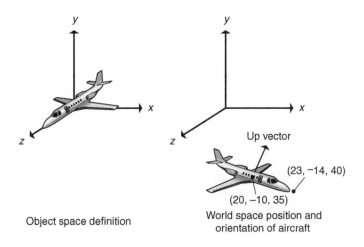

Object space definition

World space position and orientation of aircraft

FIGURE 2.6

Desired position and orientation.

The task is to determine the series of transformations that takes the aircraft from its original object space definition to its desired position and orientation in world space. This series of transformations will be one or more rotations about the principal axes followed by a translation of $(20, -10, 35)$. The rotations will transform the aircraft to an orientation so that, with its center at the origin, its nose is oriented toward $(23-20, -14 + 10, 40-35) = (3, -4, 5)$; this will be referred to as the aircraft's desired orientation vector.

In general, any such orientation can be effected by a rotation about the z-axis to tilt the object, followed by a rotation about the x-axis to tip the nose up or down, followed by a rotation about the y-axis to swing the plane around to point to the correct direction. This sequence is not unique; others could be constructed as well.

In this particular example, there is no tilt necessary because the desired up vector is already in the plane of the y-axis and orientation vector. We need to determine the x-axis rotation that will dip the nose down the right amount and the y-axis rotation that will swing it around the right amount. We do this by looking at the transformations needed to take the plane's initial orientation in object space aligned with the z-axis to its desired orientation.

The transformation that takes the aircraft to its desired orientation can be formed by determining the sines and cosines necessary for the x-axis and y-axis rotation matrices. Note that the length of the orientation vector is $\sqrt{3^2 + (-4)^2 + 5^2} = \sqrt{50}$. In first considering the x-axis rotation, initially position the orientation vector along the z-axis so that its endpoint is at $(0, 0, \sqrt{50})$. The x-axis rotation must rotate the endpoint of the orientation vector so that it is -4 in y. By the Pythagorean Rule, the z-coordinate of the endpoint would then be $\sqrt{50 - 4^2} = \sqrt{34}$ after the rotation. The sines and cosines can be read from the triangle formed by the rotated orientation vector, the vertical line segment that extends from the end of the orientation vector up to intersect the x-z plane, and the line segment from that intersection point to the origin (Figure 2.7). Observing that a positive x-axis rotation will

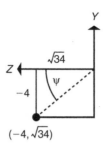

FIGURE 2.7

Projection of desired orientation vector onto *y-z* plane.

rotate the orientation vector down in y, we have $\sin\psi = 4/\sqrt{50}$ and $\cos\psi = \sqrt{34}/\sqrt{50}$. The x-axis rotation matrix looks like this:

$$R_x = \begin{bmatrix} 1 & 0 & 0 \\ 0 & \cos\psi & -\sin\psi \\ 0 & \sin\psi & \cos\psi \end{bmatrix} = \begin{bmatrix} 1 & 0 & 0 \\ 0 & \dfrac{\sqrt{34}}{\sqrt{50}} & \dfrac{-4}{\sqrt{50}} \\ 0 & \dfrac{4}{\sqrt{50}} & \dfrac{\sqrt{34}}{\sqrt{50}} \end{bmatrix} = \begin{bmatrix} 1 & 0 & 0 \\ 0 & \dfrac{\sqrt{17}}{5} & \dfrac{-2\sqrt{2}}{5} \\ 0 & \dfrac{2\sqrt{2}}{5} & \dfrac{\sqrt{17}}{5} \end{bmatrix} \quad (2.16)$$

After the pitch rotation has been applied, a y-axis rotation is required to spin the aircraft around (yaw) to its desired orientation. The sine and cosine of the y-axis rotation can be determined by looking at the projection of the desired orientation vector in the x-z plane. This projection is $(3, 0, 5)$. Thus, a positive y-axis rotation with $\sin\varphi = 3/\sqrt{34}$ and $\cos\varphi = 5/\sqrt{34}$ is required (Figure 2.8). The y-axis rotation matrix looks like this:

$$R_y = \begin{bmatrix} \cos\varphi & 0 & \sin\varphi \\ 0 & 1 & 0 \\ -\sin\varphi & 0 & \cos\varphi \end{bmatrix} = \begin{bmatrix} \dfrac{5}{\sqrt{34}} & 0 & \dfrac{3}{\sqrt{34}} \\ 0 & 1 & 0 \\ \dfrac{-3}{\sqrt{34}} & 0 & \dfrac{5}{\sqrt{34}} \end{bmatrix} \quad (2.17)$$

The final transformation of a point P would be $P' = R_y R_x P$.

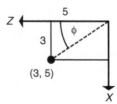

FIGURE 2.8

Projection of desired orientation vector onto *x-z* plane.

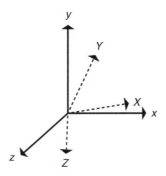

FIGURE 2.9

Global coordinate system and unit coordinate system to be transformed.

An alternative way to represent a transformation to a desired orientation is to construct what is known as the *matrix of direction cosines*. Consider transforming a copy of the global coordinate system so that it coincides with a desired orientation defined by a unit coordinate system (see Figure 2.9). To construct this matrix, note that the transformation matrix, M, should do the following: map a unit x-axis vector into the X-axis of the desired orientation, map a unit y-axis vector into the Y-axis of the desired orientation, and map a unit z-axis vector into the Z-axis of the desired orientation (see Eq. 2.18). These three mappings can be assembled into one matrix expression that defines the matrix M (Eq. 2.19).

$$X = Mx \qquad\qquad Y = My \qquad\qquad Z = Mz$$

$$\begin{bmatrix} X_x \\ X_y \\ X_z \end{bmatrix} = M \begin{bmatrix} 1 \\ 0 \\ 0 \end{bmatrix} \qquad \begin{bmatrix} Y_x \\ Y_y \\ Y_z \end{bmatrix} = M \begin{bmatrix} 0 \\ 1 \\ 0 \end{bmatrix} \qquad \begin{bmatrix} Z_x \\ Z_y \\ Z_z \end{bmatrix} = M \begin{bmatrix} 0 \\ 0 \\ 1 \end{bmatrix} \qquad (2.18)$$

$$\begin{bmatrix} X_x & Y_x & Z_x \\ X_y & Y_y & Z_y \\ X_z & Y_z & Z_z \end{bmatrix} = M \begin{bmatrix} 1 & 0 & 0 \\ 0 & 1 & 0 \\ 0 & 0 & 1 \end{bmatrix} = M \qquad (2.19)$$

Since a unit x-vector (y-vector, z-vector) multiplied by a transformation matrix will replicate the values in the first (second, third) column of the transformation matrix, the columns of the transformation matrix can be filled with the coordinates of the desired transformed coordinate system. Thus, the first column of the transformation matrix becomes the desired X-axis as described by its x-, y-, and z-coordinates in the global space, call it u; the second column becomes the desired Y-axis, call it v; and the third column becomes the desired Z-axis, call it w (Eq. 2.20). The name *matrix of direction cosines* is derived from the fact that the coordinates of a desired axis in terms of the global coordinate system are the cosines of the angles made by the desired axis with each of the global axes.

In the example of transforming the aircraft, the desired Z-axis is the desired orientation vector, $w = [3, -4, 5]$. With the assumption that there is no longitudinal rotation (roll), the desired X-axis can be formed by taking the cross-product of the original y-axis and the desired Z-axis, $u = [5, 0, -3]$. The desired Y-axis can then be formed by taking the cross-product of the desired Z-axis and the desired X-axis, $v = [12, 34, 20]$. Each of these is normalized by dividing by its length to form unit vectors. This results in the matrix of Equation 2.20, which is the same matrix formed by multiplying the x-rotation matrix, R_x of Equation 2.16, by the y-rotation matrix, R_y of Equation 2.17.

$$M = \begin{bmatrix} u^T & v^T & w^T \end{bmatrix} = \begin{bmatrix} \dfrac{5}{\sqrt{34}} & \dfrac{6}{5\sqrt{17}} & \dfrac{3}{5\sqrt{2}} \\[2mm] 0 & \dfrac{\sqrt{17}}{5} & -2\dfrac{\sqrt{2}}{5} \\[2mm] -\dfrac{3}{\sqrt{34}} & \dfrac{2}{\sqrt{17}} & \dfrac{1}{\sqrt{2}} \end{bmatrix} = R_y R_x \qquad (2.20)$$

2.1.6 Extracting transformations from a matrix

For a compound transformation matrix that represents a series of rotations and translations, a set of individual transformations can be extracted from the matrix, which, when multiplied together, produce the original compound transformation matrix. Notice that the series of transformations to produce a compound transformation is not unique, so there is no guarantee that the series of transformations so extracted will be exactly the ones that produced the compound transformation (unless something is known about the process that produced the compound matrix).

An arbitrary rigid transformation can easily be formed by up to three rotations about the principal axes (or one compound rotation represented by the direction cosine matrix) followed by a translation.

The last row of a 4 × 4 transformation matrix, if the matrix does not include a perspective transformation, will have zero in the first three entries and one as the fourth entry (ignoring the use of that element to represent uniform scale). As shown in Equation 2.21, the first three elements of the last column of the matrix, A_{14}, A_{24}, and A_{34}, represent a translation. The upper left 3 × 3 submatrix of the original 4 × 4 matrix can be viewed as the definition of the transformed unit coordinate system. It can be decomposed into three rotations around principal axes by arbitrarily choosing an ordered sequence of three axes (such as x followed by y followed by z). By using the projection of the transformed unit coordinate system to determine the sines and cosines, the appropriate rotation matrices can be formed in much the same way that transformations were determined in Section 2.1.5.

$$\begin{bmatrix} x' \\ y' \\ z' \\ 1 \end{bmatrix} = \begin{bmatrix} A_{11} & A_{12} & A_{13} & A_{14} \\ A_{21} & A_{22} & A_{23} & A_{24} \\ A_{31} & A_{32} & A_{33} & A_{34} \\ 0 & 0 & 0 & 1 \end{bmatrix} \begin{bmatrix} x \\ y \\ z \\ 1 \end{bmatrix} \qquad (2.21)$$

If the compound transformation matrix includes a uniform scale factor, the rows of the 3 × 3 submatrix will form orthogonal vectors of uniform length. The length will be the scale factor, which, when followed by the rotations and translations, forms the decomposition of the compound transformation. If the rows of the 3 × 3 submatrix form orthogonal vectors of unequal length, then their lengths represent nonuniform scale factors that precede any rotations.

2.1.7 Description of transformations in the display pipeline

Now that the basic transformations have been discussed in some detail, the previously described transformations of the display pipeline can be explained in terms of concatenating these basic transformations. It should be noted that the descriptions of eye space and the corresponding perspective transformation are not unique. They vary among the introductory graphics texts depending on where the observer is placed along the z-axis to define eye space, whether the eye space coordinate system is left- or right-handed, exactly what information is required from the user in describing the perspective transformation, and the range of visible z-values in image space. While functionally equivalent, the various approaches produce transformation matrices that differ in the values of the individual elements.

Object space to world space transformation

In a simple implementation, the transformation of an object from its object space into world space is a series of rotations, translations, and scales (i.e., an affine transformation) that are specified by the user (by explicit numeric input or by some interactive technique) to place a transformed copy of the object data into a world space data structure. In some systems, the user is required to specify this transformation in terms of a predefined order of basic transformations such as scale, rotation around the x-axis, rotation around the y-axis, rotation around the z-axis, and translation. In other systems, the user may be able to specify an arbitrarily ordered sequence of basic transformations. In either case, the series of transformations can be compounded into a single object space to world space transformation matrix.

The object space to world space transformation is usually the transformation that is modified over time to produce motion. In more complex animation systems, this transformation may include manipulations of arbitrary complexity not suitable for representation in a matrix, such as nonlinear shape deformations.

World space to eye space transformation

In preparation for the perspective transformation, a rigid transformation is performed on all of the object data in world space. The transformation is designed so that, in eye space, the observer is positioned at the origin, the view vector aligns with the positive z-axis in left-handed space, and the up vector aligns with the positive y-axis. The transformation is formed as a series of basic transformations.

First, the data is translated so that the observer is moved to the origin. Then, the observer's coordinate system (view vector, up vector, and the third vector required to complete a left-handed coordinate system) is transformed by up to three rotations so as to align the view vector with the global negative z-axis and the up vector with the global y-axis. Finally, the z-axis is flipped by negating the z-coordinate. All of the individual transformations can be represented by 4×4 transformation matrices, which are multiplied together to produce a single compound world space to eye space transformation matrix. This transformation prepares the data for the perspective transformation by putting it in a form in which the perspective divide is simply dividing by the point's z-coordinate.

Perspective matrix multiply

The perspective matrix multiplication is the first part of the perspective transformation. The fundamental computation performed by the perspective transformation is that of dividing the x- and y-coordinates by their z-coordinate and normalizing the visible range in x and y to $[-1, +1]$. This is accomplished by using a homogeneous representation of a point and, as a result of the perspective matrix multiplication, producing a representation in which the fourth element is $Z_e \tan \varphi$. Z_e is the point's z-coordinate in eye space and φ is the half angle of view in the vertical or horizontal direction (assuming a square viewport for now and, therefore, the vertical and horizontal angles are equal). The z-coordinate is transformed so that planarity is preserved and the visible range in z is mapped into $[0, +1]$. (These ranges are arbitrary and can be set to anything by appropriately forming the perspective matrix. For example, sometimes the visible range in z is set to $[-1, +1]$.) In addition, the aspect ratio of the viewport can be used in the matrix to modify the horizontal or vertical half angle of view so that no distortion results in the viewed data.

Perspective divide

Each point produced by the perspective matrix multiplication has a nonunitary fourth component that represents the perspective divide by z. Dividing each point by its fourth component completes the perspective transformation. This is considered a separate step from the perspective matrix multiply because a commonly used clipping procedure operates on the homogeneous representation of points produced by the perspective matrix multiplication but before perspective divide.

Clipping, the process of removing data that are outside the view frustum, can be implemented in a variety of ways. It is computationally simpler if clipping is performed after the world space to eye space transformation. It is important to perform clipping in z using the near clipping distance before perspective divide to prevent divide by zero and to avoid projecting objects behind the observer onto the picture plane. However, the details of clipping are not relevant to the discussion here. Interested readers should refer to one of the standard computer graphics texts (e.g., [2]) for the details of clipping procedures.

Image to screen space mapping

The result of the perspective transformation (the perspective matrix multiply followed by perspective divide) maps visible elements into the range of minus one to plus one ($[-1, +1]$) in x and y. This range is now mapped into the user-specified viewing area of the screen-based pixel coordinate system. This is a simple linear transformation represented by a scale and a translation and thus can be easily represented in a 4×4 transformation matrix.

2.1.8 Error considerations

Accumulated round-off error

Once the object space to world space transformation matrix has been formed for an object, the object is transformed into world space by simply multiplying all of the object's object space points by the transformation matrix. When an object's position and orientation are animated, its points will be repeatedly transformed over time—as a function of time. One way to do this is to repeatedly modify the object's world space points. However, incremental transformation of world space points can lead to the accumulation of round-off errors. For this reason, it is almost always better to modify the transformation from object to world space and reapply the transformation to the object space points rather than to repeatedly transform the world space coordinates. To further transform an object that already has a transformation matrix associated with it, one simply has to form a transformation matrix and premultiply it by the existing transformation matrix to produce a new one. However, round-off errors can also accumulate when one repeatedly modifies a transformation matrix. The best way is to build the transformation matrix anew each time it is to be applied.

An affine transformation matrix can be viewed as a 3×3 rotation/scale submatrix followed by a translation. Most of the error accumulation occurs because of the operations resulting from multiplying the x-, y-, and z-coordinates of the point by the 3×3 submatrix. Therefore, the following round-off error example will focus on the errors that accumulate as a result of rotations.

Consider the case of the moon orbiting the earth. For the sake of simplicity, the assumption is that the center of the earth is at the origin and, initially, the moon data are defined with the moon's center at the origin. The moon data are first transformed to an initial position relative to the earth, for example $(r, 0, 0)$ (see Figure 2.10). There are three approaches that could be taken to animate the rotation of the moon around the earth, and these will be used to illustrate various effects of round-off error.

The first approach is, for each frame of the animation, to apply a delta z-axis transformation matrix to the moon's points, in which each delta represents the angle it moves in one frame time (see Figure 2.11). Round-off errors will accumulate in the world space object points. Points that began as coplanar will no longer be coplanar. This can have undesirable effects, especially in display algorithms that linearly interpolate values to render a surface.

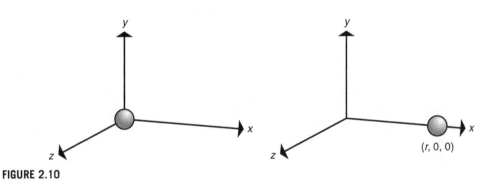

FIGURE 2.10

Translation of moon to its initial position on the x-axis.

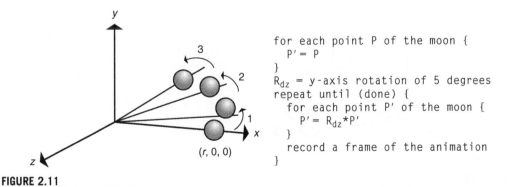

```
for each point P of the moon {
    P' = P
}
Rdz = y-axis rotation of 5 degrees
repeat until (done) {
    for each point P' of the moon {
        P' = Rdz*P'
    }
    record a frame of the animation
}
```

FIGURE 2.11

Rotation by applying incremental rotation matrices to points.

The second approach is, for each frame, to incrementally modify the transformation matrix that takes the object space points into the world space positions. In the example of the moon, the transformation matrix is initialized with the x-axis translation matrix. For each frame, a delta z-axis transformation matrix multiplies the current transformation matrix and then that resultant matrix is applied to the moon's object space points (see Figure 2.12). Round-off error will accumulate in the transformation matrix. Over time, the matrix will deviate from representing a rigid transformation. Shearing effects will begin to creep into the transformation and angles will cease to be preserved. While a square may begin to look like something other than a square, coplanarity will be preserved (because any matrix multiplication is, by definition, a linear transformation that preserves planarity), so that rendering results will not be compromised.

The third approach is to add the delta value to an accumulating angle variable and then build the z-axis rotation matrix from that angle parameter. This would then be multiplied with the x-axis translation matrix, and the resultant matrix would be applied to the original moon points in object space (see Figure 2.13). In this case, any round-off error will accumulate in the angle variable so that, over time, it may begin to deviate from what is desired. This may have unwanted effects when one tries to coordinate motions, but the transformation matrix, which is built anew every frame, will not accumulate any errors itself. The transformation will always represent a valid rigid transformation with both planarity and angles being preserved.

Orthonormalization

The rows of a matrix that represent a rigid transformation are perpendicular to each other and are of unit length (orthonormal). The same can be said of the matrix columns. If values in a rigid transformation matrix have accumulated errors, then the rows cease to be orthonormal and the matrix ceases to represent a rigid transformation; it will have the effect of introducing shear into the transformation. However, if it is known that the matrix is supposed to represent a rigid transformation, it can be massaged back into a rigid transformation matrix.

A rigid transformation matrix has an upper 3×3 submatrix with specific properties: the rows (columns) are unit vectors orthogonal to each other. A simple procedure to reformulate the transformation

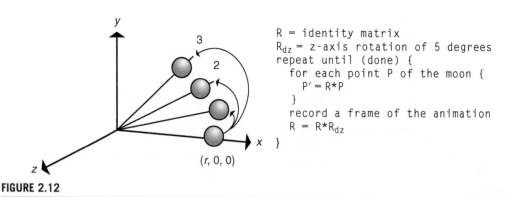

```
R = identity matrix
Rdz = z-axis rotation of 5 degrees
repeat until (done) {
    for each point P of the moon {
        P' = R*P
    }
    record a frame of the animation
    R = R*Rdz
}
```

$(r, 0, 0)$

FIGURE 2.12

Rotation by incrementally updating the rotation matrix.

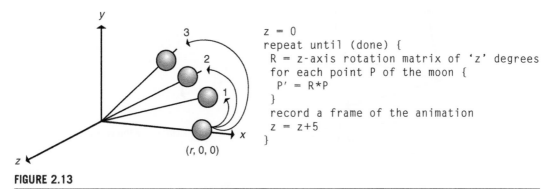

```
z = 0
repeat until (done) {
    R = z-axis rotation matrix of 'z' degrees
    for each point P of the moon {
        P' = R*P
    }
    record a frame of the animation
    z = z+5
}
```

FIGURE 2.13

Rotation by forming the rotation matrix anew for each frame.

matrix to represent a rigid transformation is to take the first row (column) and normalize it. Take the second row (column), compute the cross-product of this row (column) and the first row (column), normalize it, and place it in the third row (column). Take the cross-product of the third row (column) and the first row (column) and put it in the second row (column) (see Figure 2.14). Notice that this does not necessarily produce the correct transformation; it merely forces the matrix to represent a rigid transformation. The error has just been shifted around so that the columns of the matrix are orthonormal and the error may be less noticeable.

If the transformation might contain a uniform scale, then take the length of one of the rows, or the average length of the three rows, and, instead of normalizing the vectors by the steps described above, make them equal to this length. If the transformation might include nonuniform scale, then the difference between shear and error accumulation cannot be determined unless something more is known about the transformations represented. For example, if it is known that nonuniform scale was applied before any rotation (i.e., no shear), then Gram-Schmidt Orthonormalization [6] can be performed, without the normalization step, to force orthogonality among the vectors. Gram-Schmidt processes the vectors in any order. To process a vector, project it onto each previously processed vector. Subtract the projections from the vector currently being processed, then process the next vector. When all vectors have been processed, they are orthogonal to each other.

Considerations of scale

When constructing a large database (e.g., flight simulator), there may be such variations in the magnitude of various measures as to create precision problems. For example, you may require detail on the order of a fraction of an inch may be required for some objects, while the entire database may span thousands of miles. The scale of values would range from to 10^{-1} inches to 5000 miles·5280 feet/mile·12 inches/foot $= 3.168 \cdot 10^8$ inches. This exceeds the precision of 32-bit single-precision representations. Using double-precision will help eliminate the problem. However, using double-precision representations may also increase storage space requirements and decrease the speed of the computations. Alternatively, subdividing the database into local data, such as airports in the example of a flight simulator, and switching between these localized databases might provide an acceptable solution.

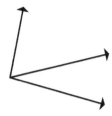

The original unit orthogonal vectors have ceased to be orthogonal to each other due to repeated transformations.

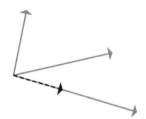

Step 1: Normalize one of the vectors.

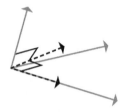

Step 2:
Form a vector perpendicular (orthogonal) to the vector just normalized and to one of the other two original vectors by taking the cross product of the two. Normalize it.

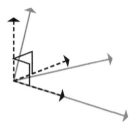

Step 3:
Form the final orthogonal vector by taking the cross product of the two just generated.

FIGURE 2.14

Orthonormalization of a set of three vectors.

2.2 Orientation representation

A common issue that arises in computer animation is deciding the best way to represent the position and orientation of an object in space and how to interpolate the represented transformations over time to produce motion. A typical scenario is one in which the user specifies an object in two transformed states and the computer is used to interpolate intermediate states, thus producing animated key-frame motion. Another scenario is when an object is to undergo two or more successive transformations and it would be efficient to concatenate these transformations into a single representation before applying it to a multitude of object vertices. This section discusses possible orientation representations and identifies strengths and weaknesses; the next chapter addresses the best way to interpolate orientations using these representations. In this discussion, it is assumed that the final transformation applied to the object is a result of rotations and translations only, so that there is no scaling involved, nonuniform or otherwise; that is, the transformations considered are *rigid body*.

The first obvious choice for representing the orientation and position of an object is by a 4×4 transformation matrix. For example, a user may specify a series of rotations and translations to apply to an object. This series of transformations is compiled into 4 × 4 matrices and is multiplied together to produce a compound 4 × 4 transformation matrix. In such a matrix, the upper left 3 × 3 submatrix represents a rotation to apply to the object, while the first three elements of the fourth column represent the translation (assuming points are represented by column vectors that are premultiplied by the transformation matrix). No matter how the 4 × 4 transformation matrix was formed (no matter in what order the transformations were given by the user, such as "rotate about x, translate, rotate about x, rotate about y, translate, rotate about y"), the final 4 × 4 transformation matrix produced by multiplying all of the individual transformation matrices in the specified order will result in a matrix that specifies the final position of the object by a 3 × 3 rotation matrix followed by a translation. The conclusion is that the rotation can be interpolated independently from the translation. (For now, consider that the interpolations are linear, although higher order interpolations are possible; see Appendix B.5.)

Consider two such transformations that the user has specified as key states with the intention of generating intermediate transformations by interpolation. While it should be obvious that interpolating the translations is straightforward, it is not at all clear how to go about interpolating the rotations. In fact, it is the objective of this discussion to show that interpolation of orientations is not nearly as straightforward as interpolation of translation. A property of 3 × 3 rotation matrices is that the rows and columns are orthonormal (unit length and perpendicular to each other). Simple linear interpolation between the nine pairs of numbers that make up the two 3 × 3 rotation matrices to be interpolated will not produce intermediate 3 × 3 matrices that are orthonormal and are therefore not rigid body rotations. It should be easy to see that interpolating from a rotation of +90 degrees about the y-axis to a rotation of −90 degrees about the y-axis results in intermediate transformations that are nonsense (Figure 2.15).

So, direct interpolation of transformation matrices is not acceptable. There are alternative representations that are more useful than transformation matrices in performing such interpolations including fixed angle, Euler angle, axis–angle, quaternions, and exponential maps.

$$\begin{bmatrix} 0 & 0 & 1 \\ 0 & 1 & 0 \\ -1 & 0 & 0 \end{bmatrix}$$

Positive 90-degree y-axis rotation

$$\begin{bmatrix} 0 & 0 & -1 \\ 0 & 1 & 0 \\ 1 & 0 & 0 \end{bmatrix}$$

Negative 90-degree y-axis rotation

$$\begin{bmatrix} 0 & 0 & 0 \\ 0 & 1 & 0 \\ 0 & 0 & 0 \end{bmatrix}$$

Interpolated matrix halfway between the orientation representations above

FIGURE 2.15

Direct interpolation of transformation matrix values can result in nonsense—transformations.

2.2.1 Fixed-angle representation

A *fixed-angle* representation[4] really refers to "angles used to rotate about fixed axes." A fixed order of three rotations is implied, such as *x-y-z*. This means that orientation is given by a set of three ordered parameters that represent three ordered rotations about fixed axes: first around *x*, then around *y*, and then around *z*. There are many possible orderings of the rotations, and, in fact, it is not necessary to use all three coordinate axes. For example, *x-y-x* is a feasible set of rotations. The only orderings that do not make sense are those in which an axis immediately follows itself, such as in *x-x-y*. In any case, the main point is that the orientation of an object is given by three angles, such as (10, 45, 90). In this example, the orientation represented is obtained by rotating the object first about the *x*-axis by 10 degrees, then about the *y*-axis by 45 degrees, and then about the *z*-axis by 90 degrees. In Figure 2.16, the aircraft is shown in its initial orientation and in the orientation represented by the values of (10, 45, 90).

The following notation will be used to represent such a sequence of rotations: $R_z(90)R_y(45)R_x(10)$ (in this text, transformations are implemented by premultiplying column vectors by transformation matrices; thus, the rotation matrices appear in right to left order).

From this orientation, changing the *x*-axis rotation value, which is applied first to the data points, will make the aircraft's nose dip more or less in the *y-z* plane. Changing the *y*-axis rotation will change the amount the aircraft, which has been rotated around the *x*-axis, rotates out of the *y-z* plane. Changing the *z*-axis rotation value, the rotation applied last, will change how much the twice-rotated aircraft will rotate about the *z*-axis.

The problem with using this scheme is that two of the axes of rotation can effectively line up on top of each other when an object can rotate freely in space (or around a 3 degree of freedom[5] joint). Consider an object in an orientation represented by (0, 90, 0), as shown in Figure 2.17. Examine the effect a slight change in the first and third parametric values has on the object in that orientation. A slight change of the third parameter will rotate the object slightly about the global *z*-axis because that is the rotation applied last to the data points. However, note that the effect of a slight change of the first

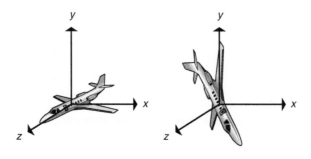

FIGURE 2.16

Fixed-angle representation.

---

[4]Terms referring to rotational representations are not used consistently in the literature. This book follows the usage found in *Robotics* [1], where fixed angle refers to rotation about the fixed (global) axes and Euler angle refers to rotation about the rotating (local) axes.

[5]The degrees of freedom that an object possesses is the number of independent variables that have to be specified to completely locate that object (and all of its parts).

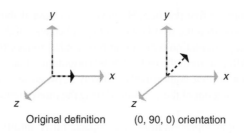

Original definition   (0, 90, 0) orientation

FIGURE 2.17

Fixed-angle representation of (0, 90, 0).

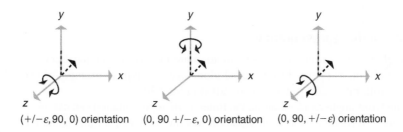

$(+/-\varepsilon, 90, 0)$ orientation $(0, 90 +/-\varepsilon, 0)$ orientation $(0, 90, +/-\varepsilon)$ orientation

FIGURE 2.18

Effect of slightly altering values of fixed-angle representation (0, 90, 0).

parameter, which rotates the original data points around the x-axis, will also have the effect of rotating the transformed object slightly about the z-axis (Figure 2.18). This results because the 90-degree y-axis rotation has essentially made the first axis of rotation align with the third axis of rotation. The effect is called *gimbal lock*. From the orientation (0, 90, 0), the object can no longer be rotated about the global x-axis by a small change in its orientation representation. Actually, the representation that will perform an incremental rotation about the x-axis from the (0, 90, 0) orientation is $(90, 90 +\varepsilon, 90)$, which is not very intuitive.

The cause of this problem can often make interpolation between key positions problematic. Consider the key orientations $(0, 90, 0)$ and $(90, 45, 90)$, as shown in Figure 2.19. The second orientation is a

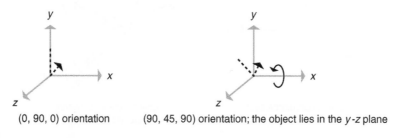

(0, 90, 0) orientation   (90, 45, 90) orientation; the object lies in the y-z plane

FIGURE 2.19

Example orientations to interpolate.

45-degree x-axis rotation from the first position. However, as discussed above, the object can no longer directly rotate about the x-axis from the first key orientation because of the 90-degree y-axis rotation. Direct interpolation of the key orientation representations would produce $(45, 67.5, 45)$ as the halfway orientation, which is very different from the $(90, 22.5, 90)$ orientation that is desired (because that is the representation of the orientation that is intuitively halfway between the two given orientations). The result is that the object will swing out of the y-z plane during the interpolation, which is not the behavior one would expect.

In its favor, the fixed-angle representation is compact, fairly intuitive, and easy to work with because the implied operations correspond to what we know how to do mathematically—rotate data around the global axes. However, it is often not the most desirable representation to use because of the gimbal lock problem.

2.2.2 Euler angle representation

In a *Euler angle* representation, the axes of rotation are the axes of the local coordinate system that rotate with the object, as opposed to the fixed global axes. A typical example of using Euler angles is found in the roll, pitch, and yaw of an aircraft (Figure 2.20).

As with the fixed-angle representation, the Euler angle representation can use any of various orderings of three axes of rotation as its representation scheme. Consider a Euler angle representation that uses an x-y-z ordering and is specified as (α, β, γ). The x-axis rotation, represented by the transformation matrix $R_x(\alpha)$, is followed by the y-axis rotation, represented by the transformation matrix $R_y(\beta)$, around the y-axis of the local, rotated coordinate system. Using a prime symbol to represent rotation about a rotated frame and remembering that points are represented as column vectors and are premultiplied by transformation matrices, one achieves a result of $R_y'(\beta)R_x(\alpha)$. Using global axis rotation matrices to implement the transformations, the y-axis rotation around the rotated frame can be effected by $R_x(\alpha)R_y(\beta)R_x(-\alpha)$. Thus, the result after the first two rotations is shown in Equation 2.22.

$$R_y'(\beta)R_x(\alpha) = R_x(\alpha)R_y(\beta)R_x(-\alpha)R_x(\alpha) = R_x(\alpha)R_y(\beta) \tag{2.22}$$

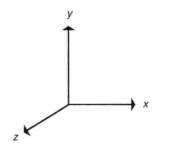
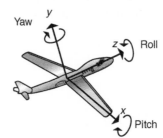

Global coordinate system   Local coordinate system
attached to object

FIGURE 2.20

Euler angle representation.

The third rotation, $R_z(\gamma)$, is around the now twice-rotated frame. This rotation can be effected by undoing the previous rotations with $R_x(-\alpha)$ followed by $R_y(-\beta)$, then rotating around the global z-axis by $R_z(\gamma)$, and then reapplying the previous rotations. Putting all three rotations together, and using a double prime to denote rotation about a twice-rotated frame, results in Equation 2.23.

$$
\begin{aligned}
R_z''(\gamma)R_y'(\beta)R_x(\alpha) &= R_x(\alpha)R_y(\beta)R_z(\gamma)R_y(-\beta)R_x(-\alpha)R_x(\alpha)R_y(\beta) \\
&= R_x(\alpha)R_y(\beta)R_z(\gamma)
\end{aligned}
\tag{2.23}
$$

Thus, this system of Euler angles is precisely equivalent to the fixed-angle system in reverse order. This is true for any system of Euler angles. For example, z-y-x Euler angles are equivalent to x-y-z fixed angles. Therefore, the Euler angle representation has exactly the same advantages and disadvantages (i.e., gimbal lock) as those of the fixed-angle representation.

2.2.3 Angle and axis representation

In the mid-1700s, Leonhard Euler showed that one orientation can be derived from another by a single rotation about an axis. This is known as the Euler Rotation Theorem [1]. Thus, any orientation can be represented by three numbers: two for the axis, such as longitude and latitude, and one for the angle (Figure 2.21). The axis can also be represented (somewhat inefficiently) by a three-dimensional vector. This can be a useful representation. Interpolation between representations (A_1, θ_1) and (A_2, θ_2), where A is the axis of rotation and θ is the angle, can be implemented by interpolating the axes of rotation and the angles separately (Figure 2.22). An intermediate axis can be determined by rotating one axis partway toward the other. The axis for this rotation is formed by taking the cross product of two axes, A_1 and A_2. The angle between the two axes is determined by taking the inverse cosine of the dot product of normalized versions of the axes. An interpolant, k, can then be used to form an intermediate axis and angle pair. Note that the axis–angle representation does not lend itself to easily concatenating a series of rotations. However, the information contained in this representation can be put in a form in which these operations are easily implemented: quaternions.

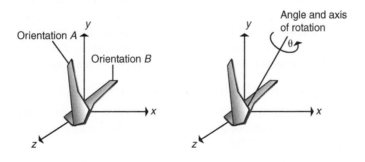

FIGURE 2.21

Euler's rotation theorem implies that for any two orientations of an object, one can be produced from the other by a single rotation about an arbitrary axis.

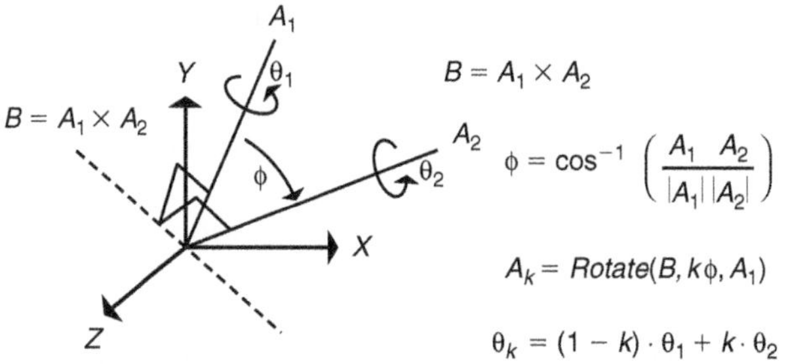

FIGURE 2.22

Interpolating axis-angle representations of (A_1, θ_1) and (A_2, θ_2) by k to get (A_k, θ_k), where 'Rotate(a,b,c)' rotates 'c' around 'a' by 'b' degrees.

2.2.4 Quaternion representation

As discussed earlier, the representations covered so far have drawbacks when interpolating intermediate orientations when an object or joint has three degrees of rotational freedom. A better approach is to use *quaternions* to represent orientation [5]. A quaternion is a four-tuple of real numbers, $[s, x, y, z]$, or, equivalently, $[s, v]$, consisting of a scalar, s, and a three-dimensional vector, v.

The quaternion is an alternative to the axis and angle representation in that it contains the same information in a different, but mathematically convenient, form. Importantly, it is in a form that can be interpolated as well as used in concatenating a series of rotations into a single representation. The axis and angle information of a quaternion can be viewed as an orientation of an object relative to its initial object space definition, or it can be considered as the representation of a rotation to apply to an object definition. In the former view, being able to interpolate between represented orientations is important in generating key-frame animation. In the latter view, concatenating a series of rotations into a simple representation is a common and useful operation to perform to apply a single, compound transformation to an object definition.

Basic quaternion math

Before interpolation can be explained, some basic quaternion math must be understood. In the equations that follow, a bullet operator represents dot product, and "×" denotes cross-product. *Quaternion addition* is simply the four-tuple addition of quaternion representations, $[s_1, v_1] + [s_2, v_2] = [s_1 + s_2, v_1 + v_2]$. *Quaternion multiplication* is defined as Equation 2.24. Notice that quaternion multiplication is associative, $(q_1 q_2)q_3 = q_1(q_2 q_3)$, but is not commutative, $q_1 q_2 \neq q_2 q_1$.

$$[s_1, v_1][s_2, v_2] = [s_1 s_2 - v_1 \cdot v_2, s_1 v_2 + s_2 v_1 + v_1 \times v_2] \tag{2.24}$$

A point in space, v, or, equivalently, the vector from the origin to the point, is represented as $[0, v]$. It is easy to see that quaternion multiplication of two orthogonal vectors $(v_1 \bullet v_2 = 0)$ computes the cross-product of those vectors (Eq. 2.25).

$$[0, v_1][0, v_2] = [0, v_1 \times v_2] \quad \text{if} \quad v_1 \cdot v_2 = 0 \tag{2.25}$$

The quaternion $[1, (0, 0, 0)]$ is the multiplicative identity; that is,

$$[s, \ v][1, (0, 0, 0)] = [s, v] \tag{2.26}$$

The *inverse of a quaternion*, $[s, v]^{-1}$, is obtained by negating its vector part and dividing both parts by the magnitude squared (the sum of the squares of the four components), as shown in Equation 2.27.

$$q^{-1} = (1/||q||)^2[s, -v] \quad \text{where } ||q|| = \sqrt{s^2 + x^2 + y^2 + z^2} \tag{2.27}$$

Multiplication of a quaternion, q, by its inverse, q^{-1}, results in the multiplicative identity $[1, (0, 0, 0)]$. A unit-length quaternion (also referred to here as a unit quaternion), $\hat{q}$, is created by dividing each of the four components by the square root of the sum of the squares of those components (Eq. 2.28).

$$\hat{q} = q/(||q||) \tag{2.28}$$

Representing rotations using quaternions

A rotation is represented in a quaternion form by encoding axis–angle information. Equation 2.29 shows a unit quaternion representation of a rotation of an angle, u, about a unit axis of rotation (x, y, z).

$$QuatRot(\theta, (x, y, z)) \equiv [\cos(\theta/2), \sin(\theta/2)(x, y, z)] \tag{2.29}$$

Notice that rotating some angle around an axis is the same as rotating the negative angle around the negated axis. In quaternions, this is manifested by the fact that a quaternion, $q = [s, v]$, and its negation, $-q = [-s, -v]$, represent the same rotation. The two negatives in this case cancel each other out and produce the same rotation. In Equation 2.30, the quaternion q represents a rotation of u about a unit axis of rotation (x, y, z), i.e.,

$$\begin{aligned}
-q &= [-\cos(\theta/2), -\sin(\theta/2)(x, y, z)] \\
&= [\cos(180-(\theta/2)), \sin(\theta/2)(-(x, y, z))] \\
&= [\cos((360-\theta)/2), \sin(180-\theta/2)(-(x, y, z))] \\
&= [\cos((360-\theta)/2), \sin(360-\theta/2)(-(x, y, z))] \\
&\equiv QuatRot(-\theta, -(x, y, z)) \\
&\equiv QuatRot(\theta, (x, y, z))
\end{aligned} \tag{2.30}$$

Negating q results in a negative rotation around the negative of the axis of rotation, which is the same rotation represented by q (Eq. 2.30).

Rotating vectors using quaternions

To rotate a vector, v, using quaternion math, represent the vector as $[0, v]$ and represent the rotation by a quaternion, q. The vector is rotated according to Equation 2.31.

$$Rot_q(v) \equiv qvq^{-1} = v' \tag{2.31}$$

A series of rotations can be concatenated into a single representation by quaternion multiplication. Consider a rotation represented by a quaternion, p, followed by a rotation represented by a quaternion, q, on a vector, v (Eq. 2.32).

$$Rot_q(Rot_p(v)) = q(pvp^{-1})q^{-1} = (qp)v(qp)^{-1} = Rot_{qp}(v) \tag{2.32}$$

The inverse of a quaternion represents rotation about the same axis by the same amount but in the reverse direction. Equation 2.33 shows that rotating a vector by a quaternion, q, followed by rotating the result by the inverse of that same quaternion produces the original vector.

$$Rot_{q^{-1}}(Rot_q(v)) = q^{-1}(qvp^{-1})q = v \tag{2.33}$$

Also, notice that in performing rotation, qvq^{-1}, all effects of magnitude are divided out due to the multiplication by the inverse of the quaternion. Thus, any scalar multiple of a quaternion represents the same rotation as the corresponding unit quaternion (similar to how the homogeneous representation of points is scale invariant).

A concise list of quaternion arithmetic and conversions to and from other representations can be found in Appendix B.3.4.

2.2.5 Exponential map representation

Exponential maps, similar to quaternions, represent an orientation as an axis of rotation and an associated angle of rotation as a single vector [3]. The direction of the vector is the axis of rotation and the magnitude is the amount of rotation. In addition, a zero rotation is assigned to the zero vector, making the representation continuous at the origin. Notice that an exponential map uses three parameters instead of the quaternion's four. The main advantage is that it has well-formed derivatives. These are important, for example, when dealing with angular velocity.

This representation does have some drawbacks. Similar to Euler angles, it has singularities. However, in practice, these can be avoided. Also, it is difficult to concatenate rotations using exponential maps and is best done by converting to rotation matrices.

2.3 Summary

Linear transformations represented by 4×4 matrices are a fundamental operation in computer graphics and animation. Understanding their use, how to manipulate them, and how to control round-off error is an important first step in mastering graphics and animation techniques.

There are several orientation representations to choose from. The most robust representation of orientation is quaternions, but fixed angle, Euler angle, and axis–angle are more intuitive and easier to implement. Fixed angles and Euler angles suffer from gimbal lock and axis–angle is not easy to composite, but they are useful in some situations. Exponential maps also do not concatenate well but offer some advantages when working with derivatives of orientation. Appendix B.3.4 contains useful conversions between quaternions and other representations.

References

[1] Craig J. Robotics. New York: Addison-Wesley; 1989.
[2] Foley J, van Dam A, Feiner S, Hughes J. Computer Graphics: Principles and Practice. 2nd ed. New York: Addison-Wesley; 1990.
[3] Grassia FS. Practical Parameterization of Rotations Using the Exponential Map. The Journal of Graphics Tools 1998;3.3.
[4] Mortenson M. Geometric Modeling. New York: John Wiley & Sons; 1997.
[5] Shoemake K. Animating Rotation with Quaternion Curves. In: Barsky BA, editor. Computer Graphics. Proceedings of SIGGRAPH 85, vol. 19(3). San Francisco, Calif; August 1985. p. 143–52.
[6] Strong G. Linear Algebra and Its Applications. New York: Academic Press; 1980.
[7] Watt A, Watt M. Advanced Animation and Rendering Techniques. New York: Addison-Wesley; 1992.

Interpolating Values

3

This chapter presents the foundation upon which much of computer animation is built—that of interpolating values. Changing values over time that somehow affect the visuals produced is, essentially, animation. Interpolation implies that the boundary values have been specified and all that remains is to fill in the intermediate details. But even filling in the values requires skill and knowledge. This chapter starts by presenting the basics of various types of interpolating and approximating curves followed by techniques to control the motion of a point along a curve. The interpolation of orientation is then considered. The chapter concludes with a section on working with paths.

3.1 Interpolation

The foundation of most computer animation is the interpolation of values. One of the simplest examples of animation is the interpolation of the position of a point in space. But even to do this correctly is nontrivial and requires some discussion of several issues: the appropriate interpolating function, the parameterization of the function based on distance traveled, and maintaining the desired control of the interpolated position over time.

Most of this discussion will be in terms of interpolating spatial values. The reader should keep in mind that any changeable value involved in the animation (and display) process such as an object's transparency, the camera's focal length, or the color of a light source could be subjected to interpolation over time.

Often, an animator has a list of values associated with a given parameter at specific frames (called *key frames* or *extremes*) of the animation. The question to be answered is how best to generate the values of the parameter for frames between the key frames. The parameter to be interpolated may be a coordinate of the position of an object, a joint angle of an appendage of a robot, the transparency attribute of an object, or any other parameter used in the manipulation and display of computer graphics elements. Whatever the case, values for the parameter of interest must be generated for all of the frames between the key frames.

For example, if the animator wants the position of an object to be $(-5, 0, 0)$ at frame 22 and the position to be $(5, 0, 0)$ at frame 67, then values for the position need to be generated for frames 23 to 66. Linear interpolation could be used. But what if the object should appear to be stopped at frame 22 and needs to accelerate from that position, reach a maximum speed by frame 34, start to decelerate at frame 50, and come to a stop by frame 67? Or perhaps instead of stopping at frame 67, the object should continue to position $(5, 10, 0)$ and arrive there at frame 80 by following a nice curved path. The next several sections address these issues of generating points along a path defined by control points and distributing the points along the path according to timing considerations.

3.1.1 The appropriate function

Appendix B.5 contains a discussion of various specific interpolation techniques. In this section, the discussion covers general issues that determine how to choose the most appropriate interpolation technique and, once it is chosen, how to apply it in the production of an animated sequence.

The following issues need to be considered in order to choose the most appropriate interpolation technique: interpolation versus approximation, complexity, continuity, and global versus local control.

Interpolation versus approximation

Given a set of points to describe a curve, one of the first decisions an animator must make is whether the given values represent actual positions that the curve should pass through (*interpolation*) or whether they are meant merely to control the shape of the curve and do not represent actual positions that the curve will intersect (*approximation*) (see Figure 3.1). This distinction is usually dependent on whether the data points are sample points of a desired curve or whether they are being used to design a new curve. In the former case, the desired curve is assumed to be constrained to travel through the sample points, which is, of course, the definition of an *interpolating spline*.[1] In the latter case, an approximating spline can be used as the animator quickly gets a feel for how repositioning the control points influences the shape of the curve.

Commonly used interpolating functions are the Hermite formulation and the Catmull-Rom spline. The Hermite formulation requires tangent information at the endpoints, whereas Catmull-Rom uses only positions the curve should pass through. Parabolic blending, similar to Catmull-Rom, is another useful interpolating function that requires only positional information. Functions that approximate some or all of the control information include Bezier and B-spline curves. See Appendix B.5 for a more detailed discussion of these functions.

An interpolating spline in which the spline passes through the interior control points

An approximating spline in which only the endpoints are interpolated; the interior control points are used only to design the curve

FIGURE 3.1

Comparing interpolation and approximating splines.

---

[1]The term *spline* comes from flexible strips used by shipbuilders and draftsmen to draw smooth curves. In computer graphics, it generally refers to a wide class of interpolating or smoothing functions.

Complexity

The complexity of the underlying interpolation equation is of concern because this translates into computational efficiency. The simpler the underlying equations of the interpolating function, the faster its evaluation. In practice, polynomials are easy to compute, and piecewise cubic polynomials are the lowest degree polynomials that provide sufficient smoothness while still allowing enough flexibility to satisfy other constraints such as beginning and ending positions and tangents. A polynomial whose degree is lower than cubic does not provide for a point of inflection between two endpoints; therefore, it might not fit smoothly to certain data points. Using a polynomial whose degree is higher than cubic typically does not provide any significant advantages and is more costly to evaluate.

Continuity

The smoothness in the resulting curve is a primary consideration. Mathematically, smoothness is determined by how many of the derivatives of the curve equation are continuous. *Zero-order continuity* refers to the continuity of values of the curve itself. Does the curve make any discontinuous jumps in its values? If a small change in the value of the parameter always results in a small change in the value of the function, then the curve has zero-order, or positional, continuity. If the same can be said of the first derivative of the function (the instantaneous change in values of the curve), then the function has *first-order*, or *tangential*, *continuity*. *Second-order continuity* refers to continuous curvature or instantaneous change of the tangent vector (see Figure 3.2). In some geometric design environments, second-order continuity of curves and surfaces may be needed, but often in animation applications, first-order continuity suffices for spatial curves. As explained later in this chapter, when dealing with time-distance curves, second-order continuity can be important.

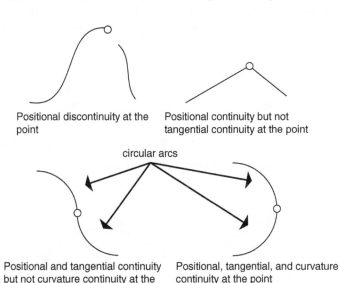

Positional discontinuity at the point

Positional continuity but not tangential continuity at the point

circular arcs

Positional and tangential continuity but not curvature continuity at the point

Positional, tangential, and curvature continuity at the point

FIGURE 3.2

Continuity (at the point indicated by the small circle).

In most applications, a curve interpolating or approximating more than a few points is defined piecewise; the curve is defined by a sequence of segments where each segment is defined by a single, vector-valued parametric function and shares its endpoints with the functions of adjacent segments. For the many types of curve segments considered here, the segment functions are cubic polynomials. The issue then becomes, not the continuity within a segment (which in the case of polynomial functions is of infinite order), but the continuity enforced at the junction between adjacent segments. Hermite, Catmull-Rom, parabolic blending, and cubic Bezier curves (see Appendix B.5) can all produce first-order continuity between curve segments. There is a form of compound Hermite curves that produces second-order continuity between segments at the expense of local control (see the following section). A cubic B-spline is second-order continuous everywhere. All of these curves provide sufficient continuity for many animation applications. Appendix B.5 discusses continuity in more mathematical terms, with topics including the difference between *parametric continuity* and *geometric continuity*.

Global versus local control

When designing a curve, a user often repositions one or just a few of the points that control the shape of the curve in order to tweak just part of the curve. It is usually considered an advantage if a change in a single control point has an effect on a limited region of the curve as opposed to affecting the entire curve. A formulation in which control points have a limited effect on the curve is referred to as providing *local control*. If repositioning one control point redefines the entire curve, however slightly, then the formulation provides only *global control*. The well-known Lagrange interpolation [5] is an example of an interpolating polynomial with global control. Figure 3.3 illustrates the difference between local control and global control. Local control is almost always viewed as being the more desirable of the two. Almost all of the composite curves provide local control: parabolic blending, Catmull-Rom splines, composite cubic Bezier, and cubic B-spline. The form of Hermite curves that enforces second-order continuity at the segment junctions does so at the expense of local control. Higher order continuity Bezier and B-spline curves have less localized control than their cubic forms.

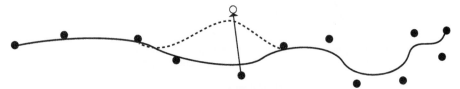

Local control: moving one control point only changes the curve over a finite bounded region

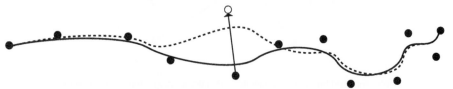

Global control: moving one control point changes the entire curve; distant sections may change only slightly

FIGURE 3.3

Comparing local and global effect of moving a control point.

3.1.2 Summary

There are many formulations that can be used to interpolate values. The specific formulation chosen depends on the desired continuity, whether local control is needed, the degree of computational complexity involved, and the information required from the user. The Catmull-Rom spline is often used in creating a path through space because it is an interpolating spline and requires no additional information from the user other than the points that the path is to pass through. Bezier curves that are constrained to pass through given points are also often used. Parabolic blending is an often overlooked technique that also affords local control and is interpolating. Formulations for these curves appear in Appendix B.5. See Mortenson [5] and Rogers and Adams [6] for more in-depth discussions.

3.2 **Controlling the motion of a point along a curve**

Designing the shape of a curve is only the first step in creating an animation path. The speed at which the curve is traced out as the parametric value is increased has to be under the direct control of the animator to produce predictable results. If the relationship between a change in parametric value and the corresponding distance along the curve is not known, then it becomes much more difficult for the animator to produce desired effects. The first step in giving the animator control is to establish a method for stepping along the curve in equal increments. Once this is done, methods for speeding up and slowing down along the curve can be made available to the animator.

For this discussion, it is assumed that an interpolating technique has been chosen and that a function $P(u)$ has been selected that, for a given value of u, will produce a value that is a point in space, that is, $p = P(u)$. Such vector valued functions will be shown in bold. Because a position in three-dimensional space is being interpolated, $P(u)$ can be considered to represent three functions. The x-, y-, and z-coordinates for the positions at the key frames are specified by the user. Key frames are associated with specific values of the time parameter, u. The x-, y-, and z-coordinates are considered independently so that, for example, the x-coordinates of the points are used as control values for the interpolating curve so that $X = P_x(u)$, where P_x denotes an interpolating function; the subscript x is used to denote that this specific curve was formed using the x-coordinates of the key positions. Similarly, $Y = P_y(u)$ and $Z = P_z(u)$ are formed so that for any specified time, u, a position (X, Y, Z) can be produced as $(P_x(u), P_y(u), P_z(u)) = P(u)$.

It is very important to note that varying the parameter of interpolation (in this case u) by a constant amount does not mean that the resulting values (in this case Euclidean position) will vary by a constant amount. Thus, if positions are being interpolated by varying u at a constant rate, the positions that are generated will not necessarily, in fact will seldom, represent a constant speed (e.g., see Figure 3.4).

To ensure a constant speed for the interpolated value, the interpolating function has to be parameterized by *arc length*, that is, distance along the curve of interpolation. Some type of reparameterization by arc length should be performed for most applications. Usually this reparameterization can be approximated without adding undue overhead or complexity to the interpolating function.

Three approaches to establishing the reparameterization by arc length are discussed here. One approach is to analytically compute arc length. Unfortunately, many curve formulations useful in animation do not lend themselves to the analytic method, so numeric methods must be applied. Two numeric methods are presented, both of which create a table of values to establish a relationship between parametric value and approximate arc length. This table can then be used to approximate

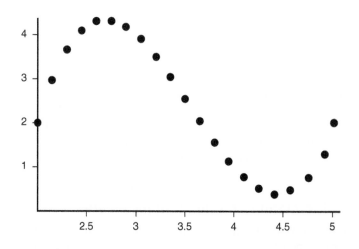

FIGURE 3.4

Example of points produced by equal increments of an interpolating parameter for a typical cubic curve. Notice the variable distance between adjacent points.

parametric values at equal-length steps along the curve. The first of these numeric methods constructs the table by supersampling the curve and uses summed linear distances to approximate arc length. The second numeric method uses Gaussian quadrature to numerically estimate the arc length. Both methods can benefit from an adaptive subdivision approach to controlling error.

3.2.1 Computing arc length

To specify how fast the object is to move along the path defined by the curve, an animator may want to specify the time at which positions along the curve should be attained. Referring to Figure 3.5 as an example in two-dimensional space, the animator specifies the following frame number and position pairs: $(0, A)$, $(10, B)$, $(35, C)$, and $(60, D)$.

Alternatively, instead of specifying time constraints, the animator might want to specify the relative velocities that an object should have along the curve. For example, the animator might specify that an object, initially at rest at position A, should smoothly accelerate until frame 20, maintain a constant

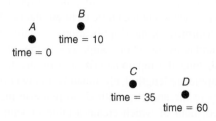

FIGURE 3.5

Position-time pairs constraining the motion.

speed until frame 35, and then smoothly decelerate until frame 60 at the end of the curve at position D. These kinds of constraints can be accommodated in a system that can compute the distance traveled along any span of the curve.

Assume that the position of an object in three-dimensional space (also referred to as three-space) is being interpolated. The objective is to define a parameterized function that evaluates to a point in three-dimensional space as a function of the parametric value; this defines a *space curve*. Assume for now that the function is a cubic polynomial as a function of a single parametric variable (as we will see, this is typically the case), that is, Equation 3.1.

$$P(u) = au^3 + bu^2 + cu + d \tag{3.1}$$

Remember that in three-space this really represents three equations: one for the x-coordinate, one for the y-coordinate, and one for the z-coordinate. Each of the three equations has its own constants a, b, c, and d. The equation can also be written explicitly representing these three equations, as in Equation 3.2.

$$\begin{aligned}
P(u) &= (x(u),\ y(u),\ z(u)) \\
x(u) &= a_x u^3 + b_x u^2 + c_x u + d_x \\
y(u) &= a_y u^3 + b_y u^2 + c_y u + d_y \\
z(u) &= a_z u^3 + b_z u^2 + c_z u + d_z
\end{aligned} \tag{3.2}$$

Each of the three equations is a cubic polynomial of the form given in Equation 3.1. The curve itself can be specified using any of the standard ways of generating a spline (see Appendix B.5 or texts on the subject, e.g., [6]). Once the curve has been specified, an object is moved along it by choosing a value of the parametric variable, and then the x-, y-, and z-coordinates of the corresponding point on the curve are calculated.

It is important to remember that in animation the path swept out by the curve in space is not the only important thing. Equally important is how the path is swept out over time. A very different effect will be evoked by the animation if an object travels over the curve at a strictly constant speed instead of smoothly accelerating at the beginning and smoothly decelerating at the end. As a consequence, it is important to discuss both the curve that defines the path to be followed by the object and the function that relates time to distance traveled. The former is the previously mentioned *space curve* and the term *distance-time function* will be used to refer to the latter. In discussing the distance-time function, the curve that represents the function will be referred to often. As a result, the terms *curve* and *function* will be used interchangeably in some contexts.

Notice that a function is desired that relates time to a position on the space curve. The user supplies, in one way or another (to be discussed in the sections that follow), the distance-time function that relates time to the distance traveled along the curve. The distance along a curve is defined as *arc length* and, in this text, is denoted by s. When the arc length computation is given as a function of a variable u and this dependence is noteworthy, then $s = S(u)$ is used. The arc length at a specific parametric value, such as $S(u_i)$, is often denoted as s_i. If the arc length computation is specified as a function of time, then $S(t)$ is used. Similarly, the function of arc length that computes a parametric value is specified as $u = U(s)$.

The interpolating function relates parametric value to position on the space curve. The relationship between distance along the curve and parametric value needs to be established. This relationship is the *arc length parameterization* of the space curve. It allows movement along the curve at a constant speed by evaluating the curve at equal arc length intervals. Further, it allows acceleration and deceleration along the curve by controlling the distance traveled in a given time interval.

For an arbitrary curve, it is usually not the case that a constant change in the parameter will result in a constant distance traveled. Because the value of the parameterizing variable is not the same as arc length, it is difficult to choose the values of the parameterizing variable so that the object moves along the curve at a desired speed. The relationship between the parameterizing variable and arc length is usually nonlinear. In the special case when a unit change in the parameterizing variable results in a unit change in curve length, the curve is said to be parameterized by arc length. Many seemingly difficult problems in controlling the motion along a path become very simple if the curve can be parameterized by arc length or, if arc length can be numerically computed, given a value for the parameterizing variable.

Let the function $LENGTH(u_a, u_b)$ be the length along the space curve from the point $P(u_a)$ to the point $P(u_b)$ (see Figure 3.6). Then the two problems to solve are

1. Given parameters u_a and u_b, find $LENGTH(u_a, u_b)$.
2. Given an arc length s and a parameter value u_a, find u_b such that $LENGTH(u_a, u_b) = \text{s}$.
   This is equivalent to finding the solution to the equation $s - LENGTH(u_a, u_b) = 0$.

The first step in controlling the timing along a space curve is to establish the relationship between parametric values and arc length. This can be done by specifying the function $s = S(u)$, which computes, for any given parametric value, the length of the curve from its starting point to the point that corresponds to that parametric value. Then if the inverse, $u = S^{-1}(s) = U(s)$, can be computed (or estimated), the curve can be effectively parameterized by arc length, that is, $P(U(s))$. Once this is done, the second step is for the animator to specify, in one way or another, the distance the object should move along the curve for each time step.

In general, neither of the problems above has an analytic solution, so numerical solution techniques must be used. As stated previously, the first step in controlling the motion of an object along a space curve is to establish the relationship between the parametric value and arc length. If possible, the curve should be explicitly parameterized by arc length by analyzing the space curve equation. Unfortunately, for most types of parametric space curves, it is difficult or impossible to find a closed-form algebraic formula to describe an arc length parameter. For example, it has been shown that B-spline curves cannot, in general, be parameterized this way [3]. As a result, several approximate parameterization techniques have been developed. But first, it is useful to look at the analytic approach.

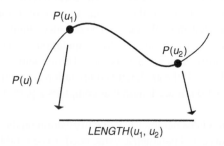

FIGURE 3.6

$LENGTH(u_1, u_2)$.

The analytic approach to computing arc length

The length of the curve from a point $P(u_a)$ to any other point $P(u_b)$ can be found by evaluating the arc length integral [4] by

$$s = \int_{u_a}^{u_b} |dP/du| \, du \qquad (3.3)$$

where the derivative of the space curve with respect to the parameterizing variable is defined to be that shown in Equations 3.4 and 3.5.

$$dP/du = ((dx(u)/(du)), (dy(u)/(du)), (dz(u)/(du))) \qquad (3.4)$$

$$|dP/du| = \sqrt{(dx(u)/du)^2 + (dy(u)/du)^2 + (dz(u)/du)^2} \qquad (3.5)$$

For a cubic curve in which $P(u) = au3 + bu2 + cu + d$, the derivative of the x-coordinate equation with respect to u is $dx(u)/du = 3a_x u2 + 2b_x u + c_x$. After squaring this and collecting the terms similarly generated by the y- and z-coordinate equations, the expression inside the radical takes the form of Equation 3.6.

$$Au^4 + Bu^3 + Cu^2 + Du + E \qquad (3.6)$$

For the two-dimensional case, the coefficients are given in Equation 3.7; the extension to three-dimensional is straightforward.

$$\begin{aligned}
A &= 9(a_x^2 + a_y^2) \\
B &= 12(a_x b_x + a_y b_y) \\
C &= 6(a_x c_x + a_y c_y) + 4(b_x^2 + b_y^2) \\
D &= 4(b_x c_x + b_y c_y) \\
E &= c_x^2 + c_y^2
\end{aligned} \qquad (3.7)$$

Estimating arc length by forward differencing

As mentioned previously, analytic techniques are not tractable for many curves of interest to animation. The easiest and conceptually simplest strategy for establishing the correspondence between parameter value and arc length is to sample the curve at a multitude of parametric values. Each parametric value produces a point along the curve. These sample points can then be used to approximate the arc length by computing the linear distance between adjacent points. As this is done, a table is built up of arc lengths indexed by parametric values. For example, given a curve $P(u)$, approximate the positions along the curve for $u = 0.00, 0.05, 0.10, 0.15, \ldots, 1.0$. The table, recording summed distances and indexed by entry number, is represented here as $G[i]$. It would be computed as follows:

$G[0] = 0.0$
$G[1] =$ the distance between $P(0.00)$ and $P(0.05)$
$G[2] = G[1]$ plus the distance between $P(0.05)$ and $P(0.10)$
$G[3] = G[2]$ plus the distance between $P(0.10)$ and $P(0.15)$
$\ldots$
$G[20] = G[19]$ plus the distance between $P(0.95)$ and $P(1.00)$

Table 3.1 Parameter, Arc Length Pairs

Index	Parametric Value (V)	Arc Length (G)
0	0.00	0.000
1	0.05	0.080
2	0.10	0.150
3	0.15	0.230
4	0.20	0.320
5	0.25	0.400
6	0.30	0.500
7	0.35	0.600
8	0.40	0.720
9	0.45	0.800
10	0.50	0.860
11	0.55	0.900
12	0.60	0.920
13	0.65	0.932
14	0.70	0.944
15	0.75	0.959
16	0.80	0.972
17	0.85	0.984
18	0.90	0.994
19	0.95	0.998
20	1.00	1.000

For example, consider the table of values, V, for u and corresponding values of the function G in Table 3.1.

As a simple example of how such a table could be used, consider the case in which the user wants to know the distance (arc length) from the beginning of the curve to the point on the curve corresponding to a parametric value of 0.73. The parametric entry closest to 0.73 must be located in the table. Because the parametric entries are evenly spaced, the location in the table of the closest entry to the given value can be determined by direct calculation. Using Table 3.1, the index, i, is determined by Equation 3.8 using the distance between parametric entries, $d = 0.05$ in this case, and the given parametric value, $v = 0.73$ in this case.

$$i = \left\lfloor \frac{v}{d} + 0.5 \right\rfloor = \left\lfloor \frac{0.73}{0.05} + 0.5 \right\rfloor = 15 \tag{3.8}$$

A crude approximation to the arc length is obtained by using the arc length entry located in the table at index 15, that is, 0.959. A better approximation can be obtained by interpolating between the arc lengths corresponding to entries on either side of the given parametric value. In this case, the index of the largest parametric entry that is less than the given value is desired (Eq. 3.9).

$$i = \left\lfloor \frac{v}{d} \right\rfloor = \left\lfloor \frac{0.73}{0.05} \right\rfloor = 14 \tag{3.9}$$

An arc length, s, can be linearly interpolated from the approximated arc lengths, G, in the table by using the differences between the given parametric value and the parametric values on either side of it in the table, as in Equation 3.10.

$$s = G[i] + \frac{v - V[i]}{V[i+1] - V[i]}(G[i+1] - G[i])$$

$$= 0.944 + \frac{0.73 - 0.70}{0.75 - 0.70}(0.959 - 0.944) \qquad (3.10)$$

$$= 0.953$$

The solution to the first problem cited above (given two parameters u_a and u_b, find the distance between the corresponding points) can be found by applying this calculation twice and subtracting the respective distances.

The converse situation, that of finding the value of u given the arc length, is dealt with in a similar manner. The table is searched for the closest arc length entry to the given arc length value, and the corresponding parametric entry is used to estimate the parametric value. This situation is a bit more complicated because the arc length entries are not evenly spaced; the table must actually be searched for the closest entry. Because the arc length entries are monotonically increasing, a binary search is an effective method of locating the closest entry. As before, once the closest arc length entry is found the corresponding parametric entry can be used as the approximate parametric value, or the parametric entries corresponding to arc length entries on either side of the given arc length value can be used to linearly interpolate an estimated parametric value.

For example, if the task is to estimate the location of the point on the curve that is 0.75 unit of arc length from the beginning of the curve, the table is searched for the entry whose arc length is closest to that value. In this case, the closest arc length is 0.72 and the corresponding parametric value is 0.40. For a better estimate, the values in the table that are on either side of the given distance are used to linearly interpolate a parametric value. In this case, the value of 0.75 is three-eighths of the way between the table values 0.72 and 0.80. Therefore, an estimate of the parametric value would be calculated as in Equation 3.11.

$$u = 0.40 + \frac{3}{8}(0.45 - 0.40) = 0.41875 \qquad (3.11)$$

The solution to the second problem (given an arc length s and a parameter value u_a, find u_b such that $LENGTH(u_a, u_b) = s$) can be found by using the table to estimate the arc length associated with u_a, adding that to the given value of s, and then using the table to estimate the parametric value of the resulting length.

The advantages of this approach are that it is easy to implement, intuitive, and fast to compute. The downside is that both the estimate of the arc length and the interpolation of the parametric value introduce errors into the calculation. These errors can be reduced in a couple of ways.

The curve can be supersampled to help reduce errors in forming the table. For example, ten thousand equally spaced values of the parameter could be used to construct a table consisting of a thousand entries by breaking down each interval into ten subintervals. This is useful if the curve is given beforehand and the table construction can be performed as a preprocessing step.

Better methods of interpolation can be used to reduce errors in estimating the parametric value. Instead of linear interpolation, higher-degree interpolation procedures can be used in computing the parametric value. Of course, higher-degree interpolation slows down the computation somewhat, so a decision about the speed/accuracy trade-off has to be made.

These techniques reduce the error somewhat blindly. There is no measure for the error in the calculation of the parametric value; these techniques only reduce the error globally instead of investing more computation in the areas of the curve in which the error is highest.

Adaptive approach

To better control error, an *adaptive forward differencing* approach can be used that invests more computation in areas of the curve that are estimated to have large errors. The approach does this by considering a section of the curve and testing to see if the estimate of the segment's length is within some error tolerance of the sum of the estimates for each of the segment's halves. If the difference is greater than can be tolerated, then the segment is split in half and each half is put on the list of segments to be tested. In addition, at each level of subdivision the tolerance is reduced by half to reflect the change in scale. If the difference is within the tolerance, then its estimated length is taken to be a good enough estimate and is used for that segment. Initially, the entire curve is the segment to be tested.

As before, a table is to be constructed that relates arc length to parametric value. Each element of the table records a parametric value and the associated arc length from the start of the curve. It is also useful to record the coordinates of the associated point on the curve. A linked list is an appropriate structure to hold this list because the number of entries that will be generated is not known beforehand; after all points are generated, the linked list can then be copied to a linear list to facilitate a binary search. In addition to the list of entries, a sorted list of segments to be tested is maintained. A segment on the list is defined and sorted according to its range of parametric values.

The adaptive approach begins by initializing the table with an entry for the first point of the curve, $<0.0, P(0)>$, and initializing the list of segments to be tested with the entire curve, $<0.0, 1.0>$. The procedure operates on segments from the list to be tested until the list is empty. The first segment on the list is always the one tested next. The segment's midpoint is computed by evaluating the curve at the middle of the range of its parametric value. The curve is also evaluated at the endpoint of the segment; the position of the start of the segment is already in the table and can be retrieved from there. The length of the segment and the lengths of each of its halves are estimated by the linear distance between the points on the curve. The sum of the estimated lengths of the two halves of the segment is compared to the estimated length of the segment. Equation 3.12 shows the test for the initial entire-curve segment.

$$\left| \|P(0.0) - P(1.0)\| - \left(\|P(0.0) - P(0.5)\| + \|P(0.5) - P(1.0)\| \right) \right| < \varepsilon \qquad (3.12)$$

If the difference between these two values is above some user-specified threshold, then both halves, in order, are added to the list of segments to be tested along with their error tolerance ($\varepsilon/2^n$ where n is the level of subdivision). If the values are within tolerance, then the parametric value of the midpoint is recorded in the table along with the arc length of the first point of the segment plus the distance from the first point to the midpoint. Also added to the table is the last parametric value of the segment along with the arc length to the midpoint plus the distance from the midpoint to the last point. When the list of segments to be tested becomes empty, a list of $<$parametric value, arc length$>$ has been generated for the entire curve.

One problem with this approach is that at a premature stage in the procedure two half segments might indicate that the subdivision can be terminated (Figure 3.7). It is usually wise to force the subdivision down to a certain level and then embark on the adaptive subdivision.

Because the table has been adaptively built, it is not possible to directly compute the index of a given parametric entry as it was with the nonadaptive approach. Depending on the data structure used

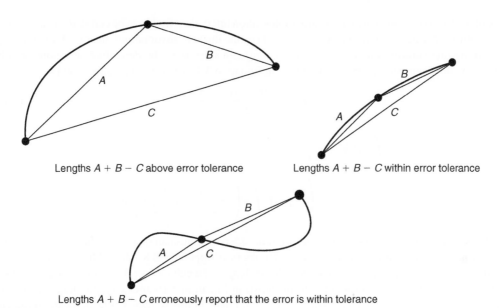

Lengths $A + B - C$ above error tolerance   Lengths $A + B - C$ within error tolerance

Lengths $A + B - C$ erroneously report that the error is within tolerance

FIGURE 3.7

Tests for adaptive subdivision.

to hold the final table, a sequential search or, possibly, a binary search must be used. Once the appropriate entries are found, then, as before, a corresponding table entry can be used as an estimate for the value or entries can be interpolated to produce better estimates.

Estimating the arc length integral numerically

For cases in which efficiency of both storage and time are of concern, calculating the arc length function numerically may be desirable. Calculating the length function involves evaluating the arc length integral (refer to Eq. 3.3). Many numerical integration techniques approximate the integral of a function with the weighted sum of values of the function at various points in the interval of integration. Techniques such as Simpson's and trapezoidal integration use evenly spaced sample intervals. Gaussian quadrature [1] uses unevenly spaced intervals in an attempt to get the greatest accuracy using the smallest number of function evaluations. Because evaluation of the derivatives of the space curve accounts for most of the processing time in this algorithm, minimizing the number of evaluations is important for efficiency. Also, because the higher derivatives are not continuous for some piecewise curves, this should be done on a per segment basis.

Gaussian quadrature, as commonly used, is defined over an integration interval from -1 to 1. The function to be integrated is evaluated at fixed points in the interval -1 to $+1$, and each function evaluation is multiplied by a precalculated weight (see Eq. 3.13).

$$\int_{-1}^{1} f(u) = \sum_{i} w_i f(u_i)$$

(3.13)

A function, $g(t)$, defined over an arbitrary integration interval $t \in [a, b]$ can be converted to a function, $h(u) = g(f(u))$, defined over the interval $u \in [-1, 1]$ by the linear transformation shown in Equation 3.14. By making this substitution and the corresponding adjustments to the derivative, any range can be handled by the Gaussian integration as shown in Equation 3.15 for arbitrary limits, a and b.

$$t = f(u) = \frac{(b-a)}{2} u + \frac{b+a}{2} \tag{3.14}$$

$$\int_a^b g(t)dt = \int_{-1}^{+1} g(f(u))f'(u)du$$

$$= \left(\frac{b-a}{2}\right) \int_{-1}^{+1} g\left(\frac{(b-a)}{2}u + \frac{b+a}{2}\right) du \tag{3.15}$$

The weights and evaluation points for different orders of Gaussian quadrature have been tabulated and can be found in many mathematical handbooks (see Appendix B.8.1 for additional details).

In using Gaussian quadrature to compute the arc length of a cubic curve, Equations 3.5–3.7 are used to define the arc length function in the form shown in Equation 3.16 after common terms have been collected into the new coefficients. Sample code to implement this is given below in the discussion on adaptive Gaussian integration.

$$\int_{-1}^{1} \sqrt{Au^4 + Bu^3 + Cu^2 + Du + E} \tag{3.16}$$

Adaptive Gaussian integration

Some space curves have derivatives that vary rapidly in some areas and slowly in others. For such curves, Gaussian quadrature will undersample some areas of the curve or oversample some other areas. In the former case, unacceptable error will accumulate in the result. In the latter case, time is wasted by unnecessary evaluations of the function. This is similar to what happens when using nonadaptive forward differencing.

To address this problem, an adaptive approach can be used [2]. Adaptive Gaussian integration is similar to the previously discussed adaptive forward differencing. Each interval is evaluated using Gaussian quadrature. The interval is then divided in half, and each half is evaluated using Gaussian quadrature. The sum of the two halves is compared with the value calculated for the entire interval. If the difference between these two values is less than the desired accuracy, then the procedure returns the sum of the halves; otherwise, as with the forward differencing approach, the two halves are added to the list of intervals to be subdivided along with the reduced error tolerance, $\varepsilon/2^n$, where, as before, ε is the user-supplied error tolerance and n is the level of subdivision.

Initially the length of the entire space curve is calculated using adaptive Gaussian quadrature. During this process, a table of the subdivision points is created. Each entry in the table is a pair (u, s) where s is the arc length at parameter value u. When calculating $LENGTH(0, u)$, the table is searched to find the values u_i, u_{i+1} such that $u_i < u < u_{i+1}$. The arc length from u_i to u is then calculated using nonadaptive Gaussian quadrature. This can be done because the space curve has been subdivided in such a way that nonadaptive Gaussian quadrature will give the required accuracy over the interval from u_i to $u_i + 1$. $LENGTH(0, u)$ is then equal to $s_i + LENGTH(u_i, u)$. $LENGTH(u_a, u_b)$ can be found by calculating $LENGTH(0, u_a)$ and $LENGTH(0, u_b)$ and then subtracting. The code for the adaptive integration and table-building procedure is shown in Figure 3.8.

```
/* ------------------------------------------------------------
 * STRUCTURES
 */


// the structure to hold entries of the table consisting of
// parameter value (u) and estimated length (length)
typedef struct table_entry_structure {
  double u,length;
} table_entry_td;


// the structure to hold an interval of the curve, defined by
// starting and ending parameter values and the estimated
// length (to be filled in by the adaptive integration
// procedure)
typedef struct interval_structure {
  double ua,ub;
  double length;
} interval_td;


// coefficients for a 2D cubic curve
typedef struct cubic_curve_structure {
  double ax,bx,cx,dx;
  double ay,by,cy,dy;
} cubic_curve_td;


// polynomial function structure; a quadric function is generated
// from a cubic curve during the arclength computation
typedef struct polynomial_structure {
  double *coeff;
  int  degree;
} polynomial_td;


/* ------------------------------------------------------------------
 * ADAPTIVE INTEGRATION
 * this is the high-level call used whenever a curve's length is to be computed
 */
void adaptive_integration(cubic_curve_td *curve, double ua, double ub,double
tolerance)
{
   double subdivide();
   polynomial_td func;
   interval_td full_interval;
   double total_length;
   double integrate_func();
   double temp;

   func.degree = 4;
   func.coeff = (double *)malloc(sizeof(double)*5);
```

FIGURE 3.8

Adaptive Gaussian integration of arc length.   *(Continued)*

```
func.coeff[4] = 9*(curve->ax*curve->ax + curve->ay*curve->ay);
   func.coeff[3] = 12*(curve->ax*curve->bx + curve->ay*curve->by);
   func.coeff[2] = (6*(curve->ax*curve->cx + curve->ay*curve->cy) +
              4*(curve->bx*curve->bx + curve->by*curve->by)   );
   func.coeff[1] = 4*(curve->bx*curve->cx + curve->by*curve->cy);
   func.coeff[0] = curve->cx*curve->cx + curve->cy*curve->cy;
   full_interval.ua = ua; full_interval.ub = ub;
   temp = integrate_func(&func,&full_interval);
   printf("\nInitial guess = %lf; %lf:%lf",temp,ua,ub);
   full_interval.length = temp;
   total_length = subdivide(&full_interval,&func,0.0,tolerance);
   printf("\n total length = %lf\n",total_length);
}


/* ------------------------------------------------------------
 * SUBDIVIDE
 * 'total_length' is the length of the curve up to, but not including,
 *          the 'full interval'
 * if the difference between the interval and the sum of its halves is
 *          less than 'tolerance,' stop the recursive  subdivision
 * 'func' is a polynomial function
 */
double subdivide(interval_td *full_interval, polynomial_td *func, double
total_length, double tolerance)
{
   interval_td left_interval, right_interval;
   double left_length,right_length;
   double midu;
   double subdivide();
   double integrate_func();
   double temp;
   void add_table_entry();

   midu = (full_interval->u1+full_interval->u2)/2;
   left_interval.ua = full_interval->ua; left_interval.ub = midu;
   right_interval.ua = midu; right_interval.ub = full_interval->ub;
   left_length = integrate_func(func, & left_interval);
   right_length = integrate_func(func, & right_interval);
   temp = fabs(full_interval->length - (left_length+right_length));
   if (temp > tolerance) {
     left_interval.length = left_length;
     right_interval.length = right_length;
     total_length = subdivide(&left_interval, func, total_length, tolerance/2.0);
     total_length = subdivide(&right_interval, func, total_length, tolerance/2.0);
     return(total_length);
   }
   else {
     total_length = total_length + left_length;
     add_table_entry(midu,total_length);
     total_length = total_length + right_length;
     add_table_entry(full_interval->ub,total_length);
     return(total_length);
   }
}
}
```

FIGURE 3.8—Cont'd

```
/* ------------------------------------------------------------------
 * ADD TABLE ENTRY
 * In an actual implementation, this code would create and add an
 *      entry to a linked list.
 * The code below simply prints out the values.
 */
void add_table_entry(double u, double length)
{
   /* add entry of (u, length) */
   printf("\ntable entry: u: %lf, length: %lf",u,length);
}


/* ------------------------------------------------------------------
 * INTEGRATE FUNCTION
* use Gaussian quadrature to integrate square root of given function
 *           in the given interval
 */
double integrate_func(polynomial_td *func,interval_td *interval)
{
   double x[5] = {.1488743389,.4333953941,.6794095682,.8650633666,
   .9739065285};
   double w[5] =
   {.2955242247,.2692667193,.2190863625,.1494513491,.0666713443};
   double length, midu, dx, diff;
   int  i;
   double evaluate_polynomial();
   double ua,ub;
   ua = interval->ua;
   ub = interval->ub;
   midu = (ua+ub)/2.0;
   diff = (ub-ua)/2.0;
   length = 0.0;
   for (i=0; i<5; i++) {
      dx = diff*x[i];
      length += w[i]*(sqrt(evaluate_polynomial(func,midu+dx)) +
           sqrt(evaluate_polynomial(func,midu-dx)));
   }
   length *= diff;
   return (length);
}


/* ------------------------------------------------------------------
 * EVALUATE POLYNOMIAL
 * evaluate a polynomial using the Horner scheme
 */
double evaluate_polynomial(polynomial_td *poly, double u)
{
   int   i;
   double value;
   value = 0.0;
   for (i=poly->degree; i<=0; i--) {
      value = value*u+poly->coeff[i];
   }
   return value;
}
```

FIGURE 3.8—Cont'd

This solves the first problem posed above; that is, given u_a and u_b find $LENGTH(u_a, u_b)$. To solve the second problem of finding u, which is a given distance, s, away from a given u_a, numerical root-finding techniques must be used as described next.

Find a point that is a given distance along a curve

The solution of the equation $s - LENGTH(u_a, u) = 0$ gives the value of u such that the point $P(u)$ is at arc length s from the point $P(u_a)$ along the curve. Since the arc length is a strictly monotonically increasing function of u, the solution is unique provided that the length of $dP(u)/du$ is not identically 0 over some interval. Newton-Raphson iteration can be used to find the root of the equation because it converges rapidly and requires very little calculation at each iteration. Newton-Raphson iteration consists of generating the sequence of points $\{p_n\}$ as in Equation 3.17.

$$p_n = p_{n-1} - f(p_{n-1})/f'(p_{n-1}) \qquad (3.17)$$

In this case, f is equal to $s - LENGTH(u_1, U(p_{n-1}))$ where $U(p)$ is a function that computes the parametric value of a point on a parameterized curve. The function $U()$ can be evaluated at p_{n-1} using precomputed tables of parametric value and points on the curve similar to the techniques discussed above for computing arc length; f' is dP/du evaluated at p_{n-1}.

Two problems may be encountered with Newton-Raphson iteration: some of the p_n may not lie on the space curve at all (or at least within some tolerance) and dP/du may be 0, or very nearly 0, at some points on the space curve. The first problem will cause all succeeding elements $p_{n+1}, p_{n+2}, \ldots$ to be undefined, while the latter problem will cause division by 0 or by a very small number. A zero parametric derivative is most likely to arise when two or more control points are placed in the same position. This can easily be detected by examining the derivative of f in Equation 3.17. If it is small or 0, then binary subdivision is used instead of Newton-Raphson iteration. Binary subdivision can be used in a similar way to handle the case of undefined p_n. When finding u such that $LENGTH(0, u) = s$, the subdivision table is searched to find the values s_i, s_{i+1} such that $s_i < s < s_{i+1}$. The solution u lies between u_i and u_{i+1}. Newton-Raphson iteration is then applied to this subinterval.

An initial approximation to the solution is made by linearly interpolating s between the endpoints of the interval and using this as the first iterate. Although Newton-Raphson iteration can diverge under certain conditions, it does not happen often in practice. Since each step of Newton-Raphson iteration requires evaluation of the arc length integral, eliminating the need to do adaptive Gaussian quadrature, the result is a significant increase in speed. At this point it is worth noting that the integration and root-finding procedures are completely independent of the type of space curve used. The only procedure that needs to be modified to accommodate new curve types is the derivative calculation subroutine, which is usually a short program.

3.2.2 Speed control

Once a space curve has been parameterized by arc length, it is possible to control the speed at which the curve is traversed. Stepping along the curve at equally spaced intervals of arc length will result in a constant speed traversal. More interesting traversals can be generated by speed control functions that relate an equally spaced parametric value (e.g., *time*) to arc length in such a way that a controlled traversal of the curve is generated. The most common example of such speed control is *ease-in/ease-out* traversal. This type of speed control produces smooth motion as an object accelerates from a stopped position, reaches a maximum velocity, and then decelerates to a stopped position.

In this discussion, the speed control function's input parameter is referred to as t, for *time*, and its output is *arc length*, referred to as *distance* or simply as s. Constant-velocity motion along the space curve can be generated by evaluating it at equally spaced values of its arc length where arc length is a linear function of t. In practice, it is usually easier if, once the space curve has been reparameterized by arc length, the parameterization is then normalized so that the parametric variable varies between 0 and 1 as it traces out the space curve; the *normalized arc length parameter* is just the arc length parameter divided by the total arc length of the curve. For this discussion, the normalized arc length parameter will still be referred to simply as the arc length.

Speed along the curve can be controlled by varying the arc length values at something other than a linear function of t; the mapping of time to arc length is independent of the form of the space curve itself. For example, the space curve might be linear, while the arc length parameter is controlled by a cubic function with respect to time. If t is a parameter that varies between 0 and 1 and if the curve has been parameterized by arc length and normalized, then ease-in/ease-out can be implemented by a function $s = ease(t)$ so that as t varies uniformly between 0 and 1, s will start at 0, slowly increase in value and gain speed until the middle values, and then decelerate as it approaches 1 (see Figure 3.9). Variable s is then used as the interpolation parameter in whatever function produces spatial values.

The control of motion along a parametric space curve will be referred to as *speed control*. Speed control can be specified in a variety of ways and at various levels of complexity. But the final result is to produce, explicitly or implicitly, a distance-time function $S(t)$, which, for any given time t, produces the distance traveled along the space curve (arc length). The space curve defines *where* to go, while the distance-time function defines *when*.

Such a function $S(t)$ would be used as follows. At a given time t, $S(t)$ is the desired distance to have traveled along the curve at time t. The arc length of the curve segment from the beginning of the curve to the point is $s = S(t)$ (within some tolerance). If, for example, this position is used to translate an object through space at each time step, it translates along the path of the space curve at a speed indicated by $S'(t)$. An arc length table (see Section 3.2.1) can then be used to find the corresponding parametric value $u = U(s)$ that corresponds to that arc length. The space curve is then evaluated at u to produce a point on the space curve, $p = P(u)$. Thus, $p = P(U(S(t)))$.

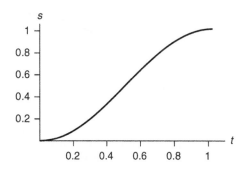

FIGURE 3.9

An example of an *ease(t)* function.

There are various ways in which the distance-time function can be specified. It can be explicitly defined by giving the user graphical curve-editing tools. It can be specified analytically. It can also be specified by letting the user define the velocity-time curve, or by defining the acceleration-time curve. The user can even work in a combination of these spaces. In the following discussion, the common assumption is that the entire arc length of the curve is to be traversed during the given total time. Two additional optional assumptions (constraints) that are typically used are listed below. In certain situations, it might be desirable to violate one or both of these constraints.

1. The distance-time function should be monotonic in t, that is, the curve should be traversed without backing up along the curve.
2. The distance-time function should be continuous, that is, there should be no instantaneous jumps from one point on the curve to a nonadjacent point on the curve.

Following the assumptions stated above, the entire space curve is to be traversed in the given total time. This means that $0.0 = S(0.0)$ and *total_distance* $= S(total\_time)$. As mentioned previously, the distance-time function may also be normalized so that all such functions end at $1.0 = S(1.0)$. Normalizing the distance-time function facilitates its reuse with other space curves. An example of an analytic definition is $S(t) = (2 - t)*t$ (although this does not have the shape characteristic of ease-in/ease-out motion control; see Figure 3.10).

3.2.3 Ease-in/ease-out

Ease-in/ease-out is one of the most useful and most common ways to control motion along a curve. There are several ways to incorporate ease-in/ease-out control. The standard assumption is that the motion starts and ends in a complete stop and that there are no instantaneous jumps in velocity (first-order continuity). There may or may not be an intermediate interval of constant speed, depending on the technique used to generate the speed control. The speed control function will be referred to as $s = ease(t)$, where t is a uniformly varying input parameter meant to represent time and s is the output parameter that is the distance (arc length) traveled as a function of time.

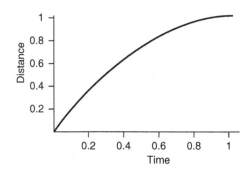

FIGURE 3.10

Graph of sample analytic distance-time function $(2 - t)*t$.

Sine interpolation

One easy way to implement ease-in/ease-out is to use the section of the sine curve from $-\pi/2$ to $+\pi/2$ as the *ease*(t) function. This is done by mapping the parameter values of 0 to $+1$ into the domain of $-\pi/2$ to $+\pi/2$ and then mapping the corresponding range of the sine functions of -1 to $+1$ in the range 0 to 1. See Equation 3.18 and the corresponding Figure 3.11.

$$s = ease(t) = \frac{\sin\left(t_\pi - \frac{\pi}{2}\right) + 1}{2} \tag{3.18}$$

In this function, t is the input that is to be varied uniformly from 0 to 1 (e.g., 0.0, 0.1, 0.2, 0.3, . . .). The output, s, also goes from 0 to 1 but does so by starting off slowly, speeding up, and then slowing down. For example, at $t = 0.25$, $ease(t) = 0.1465$, and at $t = 0.75$, $ease(t) = 0.8535$. With the sine/cosine ease-in/ease-out function presented here, the "speed" of s with respect to t is never constant over an interval, rather it is always accelerating or decelerating, as can be seen by the ever-changing slope of the curve. Notice that the derivative of this ease function is 0 at $t = 0$ and at $t = 1$. Zero derivatives at 0 and 1 indicate a smooth acceleration from a stopped position at the beginning of the motion and a smooth deceleration to a stop at the end.

Using sinusoidal pieces for acceleration and deceleration

To provide an intermediate section of the distance-time function that has constant speed, pieces of the sine curve can be constructed at each end of the function with a linear intermediate segment. Care must be taken so that the tangents of the pieces line up to provide first-order continuity. There are various ways to approach this, but as an example, assume that the user specifies fragments of the unit interval that should be devoted to initial acceleration and final deceleration. For example, the user may specify that acceleration ceases at 0.3 and the deceleration starts at 0.75. That is, acceleration occurs from time 0 to 0.3 and deceleration occurs from time 0.75 to the end of the interval at 1.0. Referring to the user-specified values as k_1 and k_2 respectively, a sinusoidal curve segment is used to implement an acceleration from time 0 to k_1. A sinusoidal curve is also used for velocity to implement deceleration from time k_2 to 1. Between times k_1 and k_2, a constant velocity is used.

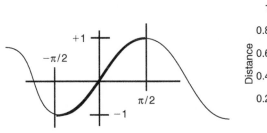

Sine curve segment to use as ease-in/ease-out control

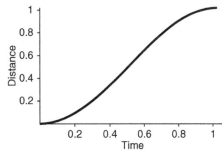

Sine curve segment mapped to useful values

FIGURE 3.11

Using a sine curve segment as the ease-in/ease-out distance-time function.

The solution is formed by piecing together a sine curve segment from $-\pi/2$ to 0 with a straight line segment (indicating constant speed) inclined at 45 degrees (because the slope of the sine curve at 0 is equal to 1) followed by a sine curve segment from 0 to $\pi/2$. The initial sine curve segment is uniformly scaled by $k_1/(\pi/2)$ so that the length of its domain is k_1. The length of the domain of the line segment is $k_2 - k_1$. And the final sine curve segment is uniformly scaled by $1.0 - k_2/(\pi/2)$ so that the length of its domain is $1.0 - k_2$. The sine curve segments must be uniformly scaled to preserve C^1 (first-order) continuity at the junction of the sine curves with the straight line segment. This means that for the first sine curve segment to have a domain from 0 to k_1, it must have a range of length $k_1/(\pi/2)$; similarly, the ending sine curve segment must have a range of length $1 - k_2/(\pi/2)$. The middle straight line segment, with a slope of 1, will travel a distance of $k_2 - k_1$.

To normalize the distance traveled, the resulting three-segment curve must be scaled down by a factor equal to the total distance traveled as computed by $k_1/(\pi/2) + k_2 - k_1 + 1 - k_2/(\pi/2)$ (see Figure 3.12).

The ease function described above and shown in Equation 3.19 is implemented in the code of Figure 3.13.

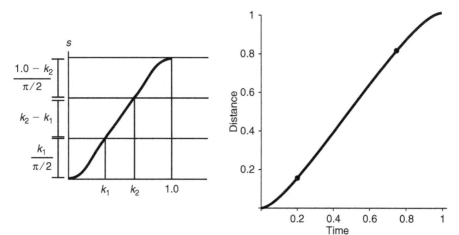

Ease-in/ease-out curve as it is initially pieced together

Curve segments scaled into useful values with points marking segment junctions

FIGURE 3.12

Using sinusoidal segments with constant speed intermediate interval.

```
double ease(float t, float k1, float k2)
{
    double f,s;
    f = k1*2/PI + k2 - k1 + (1.0 - k2)*2/PI;
    if(t < k1)          s = k1*(2/PI)*(sin((t/k1)*PI/2 - PI/2)+1);
    else if (t < k2)    s = (2*k1/PI + t - k1);
        else            s = 2*k1/PI + k2 - k1 + ((1-k2)*(2/PI)) *
                            sin(((t - k2)/(1.0 - k2))*PI/2);
    return (s/f);
}
```

FIGURE 3.13

Code to implement ease-in/ease-out using sinusoidal ease-in, followed by constant velocity and sinusoidal ease-out.

$$\left.\begin{array}{rl} = & \left(k_1 \dfrac{2}{\pi}\left(\sin\left(\dfrac{t}{k_1}\dfrac{\pi}{2}-\dfrac{\pi}{2}\right)+1\right)\right)/f \\[2em] ease(t) \;\;\; = & \left(\dfrac{k_1}{\pi/2}+t-k_1\right)/f \\[2em] = & \left(\dfrac{k_1}{\pi/2}+k_2-k_1+\left((1-k_2)\dfrac{2}{\pi}\right)\sin\left(\left(\dfrac{t-k_2}{1-k_2}\right)\dfrac{\pi}{2}\right)\right)/f \end{array}\right\} \quad \begin{array}{l} t \le k_1 \\[2em] k_1 \le t \le k_2 \qquad (3.19) \\[2em] k_2 \le t \end{array}$$

where $f = k_1(2/\pi + k_2 - k_1 + (1-k_2)(2/\pi)$

Single cubic polynomial ease-in/ease-out

A single polynomial can be used to approximate the sinusoidal ease-in/ease-out control (Eq. 3.20). It provides accuracy within a couple of decimal points while avoiding the transcendental function[2] invocations. It passes through the points (0,0) and (1,1) with horizontal beginning and ending tangents of 0. Its drawback is that there is no intermediate segment of constant speed (see Figure 3.14).

$$s = -2t^3 + 3t^2 \qquad\qquad (3.20)$$

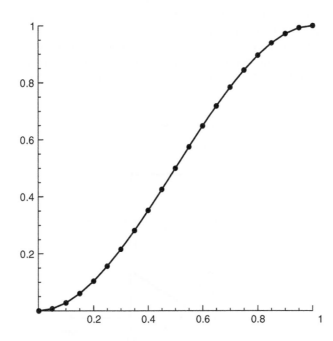

FIGURE 3.14

Ease-in/ease-out polynomial $s = -2t^3 + 3t^2$.

---

[2]A transcendental function is one that cannot be expressed algebraically. That is, it transcends expression in algebra. Transcendental functions include the exponential function as well as the trigonometric functions such as sine.

Constant acceleration: parabolic ease-in/ease-out

To avoid the transcendental function evaluation while still providing for a constant speed interval between the ease-in and ease-out intervals, an alternative approach for the ease function is to establish basic assumptions about the acceleration that in turn establish the basic form that the velocity-time curve can assume. The user can then set parameters to specify a particular velocity-time curve that can be integrated to get the resulting distance-time function.

The simple case of no ease-in/ease-out would produce a constant zero acceleration curve and a velocity curve that is a horizontal straight line at some value v_0 over the time interval from 0 to total time, t_{total}. The actual value of v_0 depends on the total distance covered, d_{total}, and is computed using the relationship

$$distance = speed \cdot time$$

so that

$$v_0 = \frac{d_{total}}{t_{total}}$$

In the case where normalized values of 1 are used for total distance covered and total time, $v_0 = 1$ (see Figure 3.15).

The distance-time curve is defined by the integral of the velocity-time curve and relates time and distance along the space curve through a function $S(t)$. Similarly, the velocity-time curve is defined by the integral of the acceleration-time curve and relates time and velocity along the space curve.

To implement an ease-in/ease-out function, constant acceleration and deceleration at the beginning and end of the motion and zero acceleration during the middle of the motion are assumed. The assumptions of beginning and ending with stopped positions mean that the velocity starts out at 0 and ends at 0. In order for this to be reflected in the acceleration/deceleration curve, the area under the curve marked

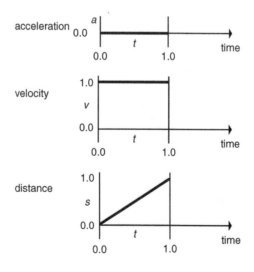

FIGURE 3.15

Acceleration, velocity, and distance curves for constant speed.

"acc" must be equal to the area above the curve marked "dec," but the actual values of the acceleration and deceleration do not have to be equal to each other. Thus, three of the four variables (acc, dec, t_1, t_2) can be specified by the user and the system can solve for the fourth to enforce the constraint of equal areas (see Figure 3.16).

This piecewise constant acceleration function can be integrated to obtain the velocity function. The resulting velocity function has a linear ramp for accelerating, followed by a constant velocity interval, and ends with a linear ramp for deceleration (see Figure 3.17). The integration introduces a constant into the velocity function, but this constant is 0 under the assumption that the velocity starts out at 0 and ends at 0. The constant velocity attained during the middle interval depends on the total distance that

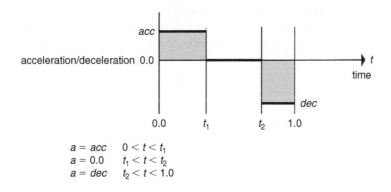

$$a = acc \quad 0 < t < t_1$$
$$a = 0.0 \quad t_1 < t < t_2$$
$$a = dec \quad t_2 < t < 1.0$$

FIGURE 3.16

Constant acceleration/deceleration graph.

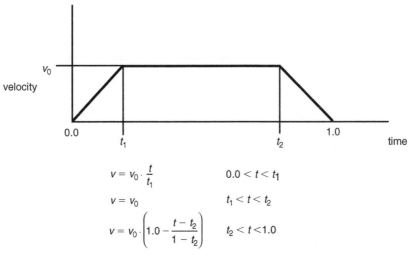

$$v = v_0 \cdot \frac{t}{t_1} \qquad 0.0 < t < t_1$$

$$v = v_0 \qquad t_1 < t < t_2$$

$$v = v_0 \cdot \left(1.0 - \frac{t - t_2}{1 - t_2}\right) \qquad t_2 < t < 1.0$$

FIGURE 3.17

Velocity-time curve for constant acceleration. Area under the curve equals the distance traveled.

must be covered during the time interval; the velocity is equal to the area below (above) the *acc* (*dec*) segment in Figure 3.16. In the case of normalized time and normalized distance covered, the total time and total distance are equal to 1. The total distance covered will be equal to the area under the velocity curve (Figure 3.17). The area under the velocity curve can be computed as in Equation 3.21.

$$1 = A_{acc} + A_{constant} + A_{dec}$$
$$1 = \frac{1}{2}v_0 t_1 + v_0(t_2 - t_1) + \frac{1}{2}v_0(1 - t_2) \tag{3.21}$$

Because integration introduces arbitrary constants, the acceleration-time curve does not bear any direct relation to total distance covered. Therefore, it is often more intuitive for the user to specify ease-in/ease-out parameters using the velocity-time curve. In this case, the user can specify two of the three variables, t_1, t_2, and v_0, and the system can solve for the third in order to enforce the "total distance covered" constraint. For example, if the user specifies the time over which acceleration and deceleration take place, then the maximum velocity can be found by using Equation 3.22.

$$v_0 = \frac{2}{(t_2 - t_2 + 1)} \tag{3.22}$$

The velocity-time function can be integrated to obtain the final distance-time function. Once again, the integration introduces an arbitrary constant, but, with the assumption that the motion begins at the start of the curve, the constant is constrained to be 0. The integration produces a parabolic ease-in segment, followed by a linear segment, followed by a parabolic ease-out segment (Figure 3.18).

The methods for ease control based on one or more sine curve segments are easy to implement and use but are less flexible than the acceleration-time and velocity-time functions. These last two functions allow the user to have even more control over the final motion because of the ability to set various parameters.

3.2.4 General distance-time functions

When working with velocity-time curves or acceleration-time curves, one finds that the underlying assumption that the total distance is to be traversed during the total time presents some interesting issues. Once the total distance and total time are given, the average velocity is fixed. This average

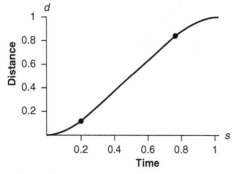

$$d = v_0 \frac{t^2}{2t_1} \qquad\qquad 0.0 < t < t_1$$

$$d = v_0 \frac{t_1}{2} + v_0(t - t_1) \qquad\qquad t_1 < t < t_2$$

$$d = v_0 \frac{t_1}{2} + v_0(t_2 - t_1) + \left[v_0 - \frac{\left(v_0 \frac{t - t_2}{1 - t_2}\right)}{2}\right](t - t_2) \qquad t_2 < t < 1.0$$

FIGURE 3.18

Distance-time function for constant acceleration.

velocity must be maintained as the user modifies, for example, the shape of the velocity-time curve. This can create a problem if the user is working only with the velocity-time curve. One solution is to let the absolute position of the velocity-time curve float along the velocity axis as the user modifies the curve. The curve will be adjusted up or down in absolute velocity in order to maintain the correct average velocity. However, this means that if the user wants to specify certain velocities, such as starting and ending velocities, or a maximum velocity, then other velocities must change in response in order to maintain total distance covered.

An alternative way to specify speed control is to fix the absolute velocities at key points and then change the interior shape of the curve to compensate for average velocity. However, this may result in unanticipated (and undesirable) changes in the shape of the velocity-time curve. Some combinations of values may result in unnatural spikes in the velocity-time curve in order to keep the area under the curve equal to the total distance (equal to 1, in the case of normalized distance). Consequently, undesirable accelerations may be produced, as demonstrated in Figure 3.19.

Notice that negative velocities mean that a point traveling along the space curve backs up along the curve until the time is reached when velocity becomes positive again. Usually, this is not desirable behavior.

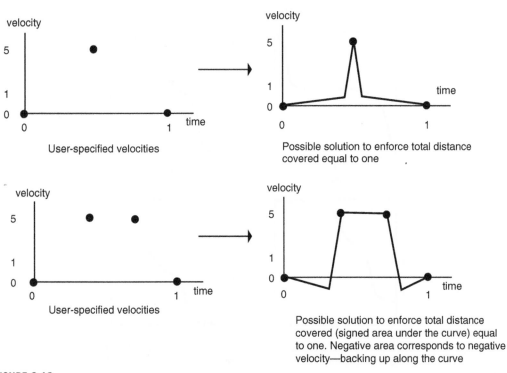

FIGURE 3.19

Some non-intuitive results of user-specified values on the velocity-time curve.

The user may also work directly with the distance-time curve. For some users, this is the most intuitive approach. Assuming that a point starts at the beginning of the curve at $t = 0$ and traverses the curve to arrive at the end of the curve at $t = 1$, then the distance-time curve is constrained to start at $(0, 0)$ and end at $(1, 1)$. If the objective is to start and stop with zero velocity, then the slope at the start and end should be 0. The restriction that a point traveling along the curve may not back up any time during the traversal means that the distance-time curve must be monotonically increasing (i.e., always have a non-negative slope). If the point may not stop along the curve during the time interval, then there can be no horizontal sections in the distance-time curve (no internal zero slopes) (see Figure 3.20).

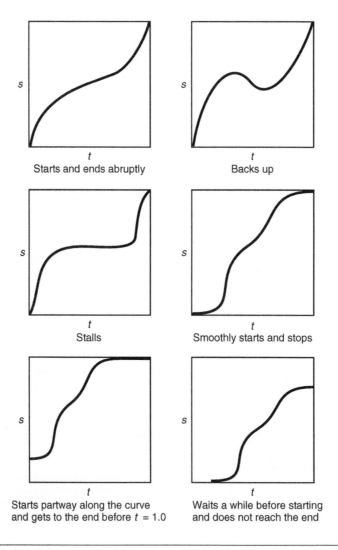

FIGURE 3.20

Sample distance-time functions.

As mentioned before, the space curve that defines the path along which the object moves is independent of the speed control curves that define the relative velocity along the path as a function of time. A single space curve could have several velocity-time curves defined along it. A single distance-time curve, such as a standard ease-in/ease-out function, could be applied to several different space curves. Reusing distance-time curves is facilitated if normalized distance and time are used.

Motion control frequently requires specifying positions and speeds along the space curve at specific times. An example might be specifying the motion of a hand as it reaches out to grasp an object; initially the hand accelerates toward the object and then, as it comes close, it slows down to almost zero speed before picking up the object. The motion is specified as a sequence of constraints on time, position (in this case, arc length traveled along a space curve), velocity, and acceleration. Stating the problem more formally, each point to be constrained is an n-tuple, $<t_i, s_i, v_i, a_i, \ldots,>$, where s_i is position, v_i is velocity, a_i is acceleration, and t_i is the time at which all the constraints must be satisfied (the ellipses, $\ldots$, indicate that higher order derivatives may be constrained). Define the zero-order constraint problem to be that of satisfying sets of two-tuples, $<t_i, s_i>$, while velocity, acceleration, and so on are allowed to take on any values necessary to meet the position constraints at the specified times. Zero-order constrained motion is illustrated at the top of Figure 3.21. Notice that there is continuity of position but not of speed. By extension, the first-order constraint problem requires satisfying sets of three-tuples, $<s_i, v_i, t_i>$, as shown in the bottom illustration in Figure 3.21. Standard interpolation techniques (see Appendix B.5) can be used to aid the user in generating distance-time curves.

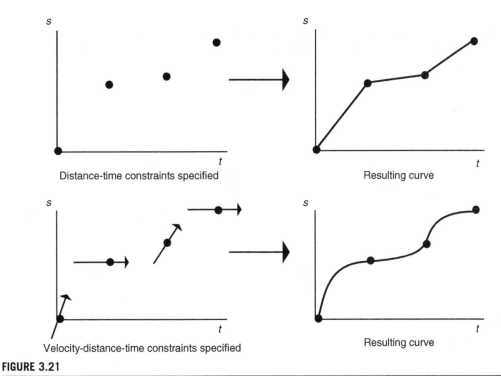

Distance-time constraints specified   Resulting curve

Velocity-distance-time constraints specified   Resulting curve

FIGURE 3.21

Specifying motion constraints.

3.2.5 Curve fitting to position-time pairs

If the animator has specific positional constraints that must be met at specific times, then the time-parameterized space curve can be determined directly. Position-time pairs can be specified by the animator, as in Figure 3.22, and the control points of the curve that produce an interpolating curve can be computed from this information [6].

For example, consider the case of fitting a B-spline[3] curve to values of the form $(P_i, t_i), i = 1, \ldots, j$. Given the form of B-spline curves shown in Equation 3.23 of degree k with $n + 1$ defining control vertices, and expanding in terms of the j given constraints $(2 \leq k \leq n + 1 \leq j)$, Equation 3.24 results. Put in matrix form, it becomes Equation 3.25, in which the given points are in the column matrix P, the unknown defining control vertices are in the column matrix B, and N is the matrix of basis functions evaluated at the given times $(t_1, t_2, \ldots, t_j)$.

$$P(t) = \sum_{i=1}^{n+1} B_i N_{i,k}(t) \tag{3.23}$$

$$\begin{aligned}
P_1 &= N_{1,k}(t_1)B_1 + N_{2,k}(t_1)B_2 + \ldots + N_{n+1,k}(t_1)B_{n+1} \\
P_2 &= N_{1,k}(t_2)B_1 + N_{2,k}(t_2)B_2 + \ldots + N_{n+1,k}(t_2)B_{n+1} \\
&\ldots \\
P_j &= N_{1,k}(t_j)B_1 + N_{2,k}(t_j)B_2 + \ldots + N_{n+1,k}(t_j)B_{n+1}
\end{aligned} \tag{3.24}$$

$$P = NB \tag{3.25}$$

If there are the same number of given data points as there are unknown control points $(2 \leq k \leq n + 1 = j)$, then N is square and the defining control vertices can be solved by inverting the matrix, as in Equation 3.26.

$$B = N^{-1}P \tag{3.26}$$

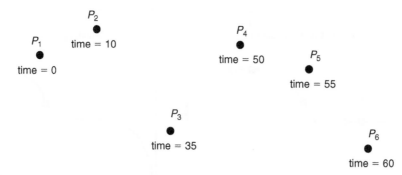

FIGURE 3.22

Position-time constraints.

---

[3]Refer to Appendix B.5.12 for more information on B-spline curves.

The resulting curve is smooth but can sometimes produce unwanted wiggles. Specifying fewer control points ($2 \leq k \leq n + 1 < j$) can remove these wiggles but N is no longer square. To solve the matrix equation, the use of the pseudoinverse of N is illustrated in Equation 3.27, but in practice a more numerically stable and robust algorithm for solving linear least-squares problems should be used (e.g., see [1]).

$$
\begin{aligned}
P &= NB \\
N^T P &= N^T NB \\
[N^T N]^{-1} N^T P &= B
\end{aligned}
\tag{3.27}
$$

3.3 Interpolation of orientations

The previous section discussed the interpolation of position through space and time. However, when dealing with objects, animators are concerned not only with the object's position, but also with the object's orientation. An object's orientation can be interpolated just like its position can be interpolated and the same issues apply. Instead of interpolating x, y, z positional values, an object's orientation parameters are interpolated. Just as an object's position is changed over time by applying translation transformations, an object's orientation is changed over time by applying rotational transformations. As discussed in the previous chapter, quaternions are a very useful representation for orientation and they will be the focus here.

3.3.1 Interpolating quaternions

One of the most important reasons for choosing quaternions to represent orientation is that they can be easily interpolated. Most important, the quaternion representation avoids the condition known as *gimbal lock*, which can trouble the other commonly used representations such as fixed angles and Euler angles.

Because the magnitude of a quaternion does not affect the orientation it represents, unit quaternions are typically used as the canonical representation for an orientation. Unit quaternions can be considered as points on a four-dimensional unit sphere.

Given two orientations represented by unit quaternions, intermediate orientations can be produced by linearly interpolating the individual quantities of the quaternion four-tuples. These orientations can be viewed as four-dimensional points on a straight-line path from the first quaternion to the second quaternion. While equal-interval, linear interpolation between the two quaternions will produce orientations that are reasonably seen as between the two orientations to be interpolated, it will not produce a constant speed rotation. This is because the unit quaternions to which the intermediate orientations map will not create equally spaced intervals on the unit sphere. Figure 3.23 shows the analogous effect when interpolating a straight-line path between points on a two-dimensional circle.

Intermediate orientations representing constant-speed rotation can be calculated by interpolating directly on the surface of the unit sphere, specifically along the great arc between the two quaternion points. In the previous chapter, it was pointed out that a quaternion, $[s, v]$, and its negation, $[-s, -v]$, represent the same orientation. That means that interpolation from one orientation, represented by the quaternion q_1, to another orientation, represented by the quaternion q_2, can also be carried out from q_1 to $-q_2$. The difference is that one interpolation path will be longer than the other. Usually, the shorter path is the more desirable because it represents the more direct way to get from one orientation to the other. The shorter path is the one indicated by the smaller angle between the four-dimensional

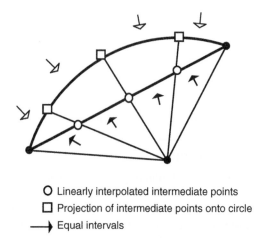

O Linearly interpolated intermediate points

□ Projection of intermediate points onto circle

⟶ Equal intervals

⟶ Unequal intervals

FIGURE 3.23

Equally spaced linear interpolations of straight-line path between two points on a circle generate unequal spacing of points after projecting onto a circle. Interpolating quaternion four-tuples exhibit the same problem.

quaternion vectors. This can be determined by using the four-dimensional dot product of the quaternions to compute the cosine of the angle between q_1 and q_2 (Eq. 3.28). If the cosine is positive, then the path from q_1 to q_2 is shorter; otherwise the path from q_1 to $-q_2$ is shorter (Figure 3.24).

$$\cos(\theta) = q_1 \cdot q_2 = s_1 s_2 + v_1 \cdot v_2 \qquad (3.28)$$

The formula for spherical linear interpolation (*slerp*) between unit quaternions q_1 and q_2 with parameter u varying from 0 to 1 is given in Equation 3.29, where $q_1 \cdot q_2 = \cos(\theta)$. Notice that this does

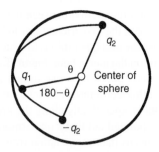

FIGURE 3.24

The closer of the two representations of orientation, q_2, is the better choice to use in interpolation. In this case $-q_2$ is closer to q_1 than q_2. The angle between q_1 and q_2, as measured from the center of the sphere, is computed by taking the dot product of q_1 and q_2. The same is done for $-q_2$.

not necessarily produce a unit quaternion, so the result must be normalized if a unit quaternion is desired.

$$slerp(q_1, q_2, u) = \frac{sin((1-u)\theta)}{sin(\theta)}q_1 + \frac{sin(u\theta)}{sin(\theta)}q_2 \tag{3.29}$$

Notice that in the case $u = 1/2$, $slerp(q_1, q_2, u)$ can be easily computed within a scale factor, as $q_1 + q_2$, which can then be normalized to produce a unit quaternion.

When interpolating between a series of orientations, slerping between points on a spherical surface has the same problem as linear interpolation between points in Euclidean space: that of first-order discontinuity (see Appendix B.5). Shoemake [7] suggests using cubic Bezier interpolation to smooth the interpolation between orientations. Reasonable interior control points are automatically calculated to define cubic segments between each pair of orientations.

To introduce this technique, assume for now that there is a need to interpolate between a sequence of two-dimensional points, $[\ldots, p_{n-1}, p_n, p_{n+1}, \ldots]$; these will be referred to as the *interpolated points*. (How to consider the calculations using quaternion representations is discussed below.) Between each pair of points, two control points will be constructed. For each of the interpolation points, p_n, two control points will be associated with it: the one immediately before it, b_n, and the one immediately after it, a_n.

To calculate the control point following any particular point p_n, take the vector defined by p_{n-1} to p_n and add it to the point p_n. Now, take this point (marked "1" in Figure 3.25) and find the average of it and p_{n+1}. This becomes one of the control points (marked "a_n" in Figure 3.25).

Next, take the vector defined by a_n to p_n and add it to p_n to get b_n (Figure 3.26). Points b_n and a_n are the control points immediately before and after the point p_n. This construction ensures first-order

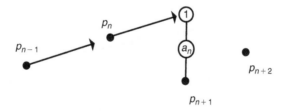

FIGURE 3.25

Finding the control point after p_n: to p_n, add the vector from p_{n-1} to p_n and average that with p_{n+1}.

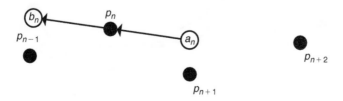

FIGURE 3.26

Finding the control point before p_n: to p_n, add the vector from a_n to p_n.

continuity at the junction of adjacent curve segments (in this case, p_n) because the control points on either side of the point are colinear with, and equidistant to, the control point itself.

The end conditions can be handled by a similar construction. For example, the first control point, a_0, is constructed as the vector from the third interpolated point to the second point ($p_1 - p_2$) is added to the second point (Figure 3.27).

In forming a control point, the quality of the interpolated curve can be affected by adjusting the distance the control point is from its associated interpolated point while maintaining its direction from that interpolated point. For example, a new b_n' can be computed with a user-specified constant, k, as in Equation 3.30 (and optionally computing a corresponding new a_n').

$$b_n' = p_n + k(b_n - p_n) \tag{3.30}$$

Between any two interpolated points, a cubic Bezier curve segment is then defined by the points p_n, a_n, b_{n+1}, and p_{n+1}. The control point b_{n+1} is defined in exactly the same way that b_n is defined except for using p_n, p_{n+1}, and p_{n+2}. The cubic curve segment is then generated between p_n and p_{n+1} (see Figure 3.28).

It should be easy to see how this procedure can be converted into the four-dimensional spherical world of quaternions. Instead of adding vectors, rotations are concatenated using quaternion multiplication. Averaging of orientations can easily be done by slerping to the halfway orientation, which is implemented by adding quaternions (and optionally normalizing). Adding the difference between two orientations, represented as unit quaternions p and q, to the second orientation, q, can be affected by constructing the representation of the result, r, on the four-dimensional unit sphere using Equation 3.31. See Appendix B.3.4 for further discussion.

$$r = 2\,(p.q)\,q - p \tag{3.31}$$

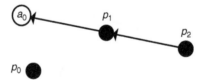

FIGURE 3.27

Constructing the first interior control point.

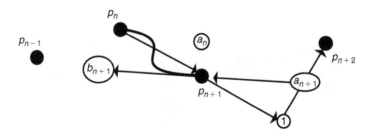

FIGURE 3.28

Construction of b_{n+1} and the resulting curve.

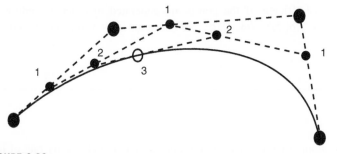

Interpolation steps

1. 1/3 of the way between pairs of points
2. 1/3 of the way between points of step 1
3. 1/3 of the way between points of step 2

FIGURE 3.29

de Casteljau construction of point on cubic Bezier segment at 1/3 (the point labeled "3").

Once the internal control points are computed, the de Casteljau algorithm can be applied to interpolate points along the curve. An example of the de Casteljau construction procedure in the two-dimensional Euclidean case of Bezier curve interpolation is shown in Figure 3.29. See Appendix B.5.10 for a more complete discussion of the procedure.

The same procedure can be used to construct the Bezier curve in four-dimensional spherical space. For example, to obtain an orientation corresponding to the $u = 1/3$ position along the curve, the following orientations are computed:

```
p1  = slerp (qₙ,aₙ,1/3)
p2  = slerp (aₙ,bₙ₊₁,1/3)
p3  = slerp (bₙ₁₁,qₙ₁₁,1/3)
p12 = slerp (p₁,p₂,1/3)
p23 = slerp (p₂,p₃,1/3)
p   = slerp (p12,p23,1/3)
```

where p is the quaternion representing an orientation 1/3 along the spherical cubic spline.

The procedure can be made especially efficient in the case of quaternion representations when calculating points at positions along the curve corresponding to u values that are powers of 1/2. For example, consider calculating a point at $u = 1/4$.

```
temp = slerp (qₙ,aₙ,1/2) = (qₙ + aₙ)/||qₙ+aₙ||
p₁ = slerp (qₙ,temp,1/2) = (qₙ + temp)/||qₙ+ temp||
temp = slerp (aₙ,bₙ₁₁,1/2) = (aₙ + bₙ₁₁)/|| aₙ+ bₙ₁₁||
p₂ = slerp (aₙ,temp,1/2) = (aₙ + temp)/|| aₙ+ temp||
temp = slerp (bₙ₁₁,qₙ₁₁,1/2) = (bₙ₊₁ + qₙ₁₁)/|| bₙ₊₁ + qₙ₁₁||
p₃ = slerp (bₙ₁₁,temp,1/2) = (bₙ₊₁ + temp)/|| bₙ₊₁ + temp ||
temp = slerp (p₁,p₂,1/2) = (p₁ + p₂)/|| p₁ + p₂||
p₁₂ = slerp (p₁,temp,1/2) = (p₁ + temp)/|| p₁ + temp ||
temp = slerp (p₂,p₃,1/2) = (p₂ + p₃)/|| p₂ + p₃||
p₂₃ = slerp (p₂,temp,1/2) = (p₂ + temp)/|| p₂ + temp ||
temp = slerp (p₁₂,p₂₃,1/2) = (p₁₂ + p₂₃)/|| p₁₂ + p₂₃||
p = slerp (p₁₂,temp,1/2) = (p₁₂ + temp)/|| p₁₂ + temp ||
```

The procedure can be made more efficient if the points are generated in order according to binary subdivision (1/2, 1/4, 3/4, 1/8, 3/8, 5/8, 7/8, 1/16, …) and temporary values are saved for use in subsequent calculations.

3.4 Working with paths
3.4.1 Path following

Animating an object by moving it along a path is a very common technique and usually one of the simplest to implement. As with most other types of animation, however, issues may arise that make the task more complicated than first envisioned. An object (or camera) following a path requires more than just translating along a space curve, even if the curve is parameterized by arc length and the speed is controlled using ease-in/ease-out. Changing the orientation of the object also has to be taken into consideration. If the path is the result of a digitization process, then often it must be smoothed before it can be used. If the path is constrained to lie on the surface of another object or needs to avoid other objects, then more computation is involved. These issues are discussed below.

3.4.2 Orientation along a path

Typically, a local coordinate system (u, v, w) is defined for an object to be animated. In this discussion, a right-handed coordinate system is assumed, with the origin of the coordinate system determined by a point along the path $P(s)$. As previously discussed, this point is generated based on the frame number, arc length parameterization, and possibly ease-in/ease-out control. This position will be referred to as *POS*. The direction the object is facing is identified with the w-axis, the up vector is identified with the v-axis, and the local u-axis is perpendicular to these two, completing the coordinate system. To form a right-handed coordinate system, for example, the u-axis points to the left of the object as someone at the object's position (*POS*) looks down the w-axis with the head aligned with the v-axis (Figure 3.30).

There are various ways to handle the orientation of the camera as it travels along a path. Of course, which method to use depends on the desired effect of the animation. The orientation is specified by determining the direction it is pointing (the w-axis) and the direction of its up vector (the v-axis); the u-axis is then fully specified by completing the left-handed camera coordinate system (eye space).

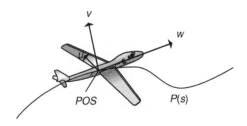

FIGURE 3.30

Object-based, right-handed, local coordinate system.

Use of the Frenet frame

If an object or camera is following a path, then its orientation can be made directly dependent on the properties of the curve. The Frenet frame[4] can be defined along the curve as a moving coordinate system, (u, v, w), determined by the curve's tangent and curvature. The Frenet frame changes orientation over the length of the curve. It is defined as normalized orthogonal vectors with w in the direction of the first derivative ($\mathbf{P}'(s)$), v orthogonal to w and in the general direction of the second derivative ($\mathbf{P}''(s)$), and u formed to complete, for example, a left-handed coordinate system as computed in a right-handed space (Figure 3.31). Specifically, at a given parameter value s, the left-handed Frenet frame is calculated according to Equation 3.32, as illustrated in Figure 3.32. The vectors are then normalized.

$$
\begin{aligned}
w &= \mathbf{P}'(s) \\
u &= \mathbf{P}'(s) \times \mathbf{P}''(s) \\
v &= u \times w
\end{aligned}
\tag{3.32}
$$

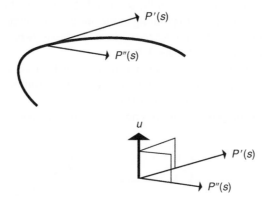

FIGURE 3.31

The derivatives at a point along the curve are used to form the u vector.

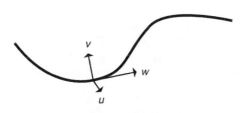

FIGURE 3.32

Left-handed Frenet frame at a point along a curve.

---

[4]Note the potential source of confusion between the use of the term frame to mean (1) a frame of animation and (2) the moving coordinate system of the Frenet frame. The context in which the term frame is used should determine its meaning.

While the Frenet frame provides useful information about the curve, there are a few problems with using it directly to control the orientation of a camera or object as it moves along a curve. First, there is no concept of "up" inherent in the formation of the Frenet frame. The v vector merely lines up with the direction of the second derivative. Another problem occurs in segments of the curve that have no curvature ($P''(u) = 0$) because the Frenet frame is undefined. These undefined segments can be dealt with by interpolating a Frenet frame along the segment from the Frenet frames at the boundary of the segment. By definition, there is no curvature along this segment, so the boundary Frenet frames must differ by only a rotation around w. Assuming that the vectors have already been normalized, the angular difference between the two can be determined by taking the arccosine of the dot product between the two v vectors so that $\theta = \text{acos}(v_1 \cdot v_2)$. This rotation can be linearly interpolated along the no-curvature segment (Figure 3.33).

The problem is more difficult to deal with when there is a discontinuity in the curvature vector as is common with piecewise cubic curves. Consider, for example, a curve composed of two semicircular segments positioned so that they form an **S** (*sigmoidal*) shape. The curvature vector, which for any point along this curve will point to the center of the semicircle that the point is on, will instantaneously switch from pointing to one center point to pointing to the other center point at the junction of the two semicircular segments. In this case, the Frenet frame is defined everywhere but has a discontinuous jump in orientation at the junction (see Figure 3.34).

However, the main problem with using the Frenet frame as the local coordinate frame to define the orientation of the camera or object following the path is that the resulting motions are usually too extreme and not natural looking. Using the w-axis (tangent vector) as the view direction of a camera can be undesirable. Often, the tangent vector does not appear to correspond to the direction of "where

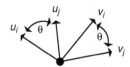

Frenet frames on the boundary of an undefined Frenet frame segment because of zero curvature

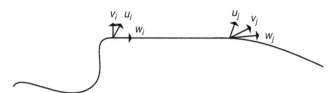

The two frames sighted down the (common) w vector

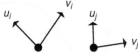

The two frames superimposed to identify angular difference

FIGURE 3.33

Interpolating Frenet frames to determine the undefined segment.

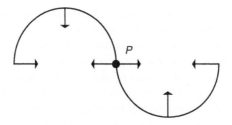

FIGURE 3.34

The curvature vector, defined everywhere but discontinuous, instantaneously switches direction at point P.

it's going" even though it is in the instantaneous (analytic) sense. The more natural orientation, for someone riding in a car or riding a bike, for example, would be to look farther ahead along the curve rather than to look tangential to the current location along the curve.

In addition, if the v-axis is equated with the up vector of an object, the object will rotate wildly about the path even when the path appears to mildly meander through an environment. With three-dimensional space curves, the problem becomes more obvious as the path bends up and down in space; the camera will flip between being upright and traveling along the path upside down. While the Frenet frame provides useful information to the animator, its direct use to control object orientation is clearly of limited value. When modeling the motion of banking into a turn, the curvature vector (u-axis) does indicate which way to bank and can be used to control the magnitude of the bank. For example, the horizontal component (the component in the x-z plane) of the u-axis can be used as the direction/magnitude indicator for the bank. For a different effect, the animator may want the object to tilt away from the curvature vector to give the impression of the object feeling a force that throws it off the path, such as a person riding a roller coaster.

Camera path following

The simplest method of specifying the orientation of a camera is to set its center of interest (COI) to a fixed point in the environment or, more elegantly, to use the center point of one of the objects in the environment. In either case, the COI is directly determined and is available for calculating the view vector, $w = COI - POS$. This is usually a good method when the path that the camera is following is one that circles some arena of action on which the camera's attention must be focused.

This still leaves one degree of freedom to be determined in order to fully specify the local coordinate system. For now, assume that the up vector, v, is to be kept "up." *Up* in this case means "generally in the positive y-axis direction," or, for a more mathematical definition, the v-axis is at right angles to the view vector (w-axis) and is to lie in the plane defined by the local w-axis and the global y-axis. The local coordinate system can be computed as shown in Equation 3.33 for a left-handed camera coordinate-system computed in right-handed space. The vectors can then be normalized to produce unit length vectors.

$$w = COI - POS$$
$$u = w \times (0,1,0)$$
$$v = u \times w$$

(3.33)

For a camera traveling down a path, the view direction can be automatically set in several ways. As previously mentioned, the COI can be set to a specific point in the environment or to the position of a specific object in the environment. This works well as long as there is a single point or single object on which the camera should focus and as long as the camera does not pass too close to the point or object. Passing close to the COI will result in radical changes in view direction (in some cases, the resulting effect may be desirable). Other methods use points along the path itself, a separate path through the environment, or interpolation between positions in the environment. The up vector can also be set in several ways. The default orientation is for the up vector to lie in the plane of the view vector and the global y-axis. Alternatively, a tilt of the up vector can be specified away from the default orientation as a user-specified value (or interpolated set of values). And, finally, the up vector can be explicitly specified by the user.

The simplest method of setting the view vector is to use a delta parametric value to define the COI. If the position of the camera on a curve is defined by $P(s)$, then the COI will be $P(s + \Delta s)$. This, of course, should be after reparameterization by arc length. Otherwise, the actual distance to the COI along the path will vary over time (although if a relatively large Δu is used, the variation may not be noticeable). At the end of the curve, once $s + \Delta s$ is beyond the path parameterization, the view direction can be interpolated to the end tangent vector as s approaches 1 (in the case that distance is normalized).

Often, updating the COI to be a specific point along the curve can result in views that appear jerky. In such cases, averaging some number of positions along the curve to be the COI point can smooth the view. However, if the number of points used is too small or the points are too close together, the view may remain jerky. If n is too large and the points are spaced out too much, the view direction may not change significantly during the traversal of the path and will appear too static. The number of points to use and their spread along the curve are dependent on the path itself and on the effect desired by the animator.

An alternative to using some function of the position path to produce the COI is to use a separate path altogether to define the COI. In this case, the camera's position is specified by $P(s)$, while the COI is specified by some $C(s)$. This requires more work by the animator but provides greater control and more flexibility.

Similarly, an up vector path, $U(s)$, might be specified so that the general up direction is defined by $U(s) - P(s)$. This is just the general direction because a valid up vector must be perpendicular to the view vector. Thus, the coordinate frame for the camera could be defined as in Equation 3.34.

$$
\begin{aligned}
w &= C(s) - P(s) \\
u &= w \times (U(s) - P(s)) \\
v &= u \times w
\end{aligned}
\tag{3.34}
$$

Instead of using a separate path for the COI, a simple but effective strategy is to fix it at one location for an interval of time and then move it to another location (using linear spatial interpolation and ease-in/ease-out temporal interpolation) and fix it there for a number of frames, and so on. The up vector can be set as before in the default "up" direction.

3.4.3 Smoothing a path

For cases in which the points making up a path are generated by a digitizing process, the resulting curve can be too jerky because of noise or imprecision. To remove the jerkiness, the coordinate values of the data can be smoothed by one of several approaches. For this discussion, the following set of data will be used: $\{(1, 1.6), (2, 1.65), (3, 1.6), (4, 1.8), (5, 2.1), (6, 2.2), (7, 2.0), (8, 1.5), (9, 1.3), (10, 1.4)\}$. This is plotted in Figure 3.35.

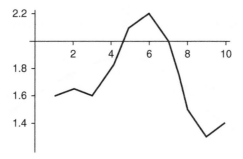

FIGURE 3.35

Sample data for path smoothing.

Smoothing with linear interpolation of adjacent values

An ordered set of points in two-space can be smoothed by averaging adjacent points. In the simplest case, the two points, one on either side of an original point, P_i, are averaged. This point is averaged with the original data point (Eq. 3.35). Figure 3.36 shows the sample data plotted with the original data. Notice how linear interpolation tends to draw the data points in the direction of local concavities. Repeated applications of the linear interpolation to further smooth the data would continue to draw the reduced concave sections and flatten out the curve.

$$P'_i = \frac{P_i + \frac{P_{i-1}+P_{i+1}}{2}}{2} = \frac{1}{4}P_{i-1} + \frac{1}{2}P_i + \frac{1}{4}P_{i+1} \tag{3.35}$$

Smoothing with cubic interpolation of adjacent values

To preserve the curvature but still smooth the data, the adjacent points on either side of a data point can be used to fit a cubic curve that is then evaluated at its midpoint. This midpoint is then averaged with the original point, as in the linear case. A cubic curve has the form shown in Equation 3.36. The two data points on either side of an original point, P_i, are used as constraints, as shown in Equation 3.37. These equations can be used to solve for the constants of the cubic curve (a, b, c, d). Equation 3.36 is then

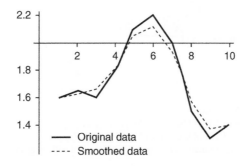

FIGURE 3.36

Sample data smoothed by linear interpolation.

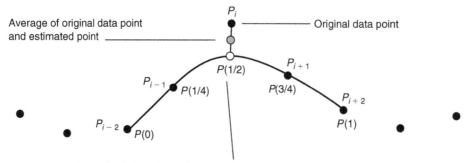

Average of original data point and estimated point

Original data point

P_i

P_{i-1}

$P(1/4)$

$P(1/2)$

P_{i+1}

$P(3/4)$

P_{i+2}

$P(1)$

P_{i-2}

$P(0)$

New estimate for P_i based on cubic curve fit through the four adjacent points

FIGURE 3.37

Smoothing data by cubic interpolation.

evaluated at $u = 1/2$; and the result is averaged with the original data point (see Figure 3.37). Solving for the coefficients and evaluating the resulting cubic curve is a bit tedious, but the solution needs to be performed only once and can be put in terms of the original data points, $P_{i-2}, P_{i-1}, P_{i+1}$, and P_{i+2}. This is illustrated in Figure 3.38.

$$P(u) = au^3 + bu^2 + cu + d \qquad (3.36)$$

$$P_{i-2} = P(0) = d$$

$$P_{i-1} = P(1/4) = a\frac{1}{64} + b\frac{1}{16} + c\frac{1}{4} + d$$

$$P_{i+1} = P(3/4) = a\frac{27}{64} + b\frac{9}{16} + c\frac{3}{4} + d$$

$$P_{i+2} = a + b + c + d$$

(3.37)

New estimate for P_i based on cubic curve fit through the four adjacent points

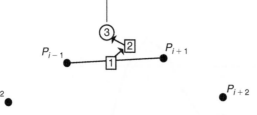

P_{i-1}

P_{i+1}

P_{i-2}

P_{i+2}

1. Average P_{i-1} and P_{i+1}
2. Add 1/6 of the vector from P_{i-2} to P_{i-1}
3. Add 1/6 of the vector from P_{i+2} to P_{i+1} to get new estimated point
4. (Not shown) Average estimated point with original data point

FIGURE 3.38

Geometric construction of a cubic estimate for smoothing a data point.

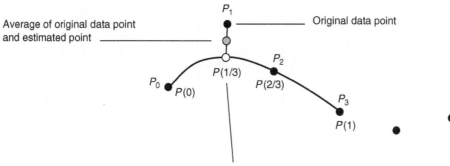

Average of original data point and estimated point ——————

Original data point

P_1

$P(1/3)$

P_0

$P(0)$

P_2

$P(2/3)$

P_3

$P(1)$

New estimate based on parabolic curve fit through the three adjacent points

FIGURE 3.39

Smoothing data by parabolic interpolation.

For the end conditions, a parabolic arc can be fit through the first, third, and fourth points, and an estimate for the second point from the start of the data set can be computed (Figure 3.39). The coefficients of the parabolic equation, $P(u) = au^2 + bu + c$, can be computed from the constraints in Equation 3.38, and the equation can be used to solve for the position $P_1 = P(1/3)$.

$$P_0 = P(0), \ P_2 = P\left(\frac{2}{3}\right), \ P_3 = P(1) \tag{3.38}$$

This can be rewritten in geometric form and the point can be constructed geometrically from the three points $P_1' = P_2 + (1/3)(P_0 - P_3)$ (Figure 3.40). A similar procedure can be used to estimate the data point second from the end. The very first and very last data points can be left alone if they represent hard constraints, or parabolic interpolation can be used to generate estimates for them as well, for

New estimate for P_1 based on parabolic curve fit through the three adjacent points

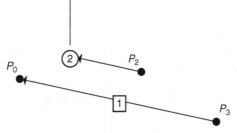

P_0

②

P_2

①

P_3

1. Construct vector from P_3 to P_0
2. Add 1/3 of the vector to P_2
3. (Not shown) Average estimated point with original data point

FIGURE 3.40

Geometric construction of a parabolic estimate for smoothing a data point.

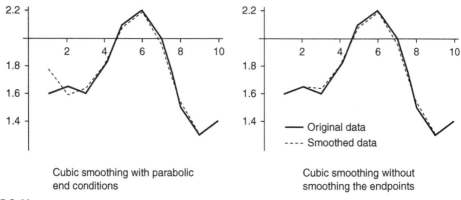

Cubic smoothing with parabolic
end conditions

Cubic smoothing without
smoothing the endpoints

FIGURE 3.41

Sample data smoothed with cubic interpolation.

example, $P'_0 = P_3 + 3(P_1 - P_2)$. Figure 3.41 shows cubic interpolation to smooth the data with and without parabolic interpolation for the endpoints.

Smoothing with convolution kernels

When the data to be smoothed can be viewed as a value of a function, $y_i = f(x_i)$, the data can be smoothed by convolution. Figure 3.42 shows such a function where the x_i are equally spaced. A smoothing kernel can be applied to the data points by viewing them as a step function (Figure 3.43). Desirable attributes of a smoothing kernel include the following: it is centered around 0, it is symmetric, it has finite support, and the area under the kernel curve equals 1. Figure 3.44 shows examples of some possibilities. A new point is calculated by centering the kernel function at the position where the new point is to be computed. The new point is calculated by summing the area under the curve that results from multiplying the kernel function, $g(u)$, by the corresponding segment of the step function, $f(x)$, beneath

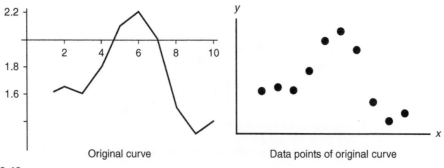

Original curve

Data points of original curve

FIGURE 3.42

Sample function to be smoothed.

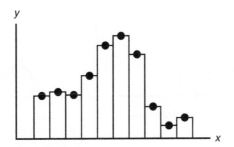

FIGURE 3.43

Step function defined by data points of original curve.

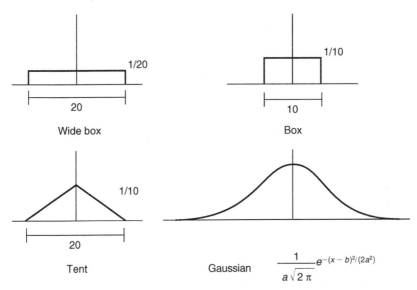

FIGURE 3.44

Sample smoothing kernels.

it (i.e., *convolution*). Figure 3.45 shows a simple tent-shaped kernel applied to a step function. In the continuous case, this becomes the integral, as shown in Equation 3.39, where $[-s, \ldots, s]$ is the extent of the support of the kernel function.

$$P(x) = \int_{-s}^{s} f(x+u)g(u)du \tag{3.39}$$

The integral can be analytically computed or approximated by discrete means. This can be done either with or without averaging down the number of points making up the path. Additional points can also be

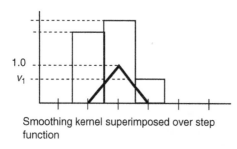 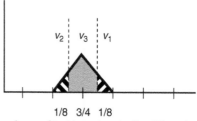

Smoothing kernel superimposed over step function

Areas of tent kernel under the different step function values

1/8 3/4 1/8

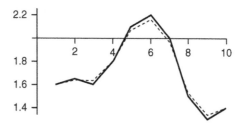

$$V = \frac{1}{8}v_1 + \frac{3}{4}v_2 + \frac{1}{8}v_3$$

Computation of value smoothed by applying area weights to step function values

FIGURE 3.45

Example of a tent-shaped smoothing filter.

FIGURE 3.46

Sample data smoothed with convolution using a tent kernel.

interpolated. At the endpoints, the step function can be arbitrarily extended so as to cover the kernel function when centered over the endpoints. Often the first and last points must be fixed because of animation constraints, so care must be taken in processing these. Figure 3.45 shows how a tent kernel is used to average the step function data; Figure 3.46 shows the sample data smoothed with the tent kernel.

Smoothing by B-spline approximation

Finally, if an approximation to the curve is sufficient, then points can be selected from the curve, and, for example, B-spline control points can be generated based on the selected points. The curve can then be regenerated using the B-spline control points, which ensures that the regenerated curve is smooth even though it no longer passes through the original points.

3.4.4 Determining a path along a surface

If one object is to move across the surface of another object, then a path across the surface must be determined. If start and destination points are known, it can be computationally expensive to find the shortest path between the points. However, it is not often necessary to find the absolute shortest path. Various alternatives exist for determining suboptimal yet reasonably direct paths.

An easy way to determine a path along a polygonal surface mesh is to determine a plane that contains the two points and is generally perpendicular to the surface. *Generally perpendicular* can be defined, for example, as the plane containing the two vertices and the average of the two vertex normals that the path is being formed between. The intersection of the plane with the faces making up the surface mesh will define a path between the two points (Figure 3.47).

If the surface is a higher order surface and the known points are defined in the u, v coordinates of the surface definition, then a straight line (or curve) can be defined in the parametric space and transferred to the surface. A linear interpolation of u, v coordinates does not guarantee the most direct route, but it should produce a reasonably short path.

Over a complex surface mesh, a greedy-type algorithm can be used to construct a path of edges along a surface from a given start vertex to a given destination vertex. For each edge emanating from the current vertex (initially the start vertex), calculate the projection of the edge onto the unit vector from the current vertex to the destination vertex. Divide this distance by the length of the edge to get the cosine of the angle between the edge and the straight line. The edge with the largest cosine is the edge most in the direction of the straight line; choose this edge to add to the path. Keep applying this until the destination edge is reached. Improvements to this approach can be made by allowing the path to cut across polygons to arrive at points along edges rather than by going vertex to vertex. For example, using the current path point and the line from the start and end vertices to form a plane, intersect the plane with the plane of the current triangle. If the line of intersection falls inside of the current triangle, use the segment inside of the triangle as part of the path. If the line of intersection leaving the current vertex is outside of the triangle, find the adjacent triangle and use it as the current triangle, and repeat. For the initial plane, the start and end vertex can be used along with some point on the surface that is judged to lie in the general direction of the path that is yet to be determined.

If a path downhill from an initial point on the surface is desired, then the surface normal and global "up" vector can be used to determine the downhill vector. The cross-product of the normal and global up vector defines a vector that lies on the surface perpendicular to the downhill direction. So the cross-product of this vector and the normal vector define the downhill (and uphill) vector on a plane (Figure 3.48). This same approach works with curved surfaces to produce the instantaneous downhill vector.

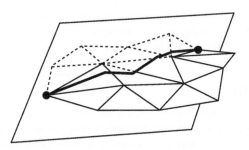

FIGURE 3.47

Determining a path along a polygonal surface mesh by using plane intersection.

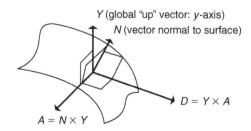

FIGURE 3.48

Calculating the downhill vector, D.

3.4.5 Path finding

Finding a collision-free path in an arbitrarily complex environment can be a complicated task. In the simple case, a point is to be moved through an otherwise stationary environment. If the obstacles are moving, but their paths are known beforehand, then the problem is manageable. If the movements of the obstructions are not known, then path finding can be very difficult. A further complication is when the object to be moved is not just a point, but has spatial extent. If the object has a nontrivial shape (i.e., not a sphere) and can be arbitrarily rotated as it moves, then the problem becomes even more interesting. A few ideas for path finding are mentioned here.

In finding a collision-free path among static obstacles, it is often useful to break the problem into simpler subproblems by introducing *via points*. These are points that the path should pass through—or at least educated guesses of points that the path might pass through. In this manner, finding the path is reduced to a number of simpler problems of finding paths between the via points.

If the size of the moving object might determine whether it can pass between obstacles, then finding a collision-free path can be a problem. However, if the moving object can be approximated by a sphere, then increasing the size of the obstacles by the size of the approximating sphere changes the problem into one of finding the path of a point through the expanded environment. More complex cases, in which the moving object has to adjust its orientation in order to pass by obstacles, are addressed in the robotics literature, for example.

Situations in which the obstacles are moving present significant challenges. A path through the static objects can be established and then movement along this path can be used to avoid moving obstacles, although there is no guarantee that a successful movement can be found. If the environment is not cluttered with moving obstacles, then a greedy-type algorithm can be used to avoid one obstacle at a time, by angling in front of the moving obstacle or angling behind it, in order to produce a path. This type of approach assumes that the motion of the obstacles, although unknown, is predictable.

3.5 Chapter summary

Interpolation is fundamental to much animation, and the ability to understand and control the interpolation process is very important in computer animation programming. Interpolation of values takes many forms, including arc length, image pixel color, and shape parameters. The control of the interpolation process can be key framed, scripted, or analytically determined. But in any case, interpolation forms the foundation upon which most computer animation takes place, including those animations that use advanced algorithms.

References

[1] Burden R, Faires J. Numerical Analysis. 8th ed. Brooks-Cole; 2005.

[2] Guenter B, Parent R. Computing the Arc Length of Parametric Curves. IEEE Comput Graph Appl May 1990; 10(3):72–8.

[3] Judd C, Hartley P. Parametrization and Shape of B-Spline Curves for CAD. Computer Aided Design September 1980; 12(5):235–8.

[4] Litwinowicz P, Williams L. Animating Images with Drawings. In: Glassner A, editor. Proceedings of SIGGRAPH 94, Computer Graphics Proceedings, Annual Conference Series (July 1994, Orlando, Fla.). Orlando, Florida: ACM Press. p. 409–12. ISBN 0-89791-667-0.

[5] Mortenson M. Geometric Modeling. New York: John Wiley & Sons; 1985.

[6] Rogers D, Adams J. Mathematical Elements for Computer Graphics. New York: McGraw-Hill; 1976.

[7] Shoemake K. Animating Rotation with Quaternion Curves. In: Barsky BA (Ed.), Computer Graphics (Proceedings of SIGGRAPH 85) (August 1985, San Francisco, Calif.), vol. 19(3). p. 143–52.

Interpolation-Based Animation

This chapter presents methods for precisely specifying the motion of objects. While the previous chapter presented basics of interpolating values, this chapter addresses how to use those basics to facilitate the production of computer animation. With these techniques, there is little uncertainty about the positions, orientations, and shapes to be produced. The computer is used only to calculate the actual values. Procedures and algorithms are used, to be sure, but in a very direct manner in which the animator has very specific expectations about the motion that will be produced on a frame-by-frame basis. These techniques are typified by the user setting precise conditions, such as beginning and ending values for position and orientation, and having the intermediate values filled in by some interpolative procedure. Later chapters cover "higher-level" algorithms in which the animator gives up some precision to produce motion with certain desired qualities. The techniques of this chapter deal with low-level, more precise, control of motion that is highly constrained by the animator.

This chapter starts off with a discussion of key-frame animation systems. These are the three-dimensional computer animation version of the main hand-drawn animation technique. The next section discusses animation languages. While animation languages can be used to control high-level algorithms, they are historically associated with their use in specifying the interpolation of values. Indeed, many of the language constructs particular to the animation domain deal with setting transformations at key frames and interpolating between them. The final sections cover techniques to deform an object, interpolating between three-dimensional object shapes, and two-dimensional morphing.

4.1 Key-frame systems

Many of the early computer animation systems were key-frame systems (e.g., [3] [4] [5] [20]). Most of these were two-dimensional systems based on the standard procedure used for hand-drawn animation, in which master animators define and draw the key frames of the sequence to be animated. In hand-drawn animation, assistant animators have the task of drawing the intermediate frames by mentally inferring the action between the keys. The key frames occur often enough in the sequence so that the intermediate action is reasonably well defined, or the keys are accompanied by additional information to indicate placement of the intermediate frames. In computer animation, the term key frame has been generalized to apply to any variable whose value is set at specific key frames and from which values for the intermediate frames are interpolated according to some prescribed procedure. These variables have been referred to in the literature as *articulation variables* (*avars*) [24], and the systems are sometimes referred to as *track based*. It is common for such systems to provide an interactive interface

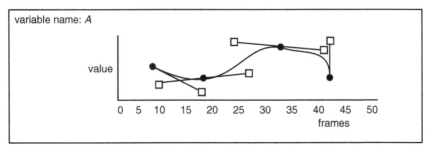

FIGURE 4.1

Sample interface for specifying interpolation of key values and tangents at segment boundaries.

with which the animator can specify the key values and can control the interpolation curve by manipulating tangents or interior control points. For example, see Figure 4.1. Early three-dimensional keyframe systems include TWIXT [15] and BBOP [33].

Because these animation systems keep the hand-drawn strategy of interpolating two-dimensional shapes, the basic operation is that of interpolating one (possibly closed) curve into another (possibly closed) curve. The interpolation is straightforward if the correspondence between lines in the frames is known in enough detail so that each pair of lines can be interpolated on a point-by-point basis to produce lines in the intermediate frames. This interpolation requires that for each pair of curves in the key frames the curves have the same number of points and that for each curve, whether open or closed, the correspondence between the points can be established.

Of course, the simplest way to interpolate the points is using linear interpolation between each pair of keys. For example, see Figure 4.2. Moving in arcs and allowing for ease-in/ease-out can be accommodated by applying any of the interpolation techniques discussed in Appendix B.5, providing that a point can be identified over several key frames.

Point-by-point correspondence information is usually not known, and even if it is, the resulting interpolation is not necessarily what the user wants. The best one can expect is for the curve-to-curve correspondence to be given. The problem is, given two arbitrary curves in key frames, to interpolate a curve as it "should" appear in intermediate frames. For example, observe the egg splatting against the wall in Figure 4.3.

For illustrative purposes, consider the simple case in which the key frames $f1$ and $f2$ consist of a single curve (Figure 4.4). The curve in frame $f1$ is referred to as $P(u)$, and the curve in frame $f2$ is referred to as $Q(v)$. This single curve must be interpolated for each frame between the key frames in which it is defined. In order to simplify the discussion, but without loss of generality, it is assumed that the curve, while it may wiggle some, is a generally vertical line in both key frames.

Some basic assumptions are used about what constitutes reasonable interpolation, such as the fact that if the curve is a single continuous open segment in frames $f1$ and $f2$, then it should remain a single continuous open segment in all the intermediate frames. Also assumed is that the top point of P, $P(0)$, should interpolate to the top point in Q, $Q(0)$, and, similarly, the bottom points should interpolate. However, what happens at intermediate points along the curve is so far left undefined (other than for the obvious assumption that the entire curve P should interpolate to the entire curve Q; that is, the mapping should be one-to-one and onto, also known as a *bijection*).

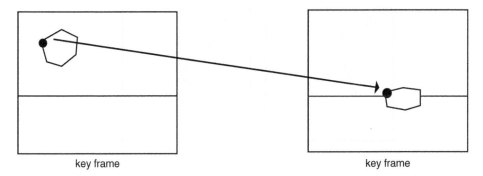

key frame   key frame

Simple key frames in which each curve of a frame has the same number of points as
its counterpart in the other frame

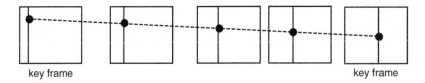

key frame   key frame

Keys and three intermediate frames with linear interpolation of a single point (with reference lines
showing the progression of the interpolation in x and y)

FIGURE 4.2

Simple key and intermediate frames. A specific point is identified in both key frames; the interpolation of the point
over intermediate frames is shown with auxiliary lines showing its relative placement in the frames.

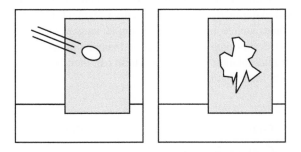

FIGURE 4.3

Object splatting against a wall. The shape must be interpolated from the initial egg shape to the splattered shape.

If both curves were generated with the same type of interpolation information, for example, each is
a single, cubic Bezier curve, then intermediate curves could be generated by interpolating the control
points and reapplying the Bezier interpolation. Another alternative would be to use interpolating functions to generate the same number of points on both curves. These points could then be interpolated on a
point-by-point basis. Although these interpolations get the job done, they might not provide sufficient
control to a user who has specific ideas about how the intermediate curves should look.

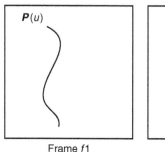
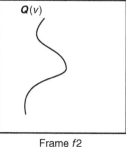

Frame *f1*   Frame *f2*

FIGURE 4.4

Two key frames showing a curve to be interpolated.

Reeves [29] proposes a method of computing intermediate curves using *moving point constraints*, which allows the user to specify more information concerning the correspondence of points along the curve and the speed of interpolation of those points. The basic approach is to use surface patch technology (two spatial dimensions) to solve the problem of interpolating a line in time (one spatial dimension, one temporal dimension).

The curve to be interpolated is defined in several key frames. Interpolation information, such as the path and speed of interpolation defined over two or more of the keys for one or more points, is also given. See Figure 4.5.

The first step is to define a segment of the curve to interpolate, bounded on top and bottom by interpolation constraints. Linear interpolation of the very top and very bottom of the curve, if not specified by a moving point constraint, is used to bind the top and bottom segments. Once a bounded segment has been formed, the task is to define an intermediate curve based on the constraints (see Figure 4.6). Various strategies can be used to define the intermediate curve segment, $C_t(u)$ in Figure 4.6, and are typically applications of surface patch techniques. For example, tangent information along the curves can be extracted from the curve definitions. The endpoint and tangent information can then be interpolated along the top and bottom interpolation boundaries. These can then be used to define an intermediate curve, $C_t(u)$, using Hermite interpolation.

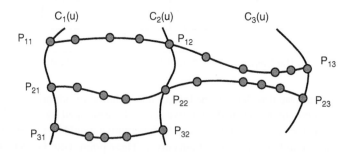

FIGURE 4.5

Moving point constraints showing the key points specified on the key curves as well as the intermediate points.

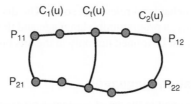

FIGURE 4.6

Patch defined by interpolation constraints made up of the boundary curve segments C_1 and C_{2q}.

4.2 Animation languages

An *animation language* is made up of structured commands that can be used to encode the information necessary to produce animations. If well-designed, an animation language can have enough expressive power to describe a wide variety of motions. Most languages are *script-based*, which means they are composed of text instructions; the instructions are recorded for later evaluation. Script-based languages can also be *interpreted* in which the animation is modified as new instructions are input by the user. Some languages are graphical in which, for example, flowchart-like diagrams encode relationships between objects and procedures. Nadia Magnenat-Thalmann and Daniel Thalmann present an excellent survey that includes many early script-based animation systems [22], and Steve May presents an excellent overview of animation languages from the perspective of his interest in procedural representations and encapsulated models [24]. Much of the discussion here is taken from these sources.

The first animation systems used general-purpose programming languages (e.g., Fortran) to produce the motion sequences. The language was used to compute whatever input values were needed by the rendering system to produce a frame of the animation. In these early systems, there were no special facilities in the language itself that were designed to support animation. Therefore, each time an animation was to be produced, there was overhead in defining graphical primitives, object data structures, transformations, and renderer output.

A language that does provide special facilities to support animation can use special libraries developed to augment a general-purpose programming language (e.g., an implementation of an applied programming interface) or it can be a specialized language designed from the ground up specifically for animation. In either case, typical features of an animation language include built-in input/output operations for graphical objects, data structures to represent objects and to support the hierarchical composition of objects, a time variable, interpolation functions, functions to animate object hierarchies, affine transformations, rendering-specific parameters, parameters for specifying camera attributes and defining the view frustum, and the ability to direct the rendering, viewing, and storing of one or more frames of animation. The program written in an animation language is often referred to as a *script*.

The advantage of using an animation language is twofold. First, the specification of an animation written in the language, the script, is a hard-coded record of the animation that can be used at any time to regenerate it. The script can be copied and transmitted easily. The script also allows the animation to be iteratively refined because it can be incrementally changed and a new animation generated. Second, the availability of programming constructs allows an algorithmic approach to motion control. The animation language, if sufficiently powerful, can interpolate values, compute arbitrarily complex numerical quantities, implement behavioral rules, and so on. The disadvantage of using an animation language is that it is, fundamentally, a programming language and thus requires the user (animator) to be a programmer. The animator must be trained not only in the arts but also in the art of programming.

4.2.1 Artist-oriented animation languages

To accommodate animators not trained in the art of computer programming, several simple animation languages were designed from the ground up with the intention of keeping the syntax simple and the semantics easy to understand (e.g., [16] [17] [23]). In the early days of computer graphics, almost all animation systems were based on such languages because there were few artists with technical backgrounds who were able to effectively program in a full-fledged programming language. These systems, as well as some of those with graphical user interfaces (e.g., [13] [26] [34]), were more accessible to artists but also had limited capabilities.

In these simple animation languages, typical statements referred to named objects, a transformation, and a time at which (or a span of time over which) to apply the transformation to the object. They also tended to have a syntax that was easy to parse (for the interpreter/compiler) and easy to read (for the animator). The following examples are from ANIMA II [16].

```
set position<name><x><y><z>at frame<number>
set rotation<name>[X,Y,Z] to<angle>at frame<number>
change position<name>to<x><y><z>from frame<number>to frame<number>
change rotation<name>[X,Y,Z] by<angle>from frame<number>to frame<number>
```

Specific values or variable names could be placed in the statements, which included an indicator as to whether the transformation was relative to the object's current position or absolute in global coordinates. The instructions operated in parallel; the script was not a traditional programming language in that sense.

As animators became more adept at writing animation scripts in such languages, they tended to demand that more capability be incorporated into the language. As Steve May [24] points out, "By eliminating the language constructs that make learning animation languages difficult, command [artist-oriented] languages give up the mechanisms that make animation languages powerful." The developers found themselves adding looping constructs, conditional control statements, random variables, procedural calls, and data structure support. An alternative to developing a full featured animation programming language from scratch was to add support for graphical objects and operations to an existing language such as C/C++, Python, Java, or LISP. For example, Steve May developed a scheme-based animation language called AL [25].

4.2.2 Full-featured programming languages for animation

Languages have been developed with simplified programming constructs, but such simplification usually reduces the algorithmic power of the language. As users become more familiar with a language, the tendency is to want more of the algorithmic power of a general-purpose programming language put back into the animation language. More recently, animation systems have been developed that are essentially user interfaces implemented by a scripting language processor. Users can use the system strictly from the interface provided, or they can write their own scripts using the scripting language.

One example of such a system is AutoDesk's Maya. The scripting language is called MEL and there is an associated animation language whose scripts are referred to as expressions [36]. Maya's interface is written in MEL so anything that can be done using the interface can be done in a MEL script (with a few exceptions). MEL scripts are typically used for scene setup. The scene is represented as a network of nodes that is traversed each time the frame is advanced in a timeline.

Expressions create nodes in the network that are therefore executed each frame. The input to, and output from, an expression node link to the attributes of other nodes in the scene. Object nodes have attributes such as geometric parameters and normals. Transformation nodes have attributes such as translation, rotation, and scale. A transformation node is typically linked to an object node in order to transform the object data into the world coordinate system. Using expression nodes, the translation of one object could, for example, affect the rotation of other objects in the scene. The sample script below creates an expression that controls the z-rotation of the object named "slab" based on the z-translation of a sphere named "sphere1." During the first frame, translation values for the objects are initialized. For each subsequent frame, two of the sphere's translation values are set as a function of the frame number and the z-rotation of slab is set as a function of the sphere's z-translation value.

```
if (frame == 1) {
   slab1.translateY = 0;
   sphere1.translateX = −25;
   sphere1.translateY = 15;
}
else {
   sphere1.translateX = 25*sin(frame/100);
   sphere1.translateZ = 25*cos(frame/100) − 25;
   slab1.rotateZ = 10*sphere1.translateZ + 90;
}
```

4.2.3 Articulation variables

A feature used by several languages is associating the value of a variable with a function, most notably of time. The function is specified procedurally or designed interactively using interpolating functions. Then a script, or other type of animation system, when it needs a value of the variable in its computation, passes the time variable to the function, which returns its value for that time to the computation. This technique goes by a variety of names, such as *track*, *channel*, or *articulation variable*. The term articulation variable, often shortened to avar, stems from its common use in some systems to control the articulation of linked appendages.

The use of avars, when an interactive system is provided to the user for inputting values or designing the functions, allows a script-based animation language to incorporate some interaction during the development of the animation. It also allows digitized data as well as arbitrarily complex functions to be easily incorporated into an animation by associating these with a particular avar.

4.2.4 Graphical languages

Graphical representations, such as those used in the commercial Houdini system [32], represent an animation by a dataflow network (see Figure 4.7). An acyclic graph is used to represent objects, operations, and the relationships among them. Nodes of the graph have inputs and outputs that connect to

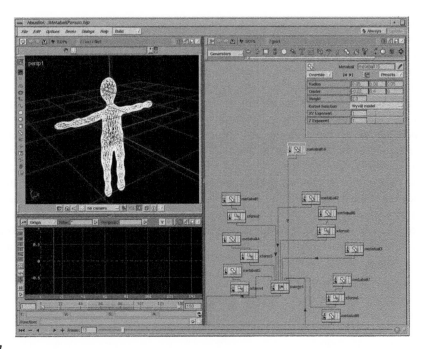

FIGURE 4.7

Sample Houdini dataflow network and the object it generates.

other nodes, and data are passed from node to node by arcs of the graphics that are interactively specified by the user. A node represents an operation to perform on the data being passed into the node, such as an object description. A transformation node will operate on an object description passed into the node and will produce a transformed object representation as output. Inputs to a node can also be used to parameterize a transformation or a data generation technique and can be set interactively, set according to an articulation variable, or set according to an arbitrary user-supplied procedure. This is a very effective way to visually design and represent the dependencies among various computations, functions, and values.

4.2.5 Actor-based animation languages

Actor-based languages are an object-oriented approach to animation in which an actor (encapsulated model [24]) is a graphical object with its associated data and procedures, including geometric description, display attributes, and motion control. Reynolds [30] popularized the use of the term *actor* in reference to the encapsulated models he uses in his ASAS system.

The basic idea is that the data associated with a graphical object should not only specify its geometry and display characteristics but also describe its motion. Thus the encapsulated model of a car includes how the doors open, how the windows roll down, and, possibly, the internal workings of the engine. Communication with the actor takes place by way of message passing; animating the object is carried out by passing requests to the actor for certain motions. The current status of the actor can be extracted by sending a request for information to the actor and receiving a message from the actor.

Actor-based systems provide a convenient way of communicating the time-varying information pertaining to a model. However, the encapsulated nature of actors with the associated message passing can result in inefficiencies when they are used with simulation systems in which all objects have the potential to affect all others.

4.3 Deforming objects

Deforming an object shape and transforming one shape into another is a visually powerful animation technique. It adds the notion of malleability and density. Flexible body animation makes the objects in an animation seem much more expressive and alive. There are physically based approaches that simulate the reaction of objects undergoing forces. However, many animators want more precise control over the shape of an object than that provided by simulations and/or do not want the computational expense of the simulating physical processes. In such cases, the animator wants to deform the object directly and define key shapes. Shape definitions that share the same edge connectivity can be interpolated on a vertex-to-vertex basis in order to smoothly change from one shape to the other. A sequence of key shapes can be interpolated over time to produce flexible body animation. Multivariate interpolation can be used to blend among a number of different shapes. The various shapes are referred to as *blend shapes* or *morph targets* and multivariate interpolation a commonly used technique in facial animation.

An immediate question that arises is "what is a shape?" Or "when are two shapes different?" It can probably be agreed that uniform scale does not change the shape of an object, but what about nonuniform scale? Does a rectangle have the same shape as a square? Most would say no. Most would agree that shearing changes the shape of an object. Elementary schools often teach that a square and a diamond are different shapes even though they may differ by only a rotation. Affine transformations are the simplest type of transformation that (sometimes) change the shape of an object. Affine transformations are defined by a 3×3 matrix followed by a translation. Affine transformations can be used to model the squash and stretch of an object, the jiggling of a block of Jello, and the shearing effect of an exaggerated stopping motion. While nonuniform scale can be used for simple squash and stretch, more interesting shape distortions are possible with nonaffine transformations. User-defined distortions are discussed in the following section; physically based approaches are discussed in Chapter 7.

4.3.1 Picking and pulling

A particularly simple way to modify the shape of an object is to displace one or more of its vertices. To do this on a vertex-by-vertex basis can be tedious for a large number of vertices. Simply grouping a number of vertices together and displacing them uniformly can be effective in modifying the shape of an object but is too restrictive in the shapes that can easily be created. An effective improvement to this approach is to allow the user to displace a vertex (the seed vertex) or group of vertices of the object and propagate the displacement to adjacent vertices along the surface while attenuating the amount of displacement. Displacement can be attenuated as a function of the distance between the seed vertex and the vertex to be displaced (see Figure 4.8). A *distance function* can be chosen to trade off quality of the results with computational complexity. One simple function uses the minimum number of edges connecting the seed vertex with the vertex to be displaced. A more accurate, but more computationally

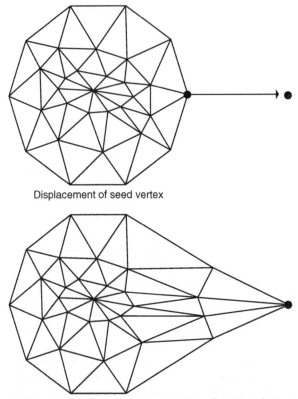

Displacement of seed vertex

Attenuated displacement propagated to adjacent vertices

FIGURE 4.8

Warping of object vertices.

expensive, function uses the minimum distance traveled over the surface of the object between the seed vertex and the vertex to be displaced.

Attenuation is typically a function of the distance metric. For example, the user can select a function from a family of power functions to control the amount of attenuation [28]. In this case, the minimum of connecting edges is used for the distance metric and the user specifies the maximum range of effect to be vertices within n edges of the seed vertex. A scale factor is applied to the displacement vector according to the user-selected integer value of k as shown in Equation 4.1.

$$S(i) = 1 - \left(\frac{i}{n+1}\right)^{k+1} \qquad k \geq 0$$

$$= \left(1 - \left(\frac{i}{n+1}\right)\right)^{-k+1} \qquad k < 0$$

(4.1)

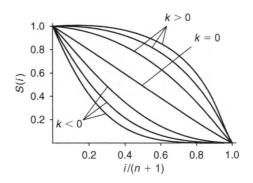

FIGURE 4.9

Power functions of Equation 4.1 for various values of k.

These attenuation functions are easy to compute and provide sufficient flexibility for many desired effects. When $k=$ zero it corresponds to a linear attenuation, while values of $k<0$ create a more elastic impression. Values of $k>0$ create the effect of more rigid displacements (Figure 4.9).

4.3.2 Deforming an embedding space

A popular technique for modifying the shape of an object is credited to Sederberg [31] and is called *free-form deformation* (FFD). FFD is only one of a number of techniques that share a similar approach, establishing a local coordinate system that encases the area of the object to be distorted. The initial configuration of the local coordinate system is such that the determination of the local coordinates of a vertex is a simple process. Typically, the initial configuration has orthogonal axes. The object to be distorted, whose vertices are defined in global coordinates, is then placed in this local coordinate space by determining the local coordinates of its vertices. The local coordinate system is distorted by the user in some way, and the local coordinates of the vertices are used to map their positions in global space. The idea behind these techniques is that it is easier or more intuitive for the user to manipulate the local coordinate system than to manipulate the vertices of the object. Of course, the trade-off is that the manipulation of the object is restricted to possible distortions of the coordinate system. The local coordinate grid is usually distorted so that the mapping is continuous (space is not torn apart), but it can be nonlinear, making this technique more powerful than affine transformations.

Two-dimensional grid deformation

Before discussing three-dimensional deformation techniques, simpler two-dimensional schemes are presented. Flexible body animation is demonstrated in the 1974 film *Hunger* [4]. In this seminal work, Peter Foldes, Nestor Burtnyk, and Marceli Wein used a two-dimensional technique that allowed for shape deformation. In this technique, the local coordinate system is a two-dimensional grid in which an object is placed. The grid is initially aligned with the global axes so that the mapping from local to global coordinates consists of a scale and a translation. For example, in Figure 4.10, assume that the local grid vertices are defined at global integer values from 20 to 28 in x and from

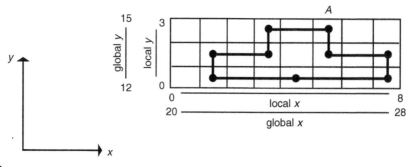

FIGURE 4.10

Two-dimensional coordinate grid used to locate initial position of vertices.

12 to 15 in y. Vertex A in the figure has global coordinates of (25.6, 14.7). The local coordinates of vertex A would be (5.6, 2.7).

The grid is then distorted by the user moving the vertices of the grid so that the local space is distorted. The vertices of the object are then relocated in the distorted grid by bilinear interpolation relative to the cell of the grid in which the vertex is located (Figure 4.11).

The bilinear interpolants used for vertex A would be 0.6 and 0.7. The positions of vertices of cell (5, 2) would be used in the interpolation. Assume the cell's vertices are named P_{00}, P_{01}, P_{10}, and P_{11}. Bilinear interpolation results in Equation 4.2.

$$P = (0.6)(0.7)P_{00} + (0.6)(1.0 - 0.7)P_{01} + (1.0 - 0.6)(0.7)P_{10} + (1.0 - 0.6)(1.0 - 0.7)P_{11} \quad (4.2)$$

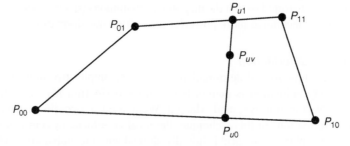

FIGURE 4.11

Bilinear interpolation used to locate vertex within quadrilateral grid.

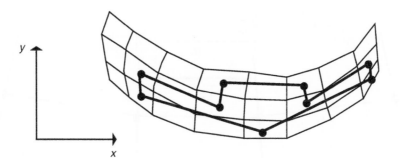

FIGURE 4.12

Two-dimensional grid deformation relocates vertices in space.

Once this is done for all vertices of the object, the object is distorted according to the distortion of the local grid (see Figure 4.12). For objects that contain hundreds or thousands of vertices, the grid distortion is much more efficient than individually repositioning each vertex. In addition, it is more intuitive for the user to specify a deformation.

Polyline deformation

A two-dimensional technique that is similar to the grid approach but lends itself to serpentine objects is based on a simple polyline (linear sequence of connected line segments) drawn by the user through the object to be deformed [28]. Polyline deformation is similar to the grid approach in that the object vertices are mapped to the polyline, the polyline is then modified by the user, and the object vertices are then mapped to the same relative location on the polyline.

The mapping to the polyline is performed by first locating the most relevant line segment for each object vertex. To do this, intersecting lines are formed at the junction of adjacent segments, and perpendicular lines are formed at the extreme ends of the polyline. These lines will be referred to as the boundary lines; each polyline segment has two boundary lines. For each object vertex, the closest polyline segment that contains the object vertex between its boundary lines is selected (Figure 4.13).

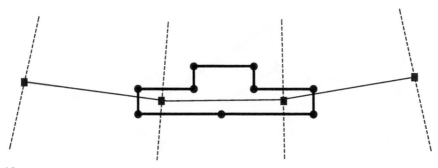

FIGURE 4.13

Polyline drawn through object; bisectors and perpendiculars are drawn as dashed lines.

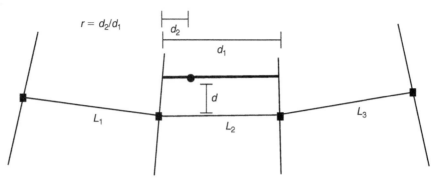

FIGURE 4.14

Measurements used to map an object vertex to a polyline.

Next, each object vertex is mapped to its corresponding polyline segment. A line segment is constructed through the object vertex parallel to the polyline segment and between the boundary lines. For a given object vertex, the following information is recorded (Figure 4.14): the closest line segment (L_2); the line segment's distance to the polyline segment (d); and the object vertex's relative position on this line segment, that is, the ratio r of the length of the line segment (d_1) and the distance from one end of the line segment to the object vertex (d_2).

The polyline is then repositioned by the user and each object vertex is repositioned relative to the polyline using the information previously recorded for that vertex. A line parallel to the newly positioned segment is constructed d units away and the vertex's new position is the same fraction along this line that it was in the original configuration (see Figure 4.15).

Global deformation

Alan Barr [1] presents a method of globally deforming the space in which an object is defined. Essentially, he applies a 3×3 transformation matrix, M, which is a function of the point being transformed, that is, $p' = M(p)\, p$, where $M(p)$ indicates the dependence of M on p. For example, Figure 4.16 shows a

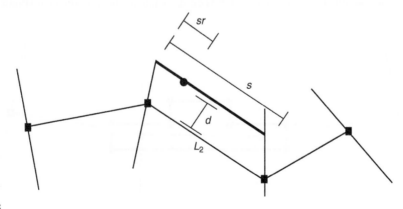

FIGURE 4.15

Remapping of an object vertex relative to a deformed polyline (see Figure 4.14).

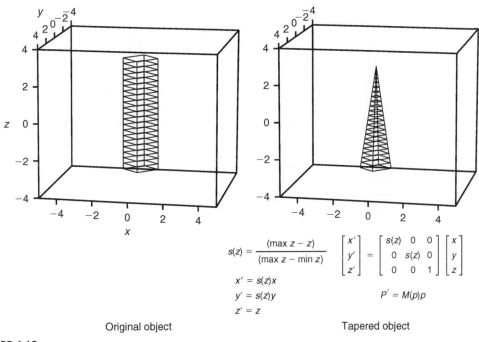

$$s(z) = \frac{(\max z - z)}{(\max z - \min z)}$$

$$\begin{bmatrix} x' \\ y' \\ z' \end{bmatrix} = \begin{bmatrix} s(z) & 0 & 0 \\ 0 & s(z) & 0 \\ 0 & 0 & 1 \end{bmatrix} \begin{bmatrix} x \\ y \\ z \end{bmatrix}$$

$$x' = s(z)x$$
$$y' = s(z)y$$
$$z' = z$$

$$P' = M(p)p$$

Original object   Tapered object

FIGURE 4.16

Global tapering.

simple linear two-dimensional tapering operation. There is no reason why the function ($s(z)$ in Figure 4.16) needs to be linear; it can be whatever function produces the desired effect. Other global deformations are possible. In addition to the taper operation, twists (Figure 4.17), bends (Figure 4.18), and various combinations of these (Figure 4.19) are possible.

FFD

An FFD is essentially a three-dimensional extension of Burtnyk's technique that incorporates higher-order interpolation. In both cases, a localized coordinate grid, in a standard configuration, is superimposed over an object. For each vertex of the object, coordinates relative to this local grid are determined that register the vertex to the grid. The grid is then manipulated by the user. Using its relative coordinates, each vertex is then mapped back into the modified grid, which relocates that vertex in global space. Instead of the linear interpolation used by Burtnyk, cubic interpolation is typically used with FFDs. In Sederberg's original paper [31], Bezier interpolation is suggested as the interpolating function, but any interpolation technique could be used.

In the first step of the FFD, vertices of an object are located in a three-dimensional rectilinear grid. Initially, the local coordinate system is defined by a not necessarily orthogonal set of three vectors (S, T, U). A vertex P is registered in the local coordinate system by determining its trilinear interpolants, as shown in Figure 4.20 and by Equations 4.3–4.5:

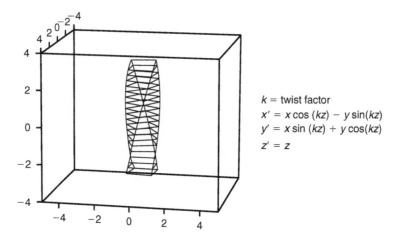

k = twist factor
$x' = x \cos(kz) - y \sin(kz)$
$y' = x \sin(kz) + y \cos(kz)$
$z' = z$

FIGURE 4.17

Twist about an axis.

$$s = (T \times U) \cdot (P - P_0)/((T \times U) \cdot S) \tag{4.3}$$

$$t = (U \times S) \cdot (P - P_0)/((U \times S) \cdot T) \tag{4.4}$$

$$u = (S \times T) \cdot (P - P_0)/((S \times T) \cdot U) \tag{4.5}$$

In these equations, the cross-product of two vectors forms a third vector that is orthogonal to the first two. The denominator normalizes the value being computed. In the first equation, for example, the projection of S onto $T \times U$ determines the distance within which points will map into the range $0 < s < 1$.

Given the local coordinates (s, t, u) of a point and the unmodified local coordinate grid, a point's position can be reconstructed in global space by simply moving in the direction of the local coordinate axes according to its local coordinates (Eq. 4.6):

$$P = P_0 + sS + tT + uU \tag{4.6}$$

To facilitate the modification of the local coordinate system, a grid of control points is created in the parallelepiped defined by the S, T, U axes. There can be an unequal number of points in the three directions. For example, in Figure 4.21, there are three in the S direction, two in the T direction, and one in the U direction.

If, not counting the origin, there are l points in the S direction, m points in the T direction, and n points in the U direction, the control points are located according to Equation 4.7.

$$P_{ijk} = P_0 + \frac{i}{l}S + \frac{j}{m}T + \frac{k}{n}U \quad \text{for} \quad \begin{pmatrix} 0 \le i \le l \\ 0 \le j \le m \\ 0 \le k \le n \end{pmatrix} \tag{4.7}$$

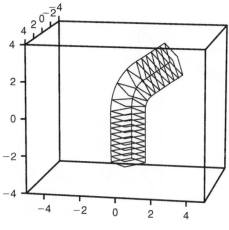

$(z_{min} : z_{max})$ − *bend region*
(y_0, z_{min}) − *center of bend*

$$x' = x$$

$$y' = \begin{cases} y & z < z_{max} \\ y_0 - (RC_\theta) & z_{min} \le z \le z_{max} \\ y_0 - (RC_\theta) + (z - z_{max})S_\theta & z > z_{max} \end{cases}$$

$$z' = \begin{cases} z & z < z_{max} \\ z_{min} - (RS_\theta) & z_{min} \le z \le z_{max} \\ z_{min} - (RS_\theta) + (z - z_{max})C_\theta & z > z_{max} \end{cases}$$

$$\theta = \begin{cases} 0 & z < z_{min} \\ z_{max} - z_{min} & z > z_{max} \\ z - z_{min} & otherwise \end{cases}$$

$$C_\theta = \cos\theta$$
$$S_\theta = \sin\theta$$
$$R = y_0 - y$$

FIGURE 4.18

Global bend operation.

The deformations are specified by moving the control points from their initial positions. The function that effects the deformation is a trivariate Bezier interpolating function. The deformed position of a point P_{stu} is determined by using its (s, t, u) local coordinates, as defined by Equations 4.3–4.5, in the Bezier interpolating function shown in Equation 4.8. In Equation 4.8, $P(s, t, u)$ represents the global coordinates of the deformed point, and P_{ijk} represents the global coordinates of the control points.

$$P(s, t, u) = \sum_{i=0}^{l} \sum_{j=0}^{m} \sum_{k=0}^{n} \binom{l}{i}\binom{m}{j}\binom{n}{k} (1 - s)^{l-i}s^i(1 - t)^{m-j}t^j(1 - u)^{n-k}u^k \qquad (4.8)$$

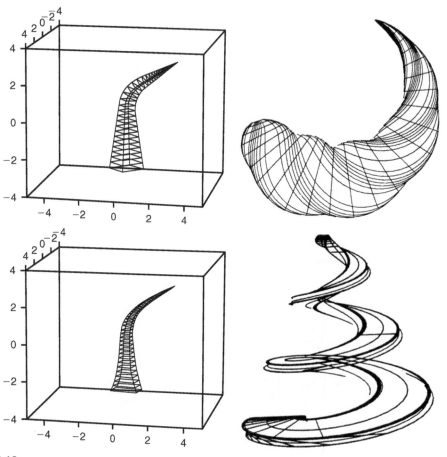

FIGURE 4.19

Examples of global deformations.

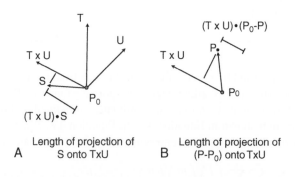

A. Length of projection of S onto TxU

B. Length of projection of (P-P₀) onto TxU

FIGURE 4.20

Computations forming the local coordinate s using an FFD coordinate system.

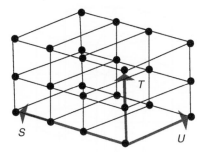

FIGURE 4.21

Grid of control points.

This interpolation function of Equation 4.8 is an example of trivariate interpolation. Just as the Bezier formulation can be used to interpolate a one-dimensional curve or a two-dimensional surface, the Bezier formulation is used here to interpolate a three-dimensional solid space.

Like Bezier curves and surfaces, multiple Bezier solids can be joined with continuity constraints across the boundaries. Of course, to enforce positional continuity, adjacent control lattices must share the control points along the boundary plane. As with Bezier curves, C^1 continuity can be ensured between two FFD control grids by enforcing colinearity and equal distances between adjacent control points across the common boundary (Figure 4.22).

Higher-order continuity can be maintained by constraining more of the control points on either side of the common boundary plane. However, for most applications, C^1 continuity is sufficient. One possibly useful feature of the Bezier formulation is that a bound on the change in volume induced by FFD can be analytically computed. See Sederberg [31] for details.

FFDs have been extended to include initial grids that are something other than a parallelepiped [11]. For example, a cylindrical lattice can be formed from the standard parallelepiped by merging the opposite boundary planes in one direction and then merging all the points along the cylindrical axis, as in Figure 4.23.

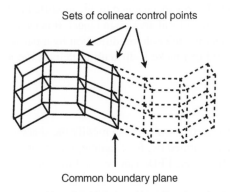

FIGURE 4.22

C^1 continuity between adjacent control grids.

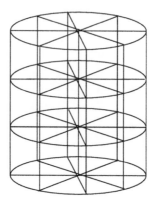

FIGURE 4.23

Cylindrical grid.

Composite FFDs—sequential and hierarchical

FFDs can be composed sequentially or hierarchically. In a sequential composition an object is modeled by progressing through a sequence of FFDs, each of which imparts a particular feature to the object. In this way, various detail elements can be added to an object in stages as opposed to trying to create one mammoth, complex FFD designed to do everything at once. For example, if a bulge is desired on a bent tube, then one FFD can be used to impart the bulge on the surface while a second is designed to bend the object (Figure 4.24).

Organizing FFDs hierarchically allows the user to work at various levels of detail. Finer-resolution FFDs, usually localized, are embedded inside FFDs higher in the hierarchy.[1] As a coarser-level FFD is used to modify the object's vertices, it also modifies the control points of any of its children FFDs that are within the space affected by the deformation. A modification made at a finer level in the hierarchy will remain well defined even as the animator works at a coarser level by modifying an FFD grid higher up in the hierarchy [18] (Figure 4.25).

If an FFD encases only part of an object, then the default assumption is that only those object vertices that are inside the initial FFD grid are changed by the modified FFD grid. Because the finer-level FFDs are typically used to work on a local area of an object, it is useful for the animator to be able to specify the part of the object that is subject to modification by a particular FFD. Otherwise, the rectangular nature of the FFD's grid can make it difficult to delimit the area that the animator actually wants to affect.

Animated FFDs

Thus far FFDs have been considered as a method to modify the shape of an object by repositioning its vertices. Animation would be performed by, for example, linear interpolation of the object's vertices on a vertex-by-vertex basis. However, FFDs can be used to control the animation in a more direct

---

[1]For this discussion, the hierarchy is conceptualized with the root node at the top, representing the coarsest level. More localized nodes with finer resolution are found lower in the hierarchy.

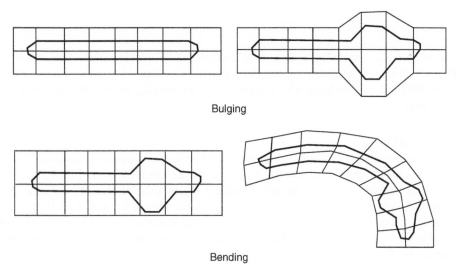

Bulging

Bending

FIGURE 4.24

Sequential FFDs.

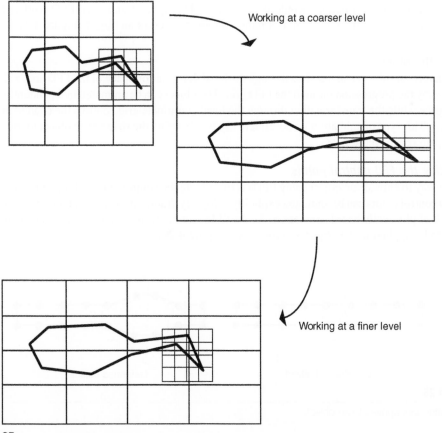

Working at a coarser level

Working at a finer level

FIGURE 4.25

Simple example of hierarchical FFDs.

manner in one of two ways. The FFD can be constructed so that traversal of an object through the FFD space results in a continuous transformation of its shape [12]. Alternatively, the control points of an FFD can be animated, which results in a time-varying deformation that then animates the object's shape.

Deformation tools

As discussed by Coquillart [12], a *deformation tool* is defined as a composition of an initial lattice and a final lattice. The initial lattice is user defined and is embedded in the region of the model to be animated. The final lattice is a copy of the initial lattice that has been deformed by the user. While the deformation tool may be defined in reference to a particular object, the tool itself is represented in an object-independent manner. This allows for a particular deformation tool to be easily applied to any object (Figure 4.26). To deform an object, the deformation tool must be associated with the object, thus forming what Coquillart calls an *AFFD object*.

Moving the tool

A way to animate the object is to specify the motion of the deformation tool relative to the object. In the example of Figure 4.26, the deformation tool could be translated along the object over time. Thus a sequence of object deformations would be generated. This type of animated FFD works effectively when a particular deformation, such as a bulge, progresses across an object (Figure 4.27).

Moving the object

Alternatively, the object can translate through the local deformation space of the FFD and, as it does, be deformed by the progression through the FFD grid. The object can be animated independently in global world space while the transformation through the local coordinate grid controls the change in shape of the object. This type of animation works effectively for changing the shape of an object to move along a certain path (e.g., Figure 4.28).

Animating the FFD control points

Another way to animate an object using FFDs is to animate the control points of the FFD. For example, the FFD control points can be animated explicitly using key-frame animation, or their movement can be the result of physically based simulation. As the FFD grid points move, they define a changing deformation to be applied to the object's vertices (see Figure 4.29).

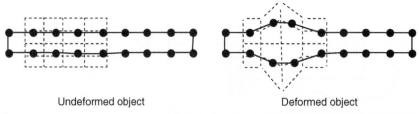

Undeformed object   Deformed object

FIGURE 4.26

Deformation tool applied to an object.

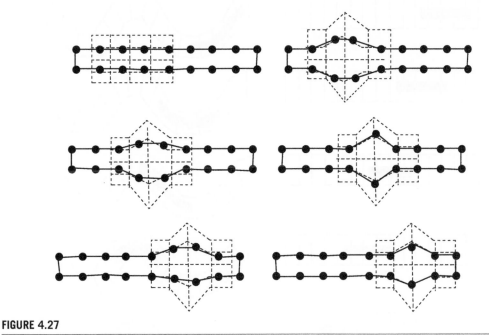

FIGURE 4.27

Deformation by translating the deformation tool relative to an object.

Chadwick et al. [8] described a layered approach to animation in which FFDs are used to animate a human form. The FFDs are animated in two ways. In the first technique, the positions of the FFD grid vertices are located relative to a wire skeleton the animator uses to move a figure. As the skeleton is manipulated, the grid vertices are repositioned relative to the skeleton automatically. The skin of the figure is then located relative to this local FFD coordinate grid. The FFDs thus play the role of muscular deformation of the skin. Joint articulation modifies the FFD grid, which in turn modifies the surface of the figure. The FFD grid is a mechanism external to the skin that effects the muscle deformation. The "muscles" in this case are meant not to be anatomical representations of real muscles but to provide for a more artistic style.

As a simple example, a hinge joint with adjacent links is shown in Figure 4.30; this is the object to be manipulated by the animator by specifying a joint angle. There is a surface associated with this structure that is intended to represent the skin. There are three FFDs: one for each of the two links and one for the joint. The FFDs associated with the links will deform the skin according to a stylized muscle, and the purpose of the FFD associated with the joint is to prevent interpenetration of the skin surface in highly bent configurations. As the joint bends, the central points of the link FFDs will displace upward and the interior panels of the joint FFD will rotate toward each other at the concave end in order to squeeze the skin together without letting it penetrate itself. Each of the grids is 5×4, and the joint grid is shown using dotted lines so that the three grids can be distinguished. Notice that the grids share a common set of control points where they meet.

Moving the FFD lattice points based on joint angle is strictly a kinematic method. The second technique employed by Chadwick [8] uses physically based animation of the FFD lattice points to animate

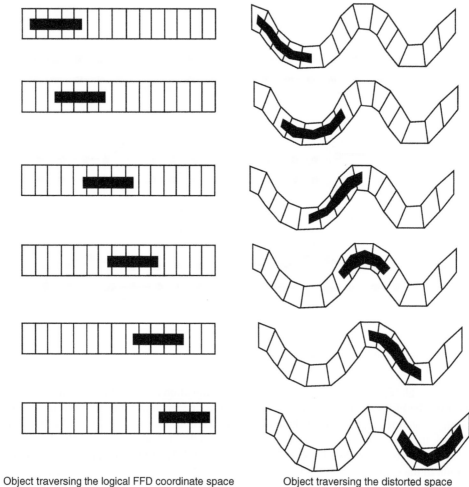

Object traversing the logical FFD coordinate space   Object traversing the distorted space

FIGURE 4.28

Deformation of an object by passing through FFD space.

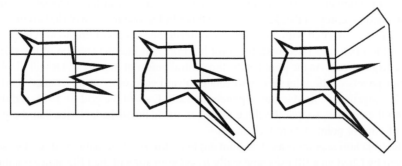

FIGURE 4.29

Using an FFD to animate a figure's head.

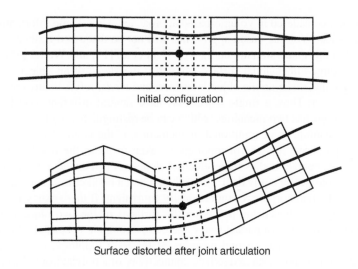

Initial configuration

Surface distorted after joint articulation

FIGURE 4.30

Using FFD to deform a surface around an articulated joint.

the figure. Animation of the FFD control points is produced by modeling the lattice with springs, dampers, and mass points. The control points of the lattice can then respond to gravity as well as kinematic motion of the figure. To respond to kinematic motion, the center of the FFD is fixed relative to the figure's skeleton. The user kinematically controls the skeleton, and the motion of the skeleton moves the center point of the FFD lattice. The rest of the FFD lattice points react to the movement of this center point via the spring-mass model, and the new positions of the FFD lattice points induce movement in the surface of the figure. This approach animates the clothes and facial tissue of a figure in animations produced by Chadwick [6] [7].

4.4 Three-dimensional shape interpolation

Changing one three-dimensional shape into another three-dimensional shape is a useful effect, but techniques are still being developed for arbitrary shapes. Several solutions exist that have various limitations. The techniques fall into one of two categories: surface based or volume based. The surface-based techniques use the boundary representation of the objects and modify one or both of them so that the vertex-edge topologies of the two objects match. Once this is done, the vertices of the object can be interpolated on a vertex-by-vertex basis. Surface-based techniques usually have some restriction on the types of objects they can handle, especially objects with holes through them. The number of holes through an object is an important attribute of an object's structure, or *topology*. The volume-based techniques consider the volume contained within the objects and blend one volume into the other. These techniques have the advantage of being less sensitive to different object topologies. However, volume-based techniques usually require volume representations of the objects and therefore tend to be more computationally intensive than surface-based approaches. In addition, the connectivity of the original shapes is typically not considered in volume-based approaches. Thus, some information, often important in animation, is lost. Volume-based approaches will not be discussed further.

The terms used in this discussion are defined by Kent et al. [19] and Weiler [35]. *Object* refers to an entity that has a three-dimensional surface geometry; the *shape* of an object refers to the set of points in object space that make up the object's surface; and *model* refers to any complete description of the shape of an object. Thus, a single object may have several different models that describe its shape. The term *topology* has two meanings, which can be distinguished by the context in which they are used. The first meaning, from traditional mathematics, is the connectivity of the surface of an object. For present purposes, this use of topology is taken to mean the number of holes an object has and the number of separate bodies represented. A doughnut and a teacup have the same topology and are said to be topologically equivalent. A beach ball and a blanket have the same topology. Two objects are said to be homeomorphic (or topologically equivalent) if there exists a continuous, invertible, one-to-one mapping between the points on the surfaces of the two objects. The *genus* of an object refers to how many holes, or passageways, there are through it. A beach ball is a genus 0 object; a teacup is a genus 1 object. The second meaning of the term topology, popular in the computer graphics literature, refers to the vertex/edge/face connectivity of a polyhedron; objects that are equivalent in this form of topology are the same except for the *x*-, *y*-, *z*-coordinate definitions of their vertices (the *geometry* of the object).

For most approaches, the shape transformation problem can be discussed in terms of the two subproblems: (1) the *correspondence problem*, or establishing the mapping from a vertex (or other geometric element) on one object to a vertex (or geometric element) on the other object, and (2) the *interpolation problem*, or creating a sequence of intermediate objects that visually represent the transformation of one object into the other. The two problems are related because the elements that are interpolated are typically the elements between which correspondences are established.

In general, it is not enough to merely come up with a scheme that transforms one object into another. An animation tool must give the user some control over mapping particular areas of one object to particular areas of the other object. This control mechanism can be as simple as aligning the object using affine transformations, or it can be as complex as allowing the user to specify an unlimited number of point correspondences between the two objects. A notable characteristic of the various algorithms for shape interpolation is the use of topological information versus geometric information. Topological information considers the logical construction of the objects and, when used, tends to minimize the number of new vertices and edges generated in the process. Geometric information considers the spatial extent of the object and is useful for relating the position in space of one object to the position in space of the other object.

While many of the techniques discussed in the following sections are applicable to surfaces defined by higher order patches, they are discussed here in terms of planar polyhedra to simplify the presentation.

4.4.1 Matching topology

The simplest case of transforming one object into another is when the two shapes to be interpolated share the same vertex-edge topology. Here, the objects are transformed by merely interpolating the positions of vertices on a vertex-by-vertex basis. For example, this case arises when one of the previous shape-modification techniques, such as FFD, has been used to modify one object, without changing the vertex-edge connectivity, to produce the second object. The correspondence between the two shapes is established by the vertex-edge connectivity structure shared by the two objects. The interpolation problem is solved, as in the majority of techniques presented here, by interpolating three-dimensional vertex positions.

4.4.2 Star-shaped polyhedra

If the two objects are both *star-shaped*[2] polyhedra, then polar coordinates can be used to induce a two-dimensional mapping between the two shapes. See Figure 4.31 for a two-dimensional example. The object surfaces are sampled by a regular distribution of rays emanating from a central point in the *kernel* of the object, and vertices of an intermediate object are constructed by interpolating between the intersection points along a ray. A surface is then constructed from the interpolated vertices by forming triangles from adjacent rays. Taken together, these triangles define the surface of the polyhedron. The vertices making up each surface triangle can be determined as a preprocessing step and are only dependent on how the rays are distributed in polar space. Figure 4.32 illustrates the sampling and interpolation for objects in two dimensions. The extension to interpolating in three dimensions is straightforward. In the three-dimensional case, polygon definitions on the surface of the object must then be formed.

4.4.3 Axial slices

Chen [9] interpolates objects that are star shaped with respect to a central axis. For each object, the user defines an axis that runs through the middle of the object. At regular intervals along this axis, perpendicular slices are taken of the object. These slices must be star shaped with respect to the point of intersection between the axis and the slice. This central axis is defined for both objects, and the part of each axis interior to its respective object is parameterized from 0 to 1. In addition, the user defines an orientation vector (or a default direction is used) that is perpendicular to the axis (see Figure 4.33).

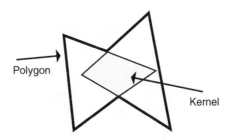

FIGURE 4.31

Star-shaped polygon and corresponding kernel from which all interior points are visible.

---

[2] A star-shaped (two-dimensional) polygon is one in which there is at least one point from which a line can be drawn to any point on the boundary of the polygon without intersecting the boundary; a star-shaped (three-dimensional) polyhedron is similarly defined. The set of points from which the entire boundary can be seen is referred to as the *kernel* of the polygon (in the two-dimensional case) or polyhedron (in the three-dimensional case).

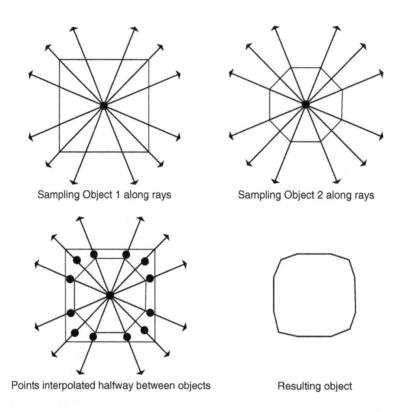

Sampling Object 1 along rays   Sampling Object 2 along rays

Points interpolated halfway between objects   Resulting object

FIGURE 4.32

Star-shaped polyhedral shape interpolation.

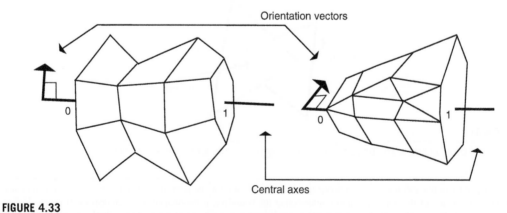

Orientation vectors

Central axes

FIGURE 4.33

Object coordinate system suitable for creating axial slices.

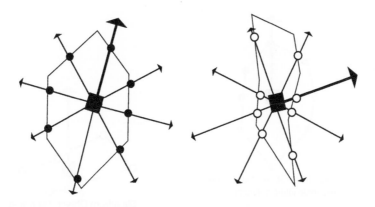

FIGURE 4.34

Projection lines for star-shaped polygons from objects in Figure 4.33.

Corresponding slices (corresponding in the sense that they use the same axis parameter to define the plane of intersection) are taken from each object. All of the slices from one object can be used to reconstruct an approximation to the original object using one of the contour-lofting techniques (e.g., [10] [14]). The two-dimensional slices can be interpolated pairwise (one from each object) by constructing rays that emanate from the center point and sample the boundary at regular intervals with respect to the orientation vector (Figure 4.34).

The parameterization along the axis and the radial parameterization with respect to the orientation vector together establish a two-dimensional coordinate system on the surface of the object. Corresponding points on the surface of the object are located in three-space. The denser the sampling, the more accurate is the approximation to the original object. The corresponding points can then be interpolated in three-space. Each set of ray–polygon intersection points from the pair of corresponding slices is used to generate an intermediate slice based on an interpolation parameter (Figure 4.35). Linear interpolation is often used, although higher-order interpolations are certainly useful. See Figure 4.36 for an example from Chen [9].

This approach can also be generalized somewhat to allow for a segmented central axis, consisting of a linear sequence of adjacent line segments. The approach may be used as long as the star-shaped restriction of any slice is maintained. The parameterization along the central axis is the same as before, except this time the central axis consists of multiple line segments.

4.4.4 **Map to sphere**

Even among genus 0 objects, more complex polyhedra may not be star shaped or allow an internal axis (single- or multi-segment) to define star-shaped slices. A more complicated mapping procedure may be required to establish the two-dimensional parameterization of objects' surfaces. One approach is to map both objects onto a common surface such as a unit sphere [19]. The mapping must be such that the entire object surface maps to the entire sphere with no overlap (i.e., it must be one-to-one and onto). Once both objects have been mapped onto the sphere, a union of their vertex-edge topologies can be constructed and then inversely mapped back onto each original object. This results in a new model for

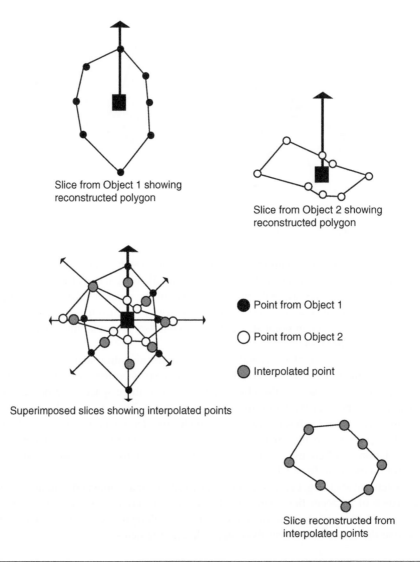

Slice from Object 1 showing
reconstructed polygon

Slice from Object 2 showing
reconstructed polygon

● Point from Object 1

○ Point from Object 2

● Interpolated point

Superimposed slices showing interpolated points

Slice reconstructed from
interpolated points

FIGURE 4.35

Interpolating slices taken from two objects along each respective central axis.

each of the original shapes, but the new models now have the same topologies. These new definitions for the objects can then be transformed by a vertex-by-vertex interpolation.

There are several different ways to map an object to a sphere. No one way has been found to work for all objects, but, taken together, most of the genus 0 objects can be successfully mapped to a sphere. The most obvious way is to project each vertex and edge of the object away from a center point of the object onto the sphere. This, of course, works fine for star-shaped polyhedra but fails for others.

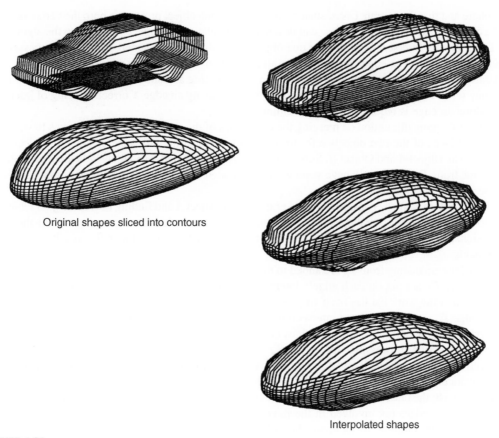

Original shapes sliced into contours

Interpolated shapes

FIGURE 4.36

Three-dimensional shape interpolation from multiple two-dimensional slices [9] (© IEEE Computer Society).

Another approach fixes key vertices to the sphere. These vertices are either selected by the user or automatically picked by being topmost, bottommost, leftmost, and so on. A spring-damper model is then used to force the remaining vertices to the surface of the sphere as well as to encourage uniform edge lengths.

If both objects are successfully mapped to the sphere (i.e., no self-overlap), the projected edges are intersected and merged into one topology. The new vertices and edges are a superset of both object topologies. They are then projected back onto both object surfaces. This produces two new object definitions, identical in shape to the original objects but now having the same vertex-edge topology, allowing for a vertex-by-vertex interpolation to transform one object into the other.

Once the vertices of both objects have been mapped onto a unit sphere, edges map to circular arcs on the sphere. The following description of the algorithm to merge the topologies follows that found in Kent et al. [19]. It assumes that the faces of the models have been triangulated prior to projection onto the sphere and that degenerate cases, such as the vertex of one object projecting onto the vertex or edge of the other object, do not occur; these can be handled by relatively simple extensions to the algorithm.

To efficiently intersect each edge of one object with edges of the second object, it is necessary to avoid a brute force edge-edge comparison for all pairs of edges. While this approach would work theoretically, it would, as noted by Kent, be very time-consuming and subject to numerical inaccuracies that may result in intersection points erroneously ordered along an edge. Correct ordering of intersections along an edge is required by the algorithm.

In the following discussion on merging the topologies of the two objects, all references to vertices, edges, and faces of the two objects refer to their projection on the unit sphere. The two objects are referred to as Object A and Object B. Subscripts on vertex, edge, and face labels indicate which object they come from. Each edge will have several lists associated with it: an intersection list, a face list, and an intersection-candidate list.

The algorithm starts by considering one vertex, V_A, of Object A and finding the face, F_B, of Object B that contains vertex V_A (see Figure 4.37). Taking into account that it is operating within the two-dimensional space of the surface of the unit sphere, the algorithm can achieve this result quite easily and quickly.

The edges emanating from V_A are added to the work list. Face F_B becomes the current face, and all edges of face F_B are put on each edge's intersection-candidate list. This phase of the algorithm has finished when the work list has been emptied of all edges.

An edge, E_A, and its associated intersection-candidate list are taken from the work list. The edge E_A is tested for any intersection with the edges on its intersection-candidate list. If no intersections are found, intersection processing for edge E_A is complete and the algorithm proceeds to the intersection-ordering phase. If an intersection, I, is found with one of the edges, E_B, then the following steps are done: I is added to the final model; I is added to both edge E_A's intersection list and edge E_B's intersection list; the face, G_B, on the other side of edge E_B becomes the current face; and the other edges of face G_B (the edges not involved in the intersection) replace the edges in edge E_A's intersection-candidate list. In addition, to facilitate the ordering of intersections along an edge, pointers to the two faces from Object A that share E_A are associated with I. This phase of the algorithm then repeats by considering the edges on the intersection-candidate list for possible intersections and, if any are

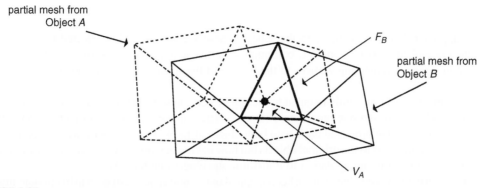

FIGURE 4.37

Locating initial vertex of Object A in the face of Object B.

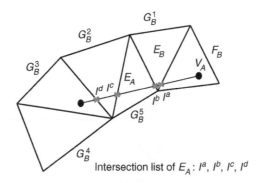

FIGURE 4.38

The intersection list for edge E_A.

found, processes subsequent intersections. When this phase is complete all edge-edge intersections have been found (see Figure 4.38).

For Object A, the intersections have been generated in sorted order along the edge. However, for Object B, the intersections have been associated with the edges but have been recorded in essentially a random order along the edge. The intersection-ordering phase uses the faces associated with intersection points and the list of intersections that have been associated with an edge of Object B to sort the intersection points along the edge. The face information is used because numerical inaccuracies can result in erroneous orderings if the numeric values of only parameters or intersection coordinates are used to order the intersections along the edge.

One of the defining vertices, V_B, for an edge, E_B, from Object B, is located within a face, F_A, of Object A. Initially, face F_A is referred to as the *current face*. As a result of the first phase, the edge already has all of its intersections recorded on its intersection list. Associated with each intersection point are the faces of Object A that share the edge E_B intersected to produce the point. The intersection points are searched to find the one that lists the current face, F_A, as one of its faces; one and only one intersection point will list F_A. This intersection point is the first intersection along the edge, and the other face associated with the intersection point will become the current face. The remaining intersection points are searched to see which lists this new current face; it becomes the next intersection point, and the other face associated with it becomes the new current face. This process continues until all intersection points have been ordered along edge E_B.

All of the information necessary for the combined models has been generated. It needs to be mapped back to each original object, and new polyhedral definitions need to be generated. The intersections along edges, kept as parametric values, can be used on the original models. The vertices of Object A, mapped to the sphere, need to be mapped onto the original Object B and vice versa for vertices of Object B. This can be done by computing the barycentric coordinates of the vertex with respect to the triangle that contains it (see Appendix B.2.9 for details). These barycentric coordinates can be used to locate the point on the original object surface.

Now all of the vertices, edges, and intersection points from both objects have been mapped back onto each object. New face definitions need to be constructed for each object. Because both models started out as triangulated meshes, there are only a limited number of configurations possible when

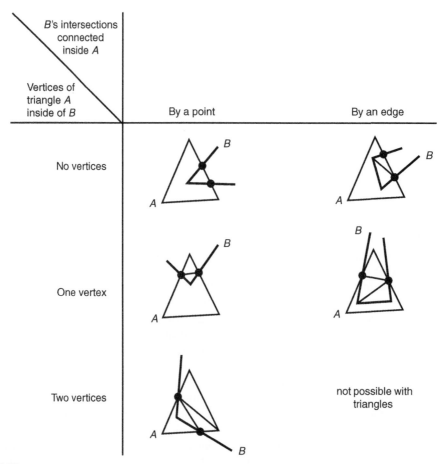

FIGURE 4.39

Configurations possible with overlapping triangles and possible triangulations.

one considers the triangulation required from the combined models (Figure 4.39). Processing can proceed by considering all of the original triangles from one of the models. For each triangle, use the intersection points along its edges and the vertices of the other object that are contained in it (along with any corresponding edge definitions) and construct a new triangulation of the triangle. When this is finished, repeat the process with the next triangle from the object.

The process of triangulation proceeds as follows. First, output any complete triangles from the other object contained in this object. Second, triangulate the fragments that include an edge or edge segment of the original triangle. Start at one of the vertices of the triangle and go to the first intersection encountered along the boundary of the triangle. The procedure progresses around the boundary of the triangle and ends when it returns to this intersection point. The next intersection along the boundary is also obtained. These two intersections are considered and the configuration identified by noting whether zero, one, or two original vertices are contained in the boundary between the two intersection points. The element inside the triangle that connects the two vertices/intersections is a vertex or an edge of the other object (along with the two edge segments involved in the intersections) (see Figure 4.39). Once the configuration is determined, triangulating the region is a simple task. The procedure then continues with the succeeding vertices or intersections in the order that they appear around the boundary of the original triangle.

This completes the triangulation of the combined topologies on the sphere. The resulting mesh can then be mapped back onto the surfaces of both objects, which establishes new definitions of the original objects but with the same vertex-edge connectivity. Notice that geometric information is used in the various mapping procedures in an attempt to map similarly oriented regions of the objects onto one another, thus giving the user some control over how corresponding parts of the objects map to one another.

4.4.5 Recursive subdivision

The main problem with the procedure above is that many new edges are created as a result of the merging operation. There is no attempt to map existing edges into one another. To avoid a plethora of new edges, a recursive approach can be taken in which each object is reduced to two-dimensional polygonal meshes [27]. Meshes from each object are matched by associating the boundary vertices and adding new ones when necessary. The meshes are then similarly split and the procedure is recursively applied until everything has been reduced to triangles. During the splitting process, existing edges are used whenever possible to reduce the number of new edges created. Edges and faces are added during subdivision to maintain topological equivalence. A data structure must be used that supports a closed, oriented path of edges along the surface of an object. Each mesh is defined by being on a particular side (e.g., right side) of such a path, and each section of a path will be shared by two and only two meshes.

The initial objects are divided into an initial number of polygonal meshes. Each mesh is associated with a mesh from the other object so that adjacency relationships are maintained by the mapping. The simplest way to do this is merely to break each object into two meshes—a front mesh and a back mesh. A front and back mesh can be constructed by searching for the shortest paths between the topmost, bottommost, leftmost, and rightmost vertices of the object and then appending these paths (Figure 4.40). On particularly simple objects, care must be taken so that these paths do not touch except at the extreme points.

This is the only place where geometric information is used. If the user wants certain areas of the objects to map to each other during the transformation process, then those areas should be the initial meshes associated with each other, providing the adjacency relationships are maintained by the associations.

When a mesh is associated with another mesh, a one-to-one mapping must be established between the vertices on the boundary of the two meshes. If one of the meshes has fewer boundary vertices than the other, then new vertices must be introduced along its boundary to make up for the difference. There are various ways to do this, and the success of the algorithm is not dependent on the method. A suggested method is to compute the normalized distance of each vertex along the boundary as measured from the first vertex of the boundary (the topmost vertex can be used as the first vertex of the boundaries of the initial meshes). For the boundary with fewer vertices, new vertices can be added one at a time by searching for the largest gap in normalized distances for successive vertices in the boundary (Figure 4.41). These vertices must be added to the original object definition. When the boundaries have the same number of vertices, a vertex on one boundary is said to be associated with the vertex on the other boundary at the same relative location.

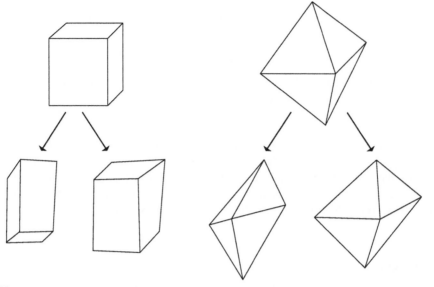

FIGURE 4.40

Splitting objects into initial front and back meshes.

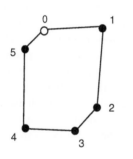

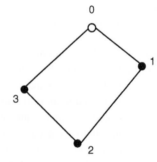

○ First vertex of boundary

Normalized distances			Normalized distances	
0	0.00		0	0.00
1	0.15		1	0.30
2	0.20		2	0.55
3	0.25		3	0.70
4	0.40			
5	0.70			

Boundary after adding additional vertices

FIGURE 4.41

Associating vertices of boundaries.

Once the meshes have been associated, each mesh is recursively divided. One mesh is chosen for division, and a path of edges is found across it. Again, there are various ways to do this and the procedure is not dependent on the method chosen for its success. However, the results will have a slightly different quality depending on the method used. One good approach is to choose two vertices across the boundary from each other and try to find an existing path of edges between them. An iterative procedure can be easily implemented that tries all pairs halfway around and then tries all pairs one less than halfway around, then two less, and so on. There will be no path only if the "mesh" is a single triangle—in which case the other mesh must be tried. There will be some path that exists on one of the two meshes unless both meshes are a single triangle, which is the terminating criterion for the recursion (and would have been previously tested for).

Once a path has been found across one mesh, then the path across the mesh it is associated with must be established between corresponding vertices. This may require creating new vertices and new edges (and, therefore, new faces) and is the trickiest part of the implementation, because minimizing the number of new edges will help reduce the complexity of the resulting topologies. When these paths (one on each mesh) have been created, the meshes can be divided along these paths, creating two pairs of new meshes. The boundary association, finding a path of edges, and mesh splitting are recursively applied to each new mesh until all of the meshes have been reduced to single triangles. At this point the new vertices and new edges have been added to one or both objects so that both objects have the same topology. Vertex-to-vertex interpolation of vertices can take place at this point in order to carry out the object interpolation.

4.5 Morphing (two-dimensional)

Two-dimensional image metamorphosis has come to be known as *morphing*. Although really an image postprocessing technique, and thus not central to the theme of this book, it has become so well known and has generated so much interest that it demands attention. Typically, the user is interested in transforming one image, called the source image, into the other image, called the destination image. There have been several techniques proposed in the literature for specifying and effecting the transformation. The main task is for the user to specify corresponding elements in the two images; these correspondences are used to control the transformation. Here, two approaches are presented. The first technique is based on user-defined coordinate grids superimposed on each image [37]. These grids impose a coordinate space to relate one image to the other. The second technique is based on pairs of user-defined feature lines. For each pair, one line in each image is marking corresponding features in the two images.

4.5.1 Coordinate grid approach

To transform one image into another, the user defines a curvilinear grid over each of the two images to be morphed. It is the user's responsibility to define the grids so that corresponding elements in the images are in the corresponding cells of the grids. The user defines the grid by locating the same number of grid intersection points in both images; the grid must be defined at the borders of the images in order to include the entire image (Figure 4.42). A curved mesh is then generated using the grid intersection points as control points for an interpolation scheme such as Catmull-Rom splines.

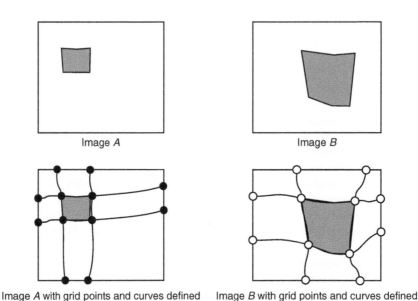

Image *A*   Image *B*

Image *A* with grid points and curves defined   Image *B* with grid points and curves defined

FIGURE 4.42

Sample grid definitions.

To generate an intermediate image, say t $(0 < t < 1.0)$ along the way from the source image to the destination image, the vertices (points of intersection of the curves) of the source and destination grids are interpolated to form an intermediate grid. This interpolation can be done linearly, or grids from adjacent key frames can be used to perform higher-order interpolation. Pixels from the source and destination images are stretched and compressed according to the intermediate grid so that warped versions of both the source image and the destination grid are generated. A two-pass procedure is used to accomplish this (described in the following section). A cross-dissolve is then performed on a pixel-by-pixel basis between the two warped images to generate the final image (see Figure 4.43).

For purposes of explaining the two-pass procedure, it will be assumed that it is the source image to be warped to the intermediate grid, but the same procedure is used to warp the destination image to the intermediate grid.

First, the pixels from the source image are stretched and compressed in the x-direction to fit the interpolated grid. These pixels are then stretched and compressed in the y-direction to fit the intermediate grid. To carry this out, an auxiliary grid is computed that, for each grid point, uses the y-coordinate from the corresponding grid point of the source image grid and the x-coordinate from the corresponding point of the intermediate grid. The source image is stretched/compressed in x by mapping it to the auxiliary grid, and then the auxiliary grid is used to stretch/compress pixels in y to map them to the intermediate grid. In the following discussion, it is assumed the curves are numbered left to right and bottom to top; a curve's number is referred to as its *index*.

Figure 4.44 illustrates the formation of the auxiliary grid from the source image grid and the intermediate grid. Once the auxiliary grid is defined, the first pass uses the source image and auxiliary grids to distort the source pixels in the x-direction. For each row of grid points in both the source and the

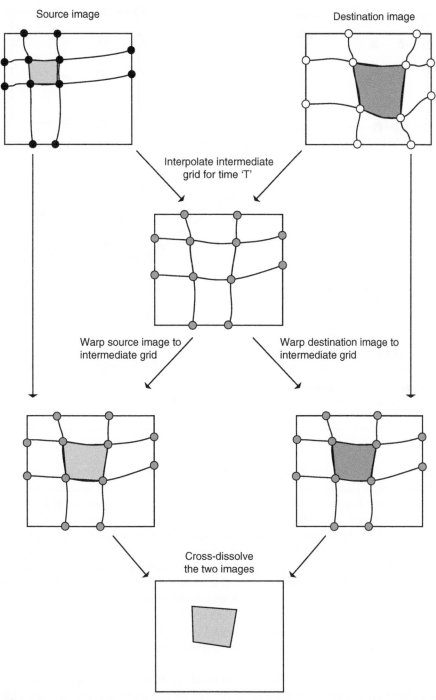

FIGURE 4.43

Interpolating to intermediate grid and cross-dissolve.

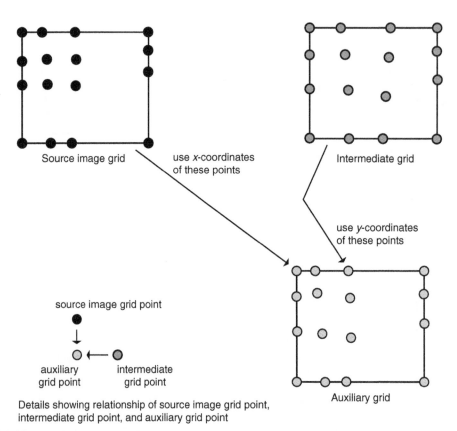

Source image grid

use *x*-coordinates
of these points

Intermediate grid

use *y*-coordinates
of these points

source image grid point

auxiliary
grid point

intermediate
grid point

Auxiliary grid

Details showing relationship of source image grid point,
intermediate grid point, and auxiliary grid point

FIGURE 4.44

Formation of auxiliary grid for two-pass warping of source image to intermediate grid.

auxiliary grid, a cubic Catmull-Rom spline is defined in pixel coordinates. The leftmost and rightmost columns define straight lines down the sides of the images; this is necessary to include the entire image in the warping process. See the top of Figure 4.45. For each scanline, the *x*-intercepts of the curves with the scanline are computed. These define a grid coordinate system on the scanline. The position of each pixel on the scanline in the source image is determined relative to the *x*-intercepts by computing the Catmull-Rom spline passing through the two-dimensional space of the (grid index, *x*-intercept) pairs. See the middle of Figure 4.45. The integer values of $x \pm 1/2$, representing the pixel boundaries, can then be located in this space and the fractional index value recorded. In the auxiliary image, the *x*-intercepts of the curves with the scanline are also computed, and for the corresponding scanline, the source image pixel boundaries can be mapped into the intermediate image by using their fractional indices and locating their *x*-positions in the auxiliary scanline. See the bottom of Figure 4.45. Once this is complete, the color of the source image pixel can be used to color in auxiliary pixels by using fractional coverage to affect anti-aliasing.

The result of the first phase generates colored pixels of an auxiliary image by averaging source image pixel colors on a scanline-by-scanline basis. The second phase repeats the same process on a

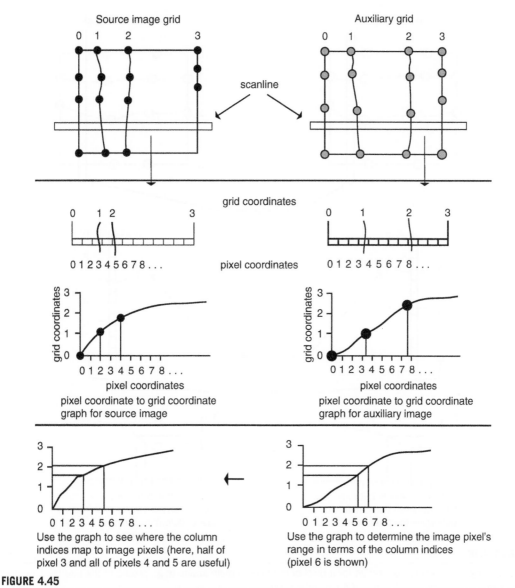

FIGURE 4.45

For a given pixel in the auxiliary image, determine the range of pixel coordinates in the source image (e.g., pixel 6 of auxiliary grid maps to pixel coordinates 3.5 to 5 of the source image).

column-by-column basis by averaging auxiliary image pixel colors to form the intermediate image. The columns are processed by using the horizontal grid curves to establish a common coordinate system between the two images (see Figure 4.46).

This two-pass procedure is applied to both the source and the destination images with respect to the intermediate grid. Once both images have been warped to the same intermediate grid, the

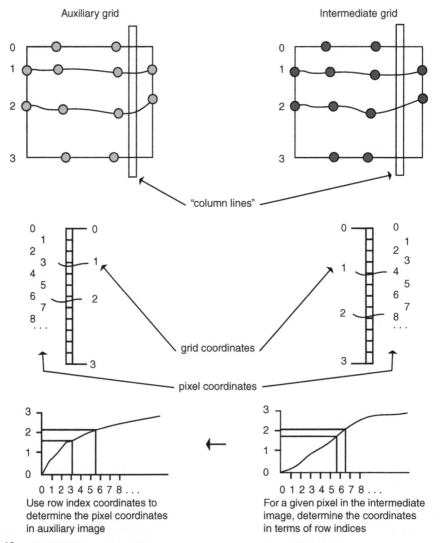

FIGURE 4.46

Establishing the auxiliary pixel range for a pixel of the intermediate image (e.g., pixel 6 of the intermediate grid maps to pixel coordinates 3.5 to 5 of the auxiliary image).

important features are presumably (if the user has done a good job of establishing the grids on the two images) in similar positions. At this point the images can be cross-dissolved on a pixel-by-pixel basis. The cross-dissolve is merely a blend of the two colors from corresponding pixels (Eq. 4.9).

$$C[i][j] = \alpha C_1[i][j] + (1 - \alpha)C_2[i][j] \tag{4.9}$$

In the simplest case, α might merely be a linear function in terms of the current frame number and the range of frame numbers over which the morph is to take place. However, as Wolberg [37] points out, a nonlinear blend is often more visually appealing. It is also useful to be able to locally control the cross-dissolve rates based on aesthetic concerns. For example, in one of the first popular commercials to use morphing, in which a car morphs into a tiger, the front of the car is morphed into the head of the tiger at a faster rate than the tail to add to the dynamic quality of the animated morph.

Animated images are morphed by the user-defined coordinate grids for various key images in each of two animation sequences. The coordinate grids for a sequence are then interpolated over time so that at any one frame in the sequence a coordinate grid can be produced for that frame. The interpolation is carried out on the x-y positions of the grid intersection points; cubic interpolation such as Catmull-Rom is typically used. Once a coordinate grid has been produced for corresponding images in the animated sequences, the morphing procedure reduces to the static image case and proceeds according to the previous description (see Figure 4.47).

4.5.2 Feature-based morphing

Instead of using a coordinate grid, the user can establish the correspondence between images by using feature lines [2]. Lines are drawn on the two images to identify features that correspond to one another and feature lines are interpolated to form an intermediate feature line set. The interpolation can be based on interpolating endpoints or on interpolating center points and orientation. In either case, a mapping for each pixel in the intermediate image is established to each interpolated feature line, and a

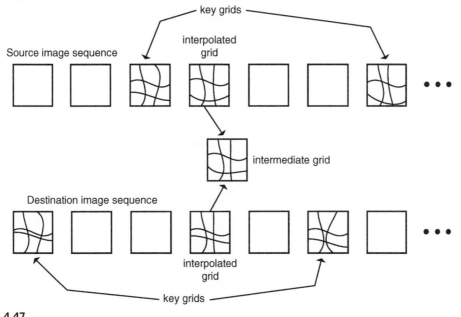

FIGURE 4.47

Morphing of animated sequences.

relative weight is computed that indicates the amount of influence that feature lines should have on the pixel. The mapping is used in the source image to locate the source image pixel that corresponds to the intermediate image pixel. The relative weight is used to average the source image locations generated by multiple feature lines into a final source image location. This location is used to determine the color of the intermediate image pixel. This same procedure is used on the destination image to form its intermediate image. These intermediate images are then cross-dissolved to form the final intermediate image.

Consider the mapping established by a single feature line, defined by two endpoints and oriented from P_1 to P_2. In effect, the feature line establishes a local two-dimensional coordinate system (U, V) over the image. For example, the first point of the line can be considered the origin. The second point of the line establishes the unit distance in the positive V-axis direction and scale. A line perpendicular to this line and of unit length extending to its right (as one stands on the first point and looks toward the second point) establishes the U-axis direction and scale. The coordinates (u, v) of a pixel relative to this feature line can be found by simply computing the pixel's position relative to the U- and V-axes of a local coordinate system defined by that feature line. Variable v is the projection of $(P - P_1)$ onto the direction of $(P_2 - P_1)$, normalized to the length of $(P_2 - P_1)$. u is calculated similarly (see Figure 4.48).

Assume that the points P_1 and P_2 are selected in the intermediate image and used to determine the (u, v) coordinates of a pixel. Given the corresponding feature line in the source image defined by the points Q_1 and Q_2, a similar two-dimensional coordinate system, (S, T), is established. Using the intermediate pixel's u-, v-coordinates relative to the feature line, one can compute its corresponding location in the source image (Figure 4.49).

To transform an image by a single feature line, each pixel of the intermediate image is mapped back to a source image position according to the equations above. The colors of the source image pixels in the neighborhood of that position are then used to color in the pixel of the intermediate image (see Figure 4.50).

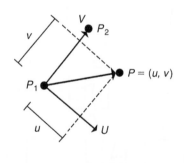

$$v = (P - P_1) \cdot \frac{(P_2 - P_1)}{|P_2 - P_1|^2}$$

$$u = \left| (P - P_1) \times \frac{(P_2 - P_1)}{|P_2 - P_1|^2} \right|$$

FIGURE 4.48

Local coordinate system of a feature in the intermediate image.

$$T = Q_2 - Q_1$$
$$S = (T_y - T_x)$$
$$Q = Q_1 + uS + vT$$

FIGURE 4.49

Relocating a point's position using local coordinates in the source image.

Source image and feature line

Intermediate feature line and resulting image

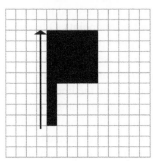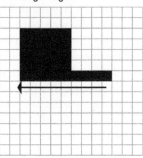

First example

Source image and feature line

Intermediate feature line and resulting image

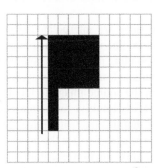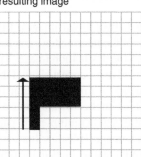

Second example

FIGURE 4.50

Two examples of single-feature line morphing.

Of course, the mapping does not typically transform an intermediate image pixel back to the center of a source image pixel. The floating point coordinates of the source image location could be rounded to the nearest pixel coordinates, which would introduce aliasing artifacts. To avoid such artifacts in the intermediate image, the corner points of the intermediate pixel could be mapped back to the source image, which would produce a quadrilateral area in the source image; the pixels, wholly or partially contained in this quadrilateral area, would contribute to the color of the destination pixel.

The mapping described so far is an affine transformation. For image pairs to be really interesting and useful, multiple line pairs must be used to establish correspondences between multiple features in the images. For a pair of images with multiple feature lines, each feature line pair produces a displacement vector from an intermediate image pixel to its source image position. Associated with this displacement is a weight based on the pixel's position relative to the feature line in the intermediate image. The weight presented by Beier and Neely [2] is shown in Equation 4.10.

$$W = \left(\frac{|Q_2 - Q_1|^p}{a + dist}\right)^b \tag{4.10}$$

The line is defined by points Q_1 and Q_2, and dist is the distance that the pixel is from the line. The distance is measured from the finite line segment defined by Q_1 and Q_2 so that if the perpendicular projection of P onto the infinite line defined by Q_1 and Q_2 falls beyond the finite line segment, then the distance is taken to be the distance to the closer of the two endpoints. Otherwise, the distance is the perpendicular distance to the finite line segment. User-supplied parameters (a, b, p in Eq. 4.10) control the overall character of the mapping. As dist increases, w decreases but never goes to zero; as a practical matter, a lower limit could be set below which w is clamped to 0 and the feature line's effect on the point is ignored above a certain distance. If a is nearly 0, then pixels on the line are rigidly transformed with the line. Increasing a makes the effect of lines over the image smoother. Increasing p increases the effect of longer lines. Increasing b makes the effect of a line fall off more rapidly. As presented here, these parameters are global; for more precise control these parameters could be set on a feature-line-by-feature-line basis. For a given pixel in the intermediate image, the displacement indicated by each feature line pair is scaled by its weight. The weights and the weighted displacements are accumulated. The final accumulated displacement is then divided by the accumulated weights. This gives the displacement from the intermediate pixel to its corresponding position in the source image. See the code segment in Figure 4.51.

When morphing between two images, the feature lines are interpolated over some number of frames. For any one of the intermediate frames, the feature line induces a mapping back to the source image and forward to the destination image. The corresponding pixels from both images are identified and their colors blended to produce a pixel of the intermediate frame. In this way, feature-based morphing produces a sequence of images that transform from the source image to the destination image.

The transformations implied by the feature lines are fairly intuitive, but some care must be taken in defining transformations with multiple line pairs. Pixels that lie on a feature line are mapped onto that feature line in another image. If feature lines cross in one image, pixels at the intersection of the feature lines are mapped to both feature lines in the other image. This situation essentially tries to pull apart the image and can produce unwanted results. Also, some configurations of feature lines can produce non-intuitive results. Other techniques in the literature (e.g., [21]) have suggested algorithms to alleviate these shortcomings.

```
//=================================================================

//XY structure
typedef struct xy_struct {
    float x,y;
}xy_td;


//FEATURE
// line in image1: p1,p2;
// line in image2: q1,q2
// weights used in mapping: a,b,p
// length of line in image2
typedef struct feature_struct {
    xy_td p1,p2,q1,q2;
    float a,b,p;
    float plength,qlength;
}feature_td;


//FEATURE LIST
typedef struct featureList_struct {
    int num;
    feature_td *features;
}featureList_td;

//□-------------------------------------------------------------
//□□□MORPH
//□-------------------------------------------------------------
void morph(featureList_td *featureList)
{
    Float a,b,p,plength,qlength;
    xy_td p1,p2,q1,q2;
    xy_td vp,wp,vq,wq,v,qq;
    int   ii,jj,indexI,indexD;
    float idisp,jdisp;
    float t,s,vx,vy;
    float weight;
    char  background[3];

    background[0] = background[1] = background[2] = 120;
    for (int i=0; i<HEIGHT; i++) {
        for (int j=0; j<WIDTH; j++) {
            weight = 0;
            idsp = jdsp = 0.0;
            for (int k=0; k<featureList->num; k++) {
                // get info about kth feature line
                a = featureList->features[k].a;
                b = featureList->features[k].b;
                p = featureList->features[k].p;
                p1 = featureList->features[k].p1;
                p2 = featureList->features[k].p2;
                q1 = featureList->features[k].q1;
                q2 = featureList->features[k].q2;
                plength = featureList->features[k].plength;
                qlength = featureList->features[k].qlength;
```

FIGURE 4.51

Code using feature lines to morph from source to destination image.

Continued

```
    // get local feature coordinate system in image1
    vp.x = p2.x-p1.x;
    vp.y = p2.y-p1.y;
    wp.x = vp.y;
    wp.y = -vp.x;
    // compute local coordinates of pixel (i,j)
    v.x = j-q1.x;
    v.y = i-q1.y;
    s = (v.x*vp.x + v.y*vp.y)/plength;
    t = (v.x*wp.y - v.y*wp.x)/plength;
    // map the pixel to the source image
    jj = (int)(q1.x + s*v1.x + t*wp.x);
    ii = (int)(q1.y + s*v1.y + t*wp.y);
    // compute distance of pixel (Iii,jj) from line segment q1q2
    If (s<0.0) {
        v.x = jj - q2.x;
        v.y = ii - w2.y;
        t = sqrt(v.x*v.x + v.y*v.y);
    }
    else t = fabs(t*qlength);
    t = pow(pow(qlength,p)/(a+t,b);
    jdisp + = (jj-j)*t;
    idisp + = (ii-i)*t;
    weight + = t;
}
jdisp /= weight;
idisp /= weight;
ii = (int)(i+idisp);
jj = (int)(j+jdisp);
indexI = (WIDTH*i+j)*3;
if ((ii<0) || (ii>=HEIGHT) || (jj<0) || (jj>=WIDTH) ) {
    image2[indexD] = background[0];
    image2[indexD+1] = background[1];
    image2[indexD+2] = background[2];
}
else {
    indexS = (WIDTH*ii+jj)*3;
    imageI[indexI] = image1[indexS];
    imageI[indexI+1] = image1[indexS+1];
    imageI[indexI+2] = image1[indexS+2];
}
            }
        }
    }
```

FIGURE 4.51—Cont'd

4.6 Chapter summary

Interpolation-based techniques form the foundation for much of computer animation. Whether interpolating images, object shapes, or parameter values, interpolation is both powerful and flexible. While this chapter has often discussed techniques in terms of linear interpolation, the reader should keep in mind that higher level techniques, such as piecewise cubic interpolation, can be much more expressive and relatively easy to implement.

References

[1] Barr A. Global and Local Deformations of Solid Primitives. In: Computer Graphics. Proceedings of SIGGRAPH 84, vol. 18(3). Minneapolis, Minn.; July 1984. p. 21–30.

[2] Beier T, Neely S. Feature Based Image Metamorphosis. In: Catmull EE, editor. Computer Graphics. Proceedings of SIGGRAPH 92, vol. 26(2). Chicago, Ill.; July 1992. p. 253–4. ISBN 0-201-51585-7.

[3] Burtnyk N, Wein M. Computer Generated Key Frame Animation. Journal of the Society of Motion Picture and Television Engineers 1971;8(3):149–53.

[4] Burtnyk N, Wein M. Interactive Skeleton Techniques for Enhancing Motion Dynamics in Key Frame Animation. Communications of the ACM October 1976;19(10):564–9.

[5] Catmull E. The Problems of Computer-Assisted Animation. In: Computer Graphics. Proceedings of SIGGRAPH 78, vol. 12(3). Atlanta, Ga.; August 1978. p. 348–53.

[6] Chadwick J. Bragger. Columbus: Ohio State University; 1989.

[7] Chadwick J. City Savvy. Columbus: Ohio State University; 1988.

[8] Chadwick J, Haumann D, Parent R. Layered Construction for Deformable Animated Characters. In: Computer Graphics. Proceedings of SIGGRAPH 89, vol. 23(3). Boston, Mass.; August 1989. p. 243–52.

[9] Chen E, Parent R. Shape Averaging and Its Application to Industrial Design. IEEE Comput Graph Appl January 1989;9(1):47–54.

[10] Christiansen HN, Sederberg TW. Conversion of Complex Contour Line Definitions into Polygonal Element Mosaics. In: Computer Graphics. Proceedings of SIGGRAPH 78, vol. 12(3). Atlanta, Ga.; August 1978. p. 187–912.

[11] Coquillart S. Extended Free-Form Deformation: A Sculpturing Tool for 3D Geometric Modeling. In: Baskett F, editor. Computer Graphics. Proceedings of SIGGRAPH 90, vol. 24(4) Dallas, Tex.; August 1990. p. 187–96. ISBN 0-201-50933-4.

[12] Coquillart S, Jancene P. Animated Free-Form Deformation: An Interactive Animation Technique. In: Sederberg TW, editor. Computer Graphics. Proceedings of SIGGRAPH 91, vol. 25(4). Las Vegas, Nev.; July 1991. p. 41–50. ISBN 0-201-56291-X.

[13] Defanti T. The Digital Component of the Circle Graphics Habitat. In: Proceedings of the National Computer Conference 76, vol. 45. New York; June 7–10. 1976. p. 195–203 AFIPS Press, Montvale, New Jersey.

[14] Ganapathy S, Denny TG. A New General Triangulation Method for Planar Contours. In: Computer Graphics. Proceedings of SIGGRAPH 82, vol. 16(3). Boston, Mass.; July 1982. p. 69–75.

[15] Gomez J. Twixt: A 3D Animation System. Computer and Graphics 1985;9(3):291–8.

[16] Hackathorn R. ANIMA II: A 3D Color Animation System. In: George J, editor. Computer Graphics. Proceedings of SIGGRAPH 77, vol. 11(2). San Jose, Calif.11.; July 1977. p. 54–64.

[17] Hackathorn R, Parent R, Marshall B, Howard M. An Interactive Micro-Computer Based 3D Animation System. In: Proceedings of Canadian Man-Computer Communications Society Conference 81. Waterloo, Ontario; June 10–12, 1981. p. 181–91.

[18] Hofer E. A Sculpting Based Solution for Three-Dimensional Computer Character Facial Animation. Master's thesis. Ohio State University; 1993.

[19] Kent J, Carlson W, Parent R. Shape Transformation for Polyhedral Objects. In: Catmull EE, editor. Computer Graphics. Proceedings of SIGGRAPH 92, vol. 26(2). Chicago, Ill.; July 1992. p. 47–54. ISBN 0-201-51585-7.

[20] Litwinowicz P. Inkwell: A $2^1/_2$-D Animation System. In: Sederberg TW, editor. Computer Graphics. Proceedings of SIGGRAPH 91, vol. 25(4). Las Vegas, Nev.; July 1991. p. 113–22.

[21] Litwinowicz P, Williams L. Animating Images with Drawings. In: Glassner A, editor. Proceedings of SIGGRAPH 94, Computer Graphics Proceedings, Annual Conference Series. Orlando, Fla.: ACM Press; July 1994. p. 409–12. ISBN 0-89791-667-0.

[22] Magnenat-Thalmann N, Thalmann D. Computer Animation: Theory and Practice. New York: Springer-Verlag; 1985.

[23] Magnenat-Thalmann N, Thalmann D, Fortin M. Miranim: An Extensible Director-Oriented System for the Animation of Realistic Images. IEEE Comput Graph Appl March 1985;5(3):61–73.

[24] May S. Encapsulated Models: Procedural Representations for Computer Animation. Ph.D. dissertation. Ohio State University; 1998.

[25] May S. AL: The Animation Language. http://accad.osu.edu/~smay/AL/scheme.html; July 2005.

[26] Mezei L, Zivian A. ARTA: An Interactive Animation System. In: Proceedings of Information Processing 71. Amsterdam: North-Holland; 1971. p. 429–34.

[27] Parent R. Shape Transformation by Boundary Representation Interpolation: A Recursive Approach to Establishing Face Correspondences, OSU Technical Report OSU-CISRC-2–91-TR7. Journal of Visualization and Computer Animation January 1992;3:219–39.

[28] Parent R. A System for Generating Three-Dimensional Data for Computer Graphics. Ph.D. dissertation. Ohio State University; 1977.

[29] Reeves W. In-betweening for Computer Animation Utilizing Moving Point Constraints. In: Computer Graphics. Proceedings of SIGGRAPH 81, vol. 15(3). Dallas, Tex.; August 1981. p. 263–70.

[30] Reynolds C. Computer Animation with Scripts and Actors. In: Computer Graphics. Proceedings of SIGGRAPH 82, vol. 16(3). Boston, Mass.; July 1982. p. 289–96.

[31] Sederberg T. Free-Form Deformation of Solid Geometric Models. In: Evans DC, Athay RJ, editors. Computer Graphics. Proceedings of SIGGRAPH 86, vol. 20(4). Dallas, Tex.; August 1986. p. 151–60.

[32] Houdini 2.0: User Guide. Toronto: Side Effects Software; September 1997.

[33] Stern G. BBOP—A program for 3-dimensional Animation. In: Nicograph 83 proceedings. Tokyo, Dec '83. p. 403–404.

[34] Talbot P, Carr III J, Coulter Jr R, Hwang R. Animator: An On-line Two-Dimensional Film Animation System. Communications of the ACM 1971;14(4):251–9.

[35] Weiler K. Edge-Based Data Structures for Solid Modeling in Curved-Surface Environments. IEEE Comput Graph Appl January 1985;5(1):21–40.

[36] Wilkins M, Kazmier C. MEL Scripting for Maya Animators. San Francisco, California: Morgan Kaufmann; 2003.

[37] Wolberg G. Digital Image Warping. Los Alamitos, California: IEEE Computer Society Press; 1988.

Kinematic Linkages

5

In describing an object's motion, it is often useful to relate it to another object. Consider, for example, a coordinate system centered at our sun in which the moon's motion must be defined. It is much easier to describe the motion of the moon relative to the earth and the earth's motion relative to the sun than it is to come up with a description of the moon's motion directly in a sun-centric coordinate system. Such sequences of relative motion are found not only in astronomy but also in robotics, amusement park rides, internal combustion engines, human figure animation, and pogo sticks.

This chapter is concerned with animating objects whose motion is relative to another object, especially when there is a sequence of objects where each object's motion can easily be described relative to the previous one. Such an object sequence forms a *motion hierarchy*. Often the components of the hierarchy represent objects that are physically connected and are referred to by the term *linked appendages* or, more simply, as *linkages*. Another common aspect of relative motion is that the motion is often restricted. The moon's position relative to the earth, for example, can be specified by a single parameter (in this case, an angle) since, at least for this discussion, it rotates around the earth in a fixed plane at a fixed distance. The plane and distance are built into the hierarchical model so the animator is only concerned with specifying one rotational parameter. This is an example of a model with *reduced dimensionality* because the hierarchical structure enforces constraints and requires fewer parameters than would be needed to specify the position of the moon otherwise.

This chapter addresses how to form data structures that support such linkages and how to animate the linkages by specifying or determining position parameters over time. As such, it is concerned with *kinematics*.

Of course, a common use for kinematic linkages is for animating human (or other) figures in which limbs are defined by a hierarchy of rotational joints connected by rigid links. The two approaches to positioning such a hierarchy are known as *forward kinematics*, in which the animator must specify rotation parameters at joints, and *inverse kinematics* (IK), in which the animator specifies the desired position of the hand, for example, and the system solves for the joint angles that satisfy that desire. Figure 5.1, demonstrating forward kinematics, shows a sample sequence of rotating a limb's joints. Figure 5.2, demonstrating IK, shows a sample sequence of positioning the hand at the desired location as a procedure automatically solves the required joint angles. These techniques are the subject of this chapter after discussing the fundamentals of modeling such hierarchies.

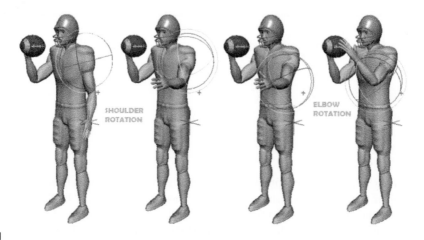

FIGURE 5.1

Sample sequence of forward kinematic specification of joint rotations.

(Image courtesy of Domin Lee.)

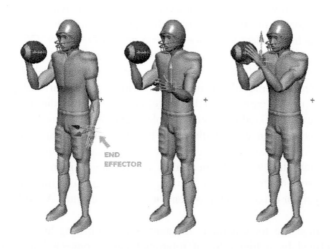

FIGURE 5.2

Sample sequence showing positioning the hand (identified as the end effector) to the desired position as a procedure solves for the required joint angles.

(Image courtesy of Domin Lee.)

5.1 Hierarchical modeling

Hierarchical modeling is the enforcement of relative location constraints among objects organized in a treelike structure. Planetary systems are one type of hierarchical model. In planetary systems, moons rotate around planets, which rotate around a sun, which moves in a galaxy. A common type of hierarchical model used in graphics has objects that are connected end to end to form multibody jointed chains. Such hierarchies are useful for modeling animals and humans so that the joints of the limbs are manipulated to produce a figure with moving appendages. Such a figure is often referred to as *articulated* and the movement of an appendage by changing the configuration of a joint is referred to as *articulation*. Because the connectivity of the figure is built into the structure of the model, the animator does not need to make sure that the objects making up the limbs stay attached to one another.

Much of the material concerning the animation of hierarchies in computer graphics comes directly from the field of robotics (e.g., [3]). The robotics literature discusses the modeling of *manipulators*, a sequence of objects connected in a chain by *joints*. The rigid objects forming the connections between the joints are called *links*, and the free end of the chain of alternating joints and links is called the *end effector*. The local coordinate system associated with each joint is referred to as the *frame*.

Robotics is concerned with all types of joints in which two links move relative to one another. Graphics, on the other hand, is concerned primarily with *revolute* joints, in which one link rotates about a fixed point of the other link. The links are usually considered to be pinned together at this point, and the link farther down the chain rotates while the other one remains fixed—at least as far as this joint is concerned. The other type of joint is the *prismatic* joint, in which one link translates relative to another (see Figure 5.3).

The joints of Figure 5.3 allow motion in one direction and are said to have one *degree of freedom* (DOF). Structures in which more than one DOF are coincident are called *complex joints*. Complex joints include the planar joint and the ball-and-socket joint. Planar joints are those in which one link slides on the planar surface of another. Sometimes when a joint has more than one ($n > 1$) DOF, such as a ball-and-socket joint, it is modeled as a set of n one-DOF joints connected by $n - 1$ links of zero length (see Figure 5.4). Alternatively, multiple DOF joints can be modeled by using a multiple-valued parameter such as Euler angles or quaternions.

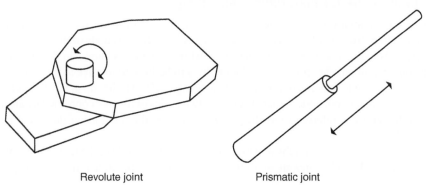

Revolute joint   Prismatic joint

FIGURE 5.3

The two basic single DOF joints.

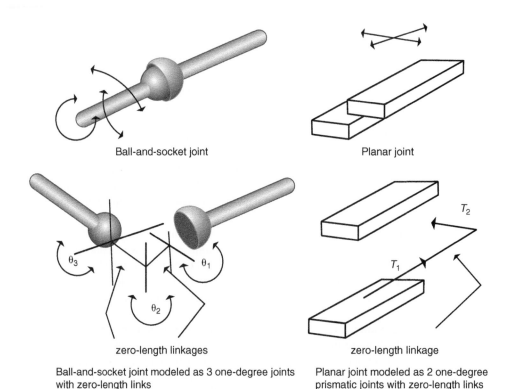

Ball-and-socket joint   Planar joint

zero-length linkages   zero-length linkage

Ball-and-socket joint modeled as 3 one-degree joints   Planar joint modeled as 2 one-degree
with zero-length links   prismatic joints with zero-length links

FIGURE 5.4

Modeling complex joints.

5.1.1 Data structure for hierarchical modeling

Human figures and animals are conveniently modeled as hierarchical linkages. Such linkages can be represented by a tree structure of *nodes* connected by *arcs*.[1] The highest node of the tree is the *root note*, which corresponds to the root object of the hierarchy whose position is known in the global coordinate system. The position of all other nodes of the hierarchy will be located relative to the root node. A node from which no arcs extend downward is referred to as a *leaf node*. "Higher up in the hierarchy" refers to a node that is closer to the root node. When discussing two nodes of the tree connected by an arc, the one higher up the hierarchy is referred to as the *parent node*, and the one farther down the hierarchy is referred to as the *child node*.

The mapping between the hierarchy and tree structure relates a node of the tree to information about the object part (the link) and relates an arc of the tree (the joint) to the transformation to apply to all of

---

[1]The connections between nodes of a tree structure are sometimes referred to as links; however, the robotics literature refers to the objects between the joints as links. To avoid overloading the term *links, arcs* is used here to refer to the connections between nodes in a tree.

the nodes below it in the hierarchy. Relating a tree arc to a figure joint may seem counterintuitive, but it is convenient because a node of the tree can have several arcs emanating from it, just as an object part may have several joints attached to it. In a discussion of a hierarchical model presented by a specific tree structure, the terms *node, object part*, and *link* are used interchangeably since all refer to the geometry to be articulated. Similarly, the terms *joint* and *arc* are used interchangeably.

In the tree structure, there is a root arc that represents a global transformation to apply to the root node (and, therefore, indirectly to all of the nodes of the tree). Changing this transformation will rigidly reposition the entire structure in the global coordinate system (see Figure 5.5).

A node of the tree structure contains the information necessary to define the object part in a position ready to be articulated. In the case of rotational joints, this means that the point of rotation on the object part is made to coincide with the origin. The object data may be defined in such a position, or there may be a transformation matrix contained in the node that, when applied to the object data, positions it so. In either case, all of the information necessary to prepare the object data for articulation is contained at the node. The node represents the transformation of the object data into a link of the hierarchical model.

Two types of transformations are associated with an arc leading to a node. One transformation rotates and translates the object into its position of attachment relative to the link one position up in the hierarchy. This defines the link's neutral position relative to its parent. The other transformation is the variable information responsible for the actual joint articulation (see Figure 5.6).

A simple example
Consider the simple, two-dimensional three-link example of Figure 5.7. In this example, there is assumed to be no transformation at any of the nodes; the data are defined in a position ready for articulation. $Link_0$, the root object, is transformed to its orientation and position in global space by T_0.

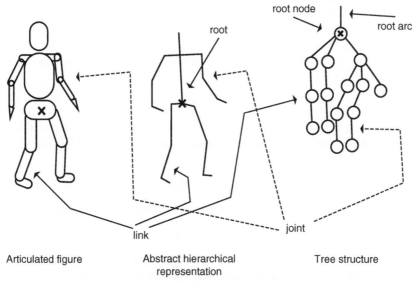

FIGURE 5.5

Example of a tree structure representing a hierarchical structure.

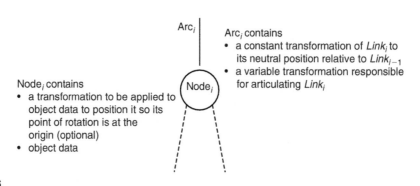

FIGURE 5.6

Arc and node definition.

Because all of the other parts of the hierarchy will be defined relative to this part, this transformation affects the entire assemblage of parts and thus will transform the position and orientation of the entire structure. This transformation can be changed over time in order to animate the position and orientation of the rigid structure. $Link_1$ is defined relative to the untransformed root object by transformation T_1. Similarly, $Link_{1.1}$ is defined relative to the untransformed $Link_1$ by transformation $T_{1.1}$. These relationships can be represented in a tree structure by associating the links with nodes and the transformations with arcs. In the example shown in Figure 5.8, the articulation transformations are not yet included in the model.

An arc in the tree representation contains a transformation that applies to the object represented by the node to which the arc immediately connects. This transformation is also applied to the rest of the linkage farther down the hierarchy. The vertices of a particular object can be transformed to their final positions by concatenating the transformations higher up the tree and applying the composite transformation matrix to the vertices. A vertex, V_0, of the root object, $Link_0$, is located in the world coordinate system, V_0' by applying the rigid transformation that affects the entire structure; see Equation 5.1.

$$V_0' = T_0 V_0 \tag{5.1}$$

A vertex of the $Link_1$ object is located in the world coordinate system by transforming it first to its location relative to $Link_0$ and then relocating it (conceptually along with $Link_0$) to world space by Equation 5.2.

$$V_1' = T_0 T_1 V_1 \tag{5.2}$$

Similarly, a vertex of the $Link_{1.1}$ object is located in world space by Equation 5.3.

$$V_{1.1}' = T_0 T_1 T_{1.1} V_{1.1} \tag{5.3}$$

Notice that as the tree is traversed farther down one of its branches, a newly encountered arc transformation is concatenated with the transformations previously encountered.

As previously discussed, when one constructs the static position of the assembly, each arc of the tree has an associated transformation that rotates and translates the link associated with the child node

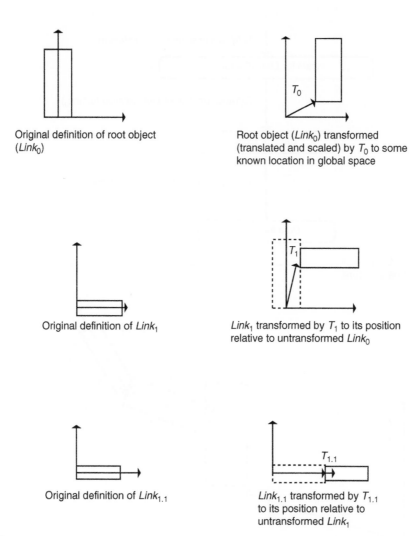

Original definition of root object
(Link₀)

Root object (*Link₀*) transformed
(translated and scaled) by T_0 to some
known location in global space

Original definition of *Link₁*

Link₁ transformed by T_1 to its position
relative to untransformed *Link₀*

Original definition of *Link₁.₁*

Link₁.₁ transformed by $T_{1.1}$
to its position relative to
untransformed *Link₁*

FIGURE 5.7

Example of a hierarchical model.

relative to the link associated with the parent node. To easily animate a revolute joint, also associated with the arc is a parameterized (variable) transformation, $R_i(\theta_i)$, which controls the rotation at the specified joint (see Figure 5.9). In the tree representation that implements a revolute joint, the rotation transformation is associated with the arc that precedes the node representing the link to be rotated (see Figure 5.10). The rotational transformation is applied to the link before the arc's constant transformation. If a transformation is present at the node (for preparing the data for articulation), then the rotational transformation is applied after the node transformation and before the arc's constant transformation.

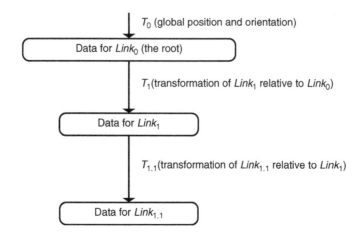

FIGURE 5.8

Example of a tree structure.

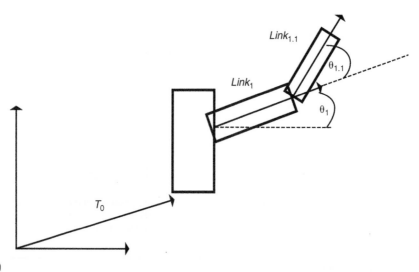

FIGURE 5.9

Variable rotations at the joints.

To locate a vertex of $Link_1$ in world space, one must first transform it via the joint rotation matrix. Once that is complete, then the rest of the transformations up the hierarchy are applied (see Eq. 5.4).

$$V'_1 = T_0 T_1 R_1(\theta_1)V_1 \qquad (5.4)$$

A vertex of $Link_{1.1}$ is transformed similarly by compositing all of the transformations up the hierarchy to the root, as in Equation 5.5.

$$V'_{1.1} = T_0 T_1 R_1(\theta_1)T_{1.1}R_{1.1}(\theta_{1.1})V_{1.1} \qquad (5.5)$$

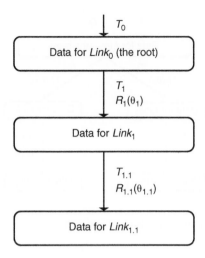

FIGURE 5.10

Hierarchy showing joint rotations.

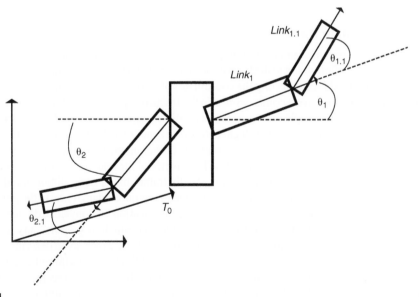

FIGURE 5.11

Hierarchy with two appendages.

In the case of a second appendage, the tree structure would reflect the bifurcations (or multiple branches if more than two). Adding another arm to our simple example results in Figure 5.11.

The corresponding tree structure would have two arcs emanating from the root node, as in Figure 5.12. Branching in the tree occurs whenever multiple appendages emanate from the same object. For example, in a simplified human figure, the root hip area (see Figure 5.5) might branch into the torso

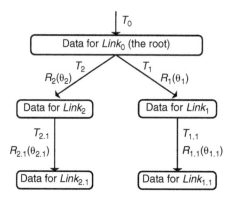

FIGURE 5.12

Tree structure corresponding to hierarchy with two appendages.

and two legs. If prismatic joints are used, the strategy is the same; the only difference is that the rotation transformation of the joint (arc) is replaced by a translation.

5.1.2 Local coordinate frames

In setting up complex hierarchies and in applying computationally involved procedures such as IK, it is convenient to be able to define points in the local coordinate system (frame) associated with a joint and to have a well-defined method for converting the coordinates of a point from one frame to another. A common use for this method is to convert points defined in the frame of a joint to the global coordinate system for display purposes. In the example above, a transformation matrix is associated with each arc to represent the transformation of a point from the local coordinate space of a child node to the local coordinate space of the parent node. Successively applying matrices farther up the hierarchy can transform a point from any position in the tree into world coordinates. The inverse of the transformation matrix can be used to transform a point from the parent's frame to the child's frame. In the three-dimensional case, 4×4 transformation matrices can be used to describe the relation of one coordinate frame to the next. Robotics has adopted a more concise and more appropriate parameterization for physical devices called the Denavit-Hartenberg notation. Although this notation is not used much in graphics anymore because it lacks flexibility appropriate for synthetic linkages, it remains an effective way to describe the physical configurations used in robotics. See Appendix B.4 for details of the Denavit-Hartenberg notation.

When implementing a kinematic linkage, especially when the linkage contains joints with multiple DOF, there are options to consider when choosing the method to implement rotational parameterization of a local coordinate frame. Part of the issue involves the user interface. What kind of information is required from the user to specify the rotational values at the joint? Users are often already comfortable with specifications such as x-y-z Euler angles. The other part of the issue involves the method used to implement the rotations in order to produce the required transformations. If the user specifies the rotation(s) interactively, then the operations are mapped directly to this internal representation. Quaternions are often used to actually implement the transformations in order to avoid the gimbal lock problem with Euler and fixed-angle representations.

5.2 **Forward kinematics**

Evaluation of a hierarchy by traversing the corresponding tree produces the figure in a position that reflects the setting of the joint parameters. Traversal follows a depth-first pattern from root to leaf node. The traversal then backtracks up the tree until an unexplored downward arc is encountered. The downward arc is then traversed, followed by backtracking up to find the next unexplored arc. This traversal continues until all nodes and arcs have been visited. Whenever an arc is followed down the tree hierarchy, its transformations are concatenated to the transformations of its parent node. Whenever an arc is traversed back up the tree to a node, the transformation of that node must be restored before traversal continues downward.

A stack of transformations is a conceptually simple way to implement the saving and restoring of transformations as arcs are followed down and then back up the tree. Immediately before an arc is traversed downward, the current composite matrix is pushed onto the stack. The arc transformation is then concatenated with the current transformation by premultiplying it with the composite matrix. Whenever an arc is traversed upward, the top of the stack is popped off of the stack and becomes the current composite transformation. (If node transformations, which prepare the data for transformation, are present, they must not be included in a matrix that gets pushed onto the stack.)

In the C-like pseudocode in Figure 5.13, each arc has associated with it the following:

- nodePtr: A pointer to a node that holds the data to be articulated by the arc.
- Lmatrix: A matrix that locates the following (child) node relative to the previous (parent) node.

```
traverse(arcPtr,matrix)
{
    ; get transformations of arc and concatenate to current matrix
    matrix = matrix*arcPtr->Lmatrix      ; concatenate location
    matrix = matrix*arcPtr->Amatrix      ; concatenate articulation
    ; process data at node
    nodePtr = arcPtr->nodePtr            ; get the node of the arc
    push(matrix)                         ; save the matrix
    matrix = matrix * nodePtr->matrix    ; ready for articulation
    articulatedData = transformData(matrix,dataPtr) ; articulate the data
    draw(articulatedData);               ; and draw it
    matrix = pop()                       ; restore matrix for node's children
    ; process children of node
    if (nodePtr->arcPtr!= NULL) {        ; if not a terminal node
        nextArcPtr = nodePtr->arcPtr     ; get first arc emanating from node
        while (nextArcPtr != NULL) {     ; while there's an arc to process
            push(matrix)                 ; save matrix at node
            traverse(nextArcPtr,matrix)  ; traverse arc
            matrix = pop()               ; restore matrix at node
            nextArcPtr = nextArcPtr->arcPtr ; set next child of node
        }
    }
}
```

FIGURE 5.13

Forward kinematics pseudocode.

- Amatrix: A matrix that articulates the node data; this is the matrix that is changed in order to animate (articulate) the linkage.
- arcPtr: A pointer to a sibling arc (another child of this arc's parent node); this is NULL if there are no more siblings.

Each node has associated with it the following:

- dataPtr: Data (possibly shared by other nodes) that represent the geometry of this segment of the figure.
- Tmatrix: A matrix to transform the node data into position to be articulated (e.g., put the point of rotation at the origin).
- ArcPtr: A pointer to a single child arc.

The code in Figure 5.13 uses `Push()` and `Pop()` calls that operate on a matrix stack in order to save and restore the transformation matrix.

There is a root arc that holds the global transformation for the figure and points to the root node of the figure. The hierarchy is traversed by passing a pointer to the root arc and a matrix initialized as the identity matrix to the traversal routine:

```
traverse(rootArcPtr,I)  ;  'I' is identity matrix
```

To animate the linkage, the parameters at the joints (rotation angles in our case) are manipulated. These parameters are used to construct the changeable transformation matrix associated with the tree arc.

A completely specified set of linkage parameters, which results in positioning the hierarchical figure, are called a pose. A pose is specified by a vector (the *pose vector*) consisting of one parameter for each DOF.

In a simple animation, a user may determine a key position interactively or by specifying numeric values and then interpolate the joint values between key positions. Specifying all of the joint parameter values for key positions is called forward kinematics and is an easy way to animate the figure. Unfortunately, getting an end effector to a specific desired position by specifying joint values can be tedious for the user. Often, it is a trial-and-error process. To avoid the difficulties in having to specify all of the joint values, IK is sometimes used, in which the desired position and orientation of the end effector are given and the internal joint values are calculated automatically.

5.3 Inverse kinematics

In IK, the desired position and possibly orientation of the end effector are given by the user along with the initial pose vector. From this, the joint values required to attain that configuration are calculated, giving the final pose vector. The problem can have zero, one, or more solutions. If there are so many constraints on the configuration that no solution exists, the system is called *overconstrained*. If there are relatively few constraints on the system and there are many solutions to the problem posed, then it is *underconstrained*. The *reachable workspace* is that volume which the end effector can reach. The *dexterous workspace* is the volume that the end effector can reach in any orientation.

Once the joint values are calculated, the figure can be animated by interpolating from the initial pose vector values to the final pose vector values calculated by IK. However, for large differences between initial and final pose vectors, this does not provide precise control over the path that the end effector follows. Alternatively, a series of intermediate end effector positions (and possibly orientations) can first be calculated and each one of these then used as input to an IK problem. In this way, the path the end effector takes is prescribed by the animator.

If the mechanism is simple enough, then the joint values (the pose vector) required to produce the final desired configuration can be calculated analytically. Given an initial pose vector and the final pose vector, intermediate configurations can be formed by interpolation of the values in the pose vectors, thus animating the mechanism from its initial configuration to the final one. However, if the mechanism is too complicated for analytic solutions, then an incremental approach can be used that employs a matrix of values (the *Jacobian*) that relates changes in the values of the joint parameters to changes in the end effector position and orientation. The end effector is iteratively nudged until the final configuration is attained within a given tolerance. In addition to the Jacobian, there are other incremental formulations that can be used to effect inverse kinematic solutions.

5.3.1 Solving a simple system by analysis

For sufficiently simple mechanisms, the joint values of a final desired position can be determined analytically by inspecting the geometry of the linkage. Consider a simple two-link arm in two-dimensional space with two rotational DOFs. Link lengths are L_1 and L_2 for the first and second link, respectively. If a position is fixed for the base of the arm at the first joint, any position beyond $|L_1 - L_2|$ units from the base of the link and within $L_1 + L_2$ of the base can be reached (see Figure 5.14).

Assume for now (without loss of generality) that the base is at the origin. In a simple IK problem, the user gives the (X, Y) coordinate of the desired position for the end effector. The joint angles, θ_1 and θ_2, can be determined by computing the distance from the base to the goal and using the law of cosines to compute the interior angles. Once the interior angles are computed, the rotation angles for the two links can be computed (see Figure 5.15). Of course, the first step is to make sure that the position of the goal is within the reach of the end effector; that is, $|L_1 - L_2| \leq \sqrt{X^2 + Y^2} \leq L_1 + L_2$.

In this simple scenario, there are only two solutions that will give the correct answer; the configurations are symmetric with respect to the line from $(0, 0)$ to (X, Y). This is reflected in the equation in

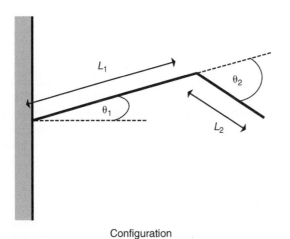

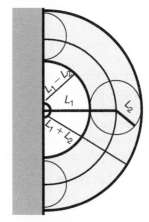

Configuration   Reachable workspace

FIGURE 5.14

Simple linkage.

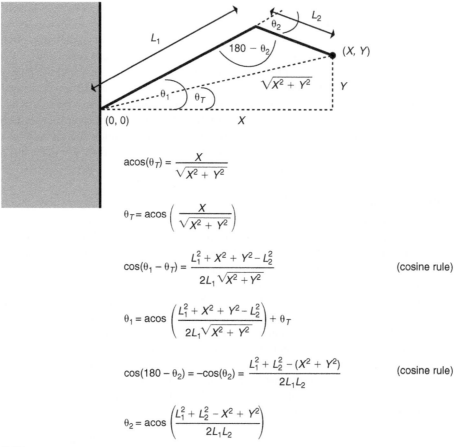

$$\operatorname{acos}(\theta_T) = \frac{X}{\sqrt{X^2 + Y^2}}$$

$$\theta_T = \operatorname{acos}\left(\frac{X}{\sqrt{X^2 + Y^2}}\right)$$

$$\cos(\theta_1 - \theta_T) = \frac{L_1^2 + X^2 + Y^2 - L_2^2}{2L_1\sqrt{X^2 + Y^2}} \qquad \text{(cosine rule)}$$

$$\theta_1 = \operatorname{acos}\left(\frac{L_1^2 + X^2 + Y^2 - L_2^2}{2L_1\sqrt{X^2 + Y^2}}\right) + \theta_T$$

$$\cos(180 - \theta_2) = -\cos(\theta_2) = \frac{L_1^2 + L_2^2 - (X^2 + Y^2)}{2L_1 L_2} \qquad \text{(cosine rule)}$$

$$\theta_2 = \operatorname{acos}\left(\frac{L_1^2 + L_2^2 - X^2 + Y^2}{2L_1 L_2}\right)$$

FIGURE 5.15

Analytic solution to a simple IK problem.

Figure 5.15; the inverse cosine is two-valued in both plus and minus theta ($\pm\theta$). However, for more complicated cases, there may be infinitely many solutions that will give the desired end effector location.

The joint values for relatively simple linkages can be solved by algebraic manipulation of the equations that describe the relationship of the end effector to the base frame. Most linkages used in robotic applications are designed to be simple enough for this analysis. However, for many cases that arise in computer animation, analytic solutions are not tractable. In such cases, iterative numeric solutions must be relied on.

5.3.2 The Jacobian

Many mechanisms of interest to computer animation are too complex to allow an analytic solution. For these, the motion can be incrementally constructed. At each time step, a computation is performed that determines a way to change each joint angle in order to direct the current position and orientation of the end effector toward the desired configuration. There are several methods used to compute the change in joint angle, but most involve forming the matrix of partial derivatives called the *Jacobian*.

To explain the Jacobian from a strictly mathematical point of view, consider the six arbitrary functions of Equation 5.6, each of which is a function of six independent variables. Given specific values for the input variables, x_i, each of the output variables, y_i, can be computed by its respective function.

$$
\begin{aligned}
y_1 &= f_1(x_1, x_2, x_3, x_4, x_5, x_6)\\
y_2 &= f_2(x_1, x_2, x_3, x_4, x_5, x_6)\\
y_3 &= f_3(x_1, x_2, x_3, x_4, x_5, x_6)\\
y_4 &= f_4(x_1, x_2, x_3, x_4, x_5, x_6)\\
y_5 &= f_5(x_1, x_2, x_3, x_4, x_5, x_6)\\
y_6 &= f_6(x_1, x_2, x_3, x_4, x_5, x_6)
\end{aligned}
\tag{5.6}
$$

These equations can also be used to describe the change in the output variables relative to the change in the input variables. The differentials of y_i can be written in terms of the differentials of x_i using the chain rule. This generates Equation 5.7.

$$
dy_i = \frac{\partial f_i}{\partial x_1} dx_1 + \frac{\partial f_i}{\partial x_2} dx_2 + \frac{\partial f_i}{\partial x_3} dx_3 + \frac{\partial f_i}{\partial x_4} dx_4 + \frac{\partial f_i}{\partial x_5} dx_5 + \frac{\partial f_i}{\partial x_6} dx_6
\tag{5.7}
$$

Equations 5.6 and 5.7 can be put in vector notation, producing Equations 5.8 and 5.9, respectively.

$$
dY = \frac{\partial F}{\partial X} dX
\tag{5.8}
$$

$$
dY = \frac{\partial F}{\partial X} dX
\tag{5.9}
$$

A matrix of partial derivatives, $\frac{\partial F}{\partial X}$, is called the *Jacobian* and is a function of the current values of x_i. The Jacobian can be thought of as mapping the velocities of X to the velocities of Y (Eq. 5.10).

$$
\dot{Y} = J(X)\dot{X}
\tag{5.10}
$$

At any point in time, the Jacobian is a function of the x_i. At the next instant of time, X has changed and so has the transformation represented by the Jacobian.

When one applies the Jacobian to a linked appendage, the input variables, x_i, become the joint values and the output variables, y_i, become the end effector position and orientation (in some suitable representation such as x-y-z fixed angles).

$$
Y = [p_x\ p_y\ p_z\ \alpha_x\ \alpha_y\ \alpha_z]^T
\tag{5.11}
$$

In this case, the Jacobian relates the velocities of the joint angles, $\dot{\theta}$, to the velocities of the end effector position and orientation, $\dot{Y}$ (Eq. 5.12).

$$
V = \dot{Y} = J(\theta)\dot{\theta}
\tag{5.12}
$$

V is the vector of linear and rotational velocities and represents the desired change in the end effector. The desired change will be based on the difference between its current position/orientation to that specified by the goal configuration. These velocities are vectors in three-space, so each has an x, y, and z component (Eq. 5.13).

$$
V = [v_x\ v_y\ v_z\ \omega_x\ \omega_y\ \omega_z]^T
\tag{5.13}
$$

$\dot{\theta}$ is a vector of joint value velocities, or the changes to the joint parameters, which are the unknowns of the equation (Eq. 5.14).

$$\dot{\theta} = \begin{bmatrix} \dot{\theta}_1 & \dot{\theta}_2 & \dot{\theta}_3 & \cdots & \dot{\theta}_n \end{bmatrix}^T \tag{5.14}$$

J, the Jacobian, is a matrix that relates the two and is a function of the current pose (Eq. 5.15).

$$J = \begin{bmatrix} \dfrac{\partial p_x}{\partial \theta_1} & \dfrac{\partial p_x}{\partial \theta_2} & \cdots & \dfrac{\partial p_x}{\partial \theta_n} \\[2mm] \dfrac{\partial p_y}{\partial \theta_1} & \dfrac{\partial p_y}{\partial \theta_2} & \cdots & \dfrac{\partial p_y}{\partial \theta_n} \\[2mm] \cdots & \cdots & \cdots & \cdots \\[2mm] \dfrac{\partial \alpha_z}{\partial \theta_1} & \dfrac{\partial \alpha_z}{\partial \theta_2} & \cdots & \dfrac{\partial \alpha_z}{\partial \theta_n} \end{bmatrix} \tag{5.15}$$

Each term of the Jacobian relates the change of a specific joint to a specific change in the end effector. For a revolute joint, the rotational change in the end effector, ω, is merely the velocity of the joint angle about the axis of revolution at the joint under consideration. For a prismatic joint, the end effector orientation is unaffected by the joint articulation. For a rotational joint, the linear change in the end effector is the cross-product of the axis of revolution and a vector from the joint to the end effector. The rotation at a rotational joint induces an instantaneous linear direction of travel at the end effector. For a prismatic joint, the linear change is identical to the change at the joint (see Figure 5.16).

The desired angular and linear velocities are computed by finding the difference between the current configuration of the end effector and the desired configuration. The angular and linear velocities of the end effector induced by the rotation at a specific joint are determined by the computations shown in Figure 5.16. The problem is to determine the best linear combination of velocities induced by the various joints that would result in the desired velocities of the end effector. The Jacobian is formed by posing the problem in matrix form.

In assembling the Jacobian, it is important to make sure that all of the coordinate values are in the same coordinate system. It is often the case that joint-specific information is given in the coordinate

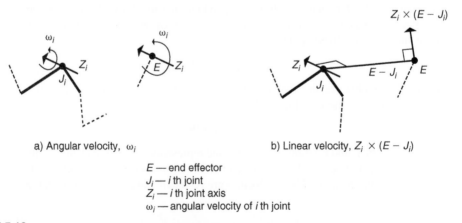

a) Angular velocity, ω_i b) Linear velocity, $Z_i \times (E - J_i)$

E — end effector
J_i — i th joint
Z_i — i th joint axis
ω_i — angular velocity of i th joint

FIGURE 5.16

Angular and linear velocities induced by joint axis rotation.

system local to that joint. In forming the Jacobian matrix, this information must be converted into some common coordinate system such as the global inertial (world) coordinate system or the end effector coordinate system. Various methods have been developed for computing the Jacobian based on attaining maximum computational efficiency given the required information in local coordinate systems, but all methods produce the derivative matrix in a common coordinate system.

A simple example

Consider the simple three-revolute-joint, planar manipulator of Figure 5.17. In this example, the objective is to move the end effector, E, to the goal position, G. The orientation of the end effector is of no concern in this example. The axis of rotation of each joint is perpendicular to the figure, coming out of the paper. The effect of an incremental rotation, g_i, of each joint can be determined by the cross-product of the joint axis and the vector from the joint to the end effector, V_i (Figure 5.18), and form the columns of the Jacobian. Notice that the magnitude of each g_i is a function of the distance between the locations of the joint and the end effector.

The desired change to the end effector is the difference between the current position of the end effector and the goal position (Eq. 5.16).

$$V = \begin{bmatrix} (G-E)_x \\ (G-E)_y \\ (G-E)_z \end{bmatrix} \tag{5.16}$$

A vector of the desired change in values is set equal to the Jacobian matrix (Eq. 5.17) multiplied by a vector of the unknown values, which are the changes to the joint angles.

$$J = \begin{bmatrix} ((0,0,1) \times E)_x & ((0,0,1) \times (E-P_1))_x & ((0,0,1) \times (E-P_2))_x \\ ((0,0,1) \times E)_y & ((0,0,1) \times (E-P_1))_y & ((0,0,1) \times (E-P_2))_y \\ ((0,0,1) \times E)_z & ((0,0,1) \times (E-P_1))_z & ((0,0,1) \times (E-P_2))_z \end{bmatrix} \tag{5.17}$$

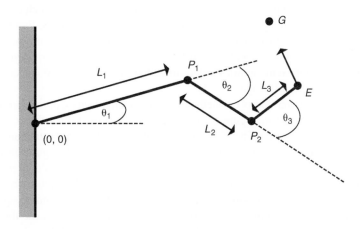

FIGURE 5.17

Planar three-joint manipulator.

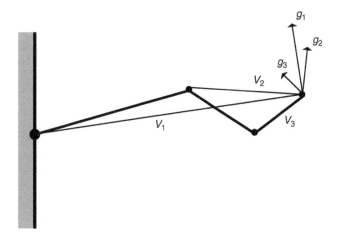

FIGURE 5.18

Instantaneous changes in position induced by joint angle rotations.

5.3.3 Numeric solutions to IK

Solution using the inverse Jacobian

Once the Jacobian has been computed, an equation in the form of Equation 5.18 is to be solved. In the case that J is a square matrix, the inverse of the Jacobian, J^{-1}, is used to compute the joint angle velocities given the end effector velocities (Eq. 5.19).

$$V = J\dot{\theta} \tag{5.18}$$

$$J^{-1}V = \dot{\theta} \tag{5.19}$$

If the inverse of the Jacobian (J^{-1}) does not exist, then the system is said to be singular for the given joint angles. A singularity occurs when a linear combination of the joint angle velocities cannot be formed to produce the desired end effector velocities. As a simple example of such a situation, consider a fully extended planar arm with a goal position somewhere on the forearm (see Figure 5.19). In such a case, a change in each joint angle would produce a vector perpendicular to the desired direction. Obviously, no linear combination of these vectors could produce the desired motion vector. Unfortunately, all of the singularities of a system cannot be determined simply by visually inspecting the possible geometric configurations of the linkage.

Additionally, a configuration that is only close to being a singularity can still present major problems. If the joints of the linkage in Figure 5.19 are slightly perturbed, then the configuration is not singular. However, in order to form a linear combination of the resulting instantaneous change vectors, very large values must be used. This results in large impulses near areas of singularities. These must be clamped to more reasonable values. Even then, numerical error can result in unpredictable motion.

Problems with singularities can be reduced if the manipulator is redundant—when there are more DOF than there are constraints to be satisfied. In this case, the Jacobian is not a square matrix and, potentially, there are an infinite number of solutions to the IK problem. Because the Jacobian is not square, a conventional inverse does not exist. However, if the columns of J are linearly independent

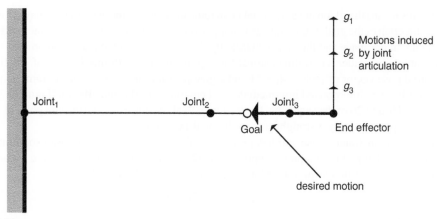

FIGURE 5.19

Simple example of a singular configuration.

(i.e., J has full column rank), then $(J^T J)^{-1}$ exists and instead, the *pseudoinverse*, J^+, can be used as in Equation 5.20. This approach is viable because a matrix multiplied by its own transpose will be a square matrix whose inverse may exist.

$$V = J\dot{\theta}$$
$$J^T V = J^T J \dot{\theta}$$
$$(J^T J)^{-1} J^T V = (J^T J)^{-1} J^T \dot{\theta} \tag{5.20}$$
$$J^+ V = \dot{\theta}$$

If the rows of J are linearly independent, then JJ_T is invertible and the equation for the pseudoinverse is $J^+ = J^T (JJ^T)^{-1}$. The pseudoinverse maps the desired velocities of the end effector to the required changes of the joint angles. After making the substitutions shown in Equation 5.21, LU decomposition can be used to solve Equation 5.22 for β. This can then be substituted into Equation 5.23 to solve for $\dot{\theta}$.

$$J^+ V = \dot{\theta}$$
$$J^T (JJ^T)^{-1} V = \dot{\theta} \tag{5.21}$$

$$\beta = (JJ^T)^{-1} V$$
$$(JJ^T)\beta = V \tag{5.22}$$

$$J^T \beta = \dot{\theta} \tag{5.23}$$

Simple Euler integration can be used at this point to update the joint angles. The Jacobian has changed at the next time step, so the computation must be performed again and another step taken. This process repeats until the end effector reaches the goal configuration within some acceptable (i.e., user-defined) tolerance.

It is important to remember that the Jacobian is only valid for the instantaneous configuration for which it is formed. That is, as soon as the configuration of the linkage changes, the Jacobian ceases to accurately describe the relationship between changes in joint angles and changes in end effector position and orientation. This means that if too big a step is taken in joint angle space, the end effector may not appear to travel in the direction of the goal. If this appears to happen during an animation sequence, then taking smaller steps in joint angle space and thus recalculating the Jacobian more often may be in order.

As an example, consider a two-dimensional three-joint linkage with link lengths of 15, 10, and 5. Using an initial pose vector of $\{\pi/8, \pi/4, \pi/4\}$ and a goal position of $\{-20, 5\}$, a 21-frame sequence is calculated[2] for linearly interpolated intermediate goal positions for the end effector. Figure 5.20 shows frames 0, 5, 10, 15, and 20 of the sequence. Notice the path of the end effector (the end point of the third link) travels in approximately a straight line to the goal position.

Even the underconstrained case still has problems with singularities. A proposed solution to such bad behavior is the damped least-squares approach [1] [2]. Referring to Equation 5.24, a user-supplied parameter is used to add in a term that reduces the sensitivity of the pseudoinverse.

$$\dot{\theta} = J^T (JJ^T + \lambda^2 I)^{-1} V \qquad (5.24)$$

It is argued that this form behaves better in the neighborhood of singularities at the expense of rate of convergence to a solution. Figure 5.21 shows solutions using the pseudoinverse of the Jacobian

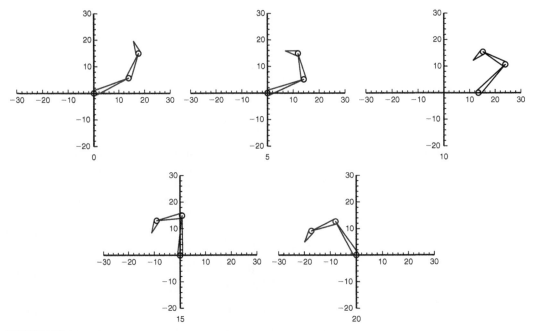

FIGURE 5.20

IK solution for a two-dimensional three-link armature of lengths 15, 10, and 5. The initial pose is $\{\pi/8, \pi/4, \pi/4\}$ and goal is $\{-20, 5\}$. Panels show frames 0, 5, 10, 15, and 20 of a 21-frame sequence in which the end effector tracks a linearly interpolated path to goal. (a) The pseudoinverse of Jacobian solution without damped least-squares. (b) The pseudoinverse of Jacobian solution with damped least-squares.

[2]The examples showing solutions of numeric IK were generated using the *LinearSolve* function of Mathematica.

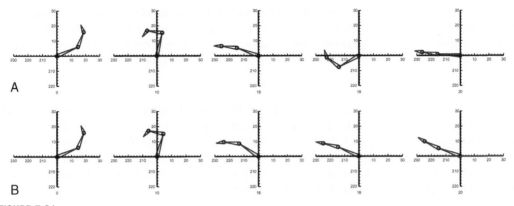

FIGURE 5.21

Example comparing the pseudoinverse of the Jacobian solution to inverse kinematics (a) without damped least-squares and (b) with damped least squares for a two-dimensional three-link armature of lengths 15, 10, and 5. The initial pose is {π/8, π/4, π/4} and goal is {−35, 5}. Panels show frames 0, 10, 18, 19, and 20 of a 21-frame sequence in which the end effector tracks a linearly interpolated path to goal.

without, and then with, damping for the linkage used in the previous example. However, in this case, the goal is at {−35, 5}—out of reach of the end effector. This example demonstrates better behavior when damped as the linkage approaches the limits of its reach.

Adding more control

The pseudoinverse computes one of many possible solutions. It minimizes joint angle rates. The configurations produced, however, do not necessarily correspond to what might be considered natural poses. To better control the kinematic model such as encouraging joint angle constraints, a control expression can be added to the pseudoinverse Jacobian solution. The control expression is used to solve for control angle rates with certain attributes. The added control expression, because of its form, contributes nothing to the desired end effector velocities.[3] The form for the control expression is shown in Equation 5.25.

$$\dot{\theta} = (J^+J - I)z \tag{5.25}$$

In Equation 5.26 it is shown that a change to pose parameters in the form of Equation 5.25 does not add anything to the velocities. As a consequence, the control expression can be added to the pseudoinverse Jacobian solution without changing the given velocities to be satisfied [4].

$$
\begin{aligned}
V &= J\dot{\theta} \\
V &= J(J^+J - I)z \\
V &= (JJ^+J - I)z \\
V &= 0z \\
V &= 0
\end{aligned}
\tag{5.26}
$$

---

[3] The columns of the control term matrix are in the null space of the Jacobian and, therefore, do not affect the end effector position [1].

To bias the solution toward specific joint angles, such as the middle angle between joint limits, z is defined as in Equation 5.27, where θ_i are the current joint angles, θ_{ci} are the desired joint angles, and a_i are the joint gains. This does not enforce joint limits as hard constraints, but the solution can be biased toward the middle values so that violating the joint limits is less probable.

$$z = \alpha_i(\theta_i - \theta_{ci})^2 \qquad (5.27)$$

The desired angles and gains are input parameters. The gain indicates the relative importance of the associated desired angle; the higher the gain, the stiffer the joint.[4] If the gain for a particular joint is high, then the solution will be such that the joint angle quickly approaches the desired joint angle. The control expression is added to the conventional pseudoinverse of the Jacobian (Eq. 5.28). If all gains are zero, then the solution will reduce to the conventional pseudoinverse of the Jacobian. Equation 5.28 can be solved by rearranging terms as shown in Equation 5.29.

$$\dot{\theta} = J^+V + (J^+J - I)z \qquad (5.28)$$

$$\dot{\theta} = J^+V + (J^+J - I)z$$

$$\dot{\theta} = J^+V + J^+Jz - Iz$$

$$\dot{\theta} = J^+(V + Jz) - z \qquad (5.29)$$

$$\dot{\theta} = J^T(JJ^T)^{-1}(V + Jz) - z$$

$$\dot{\theta} = J^T\left[(JJ^T)^{-1}(V + Jz)\right] - z$$

To solve Equation 5.29, set $\beta = (JJ^T)^{-1}(V + Jz)$ so that Equation 5.30 results. Use LU decomposition to solve for β in Equation 5.31. Substitute the solution for β in Equation 5.30 to solve for $\dot{\theta}$.

$$\dot{\theta} = J^T\beta - z \qquad (5.30)$$

$$V + Jz = (JJ^T)\beta \qquad (5.31)$$

In Figure 5.22, the difference between using a larger gain for the second joint versus using a larger gain for the third joint is shown. Notice how the joints with increased gain are kept straighter in the corresponding sequence of frames.

Alternative Jacobian

Instead of trying to push the end effector toward the goal position, a formulation has been proposed that pulls the goal to the end effector [1]. This is implemented by simply using the goal position in place of the end effector in the pseudoinverse of the Jacobian method. Comparing Figure 5.19 to Figure 5.23 for the simple example introduced earlier, the results of this approach are similar to the standard method of using the end effector position in the calculations.

---

[4]*Stiffness* refers to how much something reacts to being perturbed. A stiff spring is a strong spring. A stiff joint, as used here, is a joint that has a higher resistance to being pulled away from its desired value.

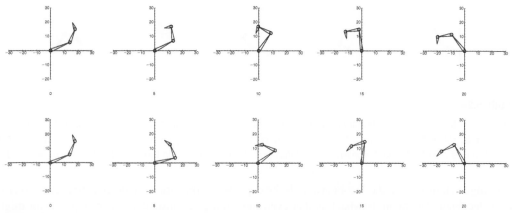

FIGURE 5.22

Inverse of the Jacobian with control term solution for a two-dimensional three-link armature of lengths 15, 10, and 5. The initial pose is {π/8, π/4, π/4} and goal is {−20, 5}. Panels show frames 0, 5, 10, 15, and 20 of a 21-frame sequence in which the end effector tracks a linearly interpolated path to goal. All joints are biased to 0; the top row uses gains of {0.1, 0.5, 0.1} and the bottom row uses gains of {0.1, 0.1, 0.5}.

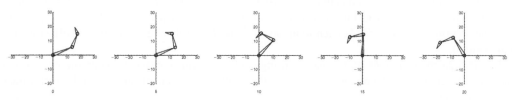

FIGURE 5.23

Inverse of the Jacobian solution formulated to pull the goal toward the end effector for a two-dimensional three-link armature of lengths 15, 10, and 5. The initial pose is {π/8, π/4, π/4} and goal is {−20, 5}. Panels show frames 0, 5, 10, 15, and 20 of a 21-frame sequence in which the end effector tracks a linearly interpolated path to goal.

Avoiding the inverse: using the transpose of the Jacobian

Solving the linear equations using the pseudoinverse of the Jacobian is essentially determining the weights needed to form the desired velocity vector from the instantaneous change vectors. An alternative way of determining the contribution of each instantaneous change vector is to form its projection onto the end effector velocity vector [5]. This entails forming the dot product between the instantaneous change vector and the velocity vector. Putting this into matrix form, the vector of joint parameter changes is formed by multiplying the transpose of the Jacobian times the velocity vector and using a scaled version of this as the joint parameter change vector.

$$\dot{\theta} = \alpha J^T V$$

Figure 5.24 shows frames from a sequence produced using the transpose of the Jacobian. Notice how the end effector tracks to the goal position while producing reasonable interior joint angles.

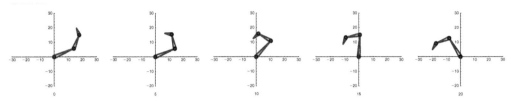

FIGURE 5.24

Transpose of the Jacobian solution for a two-dimensional three-link armature of lengths 15, 10, and 5. The initial pose is {π/8, π/4, π/4} and goal is {− 20, 5}. Panels show frames 0, 5, 10, 15, and 20 of a 21-frame sequence in which the end effector tracks a linearly interpolated path to goal. The scale term used in this example is 0.1.

Using the transpose of the Jacobian avoids the expense of computing the inverse, or pseudoinverse, of the Jacobian. The main drawback is that even though a given instantaneous change vector might contribute a small component to the velocity vector, the perpendicular component may take the end effector well away from the desired direction. Also, there is an additional parameter that must be supplied by the user.

Procedurally determining the angles: cyclic coordinate descent

Instead of relying on numerical machinery to produce desired joint velocities, a more flexible, procedural approach can be taken. Cyclic coordinate descent considers each joint one at a time, sequentially from the outermost inward [5]. At each joint, an angle is chosen that best gets the end effector to the goal position.

Figure 5.25 shows frames of a sequence using cyclic coordinate descent. While the interior joint angles are different from the previous examples, they still look reasonable and the end effector tracks the solution path well.

In a variation of this approach, each instantaneous change vector is compared to the desired change vector and the joints are rank ordered in terms of their usefulness. The highest ranked joint is then changed and the process repeats. The rank ordering can be based on how recently that joint was last modified, joint limits, how close the joint gets the end effector to the goal, or any number of other measures.

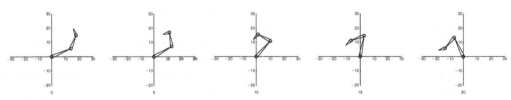

FIGURE 5.25

Cyclic coordinate descent solution for a two-dimensional three-link armature of lengths 15, 10, and 5. The initial pose is {π/8, π/4, π/4} and goal is {− 20, 5}. Panels show frames 0, 5, 10, 15, and 20 of a 21-frame sequence in which the end effector tracks a linearly interpolated path to goal.

5.3.4 Summary

Various strategies for IK have been proposed in the literature. Much of this comes from robotics literature that avoids the expense of computing an inverse [1] [5]. Several strategies are presented here. The most direct solution is by analysis, but many linkages are too complex for such analysis. A numeric approach is often needed. Trade-offs among the numeric approaches include ease of implementation, possible real-time behavior, and robustness in the face of singularities. An additional approach is targeted at linkages that are specifically human-like in their DOF and heuristically solves reaching problems by trying to model human-like reaching; this is discussed in Chapter 9.

   IK, on the surface, promises to solve many animation control problems. However, there is no single, obvious, and universally attractive solution to IK. There are options, however, and the various approaches do provide effective tools if used judiciously.

5.4 Chapter summary

Hierarchical models are extremely useful for enforcing certain relationships among the elements so that the animator can concentrate on just the DOF remaining. Forward kinematics gives the animator explicit control over each DOF but can become cumbersome when the animation is trying to attain a specific position or orientation of an element at the end of a hierarchical chain. IK, using the inverse or pseudoinverse of the Jacobian, allows the animator to concentrate only on the conditions at the end of such a chain but might produce undesirable configurations. Additional control expressions can be added to the pseudoinverse Jacobian solution to express a preference for solutions of a certain character. However, these are all kinematic techniques. Often, more realistic motion is desired and physically based simulations are needed.

References

[1] Buss S. Introduction to Inverse Kinematics with Jacobian Transpose, Pseudoinverse and Damped Least Squares Method, Unpublished survey, http://math.ucsd.edu/~sbuss/ResearchWeb/ikmethods/; July 2006.

[2] Buss S. Selectively Damped Least Squares for Inverse Kinematics. Journal of Graphics Tools 2005;10 (3):37–49.

[3] Craig J. Introduction to Robotics Mechanics and Control. New York: Addison-Wesley; 1989.

[4] Ribble E. Synthesis of Human Skeletal Motion and the Design of a Special-Purpose Processor for Real-Time Animation of Human and Animal Figure Motion. Master's thesis. Columbus, Ohio: Ohio State University; June 1982.

[5] Welman C. Inverse Kinematics and Geometric Constraints for Articulated Figure Manipulation. M.S. Thesis. Simon Frasier University; 1993.

Motion Capture

Creating physically realistic motion using techniques involving key frames, forward kinematics, or inverse kinematics is a daunting task. In the previous chapters, techniques to move articulated linkages are covered, but they still require a large amount of animator talent in order to make an object move as in the real world. There is a certain quality about physically correct motion that we are all aware of, and it only takes a small deviation from that quality to make what is supposed to be realistic movement quite distracting. As a result, it is often easier to record movement of a real object and map it onto a synthetic object than it is for the animator to craft realistic-looking motion. The recording and mapping of real motion to a synthetic object is called *motion capture*, or *mocap* for short.

Motion capture involves sensing, digitizing, and recording of an object's motion [6]. Often, the term specifically refers to capturing human motion, but some very creative animation can be accomplished by capturing the motion of structures other than animal life. However, the discussion in this chapter focuses on motion capture of a human figure.

In motion capture, a (human) subject's movement is recorded. It should be noted that locating and extracting a figure's motion directly from raw (unadulterated) video is extremely difficult and is the subject of current research. As a result, the figure whose motion is to be captured (sometimes referred to as the *talent*) is typically instrumented in some way so that positions of key feature points can be easily detected and recorded. This chapter is concerned with how this raw data is converted into data usable for animation.

There are various technologies that can be applied in this instrumentation and recording process. If the recordings are multiple two-dimensional images, then camera calibration is necessary in order to reconstruct the three-dimensional coordinates of joints and other strategic positions of the figure. Usually, joint angles are extracted from the position data and a synthetic figure, whose body dimensions match those of the mocap talent, can then be fitted to the data and animated using the extracted joint angles. In some cases, the mocap data must be edited or combined with other motion before it can be used in the final animation. These topics are discussed in the following sections.

6.1 Motion capture technologies

There are primarily two approaches to this instrumentation[1]: *electromagnetic sensors* and *optical markers*. Electromagnetic tracking, also simply called *magnetic tracking*, uses sensors placed at the joints that transmit their positions and orientations back to a central processor to record their

---

[1]There are several other technologies used to capture motion, including electromechanical suits, fiber-optic sensors, and digital armatures [6]. However, electromagnetic sensors and (passive) optical markers are by far the most commonly used technologies for capturing full-body motion.

movements. While theoretically accurate, magnetic tracking systems require an environment devoid of magnetic field distortions. To transmit their information, the sensors have to use either cables or wireless transmission to communicate with the central processor. The former requires that the subject be "tethered" with some kind of cabling harness. The latter requires that the subject also carry a power source such as a battery pack. The advantage of electromagnetic sensors is that the three-dimensional position and orientation of each sensor can be recorded and displayed in real time (with some latency). The drawbacks relate to the range and accuracy of the magnetic field and the restricted movement resulting from the instrumentation required.

Optical markers, on the other hand, have a much larger range, and the performers only have to wear reflective markers on their clothes (see Figure 6.1, Color Plate 2). The optical approach does not provide real-time feedback; however, and the data from optical systems are error prone and noisy. Optical marker systems use video technology to record images of the subject in motion. Because orientation information is not directly generated, more markers are required than with magnetic trackers. Some combination of joint and mid-segment markers is used. While industrial strength systems may use eight or more fast (up to 120 frames per second) infrared cameras, basic optical motion control can be implemented with consumer-grade video technology. Because the optical approach can be low cost, at least in some situations, it is the approach that is considered here.

6.2 Processing the images

The objective of motion control is to reconstruct the three-dimensional motion of a physical object and apply it to a synthetic model. With optical systems, three major tasks need to be undertaken. First, the two-dimensional images need to be processed so that the markers can be located, identified, and correlated in the multiple video streams. Second, the three-dimensional locations of the markers need to be reconstructed from their two-dimensional locations. Third, the three-dimensional marker locations must be constrained to a model of the physical system whose motion is being captured (e.g., a

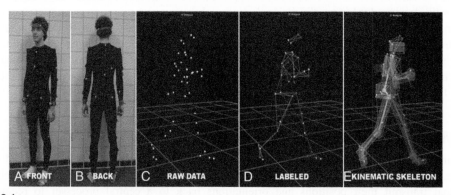

FIGURE 6.1

Images showing progression of optical motion capture session: (a) and (b) optical marker placement, (c) three-dimensional reconstruction positions of marker, (d) identified markers, (e) applied to skeleton.

(Image courtesy of Brian Windsor; data courtesy of Brent Haley.)

triangulated surface model of the performer). The first requires some basic image-processing techniques, simple logic, usually some user input, and sometimes a bit of luck. The second requires camera calibration and enough care to overcome numerical inaccuracies. The third requires satisfying constraints between relative marker positions.

Optical markers can be fashioned from table tennis balls and coated to make them stand out in the video imagery. They can be attached to the figure using Velcro strips or some other suitable method. Colored tape can also be used. The markers are usually positioned close to joints since these are the structures of interest in animating a figure. The difference between the position of the marker and that of the joint is one source of error in motion capture systems. This is further complicated if the marker moves relative to the joint during the motion of the figure. Once the video is digitized, it is simply a matter of scanning each video image for evidence of the optical markers. If the background image is static, then it can be subtracted out to simplify the processing. Finding the position of a single marker in a video frame in which the marker is visible is the easy part. This step gets messier the more markers there are, the more often some of the markers are occluded, the more often the markers overlap in an image, and the more the markers change position relative to one another. With multiple-marker systems, the task is not only to isolate the occurrence of a marker in a video frame but also to track a specific marker over a number of frames even when the marker may not always be visible.

Once all of the visible markers have been extracted from the video, each individual marker must be tracked over the video sequence. Sometimes this can be done with simple domain knowledge. For example, if the motion is constrained to be normal walking, then the ankle markers (or foot markers, if present) can be assumed to always be the lowest markers, and they can be assumed to always be within some small distance from the bottom of the image. Frame-to-frame coherence can be employed to track markers by making use of the position of a marker in a previous frame and knowing something about the limits of its velocity and acceleration. For example, knowing that the markers are on a walking figure and knowing something about the camera position relative to the figure, one can estimate the maximum number of pixels that a marker might travel from one frame to the next and thus help track it over time.

Unfortunately, one of the realities of optical motion capture systems is that periodically one or more of the markers are occluded. In situations in which several markers are used, this can cause problems in successfully tracking a marker throughout the sequence. Some simple heuristics can be used to track markers that drop from view for a few frames and that do not change their velocity much over that time. But these heuristics are not foolproof (and is, of course, why they are called heuristics). The result of failed heuristics can be markers swapping paths in mid-sequence or the introduction of a new marker when, in fact, a marker is simply reappearing again. Marker swapping can happen even when markers are not occluded. If markers pass within a small distance of each other they may swap paths because of numerical inaccuracies of the system. Sometimes these problems can be resolved when the three-dimensional positions of markers are constructed. At other times user intervention is required to resolve ambiguous cases. See Figure 6.1 for an example of mocap data processing.

As a side note, there are optical systems that use active markers. The markers are light-emitting diodes (LEDs) that flash their own unique identification code. There is no chance of marker swapping in this case, but this system has its own limitations. The LEDs are not very bright and cannot be used in bright environments. Because each marker has to take the time to flash its own ID, the system captures the motion at a relatively slow rate. Finally, there is a certain delay between the measurements of markers, so the positions of each marker are not recorded at exactly the same moment, which may present problems in animating the synthetic model.

A constant problem with motion capture systems is noise. Noise can arise from the physical system; the markers can move relative to their initial positioning and the faster the performer moves, the more the markers can swing and reposition themselves. Noise also arises from the sampling process; the markers are sampled in time and space, and errors can be introduced in all dimensions. A typical error might result in inaccurate positioning of a feature point by half a centimeter. For some animations, this can be a significant error.

To deal with the noise, the user can condition the data before they are used in the reconstruction process. Data points that are radically inconsistent with typical values can be thrown out, and the rest can be filtered. The objective is to smooth the data without removing any useful features. A simple weighted average of adjacent values can be used to smooth the curve. The number of adjacent values to use and their weights are a function of the desired smoothness. Generally, this must be selected by the user.

6.3 Camera calibration

Before the three-dimensional position of a marker can be reconstructed, it is necessary to know the locations and orientations of cameras in world space as well as the intrinsic properties of the cameras such as focal length, image center, and aspect ratio [8].

A simple pinhole camera model is used for the calibration. This is an idealized model that does not accurately represent certain optical effects often present in real cameras, but it is usually sufficient for computer graphics and image-processing applications. The pinhole model defines the basic projective geometry used to describe the imaging of a point in three-space. For example, the camera's coordinate system is defined with the origin at the center of projection and the plane of projection at a focal-length distance along the positive z-axis, which is pointed toward the camera's center of interest (Figure 6.2). Equivalently, the projection plane could be a focal length along the negative z-axis on the other side of the center of projection from the center of interest; this would produce an inverted image, but the mathematics would be the same.

The image of a point is formed by projecting a ray from the point to the center of projection (Figure 6.3). The image of the point is formed on the image (projection) plane where this ray intersects the plane. The equations for this point, as should be familiar to the reader, are formed by similar triangles. Camera calibration is performed by recording a number of image space points whose world

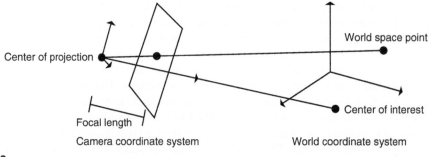

FIGURE 6.2

Camera model.

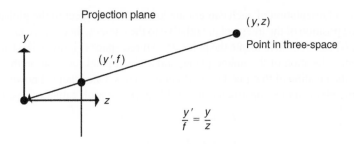

Projection plane

y

(y', f)

z

(y, z)

Point in three-space

$$\frac{y'}{f} = \frac{y}{z}$$

FIGURE 6.3

Y-Z projection of a world space point onto the image plane in the camera coordinate system.

space locations are known. These pairs can be used to create an overdetermined set of linear equations that can be solved using a least-squares solution method. This is only the most basic computation needed to fully calibrate a camera with respect to its intrinsic properties and its environment. In the most general case, there are nonlinearities in the lens, focal length, camera tilt, and other parameters that need to be determined. See Appendix B.11 for further discussion.

6.4 Three-dimensional position reconstruction

To reconstruct the three-dimensional coordinates of a marker, the user must locate its position in at least two views relative to known camera positions, C_1 and C_2. In the simplest case, this requires two cameras to be set up to record marker positions (Figure 6.4). The greater the orthogonality of the two views, the better the chance for an accurate reconstruction.

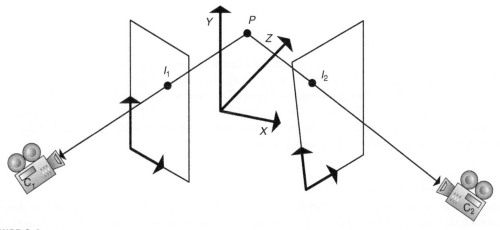

FIGURE 6.4

Two-camera view of a point.

If the position and orientation of each camera are known with respect to the global coordinate system, along with the position of the image plane relative to the camera, then the images of the point to be reconstructed (I_1, I_2) can be used to locate the point, P, in three-space (Figure 6.4). Using the location of a camera, the relative location of the image plane, and a given pixel location on the image plane, the user can compute the position of that pixel in world coordinates. Once that is known, a vector from the camera through the pixel can be constructed in world space for each camera (Eqs. 6.1 and 6.2).

$$C_1 + k_1(I_1 - C_1) = P \qquad (6.1)$$

$$C_2 + k_2(I_2 - C_2) = P \qquad (6.2)$$

By setting these equations equal to each other, $C_1 + k_1 (I_1\, C_1) = C_2 + k_2\, (I_2\, C_2)$. In three-space, this represents three equations with two unknowns, k_1 and k_2, which can be easily solved—in an ideal world. Noise tends to complicate the ideal world. In practice, these two equations will not exactly intersect, although if the noise in the system is not excessive, they will come close. So, in practice, the points of closest encounter must be found on each line. This requires that a P_1 and a P_2 be found such that P_1 is on the line from Camera 1, P_2 is on the line from Camera 2, and $P_2\, P_1$ is perpendicular to each of the two lines (Eqs. 6.3 and 6.4).

$$(P_2 - P_1) \cdot (I_1 - C_1) = 0 \qquad (6.3)$$

$$(P_2 - P_1) \cdot (I_2 - C_2) = 0 \qquad (6.4)$$

See Appendix B.2.6 on solving for P_1 and P_2. Once the points P_1 and P_2 have been calculated, the midpoint of the chord between the two points can be used as the location of the reconstructed point. In the case of multiple markers in which marker identification and tracking have not been fully established for all of the markers in all of the images, the distance between P_1 and P_2 can be used as a test for correlation between I_1 and I_2. If the distance between P_1 and P_2 is too great, then this indicates that I_1 and I_2 are probably not images of the same marker and a different pairing needs to be tried. Smoothing can also be performed on the reconstructed three-dimensional paths of the markers to further reduce the effects of noise on the system.

6.4.1 Multiple markers

The number and positioning of markers on a human figure depend on the intended use of the captured motion. A simple configuration of markers for digitizing gross human figure motion might require only 14 markers (3 per limb and 2 for positioning the head). For more accurate recordings of human motion, markers need to be added to the elbows, knees, chest, hands, toes, ankles, and spine (see Figure 6.5). Menache [6] suggests an addition of three per foot for some applications. Also refer back to Figure 6.1, which shows a sample marker set on mocap talent.

6.4.2 Multiple cameras

As the number of markers increases and the complexity of the motion becomes more involved, there is greater chance for marker occlusion. To reconstruct the three-dimensional position of a marker, the system must detect and identify the marker in at least two images. As a result, a typical system may have eight cameras simultaneously taking images. These sequences need to be synchronized either

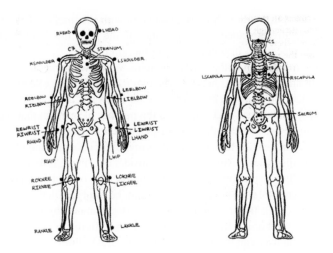

FIGURE 6.5

Complete marker set [6].

automatically or manually. This can be done manually, for example, by noting the frame in each sequence when a particular heel strike occurs. However, with manually synchronized cameras the images could be half a frame off from each other.

6.5 Fitting to the skeleton

Once the motion of the individual markers looks smooth and reasonable, the next step is to attach them to the underlying skeletal structure that is to be controlled by the digitized motion. In a straightforward approach, the position of each marker in each frame is used to absolutely position a specific joint of the skeleton. As a first approximation to controlling the motion of the skeleton, this works fine. However, on closer inspection, problems often exist. The problem with using the markers to directly specify position is that, because of noise, smoothing, and general inaccuracies, distances between joints of the skeleton will not be precisely maintained over time. This change in bone length can be significant. Length changes of 10-20 percent are common. In many cases, this is not acceptable. For example, this can cause *foot-sliding* of the skeleton (also known as *skating*). Inverse kinematics used to lock the foot to the ground can counteract this skating.

One source of the problem is that the markers are located not at the joints of the performers, but outside the joints at the surface. This means that the point being digitized is displaced from the joint itself. While a constant distance is maintained on the performer, for example, between the knee joint and the hip joint, it is not the case that a constant distance is maintained between a point to the side of the knee joint and a point to the side of the hip joint.

The one obvious correction that can be made is logically relocating the digitized positions so that they correspond more accurately to the positions of the joints. This can be done by using marker information to calculate the joint position. The distance from the marker to the actual joint can be determined easily by visual inspection. However, the problem with locating the joint relative to a marker is that there is no orientation information directly associated with the marker. This means that, given a marker location and the relative distance from the marker to the joint, the user does not know in which direction to apply the displacement in order to locate the joint.

One solution is to put markers on both sides of the joint. With two marker locations, the joint can be interpolated as the midpoint of the chord between the two markers. While effective for joints that lend themselves to this approach, the approach does not work for joints that are complex or more inaccessible (such as the hip, shoulder, and spine), and it doubles the number of markers that must be processed.

A little geometry can be used to calculate the displacement of the joint from the marker. A plane formed by three markers can be used to calculate a normal to a plane of three consecutive joints, and this normal can be used to displace the joint location. Consider the elbow. If there are two markers at the wrist (commonly used to digitize the forearm rotation) the position of the true wrist can be interpolated between them. Then the wrist-elbow-shoulder markers can be used to calculate a normal to the plane formed by those markers. Then the true elbow position is calculated by offsetting from the elbow marker in the direction of the normal by the amount measured from the performer. By recalculating the normal every frame, the user can easily maintain an accurate elbow position throughout the performance. In most cases, this technique is very effective. A problem with the technique is that when the arm straightens out the wrist-elbow-shoulder become (nearly) collinear. Usually, the normal can be interpolated during these periods of congruity from accurately computed normals on either side of the interval. This approach keeps limb lengths much closer to being constant.

Now that the digitized joint positions are more consistent with the skeleton to be articulated, they can be used to control the skeleton. To avoid absolute positioning in space and further limb-length changes, one typically uses the digitized positions to calculate joint rotations. For example, in a skeletal hierarchy, if the positions of three consecutive joints have been recorded for a specific frame, then a third of the points in the hierarchy is used to calculate the rotation of that limb relative to the limb represented by the first two points (Figure 6.6).

After posing the model using the calculated joint angles, it might still be the case that, because of inaccuracies in the digitization process, feature points of the model violate certain constraints such as avoiding floor penetration. The potential for problems is particularly high for end effectors, such as the hands or feet, which must be precisely positioned. Often, these must be independently positioned, and then the joints higher up the hierarchy (e.g., knee and hip) must be adjusted to compensate for any change to the end effector position.

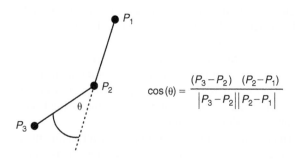

$$\cos(\theta) = \frac{(P_3 - P_2) \ (P_2 - P_1)}{|P_3 - P_2||P_2 - P_1|}$$

FIGURE 6.6

One-DOF rotation joint.

6.6 **Output from motion capture systems**

As the previous discussion has outlined, there is a lot of processing that must be done in order to convert the raw video data from the cameras to joint angles for a figure. Fortunately, motion capture software is available to enable the user to set up the marker set hierarchy, calibrate cameras, and automate as much of the marker processing as possible as well as provide interactive tools when user input is required to resolve ambiguities. The result can be used to animate a figure or saved to a file for later use.

Various de facto standard file formats exist for motion capture data. Two of the more common formats are the Acclaim two-file format (.asf/.asm) and the Biovision (a group of motion capture studios) file format (.bvh). While the various formats differ in their specifics, they all have the same basic information: the static information describing the joint hierarchy and the time-varying information containing the values for each degree of freedom (DOF) for each joint.

The joint hierarchy is described by (1) defining the joints including the number and type of each DOF, the order that DOF parameters are given, and joint limits, and (2) defining the hierarchical connectivity including the location of joint relative to previous joint. Ignoring header information and other detail not related to geometry, the static joint hierarchy information would look something like the following:

```
root
    joint Chest offset 0 5.21 0
        3 DOFs: Zrotation, Yrotation, Xrotation
        Limits: (−180,180) (−90,90) (0, 270)
        joint Neck
            offset 0 5.45 0
            3 DOFs: Zrotation, Yrotation, Xrotation
    ...
    joint LeftUpperLeg
        offset 3.91 0.0 0.0
        3 DOFs: Xrotation, Yrotation, Zrotation
        joint LeftLowerLeg
            offset 0.0 8.1 0.0
            1 DOF: Xrotation
    ...
```

In this simplified example, joint names are just for user identification, indentation indicates child–parent relationship, the offset is the location of the current joint relative to its parent, and the DOF line indicates the number of DOFs for that joint and the order they appear in the data. Optionally, joint limits can be specified.

The time-varying data records the reconstructed joint angles over some number of frames and might look something like the following:

```
total frames: 300
−2.4 40.2 3.2 2.8 22.0 −23.0...    −12.4 40.2 3.2 2.8...
−2.8 42.1 3.4 2.8 22.1 −23.2...    −12.2 42.3 3.3 3.8...
0.2 48.2 3.5 3.9 30.2 −25.2...    −14.4 45.8 3.1 3.0...
2.4 50.4 3.3 5.0 35.2 −23.3...    −20.2 44.9 3.3 2.8...
1.4 51.6 3.2 8.8 35.3 −20.2...    −22.4 45.2 3.7 2.8...
...
```

where each line is one time sample of the motion, the number of values on a line corresponds to the total DOFs of the figure, and they are given in the order they appear in the hierarchy information (the specific numbers in the previous example are completely random).

Motion capture (typically ASCII) files containing this information are the final output of a motion capture session. Many are made available on the Web for free download. It's not a difficult exercise, for example, to write a C/C++ program to read in a specific file format and visualize the motion by animating a stick figure using OpenGL.

6.7 Manipulating motion capture data

The data generated by the motion capture process are directly usable for animating a synthetic figure whose dimensions match those of the captured subject. Even then, as noted earlier, noise and numerical inaccuracies can disturb the resulting motion. In addition, it is not unusual for error-free captured motion to not quite satisfy the desires of the animator when finally incorporated into an animation. In these cases, the motion must be recaptured (an expensive and time-consuming process), or it must be manipulated to better satisfy the needs of the animator. There are various ways to manipulate mocap data. Some are straightforward applications of standard signal processing techniques while others are active areas of research. These techniques hold the promise of enabling motion libraries to be constructed along with algorithms for editing the motions, mapping the motions to new characters, and combining the motions into novel motions and/or longer motion sequences.

6.7.1 Processing the signals

The values of an individual parameter of the captured motion (e.g., a joint angle) can be thought of, and dealt with, as a time-varying signal. The signal can be decomposed into frequencies, time-warped, interpolated with other signals, and so forth. The challenge for the animator in using these techniques is in understanding that, while the original signal represents physically correct motion, there is nothing to guarantee physical correctness once the signal has been modified.

Motion signal processing considers how frequency components capture various qualities of the motion [1]. The lower frequencies of the signal are considered to represent the base activity (e.g., walking) while the higher frequencies are the idiosyncratic movements of the specific individual (e.g., walking with a limp). Frequency bands can be extracted, manipulated, and reassembled to allow the user to modify certain qualities of the signal while leaving others undisturbed. In order to do this, the signal is successively convolved with expanded versions of a filter kernel (e.g., 1/16, 1/4, 3/8, 1/4, 1/16). The frequency bands are generated by differencing the convolutions. Gains of each band are adjusted by the user and can be summed to reconstruct a motion signal.

Motion warping warps the signal in order to satisfy user-supplied key-frame-like constraints [9]. The original signal is $\theta(t)$ and the key frames are given as a set of (θ_i, t_i) pairs. In addition, there is a set of time warp constraints (t_j, t'_j). The warping procedure first creates a time warp mapping $t = g(t')$ by interpolating through the time warp constraints and then creates a motion curve warping, $\theta(t) = f(\theta, t)$. This function is created by scaling and/or translating the original curve to satisfy the key-frame constraints, $\theta'(t) = a(t)\theta(t) + b(t)$ whether to scale or translate can be user-specified. Once the functions $a(t)$

and $b(t)$ for the key frames are set, the functions $a(t)$ and $b(t)$ can be interpolated throughout the time span of interest. Combined, these functions establish a function $\theta(t')$ that satisfies the key-frame and time warp constraints by warping the original function.

6.7.2 Retargeting the motion

What happens if the synthetic character doesn't match the dimensions (e.g., limb length) of the captured subject? Does the motion have to be recaptured with a new subject that does match the synthetic figure better? Or does the synthetic figure have to be redesigned to match the dimensions of the captured subject? One solution is to map the motion onto the mismatched synthetic character and then modify it to satisfy important constraints.

This is referred to as *motion retargeting* [2]. Important constraints include such things as avoiding foot penetration of the floor, avoiding self-penetration, not letting feet slide on the floor when walking, and so forth. A new motion is constructed, as close to the original motion as possible, while enforcing the constraints. Finding the new motion is formulated as a space–time, nonlinear constrained optimization problem. This technique is beyond the scope of this book and the interested reader is encouraged to consult the work by Michael Gleicher (e.g., [2]).

6.7.3 Combining motions

Motions are usually captured in segments whose durations last a few minutes each. Often, a longer sequence of activity is needed in order to animate a specific action. The ability to assemble motion segments into longer actions makes motion capture a much more useful tool for the animator. The simplest, and least aesthetically pleasing, way to combine motion segments is to start and stop each segment in a neutral position such as standing. Motion segments can then be easily combined with the annoying attribute of the figure coming to the neutral position between each segment.

More natural transitions between segments are possible by blending the end of one segment into the beginning of the next one. Such transitions may look awkward unless the portions to be blended are similar. Similarity can be defined as the sum of the differences over all DOFs over the time interval of the blend. Both motion signal processing and motion warping can be used to isolate, overlap, and then blend two motion segments together. Automatic techniques to identify the best subsegments for transitions is the subject of current research [4].

In order to animate a longer activity, motion segments can be strung together to form a longer activity. Recent research has produced techniques such as motion graphs that identify good transitions between segments in a motion database [3] [5] [7]. When the system is confronted with a request for a motion task, the sequence of motions that satisfies the task is constructed from the segments in the motion capture database. Preprocessing can identify good points of transition among the motion segments and can cluster similar motion segments to reduce search time. Allowing minor modifications to the motion segments, such as small changes in duration and distance traveled, can help improve the usefulness of specific segments. Finally, selecting among alternative solutions usually means evaluating a motion sequence based on obstacle avoidance, distance traveled, time to completion, and so forth.

6.8 Chapter summary

Motion capture is a very powerful and useful tool. It will never replace the results produced by a skilled animator, but its role in animation will expand and increase as motion libraries are built and the techniques to modify, combine, and adapt motion capture data become more sophisticated. Current research involves extracting motion capture data from markerless video. This has the potential to free the subject from instrumentation constraints and make motion capture even more useful.

References

[1] Bruderlin A, Williams L. Motion Signal Processing. In: Cook R, editor. Proceedings of SIGGRAPH 95, Computer Graphics, Annual Conference Series. Los Angeles, Calif.: Addison-Wesley; August 1995. p. 97–104. Reading, Massachusetts. ISBN 0-201-84776-0.

[2] Gleicher M. Retargeting Motion to New Characters. In: Cohen M, editor. Proceedings of SIGGRAPH 98, Computer Graphics, Annual Conference Series. Orlando, Fla.: Addison-Wesley; July 1998. p. 33–42. ISBN 0-89791-999-8.

[3] Kovar L, Gleicher M, Pighin F. Motion Graphs, In: Proceedings of SIGGRAPH 2002, Computer Graphics, Annual Conference Series. San Antonio, Tex.; July 23–26,; 2002. p. 473–82.

[4] Kovar L, Gleicher M. Flexible Automatic Motion Blending with Registration Curves. In: Breen D, Lin M, editors. Symposium on Computer Animation. San Diego, Ca.: Eurographics Association; 2002. p. 214–24.

[5] Lee J, Chai J, Reitsma P, Hodgins J, Pollard N. Interactive Control of Avatars Animated with Human Motion Data, In: Computer Graphics, Proceedings of SIGGRAPH 2002, Annual Conference Series. San Antonio, Tex.; July 23–26, 2002. p. 491–500.

[6] Menache A. Understanding Motion Capture for Computer Animation and Video Games. New York: Morgan Kaufmann; 2000.

[7] Reitsma P, Pollard N. Evaluating Motion Graphs for Character Navigation. In: Symposium on Computer Animation. Grenoble, France. p. 89–98.

[8] Tuceryan M, Greer D, Whitaker R, Breen D, Crampton C, Rose E, et al. Calibration Requirements and Procedures for a Monitor-Based Augmented Reality System. IEEE Trans Vis Comput Graph September 1995;1 (3):255–73. ISSN 1077-2626.

[9] Witkin A, Popovic Z. Motion Warping. In: Cook R, editor. Computer Graphics, Proceedings of SIGGRAPH 95, Annual Conference Series. Los Angeles, Calif: Addison–Wesley; August 1995. p. 105–8. Reading, Massachusetts. ISBN 0-201-84776-0.

Physically Based Animation

Animators are often more concerned with the general quality of motion than with precisely controlling the position and orientation of each object in each frame. Such is the case with physically based animation. In *physically based animation*, forces are typically used to maintain relationships among geometric elements. Note that these forces may or may not be physically accurate. Animation is not necessarily concerned with accuracy, but it is concerned with believability, which is sometimes referred to as being *physically realistic*. Some forces may not relate to physics at all—they may model constraints that an animator may wish to enforce on the motion, such as directing a ball to go toward making contact with a bat. For the purposes of this chapter, the use of forces to control motion will be referred to as physically based animation whether or not the forces are actually trying to model some aspect of physics.

When modeling motion that is produced by an underlying mechanism, such as the principles of physics, a decision has to be made concerning the level at which to model the process. For example, wrinkles in a cloth can be modeled by trying to characterize common types of wrinkles and where they typically occur on a piece of cloth. On the other hand, the physics (stresses and strains) of the individual threads of cloth could be modeled in sufficient detail to give rise to naturally occurring wrinkles. Modeling the surface characteristics of the process, as in the former case, is usually computationally less expensive and easier to program, but it lacks flexibility. The only effects produced are the ones explicitly modeled. Modeling the physics of the underlying process usually incurs a higher computational expense but is more flexible—the inputs to the underlying process can usually produce a greater range of surface characteristics under a wider range of conditions. There are often several different levels at which a process can be modeled. In the case of the cloth example, a triangular mesh can be used to model the cloth as a sheet of material (which is what is usually done). The quest for the animation programmer is to provide a procedure that does as little computation as possible while still providing the desired effects and controls for the animator.

The advantage to the animator, when physical models are used, is that (s)he is relieved of low-level specifications of motions and only needs to be concerned with specifying high-level relationships or qualities of the motion. For our example of cloth, instead of specifying where wrinkles might go and the movement of individual vertices, the animator would merely be able to set parameters that indicate the type and thickness of the cloth material. For this, the animator trades off having specific control of the animation—wrinkles will occur wherever the process model produces them.

For the most part, this chapter is concerned with *dynamic control*—specifying the underlying forces that are then used to produce movement. First, a review of basic physics is given, followed by two common examples of physical simulation used in animation: spring meshes and particle systems. Rigid body dynamics and the use of constraints are then discussed.

7.1 Basic physics—a review

The physics most useful to animators is that found in high school text books. It is the physics based on Newton's laws of motion. The basics will be reviewed here; more in-depth discussion of specific animation techniques follows. More physics can be found in Appendix B.7 as well as any standard physics text (e.g., [6]).

The most basic (and most useful) equation relates force to the acceleration of a mass, Equation 7.1.

$$f = ma \tag{7.1}$$

When setting up a physically based animation, the animator will specify all of the possible forces acting in that environment. In calculating a frame of the animation, each object will be considered and all the forces acting on that object will be determined. Once that is done, the object's linear acceleration (ignoring rotational dynamics for now) can be calculated based on Equation 7.1 using Equation 7.2.

$$a = \frac{f}{m} \tag{7.2}$$

From the object's current velocity, v, and newly determined acceleration, a new velocity, v', can be determined as in Equation 7.3.

$$v' = v + a\Delta t \tag{7.3}$$

The object's new position, p', can be updated from its old position, p, by the average of its old velocity and its new velocity.[1]

$$p' = p + \frac{1}{2}(v + v')\Delta t \tag{7.4}$$

Notice that these are vector-valued equations; force, acceleration, and velocity are all vectors (indicated in bold in these equations).

There are a variety of forces that might be considered in an animation. Typically, forces are applied to an object because of its velocity and/or position in space.

The gravitational force, f, between two objects at a given distance, d, can be calculated if their masses are known (G is the gravitation constant of $6.67 \ 3 \ 10^{-11} \text{m}^3\text{s}^{-2}\text{kg}^{-1}$). This magnitude of the force, f, is applied along the direction between the two objects (Eq. 7.5).

$$f = \| \mathbf{f} \| = G\frac{m_1 m_2}{d^2} \tag{7.5}$$

If the earth is one of the objects, then Equations 7.1 and 7.5 can be simplified by estimating the distance to be the radius of the earth (r_e), using the mass of the earth (m_e), and canceling out the mass of the other object, m_o. This produces the magnitude of the gravitation acceleration, a_e, and is directed downward (see Eq. 7.6).

$$a_e = \frac{f}{m_o} = G\frac{m_e}{r_e^2} = 9.8\frac{meter}{sec^2} \tag{7.6}$$

---

[1]Better alternatives for updating these values will be discussed later in the chapter, but these equations should be familiar to the reader.

A spring is a common tool for modeling flexible objects, to keep two objects at a prescribed distance, or to insert temporary control forces into an environment. Notice that a spring may correspond to a geometric element, such as a thread of a cloth, or it may be defined simply to maintain a desired relationship between two points (sometimes called a *virtual spring*), such as a distance between the centers of two spheres. When attached to an object, a spring imparts a force on that object depending on the object's location relative to the other end of the spring. The magnitude of the spring force (f_s) is proportional to the difference between the spring's rest length (L_r) and its current length (L_c) and is in the direction of the spring. The constant of proportionality (k_s), also called the *spring constant*, determines how much the spring reacts to a change in length—its *stiffness* (see Eq. 7.7). The force is applied to both ends of the spring in opposite directions.

$$f_s = -k_s(L_c - L_r)\left(\frac{p_2 - p_1}{\|p_2 - p_1\|}\right) \tag{7.7}$$

The spring described so far is a *linear spring* because there is a linear relationship between how much the length of the spring deviates from the rest length and the resulting force. In some cases, it is useful to use something other than a linear spring. A *biphasic spring* is a spring that is essentially a linear spring that changes its spring constant at a certain length. Usually a biphasic spring is used to make the spring stiffer after it has stretched past some threshold and has found uses, for example, in facial animation and cloth animation. Other types of *nonlinear springs* have spring "constants" that are more complex functions of spring length.

A *damper*, like a spring, is attached to two points. However, the damper works against its relative velocity. The force of a damper (f_d) is negatively proportional to, and in the direction of, the velocity of the spring length (v_s). The constant of proportionality (k_d), or damper constant, determines how much resistance there is to a change in spring length. So the force at p_2 is computed as in Equation 7.8. The force on p_1 would be the negative of the force on p_2.

$$f_d = -k_d\left(\dot{p}_2 - \dot{p}_1\right) \cdot \left(\frac{p_2 - p_1}{\|p_2 - p_1\|}\right)\left(\frac{p_2 - p_1}{\|p_2 - p_1\|}\right) \tag{7.8}$$

Viscosity, similar to a damper, resists an object traveling at a velocity. However, viscosity is resisting movement through a medium, such as air. A force due to viscosity (f_v) is negatively proportional to the velocity of the object (v) (see Eq. 7.9).

$$f_v = -k_v v \tag{7.9}$$

Momentum is mass × velocity. In a closed system (i.e., a system in which energy is not changing), total momentum is conserved. This can be used to solve for unknowns in the system. For a system consisting of multiple objects in which there are no externally applied forces, the total momentum is constant (Eq. 7.10).

$$\sum m_i v_i = c \tag{7.10}$$

The rotational equivalent to force is *torque*. Think of torque, τ, as rotational force. Analogous to (linear) velocity and (linear) acceleration are *angular velocity*, ω, and *angular acceleration*, α. The mass of an object is a measure of its resistance to movement. Analogously, the *moment of inertia* is a measure of an object's resistance to change in orientation. The moment of inertia of an object is a 3×3 matrix of values, I, describing an object's distribution of mass about its center of mass. The relationship to torque, moment of inertia, and angular acceleration is similar to the familiar $f = ma$ (see Eq. 7.11).

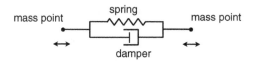

FIGURE 7.1

Spring-damper configuration.

$$\tau = I\alpha \tag{7.11}$$

More physics will be introduced as various topics are covered, but this should give the reader a starting point if physics is only a vague recollection.

7.1.1 Spring-damper pair

One of the most useful tools to incorporate some use of forces into an animation is the use of a spring-damper pair as in Figure 7.1. Consider a mass point, p_1, connected by a spring to a fixed point, p_2, with rest length L_r and current Length L_c. the resulting force on p_1 would be computed as in Equation 7.12. As mentioned before, the spring represents a force to maintain a relationship between two points or insert a control force into the system. The damper is used to restrict the motion and keep the system from reacting too violently.

$$f = \left(k_s(L_c - L_r) - k_d\left(\dot{p}_2 - \dot{p}_1\right) \cdot \left(\frac{p_2 - p_1}{\|p_2 - p_1\|}\right) \right) \left(\frac{p_2 - p_1}{\|p_2 - p_1\|}\right) \tag{7.12}$$

In physical simulations, the challenge in using a spring-damper system is in selecting an appropriate time step, setting the spring and damper constants, and setting the mass of the object so that the resulting motion is desirable. Some combination of too small a time step, a large mass, a high damping constant, and a low spring constant might result in a system that is not very lively, where the object slowly moves toward the fixed point going slower and slower. On the other hand, a larger time step, a smaller mass, a lower damping constant, and a higher spring constant might result in a system that moves too much, resulting in a mass that oscillates around the fixed point, getting further and further away from it each time step. Balancing these parameters to get the desired effect can be tricky.

7.2 Spring animation examples
7.2.1 Flexible objects

In Chapter 4, kinematic approaches to animating flexible bodies are discussed in which the animator is responsible for defining the source and target shapes as well as implying how the shapes are to be interpolated. Various physically based approaches have been proposed that model elastic and inelastic behavior, viscoelasticity, plasticity, and fracture (e.g., [11] [13] [17] [23] [24] [25]). Here, the simplest approach is presented. Flexibility is modeled by a mass-spring-damper system simulating the reaction of the body to external forces.

Mass-spring-damper modeling of flexible objects

Probably the most common approach to modeling flexible shapes is the mass-spring-damper model. The most straightforward strategy is to model each vertex of an object as a point mass and each edge of the object as a spring. Each spring's rest length is set equal to the original length of the edge. A mass can be arbitrarily assigned to an object by the animator and the mass evenly distributed among the object's vertices. If there is an uneven distribution of vertices throughout the object definition, then masses can be assigned to vertices in an attempt to more evenly distribute the mass. Spring constants are similarly arbitrary and are usually assigned uniformly throughout the object to some user-specified value. It is most common to pair a damper with each spring to better control the resulting motion at the expense of dealing with an extra set of constants for the dampers. To simplify the following discussion and diagrams, the spring-damper pairs will be referred to simply as springs with the understanding that there is usually an associated damper.

As external forces are applied to specific vertices of the object, either because of collisions, gravity, wind, or explicitly scripted forces, vertices will be displaced relative to other vertices of the object. This displacement will induce spring forces, which will impart forces to the adjacent vertices as well as reactive forces back to the initial vertex. These forces will result in further displacements, which will induce more spring forces throughout the object, resulting in more displacements, and so on. The result will be an object that is wriggling and jiggling as a result of the forces propagating along the edge springs and producing constant relative displacements of the vertices.

One of the drawbacks with using springs to model the effects of external forces on objects is that the effect has to propagate through the object, one time step at a time. This means that the number of vertices used to model an object and the length of edges used have an effect on the object's reaction to forces. Because the vertices carry the mass of the object, using a different distribution of vertices to describe the same object will result in a difference in the way the object reacts to external forces.

A simple example

In a simple two-dimensional example, an equilateral triangle composed of three vertices and three edges with uniform mass distribution and uniform spring constants will react to an initial external force applied to one of the vertices (Figure 7.2).

In the example, an external force is applied to vertex V_2, pushing it generally toward the other two vertices. Assume that the force is momentary and is applied only for one time step during the animation. At the application of the force, an acceleration ($a_2 = F/m_2$) is imparted to vertex V_2 by the force. The acceleration gives rise to a velocity at point V_2, which in turn creates a change in its position. At the next time step, the external force has already been removed, but, because vertex V_2 has been displaced, the lengths of edges E_{12} and E_{23} have been changed. As a result of this, a spring force is created along the two edges. The spring that models edge E_{12} imparts a restoring force to vertices V_1 and V_2, while the

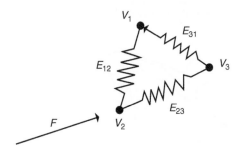

FIGURE 7.2

A simple spring-mass model of a flexible object.

spring that models edge E_{23} imparts a restoring force to vertices V_2 and V_3. Notice that the springs push back against the movement of V_2, trying to restore its position, while at the same time they are pushing the other two vertices away from V_2. In a polygonal mesh, the springs associated with mesh edges propagate the effect of the force throughout the object.

If the only springs (or spring-dampers) used to model an object are ones associated with the object edges, the model can have more than one stable configuration. For example, if a cube's edges are modeled with springs, during applications of extreme external forces, the cube can turn inside out. Additional spring dampers can help to stabilize the shape of an object. Springs can be added across the object's faces and across its volume. To help prevent such undesirable, but stable, configurations in a cube, one or more springs that stretch diagonally from one vertex across the interior of the cube to the opposite vertex can be included in the model. These springs will help to model the internal material of a solid object (Figure 7.3).

If specific angles between adjacent faces (dihedral angles) are desired, then angular springs (and dampers) can be applied. The spring resists deviation to the rest angle between faces and imparts a torque along the edge that attempts to restore the rest angle (Eq. 7.13 and Figure 7.4). The damper works to limit the resulting motion. The torque is applied to the vertex from which the dihedral angle is measured; a torque of equal magnitude but opposite direction is applied to the other vertex.

$$\hat{\tau} = k_s(\theta(t) - \theta_r) - k_d\dot{\theta}(t) \tag{7.13}$$

Alternatively, a linear spring could be defined between points A and B in Figure 7.4 in order to maintain the appropriate angle at the shared edge.

Depending on the size of forces applied to the spring's mass points, the size of the spring constant (and damper constant), and the size of the time step used to sample the system, a spring simulation may

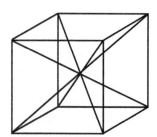

FIGURE 7.3

Interior springs to help stabilize the object's configuration.

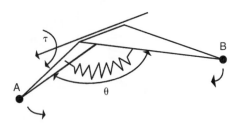

FIGURE 7.4

An angular spring imparting restoring torques.

numerically explode, resulting in computed behavior in which the motions get larger and larger over the course of time. This is because the force is (wrongly) assumed to be constant throughout the time step while, in fact, it is constantly changing as the spring length changes as it progresses from one time sample to the next. As mentioned before, modifying spring, damper, mass, and time step values can be adjusted to help control the simulation, but finding appropriate values can be tricky. Alternatively (or additionally) various simulation magnitudes, such as forces, accelerations, or velocities, can be clamped at maximum values to help control the motion.

7.2.2 Virtual springs

Virtual springs introduce forces into the system that do not directly model physical elements. These can be used to control the motion of objects in a variety of circumstances. In the previous section, it was suggested that virtual springs can be used to help maintain the shape of an object. Virtual springs with zero rest lengths can be used to constrain one object to lie on the surface of another, for example, or, with non-zero rest lengths, to maintain separation between moving objects.

Proportional derivative controllers (PDCs) are another type of virtual spring used to keep a control variable and its derivative within the neighborhood of desired values. For example, to maintain a joint angle and joint velocity close to desired values, the user can introduce a torque into a system. If the desired values are functions of time, PDCs are useful for biasing a model toward a given motion while still allowing it to react to system forces, as in Equation 7.14.

$$\tau = k_s(\theta(t) - \theta_d(t)) - k_d(\dot{\theta}(t) - \dot{\theta}_d(t)) \qquad (7.14)$$

Springs, dampers, and PDCs are an easy and handy way to control the motion of objects by incorporating forces into a physically based modeling system. The springs can model physical elements in the system as well as represent arbitrary control forces. Dampers and PDCs are commonly used to keep system parameters close to desired values. They are not without problems, however. One drawback to using springs, dampers, and PDCs is that the user-supplied constants in the equations can be difficult to choose to obtain a desired effect. A drawback to using a spring mesh is that the effect of the forces must ripple through the mesh as the force propagates from one spring to the next. Still, they are often a useful and effective tool for the animator.

7.3 Particle systems

A particle system is a collection of a large number of point-like elements. Particle systems are often animated as a simple physical simulation. Because of the large number of elements typically found in a particle system, simplifying assumptions are used in both the rendering and the calculation of their motions. Various implementations make different simplifying assumptions. Some of the common assumptions made are the following:

> Particles do not collide with other particles.
> Particles do not cast shadows, except in an aggregate sense.
> Particles only cast shadows on the rest of the environment, not on each other.
> Particles do not reflect light—they are each modeled as point light sources.

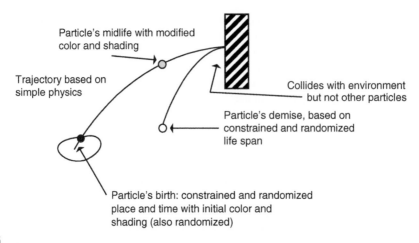

Particle's midlife with modified color and shading

Trajectory based on simple physics

Collides with environment but not other particles

Particle's demise, based on constrained and randomized life span

Particle's birth: constrained and randomized place and time with initial color and shading (also randomized)

FIGURE 7.5

The life of a particle.

Particles are often modeled as having a finite life span, so that during an animation there may be hundreds of thousands of particles used but only thousands active at any one time. Randomness is introduced into most of the modeling and processing of particles to avoid excessive orderliness. In computing a frame of motion for a particle system, the following steps are taken (see Figure 7.5):

- Any new particles that are born during this frame are generated.
- Each new particle is assigned attributes.
- Any particles that have exceeded their allocated life span are terminated.
- The remaining particles are animated and their shading parameters changed according to the controlling processes.
- The particles are rendered.

The steps are then repeated for each frame of the animation. If the user can enforce the constraint that there are a maximum number of particles active at any one time, then the data structure can be static and particle data structures can be reused as one dies and another is created in its place.

7.3.1 Particle generation

Particles are typically generated according to a controlled stochastic process. For each frame, a random number of particles are generated using some user-specified distribution centered at the desired average number of particles per frame (Eq. 7.15.). The distribution could be uniform or Gaussian or anything else the animator wants. In Equation 7.15, $Rand(\_)$ returns a random number from -1.0 to $+1.0$ in the desired distribution, and r scales it into the desired range. If the particles are used to model a fuzzy object, then it is best to make the number of particles a function of the area, A, of the screen covered by the object to control the number of particles generated (Eq. 7.16). The features of the random number generator, such as average value and variance, can be a function of time to enable the animator to have more control over the particle system.

$$\# \; of \; particles = n + Rand() * r \tag{7.15}$$

$$\# \; of \; particles = n(A) + Rand() * r \tag{7.16}$$

7.3.2 Particle attributes

The attributes of a particle determine its motion status, its appearance, and its life in the particle system. Typical attributes include the following:

> Position
> Velocity
> Shape parameters
> Color
> Transparency
> Lifetime

Each of the attributes is initialized when the particle is created. Again, to avoid uniformity, the user typically randomizes values in some controlled way. The position and velocity are updated according to the particle's motion. The shape parameters, color, and transparency control the particle's appearance. The lifetime attribute is a count of how many frames the particle will exist in.

7.3.3 Particle termination

At each new frame, each particle's lifetime attribute is decremented by one. When the attribute reaches zero, the particle is removed from the system. This completes the particle's life cycle and can be used to keep the number of particles active at any one time within some desired range of values.

7.3.4 Particle animation

Typically, each active particle in a particle system is animated throughout its life. This activation includes not only its position and velocity but also its display attributes: shape, color, and transparency. To animate the particle's position in space, the user considers forces and computes the resultant particle acceleration. The velocity of the particle is updated from its acceleration, and then the average of its old velocity and newly updated velocity is computed and used to update the particle's position. Gravity, other global force fields (e.g., wind), local force fields (vortices), and collisions with objects in the environment are typical forces modeled in the environment.

The particle's color and transparency can be a function of global time, its own life span remaining, its height, and so on. The particle's shape can be a function of its velocity. For example, an ellipsoid can be used to represent a moving particle where the major axis of the ellipsoid is aligned with the direction of travel and the ratio of the ellipsoid's length to the radius of its circular cross section is related to its speed.

7.3.5 Particle rendering

To simplify rendering, model each particle as a point light source so that it adds color to the pixel(s) it covers but is not involved in the display pipeline (except to be hidden) or shadowing. In some applications, shadowing effects have been determined to be an important visual cue. The density of particles between a position in space and a light source can be used to estimate the amount of shadowing. See

Blinn [2], Ebert et al. [4], and Reeves [20] for some of these considerations. For more information on particle systems, see Reeves and Blau [21] and Sims [22].

7.3.6 Particle system representation

A particle is represented by a tuple $[x, v, f, m]$, which holds its position, velocity, force accumulator, and mass. It can also be used to hold any other attributes of the particle such as its age and color.

```
typedef struct particle_struct {
        vector3D        p;
        vector3D        v;
        vector3D        f;
        float           mass;
} particle;
```

The state of a particle, $[x, v]$, will be updated by $[v, f/m]$ by the ordinary differential equation solver. The solver can be considered to be a black box to the extent that the actual method used by the solver does not have to be known.

A particle system is an array of particles with a time variable $[*p, n, t]$.

```
typedef struct particleSystem_struct {
        particle    *p;    // array particle information
        int          n;    // number of particles
        float        t;    // current time (age) of the particle system
} particleSystem;
```

Updating particle system status

For each time step, the particle system is updated by the following steps:

a. For each particle, zero out the force vector.
b. For each particle, sum all forces affecting the particle.
c. Form the state array of all particles.
d. Form the update array of all particles.
e. Add the update array to the state array.
f. Save the new state array.
g. Update the time variable.

7.3.7 Forces on particles

Forces can be unary or binary and typically arise from the particle's relationship to the environment. Additional forces might be imposed by the animator in order to better control the animation. Unary forces include gravity and viscous drag. Particle pair forces, if desired, can be represented by springs and dampers. However, implementing particle pair forces for each possible pair of particles can be prohibitively expensive in terms of computational requirements. Auxiliary processing, such as the use of a spatial bucket sort, can be employed to consider only n-nearest-neighbor pairs and reduce the computation required, but typically particle–particle interaction is ignored. Environmental forces arise from a particle's relationship to the rest of the environment such as gravity, wind, air viscosity, and responding to collisions; most of these may be modeled using simple physics.

7.3.8 Particle life span

Typically, each particle will have a life span. The particle data structure itself can be reused in the system so that a particle system might have tens of thousands of particles over the life of the simulation but only, for example, one thousand in existence at any one time. Initial values are set pseudorandomly so that particles are spawned in a controlled manner but with some variability.

7.4 Rigid body simulation

A common objective in computer animation is to create realistic-looking motion. A major component of realistic motion is the physically based reaction of rigid bodies to commonly encountered forces such as gravity, viscosity, friction, and those resulting from collisions. Creating realistic motion with keyframe techniques can be a daunting task. However, the equations of motion can be incorporated into an animation system to automatically calculate these reactions. This can eliminate considerable tedium—if the animator is willing to relinquish precise control over the motion of some objects.

In rigid body simulation, various forces to be simulated are modeled in the system. These forces may arise due to relative positioning of objects (e.g., gravity, collisions), object velocity (e.g., viscosity), or the absolute position of objects in user-specified vector fields (e.g., wind). When applied to objects, these forces induce linear and angular accelerations based on the mass of the object (in the linear case) and mass distribution of the object (in the angular case). These accelerations, which are the time derivative of velocities, are integrated over a delta time step to produce changes in object velocities (linear and angular). These velocities, in turn integrated over a delta time step, produce changes in object positions and orientations (Figure 7.6). The new positions and velocities of the objects give rise to new forces, and the process repeats for the next time step.

The free flight of objects through space is a simple type of rigid body simulation. The simulation of rigid body physics becomes more complex as objects collide, roll and slide over one another, and come to rest in complex arrangements. In addition, a persistent issue in rigid body simulation, as with most animation, is the modeling of a continuous process (such as physics) with discrete time steps. The trade-off of accuracy for computational efficiency is an ever-present consideration.

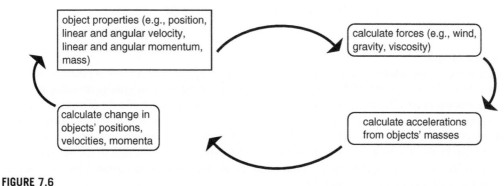

FIGURE 7.6

Rigid body simulation update cycle.

It should be noted that the difference between material commonly taught in standard physics texts and that used in computer animation is in how the equations of motion are used. In standard physics, the emphasis is usually in analyzing the equations of motion for times at which significant events happen, such as an object hitting the ground. In computer animation, the concern is with modeling the motion of objects at discrete time steps [12] as well as the significant events and their aftermath. While the fundamental principles and basic equations are the same, the discrete time sampling creates numerical issues that must be dealt with carefully. See Appendix B.7 for equations of motion from basic physics.

7.4.1 Bodies in free fall

To understand the basics in modeling physically based motion, the motion of a point in space will be considered first. The position of the point at discrete time steps is desired, where the interval between these time steps is some uniform Δt. To update the position of the point over time, its position, velocity, and acceleration are used, which are modeled as functions of time, $x(t)$, $v(t)$, $a(t)$, respectively.

If there are no forces applied to a point, then a point's acceleration is zero and its velocity remains constant (possibly non-zero). In the absence of acceleration, the point's position, $x(t)$, is updated by its velocity, $v(t)$, as in Equation 7.17. A point's velocity, $v(t)$, is updated by its acceleration, $a(t)$. Acceleration arises from forces applied to an object over time. To simplify the computation, a point's acceleration is usually assumed to be constant over the time period Δt (see Eq. 7.18). A point's position is updated by the average velocity during the time period Δt. Under the constant-acceleration assumption, the average velocity during a time period is the average of its beginning velocity and ending velocity, as in Equation 7.19. By substituting Equation 7.18 into Equation 7.19, one defines the updated position in terms of the starting position, velocity, and acceleration (Eq. 7.20).

$$x(t + \Delta t) = x(t) + v(t)\Delta t \tag{7.17}$$

$$v(t + \Delta t) = v(t) + a(t)\Delta t \tag{7.18}$$

$$x(t + \Delta t) = x(t) + \frac{v(t) + v(t + \Delta t)}{2}\Delta t \tag{7.19}$$

$$x(t + \Delta t) = x(t) + \frac{v(t) + v(t) + a(t) * \Delta t}{2}\Delta t$$

$$\tag{7.20}$$

$$x(t + \Delta t) = x(t) + v(t) * \Delta t + \frac{1}{2}a(t) * \Delta t^2$$

A simple example

Using a standard two-dimensional physics example, consider a point with an initial position of $(0, 0)$, with an initial velocity of $(100, 100)$ feet per second, and under the force of gravity resulting in a uniform acceleration of $(0, 232)$ feet per second. Assume a delta time interval of 1/30 of a second (corresponding roughly to the frame interval in the National Television Standards Committee video). In this example, the acceleration is uniformly applied throughout the sequence, and the velocity is modified at each time step by the downward acceleration. For each time interval, the average of the beginning and ending velocities is used to update the position of the point. This process is repeated for each step in time (see Eq. 7.21 and Figures 7.7 and 7.8).

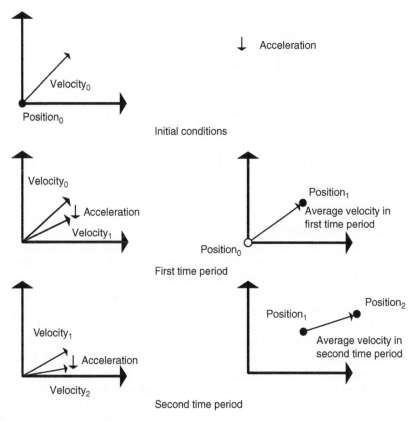

FIGURE 7.7

Modeling of a point's position at discrete time intervals (vector lengths are for illustrative purposes and are not accurate).

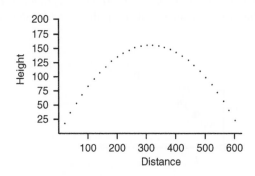

FIGURE 7.8

Path of a particle in the simple example from the text.

$$v(0) = (100, 100)$$
$$x(0) = (0, 0)$$

$$v\left(\frac{1}{30}\right) = (100, 100) + (0, -32)\frac{1}{30}$$

$$x\left(\frac{1}{30}\right) = (0, 0) + \frac{1}{2}\left(v(0) + v\left(\frac{1}{30}\right)\right)$$

(7.21)

$$v\left(\frac{2}{30}\right) = v\left(\frac{1}{30}\right) + (0, -32)\frac{1}{30}$$

etc.

A note about numeric approximation

For an object at or near the earth's surface, the acceleration of the object due to the earth's gravity can be considered constant. However, in many rigid body simulations, the assumption that acceleration remains constant over the delta time step is incorrect. Many forces continually vary as the object changes its position and velocity over time, which means that the acceleration actually varies over the time step and it often varies in nonlinear ways. Sampling the force at the beginning of the time step and using that to determine the acceleration throughout the time step is not the best approach. The multiplication of acceleration at the beginning of the time step by Δt to step to the next function value (velocity) is an example of using the Euler integration method (Figure 7.9). It is important to understand the shortcomings of this approach and the available options for improving the accuracy of the simulation. Many books have been written about the subject; *Numerical Recipes: The Art of Scientific Computing*, by Press et al. [19], is a good place to start. This short section is intended merely to demonstrate that better options exist and that, at the very least, a Runge-Kutta method should be considered.

It is easy to see how the size of the time step affects accuracy (Figure 7.10). By taking time steps that are too large, the numerically approximated path deviates dramatically from the ideal continuous path. Accuracy can be increased by taking smaller steps, but this can prove to be computationally expensive.

Accuracy can also be increased by using better methods of integration. *Runge-Kutta* is a particularly useful one. Figure 7.11 shows the advantage of using the *second-order Runge-Kutta*, or *midpoint*,

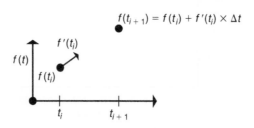

FIGURE 7.9

Euler integration.

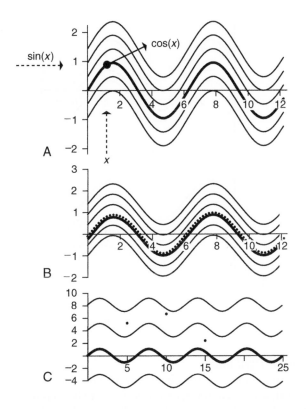

FIGURE 7.10

Approximating the sine curve by stepping in the direction of its derivative. (a) In this example, the sine function is the underlying (unknown) function. The objective is to reconstruct it based on an initial point and knowledge about the derivative function (cosine). Start out at any (x, y) location and, if the reconstruction is completely accurate, the sinusoidal curve that passes through that point should be followed. In (b) and (c), the initial point is (0, 0) so the thicker sine curve should be followed. (b) Updating the function values by taking small enough steps along the direction indicated by the derivative generates a good approximation to the function. In this example, $\Delta x = 0.2$. (c) However, if the step size becomes too large, then the function reconstructed from the sample points can deviate widely from the underlying function. In this example, $\Delta x = 5$.

method, which uses derivative information from the midpoint of the stepping interval. *Second-order* refers to the magnitude of the error term. Higher order Runge-Kutta methods are more accurate and therefore allow larger time steps to be taken, but they are more computationally expensive per time step. Generally, it is better to use a fourth- or fifth-order Runge-Kutta method.

While computer animation is concerned primarily with visual effects and not numeric accuracy, it is still important to keep an eye on the numerics that underlie any simulation responsible for producing the motion. Visual realism can be compromised if the numeric calculation is allowed to become too sloppy. One should have a basic understanding of the strengths and weaknesses of the numeric techniques used, and, in most cases, employing the Euler method should be done with caution. See Appendix B.8 for more information on numerical integration techniques.

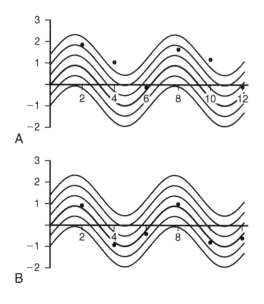

FIGURE 7.11

Euler method and midpoint method. (A) Using step sizes of 2 with the Euler method. (B) Using step sizes of 2 with the midpoint method.

Equations of motion for a rigid body

To develop the equations of motion for a rigid body, several concepts from physics are presented first [12]. The rotational equivalent of linear force, or *torque*, needs to be considered when a force is applied to an object not directly in line with its *center of mass*. To uniquely solve for the resulting motions of interacting objects, *linear momentum* and *angular momentum* have to be conserved. And, finally, to calculate the angular momentum, the distribution of an object's mass in space must be characterized by its *inertia tensor*. These concepts are discussed in the following sections and are followed by the equations of motion.

Orientation and rotational movement

Similar to linear attributes of position, velocity, and acceleration, three-dimensional objects have rotational attributes of orientation, angular velocity, and angular acceleration as functions of time. If an individual point in space is modeled, such as in a particle system, then its rotational information can be ignored. Otherwise, the physical extent of the mass of an object needs to be taken into consideration in realistic physical simulations.

For current purposes, consider an object's orientation to be represented by a rotation matrix, $R(t)$. Angular velocity is the rate at which the object is rotating irrespective of its linear velocity. It is represented by a vector, $\omega(t)$. The direction of the vector indicates the orientation of the axis about which the object is rotating; the magnitude of the angular velocity vector gives the speed of the rotation in revolutions per unit of time. For a given number of rotations in a given time period, the angular velocity of an object is the same whether the object is rotating about its own axis or rotating about an axis some distance away. Consider a point that is rotating about an axis that passes through the point and that the rate of rotation is two revolutions per minute. Now consider a second point that is rotating

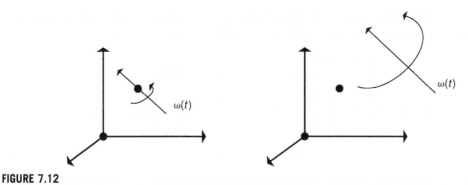

FIGURE 7.12

For a given number of rotations per unit of time, the angular velocity is the same whether the axis of rotation is near or far away.

about an axis that is parallel to the previous axis but is ten miles away, again at the rate of two revolutions per minute. The second point will have the same angular velocity as the first point. However, in the case of the second point, the rotation will also induce an instantaneous linear velocity (which constantly changes). But in both cases the object is still rotating at two revolutions per minute (Figure 7.12).

Consider a point, a, whose position in space is defined relative to a point, $b = x(t)$; a's position relative to b is defined by $r(t)$. The point a is rotating and the axis of rotation passes through the point b (Figure 7.13). The change in $r(t)$ is computed by taking the cross-product of $r(t)$ and $\omega(t)$ (Eq. 7.22). Notice that the change in $r(t)$ is perpendicular to the plane formed by $\omega(t)$ and $r(t)$ and that the magnitude of the change is dependent on the perpendicular distance between $\omega(t)$ and $r(t)$, $|r(t)|\sin\theta$, as well as the magnitude of $\omega(t)$.

$$\dot{r}(t) = \omega(t) \times r(t)$$
$$|\dot{r}(t)| = |\omega(t)||r(t)|\sin \tag{7.22}$$

Now consider an object that has an extent (distribution of mass) in space. The orientation of an object, represented by a rotation matrix, can be viewed as a transformed version of the object's local unit coordinate system. As such, its columns can be viewed as vectors defining relative positions in the object. Thus, the change in the rotation matrix can be computed by taking the cross-product of $\omega(t)$ with

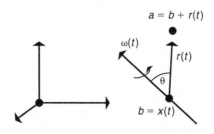

FIGURE 7.13

A point rotating about an axis.

each of the columns of $R(t)$ (Eq. 7.23). By defining a special matrix to represent cross-products (Eq.7.24), one can represent Equation 7.23 by matrix multiplication (Eq. 7.25).

$$R(t) = [R_1(t) \quad R_2(t) \quad R_3(t)]$$
$$\dot{R}(t) = [\omega \times R_1(t) \quad \omega \times R_2(t) \quad \omega \times R_3(t)] \tag{7.23}$$

$$A \times B = \begin{bmatrix} A_y B_z - A_z B_y \\ A_z B_x - A_x B_z \\ A_x B_y - A_y B_x \end{bmatrix} = \begin{bmatrix} 0 & -A_z & A_y \\ A_z & 0 & -A_x \\ -A_y & A_x & 0 \end{bmatrix} \begin{bmatrix} B_x \\ B_y \\ B_z \end{bmatrix} \equiv A^* B \tag{7.24}$$

$$\dot{R}(t) = \omega(t)^* R(t) \tag{7.25}$$

Consider a point, Q, on a rigid object. Its position in the local coordinate space of the object is q; q does not change. Its position in world space, $q(t)$, is given by Equation 7.26. The position of the body in space is given by $x(t)$, and the orientation of the body is given by $R(t)$. The velocity of the particle is given by differentiating Equation 7.26. The relative position of the particle in the rigid body is constant. The change in orientation is given by Equation 7.25, while the change in position is represented by a velocity vector. These are combined to produce Equation 7.27. The world space position of the point Q in the object, taking into consideration the object's orientation, is given by rearranging Equation 7.26 to give $R(t)q = q(t) - x(t)$. Substituting this into Equation 7.27 and distributing the cross-product produces Equation 7.28.

$$q(t) = R(t)q + x(t) \tag{7.26}$$

$$\dot{q}(t) = \omega(t)^* R(t)q + v(t) \tag{7.27}$$

$$\dot{q}(t) = \omega(t) \times (q(t) - x(t)) + v(t) \tag{7.28}$$

Center of mass

The center of mass of an object is defined by the integration of the differential mass times its position in the object. In computer graphics, the mass distribution of an object is typically modeled by individual points, which is usually implemented by assigning a mass value to each of the object's vertices. If the individual masses are given by m_i, then the total mass of the object is represented by Equation 7.29. Using an object coordinate system that is located at the center of mass is often useful when modeling the dynamics of a rigid object. For the current discussion, it is assumed that $x(t)$ is the center of mass of an object. If the location of each mass point in world space is given by $q_i(t)$, then the center of mass is represented by Equation 7.30.

$$M = \Sigma m_i \tag{7.29}$$

$$x(t) = \frac{\Sigma m_i q_i(t)}{M} \tag{7.30}$$

Forces

A linear force (a force along a straight line), f, applied to a mass, m, gives rise to a linear acceleration, $\mathbf{a}$, by means of the relationship shown in Equations 7.31 and 7.32. This fact provides a way to calculate acceleration from the application of forces. Examples of such forces are gravity, viscosity, friction, impulse forces due to collisions, and forces due to spring attachments. See Appendix B.7 for the basic equations from physics that give rise to such forces.

$$F = ma \tag{7.31}$$

$$a = F/m \tag{7.32}$$

The various forces acting on a point can be summed to form the total external force, $F(t)$ (Eq. 7.33). Given the mass of the point, the acceleration due to the total external force can be calculated and then used to modify the velocity of the point. This can be done at each time step. If the point is assumed to be part of a rigid object, then the point's location on the object must be taken into consideration, and the effect of the force on the point has an impact on the object as a whole. The rotational equivalent of linear force is *torque*, $\tau(t)$. The torque that arises from the application of forces acting on a point of an object is given by Equation 7.34.

$$F = \sum f_i(t) \tag{7.33}$$

$$\tau_i = (q_i(t) - x(t)) \times f_i(t)$$
$$\tau = \sum \tau_i(t) \tag{7.34}$$

Momentum

As with force, the momentum (mass $\times$ velocity) of an object is decomposed into a linear component and an angular component. The object's local coordinate system is assumed to be located at its center of mass. The linear component acts on this center of mass, and the angular component is with respect to this center. *Linear momentum* and *angular momentum* need to be updated for interacting objects because these values are conserved in a closed system. Saying that the linear momentum is conserved in a closed system, for example, means that the sum of the linear momentum does not change if there are no outside influences on the system. The case is similar for angular momentum. That they are conserved means they can be used to solve for unknown values in the system, such as linear velocity and angular velocity.

Linear momentum is equal to velocity times mass (Eq. 7.35). The total linear momentum $P(t)$ of a rigid body is the sum of the linear momentums of each particle (Eq. 36). For a coordinate system whose origin coincides with the center of mass, Equation 7.36 simplifies to the mass of the object times its velocity (Eq. 7.37). Further, since the mass of the object remains constant, taking the derivative of momentum with respect to time establishes a relationship between linear momentum and linear force (Eq. 7.38). This states that the force acting on a body is equal to the change in momentum. Interactions composed of equal but opposite forces result in no change in momentum (i.e., momentum is conserved).

$$P = mv \tag{7.35}$$

$$P(t) = \Sigma m_i \dot{q}_i(t) \tag{7.36}$$

$$P(t) = Mv(t) \tag{7.37}$$

$$\dot{P}(t) = M\dot{v}(t) = F(t) \tag{7.38}$$

Angular momentum, L, is a measure of the rotating mass weighted by the distance of the mass from the axis of rotation. For a mass point in an object, the angular momentum is computed by taking the cross-product of a vector to the mass and a velocity vector of that mass point $\times$ the mass of the point. These vectors are relative to the center of mass of the object. The total angular momentum of a rigid body is computed by integrating this equation over the entire object. For the purposes of computer animation, the computation is usually summed over mass points that make up the object (Eq. 7.39). Notice that angular momentum is not directly dependent on the linear components of the object's motion.

$$
\begin{aligned}
L(t) &= \sum((q_i(t) - x(t)) \times m_i(\dot{q}_i(t) - v(t))) \\
&= \sum(R(t)q \times m_i(\omega(t) \times (q(t) - x(t)))) \\
&= \sum(m_i(R(t)q \times (\omega(t) \times R(t)q)))
\end{aligned} \tag{7.39}
$$

In a manner similar to the relation between linear force and the change in linear momentum, torque equals the change in angular momentum (Eq. 7.40). If no torque is acting on an object, then angular momentum is constant. However, the angular velocity of an object does not necessarily remain constant even in the case of no torque. The angular velocity can change if the distribution of mass of an object changes, such as when an ice skater spinning on his skates pulls his arms in toward his body to spin faster. This action brings the mass of the skater closer to the center of mass. Angular momentum is a function of angular velocity, mass, and the distance the mass is from the center of mass. To maintain a constant angular momentum, the angular velocity must increase if the distance of the mass decreases.

$$
\dot{L}(t) = \tau(t) \tag{7.40}
$$

Inertia tensor

Angular momentum is related to angular velocity in much the same way that linear momentum is related to linear velocity, $P(t) = Mv(t)$ (see Eq. 7.41). However, in the case of angular momentum, a matrix is needed to describe the distribution of mass of the object in space, the *inertia tensor*, $I(t)$. The inertia tensor is a symmetric 3×3 matrix. The initial inertia tensor defined for the untransformed object is denoted as I_{object} (Eq. 7.42). Terms of the matrix are calculated by integrating over the object (e.g., Eq. 7.43). In Equation 7.43, the density of an object point, $q = (q_x, q_y, q_z)$, is $r(q)$. For the discrete case, Equation 7.44 is used. In a center-of-mass-centered object space, the inertia tensor for a transformed object depends on the orientation, $R(t)$, of the object but not on its position in space. Thus, it is dependent on time. It can be transformed according to the transformation of the object by Equation 7.45. Its inverse, which we will need later, is transformed in the same way.

$$
L(t) = I(t)\omega(t) \tag{7.41}
$$

$$
I_{object} = \begin{bmatrix} I_{xx} & I_{xy} & I_{xz} \\ I_{xy} & I_{yy} & I_{yz} \\ I_{xz} & I_{yz} & I_{zz} \end{bmatrix} \tag{7.42}
$$

$$
I_{xx} = \iiint (\rho(q)(q_y^2 + q_z^2)) dx dy dz \tag{7.43}
$$

$$
\begin{aligned}
I_{xx} &= \sum m_i(y_i^2 + z_i^2) & I_{xy} &= \sum m_i x_i y_i \\
I_{yy} &= \sum m_i(x_i^2 + z_i^2) & I_{xz} &= \sum m_i x_i z_i \\
I_{zz} &= \sum m_i(x_i^2 + y_i^2) & I_{yz} &= \sum m_i x_i z_i
\end{aligned} \tag{7.44}
$$

$$
I(t) = R(t) I_{object} R(t)^T \tag{7.45}
$$

The equations

The state of an object can be kept in a heterogeneous structure called the *state vector*, $S(t)$, consisting of its position, orientation, linear momentum, and angular momentum (Eq. 7.46). Object attributes, which do not change over time, include its mass, M, and its object-space inertia tensor, I_{object}. At any time, an object's time-varying inertia tensor, angular velocity, and linear velocity can be computed (Eqs. 7.47–7.49). The time derivative of the object's state vector can now be formed (Eq. 7.50).

$$S(t) = \begin{bmatrix} x(t) \\ R(t) \\ P(t) \\ L(t) \end{bmatrix} \tag{7.46}$$

$$I(t) = R(t)I_{object}R(t)^T \tag{7.47}$$

$$\omega(t) = I(t)^{-1}L(t) \tag{7.48}$$

$$v(t) = \frac{P(t)}{M} \tag{7.49}$$

$$\frac{d}{dt}S(t) = \frac{d}{dt}\begin{bmatrix} x(t) \\ R(t) \\ P(t) \\ L(t) \end{bmatrix} = \begin{bmatrix} v(t) \\ \omega(t)^*R(t) \\ F(t) \\ \tau(t) \end{bmatrix} \tag{7.50}$$

This is enough information to run a simulation. Once the ability to compute the derivative information is available, then a differential equation solver can be used to update the state vector. In the simplest implementation, Euler's method can be used to update the values of the state array. The values of the state vector are updated by multiplying their time derivatives by the length of the time step. In practice, Runge-Kutta methods (especially fourth-order) have found popularity because of their trade-off in speed, accuracy, and ease of implementation.

Care must be taken in updating the orientation of an object. Because the derivative information from earlier is only valid instantaneously, if it is used to update the orientation rotation matrix, then the columns of this matrix can quickly become nonorthogonal and not of unit length. At the very least, the column lengths should be renormalized after being updated. A better method is to update the orientation matrix by applying to its columns the axis-angle rotation implied by the angular velocity vector, $\omega(t)$. The magnitude of the angular velocity vector represents the angle of rotation about the axis along the angular velocity vector (see Appendix B.3.3 for the axis-angle calculation). Alternatively, quaternions can be used to represent both the object's orientation and the orientation derivative, and its use here is functionally equivalent to the axis-angle approach.

7.4.2 Bodies in collision

When an object starts to move in any kind of environment other than a complete void, chances are that sooner or later it will bump into something. If nothing is done about this in a computer animation, the object will penetrate and then pass through other objects. Other types of contact include objects sliding against and resting on each other. All of these types of contact require the calculation of forces in order to accurately simulate the reaction of one object to another.

Colliding bodies

As objects move relative to one another, there are two issues that must be addressed: (1) detecting the occurrence of collision and (2) computing the appropriate response to those collisions. The former is strictly a kinematic issue in that it has to do with the positions and orientations of objects and how they change over time. The latter is usually a dynamic issue in that forces that are a result of the collision are computed and used to produce new motions for the objects involved.

Collision detection considers the movement of one object relative to another. In its most basic form, testing for a collision amounts to determining whether there is intersection in the static position of two objects at a specific instance in time. In a more sophisticated form, the movement of one object relative to the other object during a finite time interval is tested for overlap. These computations can become quite involved when dealing with complex geometries.

Collision response is a consideration in physically based simulation. The geometric extent of the object is not of concern but rather the distribution of its mass. Localized forces at specific points on the object impart linear and rotational forces onto the other objects involved.

In dealing with the time of collision, there are two options. The first is to proceed as best as one can from this point in time by calculating an appropriate reaction to the current situation by the particle involved in the collision (the penalty method). This option allows penetration of the particle before the collision reaction takes place. Of course, if the particle is moving rapidly, this penetration might be visually significant. If multiple collisions occur in a time interval, they are treated as occurring simultaneously even though handling them in their correct sequential order may have produced different results. While more inaccurate than the second option, this is simpler to implement and often gives acceptable results.

The second option is to back up time t_i to the first instant that a collision occurred and determine the appropriate response at the time of collision. If multiple collisions occur in a time interval, then time is backed up to the point at which the first collision took place. In complex environments in which collisions happen at a high rate, this constant backing up of time and recomputing the motion of objects can become quite time-consuming.

There are three common options for collision response: a strictly kinematic response, the penalty method, and the calculation of an impulse force. The kinematic response is quick and easy. It produces good visual results for particles and spherically shaped objects. The penalty method introduces a temporary, nonphysically based force in order to restore nonpenetration. It is typically used when response to the collision occurs at the time step when penetration is detected (as opposed to backing up time). The advantage of this technique is that it is easy to compute and the force is easily incorporated into the computational mechanism that simulates rigid body movement. Calculating the impulse force is a more precise way of introducing a force into the system and is typically used when time is backed up to the point of first contact. Detecting collisions and reacting to them are discussed next.

Particle-plane collision and kinematic response

One of the simplest illustrative situations to consider for collision detection and response is that of a particle traveling at some velocity toward a stationary plane at an arbitrary angle (see Figure 7.14). The task is to detect when the particle collides with the plane and have it bounce off the plane. Because a simple plane is involved, its planar equation can be used (Eq. 7.51). $E(p)$ is the planar

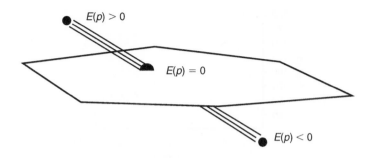

FIGURE 7.14

Point-plane collision.

function, a, b, c, d are coefficients of the planar equation, and p is, for example, a particle's position in space that has no physical extent of its own. Points on the plane satisfy the planar equation, that is, $E(p) = 0$. Assuming for now that the plane represents the ground, the planar equation can be formed so that for points above the plane the planar equation evaluates to a positive value, $E(p) > 0$; for points below the plane, $E(p) < 0$.

$$E(\mathrm{p}) = ap_x + bp_y + cp_z + d = 0 \qquad (7.51)$$

The particle travels toward the plane as its position is updated according to its average velocity over the time interval, as in Equation 7.52. At each time step t_i, the particle is tested to see if it is still above the plane, $E(p(t_i)) > 0$. As long as this evaluates to a positive value, there is no collision. The first time t at which $E(p(t_i)) < 0$ indicates that the particle has collided with the plane at some time between t_{i-21} and t_i. What to do now? The collision has already occurred and something has to be done about it.

$$\mathrm{p}(t_i) = \mathrm{p}(t_{i+1}) + \mathrm{v}_{\mathrm{ave}}(t)\Delta t \qquad (7.52)$$

In simple kinematic response, when penetration is detected, the velocity of the particle is decomposed, using the normal to the plane (N), into the normal component and the perpendicular-to-the-normal component (Figure 7.15). The normal component of the velocity vector is negated by subtracting it out of the original velocity vector and then subtracting it out again. To reduce the height of each successive bounce, a damping factor, $0 < k < 1$, can be applied when subtracting it out the second time (Eq. 7.53). This bounces the particle off a surface at a reduced velocity This approach

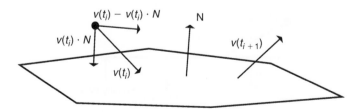

FIGURE 7.15

Kinematic solution for collision reaction.

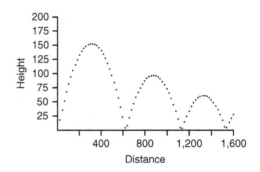

FIGURE 7.16

Kinematic response to collisions with ground using 0.8 as the damping.

is not physically based, but it produces reasonable visuals, especially for particles and spherically shaped objects. Incorporating kinematic response to collisions with the ground produces Figure 7.16.

$$
\begin{aligned}
\mathbf{v}(t_{t+1}) &= \mathbf{v}(t_i) - (\mathbf{v}(t_i) \cdot \mathbf{N})\mathbf{N} - k(\mathbf{v}(t_i) \cdot \mathbf{N})\mathbf{N} \\
&= \mathbf{v}(t_i) - (1+k)(\mathbf{v}(t_i) \cdot \mathbf{N})\mathbf{N}
\end{aligned}
\tag{7.53}
$$

The penalty method

When objects penetrate due to temporal sampling, a simple method of constructing a reaction to the implied collision is the *penalty method*. As the name suggests, a point is penalized for penetrating another object. In this case, a spring, with a zero rest length, is momentarily attached from the offending point to the surface it penetrated in such a way that it imparts a restoring force on the offending point. For now, it is assumed that the surface it penetrates is immovable and thus does not have to be considered as far as collision response is concerned. The closest point on the penetrated surface to the penetrating point is used as the point of attachment (see Figure 7.17). The spring, therefore, imparts a force on the point in the direction of the penetrated surface normal and with a magnitude according to Hooke's law ($F = -kd$). A mass assigned to the point is used to compute a resultant acceleration ($a = F/m$), which contributes an upward velocity to the point. When this upward velocity is combined with the point's original motion, the point's downward motion will be stopped and reversed by the spring, while any component of its motion tangent to the surface will be unaffected. While easy to implement, this approach is not ideal. An arbitrary mass (m) must be assigned to the point, and an arbitrary constant (k) must be determined for the spring. It is difficult to control because if the spring

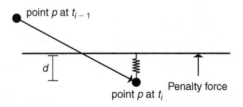

FIGURE 7.17

Penalty spring.

constant is too weak or if the mass is too large, then the collision will not be corrected immediately. If the spring constant is too strong or the mass is too small, then the colliding surfaces will be thrown apart in an unrealistic manner. If the point is moving fast relative to the surface it penetrated, then it may take a few time steps for the spring to take effect and restore a state of nonpenetration. If the desired velocity after the next time step is known, then this can be used help determine the spring constants.

Using the previous example and implementing the penalty method produces the motion traced in Figure 7.18. In this implementation, a temporary spring is introduced into the system whenever both the y component of the position and the y component of the velocity vector are negative; a spring constant of 250 and a point mass of 10 are used. While the spring force is easy to incorporate with other system forces such as gravity, the penalty method requires the user to specify control parameters that typically require a trial-and-error process to determine appropriate values because of other forces that may be present in the system.

When used with polyhedra objects, the penalty force can give rise to torque when not acting in line with the center of mass of an object. When two polyhedra collide, the spring is attached to both objects and imparts an equal but opposite force on the two to restore nonpenetration. Detecting collisions among polyhedra is discussed next, followed by a more accurate method of computing the impulse forces due to such collisions.

Testing polyhedra

In environments in which objects are modeled as polyhedra, at each time step each polyhedron is tested for possible penetration against every other polyhedron. A collision is implied whenever the environment transitions from a nonpenetration state to a state in which a penetration is detected. A test for penetration at each time step can miss some collisions because of the discrete temporal sampling, but it is sufficient for many applications.

Various tests can be used to determine whether an overlap condition exists between polyhedra, and the tests should be chosen according to computational efficiency and generality. Bounding box tests, also known as min-max tests, can be used to quickly determine if there is any chance for an intersection. A bounding box can easily be constructed by searching for the minimum and maximum values in x, y, and z. Bounding boxes can be tested for overlap. If there is no overlap of the bounding boxes, then there can be no overlap of the objects. If the object's shape does not match a rectangle well, then a bounding

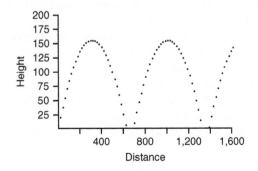

FIGURE 7.18

Penalty method with a spring constant of 250 and a point mass of 10, for example, from Section 7.4.2.

sphere or bounding slabs (pairs of parallel planes at various orientations that bound the object) can be used. A hierarchy of bounding shapes can be used on object groups, individual objects, and individual faces. If these bounding tests fail to conclusively establish that a nonoverlap situation exists, then more elaborate tests must be conducted.

Most collisions can be detected by testing each of the vertices of one object to see if any are inside another polyhedron and then testing each of the vertices of the other object to see if any are inside the first. To test whether a point is inside a convex polyhedron, each of the planar equations for the faces of the object is evaluated using the point's coordinates. If the planes are oriented so that a point to the outside of the object evaluates to a positive value and a point to the inside of the object evaluates to a negative value, then the point under consideration has to evaluate to a negative value in all of the planar equations in order to be declared inside the polyhedron.

Testing a point to determine if it is inside a concave polyhedron is more difficult. One way to do it is to construct a semi-infinite ray emanating from the point under consideration in any particular direction. For example, $y = P_y$, $z = P_z$, $x > P_x$ defines a line parallel to the x-axis going in the positive direction from the point P. This semi-infinite ray is used to test for intersection with all of the faces of the object, and the intersections are counted; an odd number of intersections means that the point is inside the object, and an even number means that the point is outside. To intersect a ray with a face, the ray is intersected with the planar equation of the face and then the point of intersection is tested to see if it is inside the polygonal face.

Care must be taken in correctly counting intersections if this semi-infinite ray intersects the object exactly at an edge or vertex or if the ray is colinear with an edge of the object. (These cases are similar to the situations that occur when scan converting concave polygons.) While this does not occur very often, it should be dealt with robustly. Because the number of intersections can be difficult to resolve in such situations, it can be computationally more efficient to simply choose a semi-infinite ray in a different direction and start over with the test.

One can miss some intersection situations if performing only these tests. The edges of each object must be tested for intersection with the faces of the other object and vice versa to capture other overlap situations. The edge–face intersection test consists of two steps. In the first step, the edge is tested for intersection with the plane of the face (i.e., the edge vertices are on opposite sides of the plane of the face) and the point of intersection is computed if it exists. In the second step, the point of intersection is tested for two-dimensional containment inside the face. The edge–face intersection tests capture almost all of the penetration situations,[2] but because it is the more expensive of the two tests, it is usually better to perform the vertex tests first (see Figure 7.19).

In any of these tests, vertices of one object that lie exactly in the plane of a face from the other object are a special case that must be handled carefully. Usually it is sufficient to logically and consistently push such a vertex to one side of the plane.

Often, a normal that defines the plane of intersection is used in the collision response calculations. For collisions that are detected by the penetration tests, one of two collision situations can usually be established by looking at the relative motion of the objects. Either a vertex has just penetrated a face, or an edge from one object has just intersected an edge from the other object. In either case, the plane of

---

[2]The edge–face intersection test will miss the case when one object is entirely inside of the other; this case is handled by testing a single vertex from each object to see if it is inside the other object.

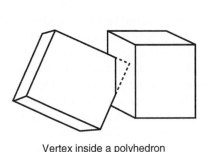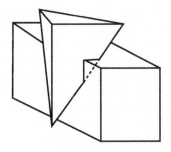

Vertex inside a polyhedron

Object penetration without a vertex
of one object contained in the other

FIGURE 7.19

Detecting polyhedra intersections.

intersection can be established. When a vertex has penetrated a face, a normal to the face is used. When there is an edge–edge intersection, the normal is defined as the cross-product of the two edges.

In certain limited cases, a more precise determination of collision can be performed. For example, if the movement of one object with respect to another is a linear path, then the volume swept by one object can be determined and intersected with the other object whose position is assumed to be static relative to the moving object. If the direction of travel is indicated by a vector, V, then front faces and back faces with respect to the direction of travel can be determined. The *boundary edges* that share a front face and a back face can be identified; *boundary vertices* are the vertices of boundary edges. The volume swept by the object is defined by the original back faces, the translated positions of the front faces, and new faces and edges that are extruded in the direction of travel from boundary edges and boundary vertices (Figure 7.20). This volume is intersected with the static object.

Impulse force of collision

To allow for more accurate handling of the moment of collision, time can be "backed up" to the point of impact, the reaction can be computed, and the time can be moved forward again. While more accurate in its reconstruction of events, this approach can become very computationally intense in complex

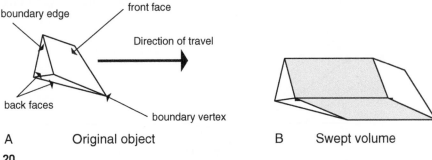

FIGURE 7.20

Sweep volume by translating front-facing faces and extruding new edges from original boundary vertices and extruding new faces from original boundary edges.

environments if many closely spaced collisions occur. However, for a point-plane collision, the impulse can be simplified because the angular velocity can be ignored.

The actual time of collision within a time interval can be searched for by using a binary search strategy. If it is known that a collision occurred between t_{i-1} and t_i, these times are used to initialize the lower ($L = t_{i-1}$) and upper ($U = t_i$) bounds on the time interval to be tested. At each iteration of the search, the point's position is tested at the middle of the bounded interval. If the test indicates that the point has not collided yet, then the lower bound is replaced with the middle of the time interval. Otherwise, the upper bound is replaced with the middle of the time interval. This test repeats with the middle of the new time interval and continues until some desired tolerance is attained. The tolerance can be based either on the length of the time interval or on the range of distances the point could be on either side of the surface.

Alternatively, a linear-path constant-velocity approximation can be used over the time interval to simplify the calculations. The position of the point before penetration and the position after penetration can be used to define a linear path. The point of intersection along the line with a planar surface can be calculated analytically. The relative position of the intersection point along this line can be used to estimate the precise moment of impact (Figure 7.21).

As previously discussed, at the time of impact, the normal component of the point's velocity can be modified as the response to the collision. A scalar, the coefficient of restitution, between zero and one can be applied to the resulting velocity to model the degree of elasticity of the collision (Figure 7.22).

Computing impulse forces

Once a collision is detected and the simulation has been backed up to the point of intersection, the reaction to the collision can be calculated. By working back from the desired change in velocity due to a collision, an equation that computes the required change in momentum can be formed. This equation uses a new term, called *impulse*, expressed in units of momentum. The impulse, J, can be viewed as a large force acting over a short time interval (Eq. 7.54). The change in momentum P, and therefore the change in velocity, can be computed once the impulse is known (Eq. 7.55).

$$J = F\Delta t \tag{7.54}$$

$$J = F\Delta t = Ma\Delta t = M\Delta a = \Delta(Mv) = \Delta P \tag{7.55}$$

To characterize the elasticity of the collision response, the user selects the coefficient of restitution, in the range $0 \le k \le 1$, which relates the relative velocity before the collision, v_{rel}^+ to the relative

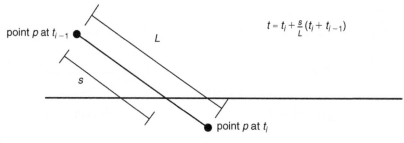

FIGURE 7.21

Linearly estimating time of impact, t.

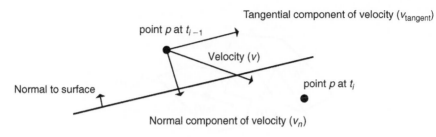

Tangential component of velocity ($v_{tangent}$)

point p at t_{i-1}

Velocity (v)

point p at t_i

Normal to surface

Normal component of velocity (v_n)

Components of a particle's velocity colliding with a plane

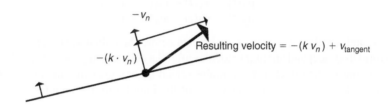

$-v_n$

$-(k \cdot v_n)$

Resulting velocity $= -(k\, v_n) + v_{tangent}$

Computing velocity resulting from a collision (k is the coefficient of restitution)

FIGURE 7.22

Impact response of a point with a plane.

velocity after the collision in the direction normal to the surface of contact, N (Eq. 7.56). Impulse is a vector quantity in the direction of the normal to the surface of contact, $J = jN$. To compute J, the equations of how the velocities must change based on the equations of motion before and after the collision are used. These equations use the magnitude of the impulse, j, and they are used to solve for its value.

$$v_{rel}^{+}(t) = -kv_{rel}^{-}(t) \qquad (7.56)$$

Assume that the collision of two objects, A and B, has been detected at time t. Each object has a position for its center of mass ($x_A(t)$, $x_B(t)$), linear velocity ($v_A(t)$, $v_B(t)$), and angular velocity ($v_A(t)$, $v_B(t)$). The points of collision (p_A, p_B) have been identified on each object (see Figure 7.23).

At the point of intersection, the normal to the surface of contact, N, is determined depending on whether it is a vertex–face contact or an edge–edge contact. The relative positions of the contact points with respect to the center of masses are r_A and r_B, respectively (Eq. 7.57). The relative velocity of the contact points of the two objects in the direction of the normal to the surface is computed by Equation 7.58. The velocities of the points of contact are computed as in Equation 7.59.

$$r_A(t) = p_A(t) - x_A(t)$$
$$r_B(t) = p_B(t) - x_B(t) \qquad (7.57)$$

$$v_{rel}(t) = ((\dot{p}_A(t) - \dot{p}_B(t)) \cdot N)N \qquad (7.58)$$

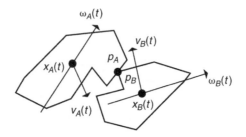

FIGURE 7.23

Configuration of colliding objects.

$$\dot{p}_A(t) = v_A(t) + \omega_A(t) \times r_A(t)$$
$$\dot{p}_B(t) = v_B(t) + \omega_B(t) \times r_B(t) \tag{7.59}$$

The linear and angular velocities of the objects before the collision $(v_A{}^-, w_B{}^-)$ are updated by the impulse to form the linear and angular velocities after the collision $(v_A{}^+, w_B{}^+)$. The linear velocities are updated by adding in the effect of the impulse scaled by one over the mass (Eq. 7.60). The angular velocities are updated by computing the effect of the impulse on the angular velocity of the objects (Eq. 7.61).

$$v_A^+(t) = v_A^-(t) + \frac{jN}{M_A}$$
$$v_B^+(t) = v_B^-(t) + \frac{jN}{M_B} \tag{7.60}$$

$$\omega_A^+(t) = \omega_A^-(t) + I_A^-(r_A(t) \times jN)$$
$$\omega_B^+(t) = \omega_B^-(t) + I_B^-(r_B(t) \times jN) \tag{7.61}$$

To solve for the impulse, form the difference between the velocities of the contact points after collision in the direction of the normal to the surface of contact (Eq. 7.62). The version of Equation 7.59 for velocities after collision is substituted into Equation 7.62. Equations 7.60 and 7.61 are then substituted into that to produce Equation 7.63. Finally, substituting into Equation 7.56 and solving for j produces Equation 7.64.

$$v_A^+(t) = ((\dot{p}_A^+(t) - \dot{p}_B^+(t)) \cdot N)N \tag{7.62}$$

$$v_{rel}^+(t) = N(v_A^+(t) + \omega_A^+(t) \times r_A(t) - (v_B^+(t) + \omega_B^+ \times r_B(t))) \tag{7.63}$$

$$j = \frac{-((1+k)v_{rel}^-)}{\frac{1}{M_A} + \frac{1}{M_B} + N(I_A^{-1}(t)(r_A(t) \times N) \times r_A(t) - I_B^{-1}(t)(r_B(t) \times N) \times r_B(t))} \tag{7.64}$$

Contact between two objects is defined by the point on each object involved in the contact and the normal to the surface of contact. A point of contact is tested to see if an actual collision is taking place. When collision is tested for, the velocity of the contact point on each object is calculated. A collision is detected if the component of the relative velocity of the two points in the direction of the normal indicates the contact points are approaching each other.

```
Compute dpA, dpB                    ;Eq. 7.60
Vrelative <- dot(N,( dpA - dpB)     ;Eq. 7.63
if Vrelative > threshold
            compute j               ;Eq. 7.65
            J <- j*N
            PA <- PA + J                ;update object A linear momentum
            PB <- PB - J                ;update object B linear momentum
            LA <- LA + rA x J           ;update object A angular momentum
            LB <- LB - rB x J           ;update object B angular momentum
else if Vrelative > -threshold
            resting contact
else
            objects are moving away from each other
```

FIGURE 7.24

Computing the impulse force of collision.

If there is a collision, then Equation 7.64 is used to compute the magnitude of the impulse (Figure 7.24). The impulse is used to scale the contact normal, which can then be used to update the linear and angular momenta of each object.

If there is more than one point of contact between two objects, then each must be tested for collision. Each time a collision point is identified, it is processed for updating the momentum as above. If any collision is processed in the list of contact points, then after the momenta have been updated, the contact list must be traversed again to see if there exist any collision points with the new object momenta. Each time one or more collisions are detected and processed in the list, the list must be traversed again; this is repeated until it can be traversed and no collisions are detected.

Friction

An object resting on the surface of another object (the *supporting object*) is referred to as being in a state of *resting contact* with respect to that supporting object. In such a case, any force acting on the first object is decomposed into a component parallel to the contact surface and a component in the direction of the normal of the contact surface, called the *normal force*, F_N (Figure 7.25). If the normal force is

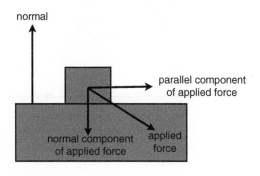

FIGURE 7.25

Normal and parallel components of applied force.

toward the supporting object, and if the supporting object is considered immovable, such as the ground or a table, the normal force is immediately and completely canceled by a force equal and opposite in direction that is supplied by the supporting object. If the supporting object is movable, then the normal force is applied transitively to the supporting object and added to the forces being applied to the supporting object. If the normal force is directed away from the supporting object, then it is simply used to move the object up and away from the supporting object.

The component of the force applied to the object that is parallel to the surface is responsible for sliding (or rolling) the object along the surface of the supporting object. If the object is initially stationary with respect to the supporting object, then there is typically some threshold force, due to *static friction*, that must be exceeded before the object will start to move. This static friction force, F_s, is a function of the normal force, F_N (Eq. 7.65). The constant μ_s is the coefficient of static friction and is a function of the two surfaces in contact.

$$F_s = \mu_s F_N \tag{7.65}$$

Once the object is moving along the surface of the supporting object, there is a *kinetic friction* that works opposite to the direction of travel. This friction creates a force, opposite to the direction of travel, which is a linear function of the normal force, F_k, on the object (Eq. 7.66). The constant μ_k is the coefficient of kinetic friction and is a function of the two surfaces in contact.

$$F_k = \mu_k F_N \tag{7.66}$$

Resting contact

Computing the forces involved in resting contact is one of the more difficult dynamics problems for computer animation. The exposition here follows that found in Baraff's work [1]. The solution requires access to quadratic programming, the implementation of which is beyond the scope of this book. An example situation in which several objects are resting on one another is shown in Figure 7.26.

For each contact point, there is a force normal to the surface of contact, just as in colliding contact. The objective is to find the magnitude of that force for a given configuration of objects. These forces (1)

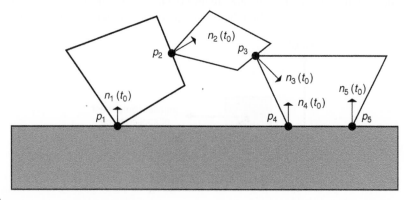

FIGURE 7.26

Multiple-object resting contacts.

have to be strong enough to prevent interpenetration; (2) must only push objects apart, not together; and (3) have to go to zero at the point of contact at the moment that objects begin to separate.

To analyze what is happening at a point of contact, use a distance function, $d_i(t)$, which evaluates to the distance between the objects at the ith point of contact. Assuming objects A and B are involved in the ith contact and the points involved in the ith contact are p_A and p_B from the respective objects, then the distance function is given by Equation 7.67.

$$d_i(t) = (p_A(t) - p_B(t)) \cdot N_i \tag{7.67}$$

If $d_i(t)$ is zero, then the objects involved are still in contact. Whenever $d_i(t) > 0$, the objects are separated. One of the objectives is to avoid $d_i(t) < 0$, which would indicate penetration.

Assume at some time t_0, the distance between objects is zero. To prevent object penetration from time t_0 onward, the relative velocity of the two objects must be greater than or equal to zero, $\dot{d}_i(t_0) \geq 0$ for $t > t_0$. The equation for relative velocity is produced by differentiating Equation 7.67 and is shown in Equation 7.68. At time t_0, the objects are touching, so $p_A(t_0) = p_B(t_0)$. This results in Equation 7.69. In addition, for resting contact, $\dot{d}_i(t_0) = 0$.

$$\dot{d}_i(t) = \dot{N}_i(t) \cdot (p_A(t) - p_B(t)) + N_i \cdot (\dot{p}_A(t) - \dot{p}_B(t)) \tag{7.68}$$

Since $d_i(t_0) = \dot{d}_i(t_0) = 0$, penetration will be avoided if the second derivative is greater than or equal to zero, $\ddot{d}_i(t) \geq 0$. The second derivative is produced by differentiating Equation 7.68 as shown in Equation 7.69. At t_0, remembering that $p_A(t_0) = p_B(t_0)$, one finds that the second derivative simplifies as shown in Equation 7.70. Notice that Equation 7.70 further simplifies if the normal to the surface of contact does not change ($\dot{n}_i(t_0) = 0$).

$$\ddot{d}_i(t) = (p_A(t) - p_B(t)) \cdot \ddot{N}_i + (\dot{p}_A(t) - \dot{p}_B(t)) \cdot \dot{N}_i + (\ddot{p}_A(t) - \ddot{p}_B(t)) \cdot N_i \tag{7.69}$$

$$\ddot{d}_i(t) = 2(\dot{p}_A(t_0) - \dot{p}_B(t_0)) \cdot \dot{N}_i + (\ddot{p}_A(t_0) - \ddot{p}_B(t_0)) \cdot N_i \tag{7.70}$$

The constraints on the forces as itemized at the beginning of this section can now be written as equations: the forces must prevent penetration (Eq. 7.71); the forces must push objects apart, not together (Eq. 7.72); and either the objects are not separating or, if the objects are separating, then the contact force is zero (Eq. 7.73).

$$d_i(t) \geq 0 \tag{7.71}$$

$$f_i \geq 0 \tag{7.72}$$

$$\ddot{d}_i(t)f_i = 0 \tag{7.73}$$

The relative acceleration of the two objects at the ith contact point, $\ddot{d}_i(t_0)$, is written as a linear combination of all of the unknown f_{ij}s (Eq. 7.74). For a given contact, j, its effect on the relative acceleration of the bodies involved in the ith contact point needs to be determined. Referring to Equation 7.70, the component of the relative acceleration that is dependent on the velocities of the points, $2 \cdot \dot{n}_i(t_0)(\dot{p}_A(t_0) - \dot{p}_B(t_0))$, is not dependent on the force f_j and is therefore part of the b_i term. The component of the relative acceleration dependent on the accelerations of the points involved in the contact, $n_i(t_0)(\ddot{p}_A(t_0) - \ddot{p}_B(t_0))$, is dependent on f_j if object A or object B is involved in the jth contact. The acceleration at a point on an object arising from a force can be determined by differentiating Equation 7.59, producing Equation 7.75 where $r_A(t) = p_A(t) - x_A(t)$.

$$\ddot{d}_i(t) = b_i + \sum_{j=1}^{n}(a_{ij}f_j) \tag{7.74}$$

$$\ddot{p}_A(t) = \dot{v}_A + \dot{\omega}_A(t) \times r_A(t) + \omega_A(t) \times (\omega_A(t) \times r_A(t)) \tag{7.75}$$

In Equation 7.75, $\dot{v}_A(t)$ is the linear acceleration as a result of the total force acting on object A divided by its mass. A force f_j acting in direction $n_j(t_0)$ produces $f_j /m_A \cdot n_j(t_0)$. Angular acceleration, $\dot{\omega}_A(t)$, is formed by differentiating Equation 7.48 and produces Equation 7.76, in which the first term contains torque and therefore relates the force f_j to $\ddot{p}_A(t)$ while the second term is independent of the force f_j and so is incorporated into a constant term, b_i.

$$\ddot{\omega}_A(t) = I_A^{-1}(t)\tau_A(t) + I^{-1}(t)(L_A(t) \times \omega_A(t)) \tag{7.76}$$

The torque from force f_j is $(p_j - x_A(t_0)) \times f_j \, n_j(t_0)$. The total dependence of $\ddot{p}_A(t)$ on f_j is shown in Equation 7.77. Similarly, the dependence of $\ddot{p}_B(t)$ on f_j can be computed. The results are combined as indicated by Equation 7.70 to form the term a_{ij} as it appears in Equation 7.74.

$$f_j\left(\frac{N_j(t_0)}{m_A} + (I_A^{-1}(t_0)(p_j - x_A(t_0) \times N_j(t_0)) \times r_A\right) \tag{7.77}$$

Collecting the terms not dependent on an f_j and incorporating a term based on the net external force, $F_A(t_0)$, and net external torque, $\tau_A(t_0)$ acting on A, the part of $\ddot{p}_A(t)$ independent of the f_j is shown in Equation 7.78. A similar expression is generated for $\ddot{p}_B(t)$. To compute b_i in Equation 7.74, the constant parts of $\ddot{p}_A(t)$ and $\ddot{p}_B(t)$ are combined and dotted with $n_i(t_0)$. To this, the term $2\dot{n}_i(t_0)(\dot{p}_A(t_0) - \dot{p}_B(t_0))$ is added.

$$\frac{F_A(t_0)}{m_A} + I_A^{-1}(t)\tau_A(t) + \omega_A(t) \times \left(\omega_A(t) \times r_A\right) + (I_A^{-1}(t)(L_A(t) \times \omega_A(t))) \times r_A \tag{7.78}$$

Equation 7.74 must be solved subject to the constraints in Equations 7.72 and 7.73. These systems of equations are solved by *quadratic programming*. It is nontrivial to implement. Baraff [1] uses a package from Stanford [7] [8] [9] [10]. Baraff notes that the quadratic programming code can easily handle $\ddot{d}_i(t) = 0$ instead of $\ddot{d}_i(t) \geq 0$ in order to constrain two bodies to never separate. This enables the modeling of hinges and pin-joints.

7.4.3 Dynamics of linked hierarchies

Applying forces to a linked figure, such as when falling under the force of gravity or taking a blow to the chest from a heavy object, results in complex reactions as the various links pull and push on each other. There are various ways to deal with the resulting motion depending on the quality of motion desired and the amount of computation tolerated. We consider two approaches here: constrained dynamics and the Featherstone equations.

Constrained dynamics

Geometric point constraints can be used as an alternative to rigid body dynamics [18]. In a simple case, particles are connected together with distance constraints. A mass particle (e.g., representing the torso of a figure) reacts to an applied force resulting in a spatial displacement. The hierarchy is traversed starting at the initially displaced particle. During traversal, each particle is repositioned in order to

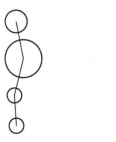

FIGURE 7.27

Frames from a sequence of constrained dynamics. The large sphere's motion is driven by outside forces; the small spheres are repositioned to satisfy distance constraints.

reestablish the distance constraint with respect to a particle already processed. Figure 7.27 shows frames from a simple example.

When an impulse force is applied to the torso of a figure, the torso reacts to the force as an independent rigid body. The appendages react to the force by enforcing distance constraints one link at a time traversing the hierarch from the root outwards as shown in Figure 7.28.

In reality, however, as the torso reacts to the applied force, the appendages exert forces and torques on the torso as it tries to move away from them. The full equations of motion must account for the dynamic interactions among all the links. The *Featherstone equations* do this.

The Featherstone equations

The actual forces of one link acting on adjacent links can be explicitly calculated using basic principles of physics. The effect of one link on any attached links must be computed. The equations presented here are referred to as the Featherstone equations of motion [5] [14] [15]. In a nutshell, the algorithm uses four hierarchy-traversal loops: initializing link velocities (from root outward), initializing link values (from root outward), updating values (from end effector inward), and computing accelerations (from root outward).

The notation to relate one link to the next is shown in Figure 7.29.

A *spatial notation*, which combines the angular and linear components, has been developed to make the computations more concise [5]. The spatial velocity vector is a six-element vector of angular velocity and the linear velocity (Eq. 7.79).

$$\hat{v} = \begin{bmatrix} \omega \\ v \end{bmatrix} \tag{7.79}$$

Similarly, the spatial acceleration vector is made of angular and linear acceleration terms (Eq. 7.80).

FIGURE 7.28

Sequence of impulse force applied to linked figure. Motion is generated by constrained dynamics.

(Image courtesy of Domin Lee.)

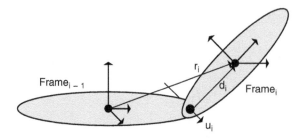

FIGURE 7.29

Vectors relating one coordinate frame to the next: u_i is the axis of revolution associated with *Frame$_i$*, r_i is the displacement vector from the center of *Frame$_{i-1}$* to the center of *Frame$_i$*; d_i is the displacement vector from the axis of revolution to the center of *Frame$_i$*.

$$\hat{a} = \begin{bmatrix} \alpha \\ a \end{bmatrix} \tag{7.80}$$

The force vector (Eq. 7.81) and mass matrix (Eq. 7.82) are defined so that $\hat{f} = \hat{M}\hat{a}$ holds in the spatial notation.

$$\hat{f} = \begin{bmatrix} f \\ \tau \end{bmatrix} \tag{7.81}$$

$$\hat{M} = \begin{bmatrix} 0 & M \\ I & 0 \end{bmatrix} \tag{7.82}$$

The spatial axis is given by combining the axis of rotation of a revolute joint and the cross-product of that axis with the displacement vector from the axis to the center of the frame (Eq. 7.83).

$$\hat{S}_i = \begin{bmatrix} u_i \\ u_i \times d_i \end{bmatrix} \tag{7.83}$$

Using Equation 7.84 to represent the matrix cross-product operator, if r is the translation vector and R is the rotation matrix between coordinate frames, then Equation 7.85 transforms a value from frame F to frame G.

$$r^* = \begin{bmatrix} 0 & -r_z & r_y \\ r_z & 0 & -r_x \\ -r_y & r_x & 0 \end{bmatrix} \tag{7.84}$$

$$G\hat{\times}F = \begin{bmatrix} R & 0 \\ r^*R & R \end{bmatrix} \tag{7.85}$$

Armed with this notation, a procedure that implements the Featherstone equations can be written succinctly [5] [15]. The interested reader is encouraged to refer to Mirtich's Ph.D. dissertation [15] for a more complete discussion.

Computing accurate dynamics is a nontrivial thing to do. However, because of the developed literature from the robotics community and the interest from the computer animation community, understandable presentations of the techniques necessary are available to the mathematically inclined. For real-time applications, approximate methods that produce visually reasonable results are very accessible.

7.5 Cloth

Cloth is a type of object that is commonly encountered in everyday living in a variety of forms— most notably in the form of clothes, but also as curtains, tablecloths, flags, and bags. Because of its ubiquity, it is a type of object worth devoting special attention to modeling and animating. Cloth is any man-made fabric made out of fibers. It can be woven, knotted, or pressed (e.g, felt). The distinguishing attributes of cloth are that it is thin, flexible, and has some limited ability to be stretched. In addition, there are other cloth-like objects, such as paper and plastic bags, that have similar attributes that can be modeled with similar techniques. However, for purposes of this discussion, woven cloth, a frequently encountered type of cloth-like object, will be considered.

Woven cloth, as well as some knotted cloth, is formed by a rectilinear pattern of threads. The two directions of the threads are typically referred to as *warp*, in the longitudinal direction, and *weft*, in the side-to-side direction and orthogonal to the warp direction. The thread patterns of cloth vary by the over-under structure of the threads. For example, in a simple weave, each thread of the warp goes over one weft thread and then under the next. The over-under of adjacent warp rows are offset by one thread. This same over-under pattern holds for the weft threads relative to the warp threads. Other patterns can be used, such as over-over-under and over-over-under-under (Figure 7.30). For more variety, different over-under patterns can be used in the weft and warp directions. Such differences in the weave pattern make the material more or less subject to shear and bend.

Because cloth is thin, it is usually modeled as a single sheet of geometric elements, that is, it is not modeled as having an interior or having any depth. Because it is flexible and distorts easily, cloth of any significant size is usually modeled by a large number of geometric elements typically on the order of hundreds, thousands, or even tens of thousands of elements. The number of elements depends on the desired quality of the resulting imagery and the amount of computation that can be tolerated. Because of the large number of elements, it is difficult for the animator to realistically animate cloth using simple positioning tools, so more automatic, and computationally intensive, approaches are employed. Major cloth features, such as simple draping, can be modeled directly or the underlying processes that give rise to those features can be modeled using physically based methods.

The main visual feature of cloth is simple draping behavior. This draping behavior can be modeled kinematically— just by modeling the fold that is generated between fixed points of the cloth (see the following section). This is possible in simple static cases where the cloth is supported at a few specific points. However, in environments where the cloth is subject to external forces, a more physically based approach is more appropriate. Because cloth can be stretched to some limited degree, a common approach to modeling cloth is the use a mass-spring-damper system. Modifying the mass-spring-damper parameters provides for some degree of modeling various types of cloth material and various types of cloth weave. The spring model, being physically based, also allows the cloth mesh to react to external forces from the environment, such as gravity, wind, and collisions. Real-time systems, such as computer games or interactive garment systems, place a heavy demand on computational efficiency not required of off-line applications and techniques have to be selected with an appropriate quality-cost

FIGURE 7.30

Example warp and weft patterns. Top: over-under. Bottom: over-over-under-under.

(Image courtesy of Di Cao.)

trade-off. At the other extreme, the approach of using finite element methods for modeling can provide very accurate simulation results, but it will not be discussed here.

7.5.1 Direct modeling of folds

With cloth, as with a lot of phenomena, the important superficial features can be modeled directly as opposed to using a more computationally expensive physically based modeling approach. In the case of cloth an important feature is the cloth fold that forms between two supported points. In the special case of a cloth supported at a fixed number of constrained points, the cloth will drape along well-defined lines. Weil [28] presents a geometric method for hanging cloth from a fixed number of points. The cloth is represented as a two-dimensional grid of points located in three-space with a certain number of the grid points constrained to fixed positions.

The procedure takes place in two stages. In the first stage, an approximation is made to the draped surface within the convex hull of the constrained points. The second stage is a relaxation process that continues until the maximum displacement during a given pass falls below a user-defined tolerance.

Vertices on the cloth grid are classified as either *interior* or *exterior*, depending on whether they are inside or outside the convex hull of the constrained points. This inside/outside determination is based on two-dimensional grid coordinates. The first step is to position the vertices that lie along the line between constrained vertices. The curve of a thread hanging between two vertices is called a *catenary curve* and has the form shown in Equation 7.86.

$$y = c + \left(a \cdot \cos h \left(\frac{x - b}{a} \right) \right) \tag{7.86}$$

Catenary curves traced between paired constrained points are generated using vertices of the cloth that lie along the line between constrained vertices. The vertices between the constrained vertices are identified in the same way that a line is drawn between points on a raster display (Figure 7.31).

If two catenary curves cross each other in grid space (Figure 7.32), then the lower of the two curves is simply removed. The reason for this is that the catenary curves essentially support the vertices along the curve. If a vertex is supported by two curves, but one curve supports it at a higher position, then the higher curve takes precedence.

A catenary curve is traced between each pair of constrained vertices. After the lower of two crossing catenary curves is removed, a triangulation is produced in the grid space of the constrained vertices

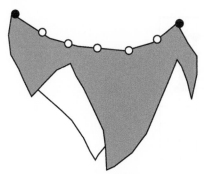

Cloth supported at two constrained points

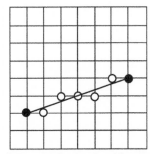

Constrained points in grid space

FIGURE 7.31

Constrained cloth and grid coordinate space.

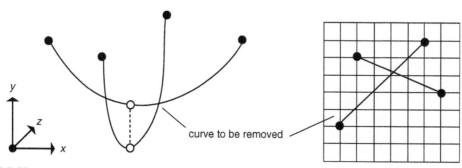

FIGURE 7.32

Two catenary curves supporting the same point.

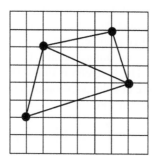

FIGURE 7.33

Triangulation of constrained points in grid coordinates.

(Figure 7.33). The vertices of the grid that fall on the lines of triangulation are positioned in three-space according to the catenary equations. To find the catenary equation between two vertices (x_1, y_1), (x_2, y_2) of length L, see Equation 7.87. Each of these triangles is repeatedly subdivided by constructing a catenary from one of the vertices to the midpoint of the opposite edge on the triangle. This is done for all three vertices of the triangle. The highest of the three catenaries so formed is kept and the others are discarded. This breaks the triangle into two new triangles (Figure 7.34). This continues until all interior vertices have been positioned.

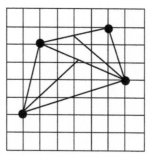

FIGURE 7.34

Subdividing triangles.

$$y_1 = c + a \cdot \cosh\left[\frac{(x_1 - b)}{a}\right]$$

$$y_2 = c + a \cdot \cosh\left[\frac{(x_2 - b)}{a}\right]$$

$$L = a \cdot \sinh\left[\frac{(x_2 - b)}{a}\right] - a \cdot \sinh\left[\frac{(x_1 - b)}{a}\right]$$

$$\sqrt{L^2 - (y_2 - y_1)^2} = 2 \cdot a \cdot \sinh\left[\frac{(x_2 - x_1)}{2 \cdot a}\right] \quad (a \text{ can be solved for numerically at this point})$$

$$M = \sinh\left(\frac{x_2}{a}\right) - \sinh\left(\frac{x_1}{a}\right)$$

$$N = \cosh\left(\frac{x_2}{a}\right) - \cosh\left(\frac{x_1}{a}\right) \tag{7.87}$$

$$\text{if } N > M \quad \mu = \tanh^{-1}\left(\frac{M}{N}\right)$$

$$Q = \frac{M}{\sinh(\mu)} = \frac{N}{\cosh(\mu)}$$

$$b = a \cdot \left[\mu - \sinh^{-1}\left(\frac{L}{Q \cdot a}\right)\right]$$

$$\text{if } M > N \quad \mu = \tanh^{-1}\left(\frac{N}{M}\right)$$

$$Q = \frac{N}{\sinh(\mu)} = \frac{M}{\cosh(\mu)}$$

$$b = a \cdot \left[\mu - \cosh^{-1}\left(\frac{L}{Q \cdot a}\right)\right]$$

A relaxation process is used as the second and final step in positioning the vertices. The effect of gravity is ignored; it has been used implicitly in the formation of the catenary curves using the interior vertices. The exterior vertices are initially positioned to affect a downward pull. The relaxation procedure repositions each vertex to satisfy unit distance from each of its neighbors. For a given vertex, displacement vectors are formed to each of its neighbors. The vectors are added to determine the direction of the repositioning. The magnitude of the repositioning is determined by taking the square root of the average of the squares of the distance to each of the neighbors. Self-intersection of the cloth is ignored. To model material stiffness, the user can add dihedral angle springs to bias the shape of the material.

Animation of the cloth is produced by animating the existence and/or position of constrained points. In this case, the relaxation process can be made more efficient by updating the position of points based on positions in the last time step. Modeling cloth this way, however, is not appropriate for many situations, such as clothing, in which there are a large number of closely packed contact points. In these situations, the formation of catenary curves is not an important aspect, and there is no provision for the wrinkling of material around contact points. Therefore, a more physically based approach is often needed.

7.5.2 Physically based modeling

In many cases, a more useful approach for modeling and animating cloth is to develop a physically based model of it. For example, see Figure 7.35. This allows the cloth to interact and respond to the environment in fairly realistic ways. The cloth weave is naturally modeled as a quadrilateral mesh that mimics the warp and weft of the threads. The characteristics of cloth are its ability to stretch, bend, and skew. These tendencies can be modeled as springs that impart forces or as energy functions that contribute to the overall energy of a given configuration. Forces are used to induce accelerations on system masses. The energy functions can be minimized to find optimal configurations or differentiated to find local force gradients. Whether forces or energies are used, the functional forms are similar and are usually based on the amount some metric deviates from a rest value.

Stretching is usually modeled by simply measuring the amount that the length of an edge deviates from its rest length. Equation 7.88 shows a commonly used form for the force equation of the corresponding spring. $v1*$ and $v2*$ are the rest positions of the vertices defining the edge; $v1$ and $v2$ are the current positions of the edge vertices. Notice that the force is a vector equation, whereas the energy equation is a scalar. The analogous energy function is given in Equation 7.89. The metric has been normalized by the rest length ($|v1* - v2*|$).

FIGURE 7.35

Cloth example.

(Image courtesy of Di Cao).

$$F_s = \left(\frac{k_s|v1 - v2| - |v1^* - v2^*|}{|v1^* - v2^*|}\right)\frac{v1 - v2}{|v1 - v2|} \tag{7.88}$$

$$E_s = k_s\frac{1}{2}\left(\frac{|v1 - v2| - |v1^* - v2^*|}{|v1^* - v2^*|}\right)^2 \tag{7.89}$$

Restricting the stretching of edges only controls changes to the surface area of the mesh. Skew is in-plane distortion of the mesh, which still maintains the length of the original edges (Figure 7.36a,b). To control such distortion (when using forces), one may employ diagonal springs (Figure 7.36c). The energy function suggested by DeRose, Kass, and Truong [29] to control skew distortion is given in Equation 7.90.

$$S(v1, v2) = \left(\frac{1}{2}\right)\left(\frac{|v1 - v2| - |v1^* - v2^*|}{|v1^* - v2^*|}\right)^2 \tag{7.90}$$
$$E_w = k_w \cdot S(v1, v3)\, S(v2, v4)$$

Edge and diagonal springs (energy functions) control in-plane distortions, but out-of-plane distortions are still possible. These include the bending and folding of the mesh along an edge that does not

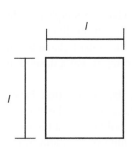

A   Original quadrilateral of mesh

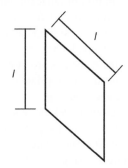

B   Skew of original quadrilateral without changing the length of edges

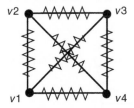

C   Diagonal springs to control skew

FIGURE 7.36

Original quadrilateral, skewed quadrilateral, and controlling skew with diagonal springs.

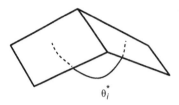 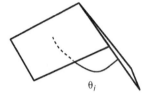

θ_i^* θ_i

Original dihedral angle   Bending along the edge that
                                            changes dihedral angle

FIGURE 7.37

Control of bending by dihedral angle.

change the length of the edges or diagonal measurements. Bending can be controlled by either restricting the dihedral angles (the angle between adjacent quadrilaterals) or controlling the separation among a number of adjacent vertices. A spring-type equation based on deviation from the rest angle can be used to control the dihedral angle: $F_b = k_b?(\theta_i - \theta_i^*)$ (Figure 7.37). Bending can also be controlled by considering the separation of adjacent vertices in the direction of the warp and weft of the material (Figure 7.38). See Equation 7.91 for an example.

The mass-spring-damper system can then be updated according to basic principles of physics and the cloth model can respond when subjected to external forces. Whether spring-like forces or energy functions are used, the constants are globally and locally manipulated to impart characteristics to the mesh. However, choosing appropriate constants can be an issue in trying to get a desired amount of stretching, bending, and shearing.

For an effective cloth model, there are several issues that deserve special attention. First, the integration method used to update the physical model has an effect on the accuracy of the model as well as on the computational efficiency. The computational efficiency is especially important in real-time systems. Second, a simple spring model can allow for unrealistic stretching of the cloth, referred to as *super-elasticity*. Cloth models typically use something other than the basic linear spring. Third, collision detection can be a large part of the computational effort in animating cloth. This is especially true for clothes where the cloth is in near-constant contact with the underlying figure geometry. Fourth, collision response has to be handled appropriately for the inelastic properties of cloth. Cloth does not bounce away from collisions as much as it distorts, often taking the shape of the underlying surface. Fifth, unlike a spring, cloth does not really compress. Instead, cloth bends, forming folds and wrinkles.

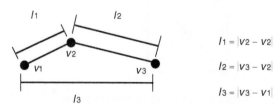

$l_1 = |v_2 - v_2|$

$l_2 = |v_3 - v_2|$

$l_3 = |v_3 - v_1|$

FIGURE 7.38

Control of bending by separation of adjacent vertices.

Special attention to bending can make a cloth simulation more efficient and more realistic. These issues are discussed in the following paragraphs.

Typical simulations of cloth use explicit, or forward, Euler integration, especially for real-time applications because it is fast to execute and easy to implement. But because of the stiffness encountered in cloth simulations, which explicit Euler integration does not handle well, other integration schemes have been used, especially for off-line applications. Implicit Euler is used but, because it is iterative, it tends to be too computationally expensive for real-time or interactive applications. However, it does provide a more stable simulation for the stiffer models. Semi-implicit Euler integration has been used with success. This integration technique is easy to implement, is fast to execute, and handles stiffness fairly well. See Appendix B.8 for details.

In modeling cloth, controlling stretching is a primary concern. When cloth is pulled, there is typically some initial stretching that occurs as the weave straightens out, but the cloth soon becomes very stiff and resistant to further stretching. If the cloth is modeled using a basic mass-spring-damper system, the pulled cloth will not become stiff and will continue to stretch in unrealistic ways because of the linear response inherent in the spring-damper equations. This is referred to as super-elasticity. This pulling can simply be the result of gravity as it pulls down on all the mass points of the cloth model when the cloth is suspended along one edge. To prevent this super-elasticity of the cloth model, either very stiff springs must be used, requiring very small time steps and limiting the amount of initial stretching, or some nonlinear effects must be captured by the model. Approaches to prevent super-elasticity include biphasic springs and velocity clamping. Biphasic springs are springs that respond differently to large displacements. Under small displacements, they allow the initial stretching, but at a certain displacement, they become very stiff. This still requires small time steps at large displacements. Velocity clamping limits the velocity of the mass points due to spring response and thus is a kinematic solution to controlling vertex displacement.

One of the major activities of cloth animation is to detect and respond to collisions. Especially in real-time environments, the efficient detection and response of collisions is a primary concern where computational shortcuts are often worth sacrificing accuracy. There are two types of collision: self-collision and collision with the environment. These two can be handled separately because knowledge about the cloth configuration can help in processing self-collisions. Collisions with cloth are handled much like collisions with elements of any complex environment; that is, they must be handled efficiently. One of the most common techniques for handling collisions in a complex environment is to organize data in a hierarchy of bounding boxes. As suggested by DeRose, Kass, and Truong [29], such a hierarchy can be built from the bottom up by first forming bounding boxes for each of the faces of the cloth. Each face is then logically merged with an adjacent face, forming the combined bounding box and establishing a node one level up in the hierarchy. This can be done repeatedly to form a complete hierarchy of bounding boxes for the cloth faces. Care should be taken to balance the resulting hierarchy by making sure all faces are used when forming nodes at each level of the hierarchy. Because the cloth is not a rigid structure, additional information can be included in the data structure to facilitate updating the bounding boxes each time a vertex is moved. Hierarchies can be formed for rigid and nonrigid structures in the environment. To test a vertex for possible collision with geometric elements, compare it with all the object-bounding box hierarchies, including the hierarchy of the object to which it belongs. Another approach takes advantage of the graphics process unit's on-board z-buffer to detect an underlying surface that shows through the cloth and, therefore, indicates a collision [30].

Cloth response to collisions does not follow the rigid body paradigm. The collision can be viewed as elastic in that the cloth will change shape as a result. It is really a damped local inelastic collision, but rigid body computations are usually too expensive, especially for real-time applications, and even for off-line simulations. The response to collision is typically to constrain the position and/or velocity of the colliding vertices. This will constrain the motion of the colliding cloth to be on the surface of the body. This avoids treating periods of extended contact with a collision detection–collision response cycle.

Folds are a particular visual feature of cloth that are important to handle in a believable way. In the case of spring-damper models of cloth, there is typically no compression because the cloth easily bends, producing folds and wrinkles. It is typical to incorporate some kind of bending spring in the cloth model. Choi and Koh [31] have produced impressive cloth animation by special handling of what they call the post-buckling instability together with the use of implicit numerical integration. A key feature of their approach is the use of a different energy function for stretching than for compression.

$$F_b = k_b \left(\frac{l3}{(l1 + l2)} - \frac{l3^*}{(l1^* + l2^*)} \right) \tag{7.91}$$

7.6 Enforcing soft and hard constraints

One of the main problems with using physically based animation is for the animator to get the object to do what he or she wants while having it react to the forces modeled in the environment. One way to solve this is to place constraints on the object that restrict some subset of the degrees of freedom (DOF) of the object's motion. The remaining DOF are subject to the physically based animation system. Constraints are simply requirements placed on the object. For animation, constraints can take the form of colocating points, maintaining a minimum distance between objects, or requiring an object to have a certain orientation in space. The problem for the animation system is in enforcing the constraints while the object's motion is controlled by some other method.

If the constraints are to be strictly enforced, then they are referred to as *hard constraints*. If the constraints are relations the system should only attempt to satisfy, then they are referred to as *soft constraints*. Satisfying hard constraints requires more sophisticated numerical approaches than satisfying soft constraints. To satisfy hard constraints, computations are made that search for a motion that reacts to forces in the system while satisfying all of the constraints. As more constraints are added to the system, this becomes an increasingly difficult problem. Soft constraints are typically incorporated into a system as additional forces that influence the final motion. One way to model flexible objects is to create soft distance constraints between the vertices of an object. These constraints, modeled by a mesh of interconnected springs and dampers, react to other forces in the system such as gravity and collisions to create a dynamic structure. Soft constraint satisfaction can also be phrased as an energy minimization problem in which deviation from the constraints increases the system's energy. Forces are introduced into the system that decrease the system's energy. These approaches are discussed in the following section.

7.6.1 Energy minimization

The concept of a system's energy can be used in a variety of ways to control the motion of objects. Used in this sense, energy is not defined solely as physically realizable, but it is free to be defined in whatever form serves the animator. Energy functions can be used to pin objects together, to restore the shape of objects, or to minimize the curvature in a spline as it interpolates points in space.

As presented by Witkin, Fleischer, and Barr [26], energy constraints induce restoring forces based on arbitrary functions of a model's parameters. The current state of a model can be expressed as a set of parameter values. These are external parameters such as position and orientation as well as internal parameters such as joint angles, the radius of a cylinder, or the threshold of an implicit function. They are any value of the object model subject to change. The set of state parameters will be referred to as ψ.

A constraint is phrased in terms of a non-negative smooth function, $E(\psi)$, of these parameters, and a local minimum of the constraint function is searched for in the space of all parametric values. The local minimum can be searched for by evaluating the gradient of the energy function and stepping in the direction of the negative of the gradient, $-\nabla E$. The gradient of a function is the parameter space vector in the direction of greatest increase of that function. Usually the parametric state set will have quite a few elements, but for illustrative purposes suppose there are only two state parameters. Taking the energy function as the surface of a height function of those two parameters, one can consider the gradient as pointing uphill from any point on that surface (see Figure 7.39).

Given an initial set of parameters, ψ_0, the parameter function at time zero is defined as $F(0) = \psi_0$. From this it follows that $(d/dt)\,F(t) = -\nabla E$. The force vector in parameter space is the negative of the gradient of the energy function, $-\nabla E$. Any of a number of numerical techniques can be used to solve the equation, but a particularly simple one is Euler's method, as shown in Equation 7.92 (see Appendix B.8 or [19] for better methods).

$$F(t_{i+1}) = F(t_i) - h\nabla E \qquad (7.92)$$

Determining the gradient, ∇E, is usually done numerically by stepping along each dimension in parameter space, evaluating the energy function, and taking differences between the new evaluations and the original value.

Three useful functions

As presented by Witkin, Fleischer, and Barr [26], three useful functions in defining an energy function are as follows:

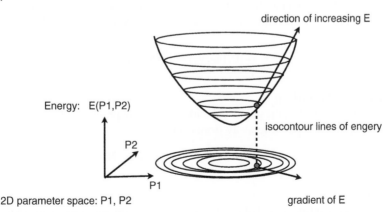

FIGURE 7.39

Sample simple energy function.

1. The parametric position function, $P(u, v)$
2. The surface normal function, $N(u, v)$
3. The implicit function, $I(x)$

The parametric position function and surface normal function are expressed as functions of surface parameters u and v (although for a given model, the surface parameters may be more complex, incorporating identifiers of object parts, for example). Given a value for u and v, the position of that point in space and the normal to the surface at that point are given by the functions. The implicit function takes as its parameter a position in space and evaluates to an approximation of the signed distance to the surface where a point on the surface returns a zero, a point outside the object returns a positive distance to the closest point on the surface, and a point inside the object returns a negative distance to the closest point on the surface. These functions will, of course, also be functions of the current state of the model, ψ.

Useful constraints
The functions above can be used to define several useful constraint functions. The methods of Witkin, Fleischer, and Barr [26] can also lead to several intuitive constraint functions. The functions typically are associated with one or more user-specified weights; these are not shown.

Point-to-fixed-point
The point Q in space is fixed and is given in terms of absolute coordinate values. The point P is a specific point on the surface of the model and is specified by the u, v parameters. The energy function will be zero when these two points coincide.

$$E = |P(u, v) - Q|^2$$

Point-to-point
A point on the surface of one object is given and the point on the surface of a second object is given. The energy function will be zero when they coincide. Notice that within a constant, this is a zero-length spring. Also notice that the orientations of the objects are left unspecified. If this is the only constraint given, then the objects could completely overlap or partially penetrate each other to satisfy this constraint.

$$E = |P^a(u_a, v_a) - P^b(u_b, v_b)|^2$$

Point-to-point locally abutting
The energy function is zero when the points coincide, and the dot product of the normals at those points is equal to -1 (i.e., they are pointing away from each other).

$$E = |P^a(u_a, v_a) - P^b(u_b, v_b)|^2 + N^a(u_a, v_a) \cdot N^b(u_b, v_b) + 1.0$$

Floating attachment
With the use of the implicit function of object b, a specific point on object a is made to lie on the surface of object b.

$$E = (I^b(P^a(u_a, v_a)))^2$$

Floating attachment locally abutting

A point of object a is constrained to lie on the surface of b, using the implicit function of object b as previously mentioned. In addition, the normal of object a and the point must be colinear to and in the opposite direction of the normal of object b at the point of contact. The normal of object b is computed as the gradient of its implicit function.

Other constraints are possible. Witkin, Fleischer, and Barr [26] present several others as well as some examples of animations using this technique.

$$E = (I^b(\mathbf{P}^a(u_a, v_a)))^2 + \mathbf{N}^a(u_a, v_a) \cdot \frac{\nabla I^b(\mathbf{P}^a(u_a, v_a))}{|\nabla I^b(\mathbf{P}^a(u_a, v_a))|} + 1.0$$

Energy constraints are not hard constraints

While such energy functions can be implemented quite easily and can be effectively used to assemble a structure defined by relationships such as the ones discussed previously, a drawback to this approach is that the constraints used are not *hard constraints* in the sense that the constraint imposed on the model is not always met. For example, if a point-to-point constraint is specified and one of the points is moved rapidly away, the other point, as moved by the constraint satisfaction system, will chase around the moving point. If the moving point comes to a halt, then the second point will, after a time, come to rest on top of the first point, thus satisfying the point-to-point constraint.

7.6.2 Space-time constraints

Space-time constraints view motion as a solution to a constrained optimization problem that takes place over time in space. Hard constraints, which include equations of motion as well as nonpenetration constraints and locating the object at certain points in time and space, are placed on the object. An objective function that is to be optimized is stated, for example, to minimize the amount of force required to produce the motion over some time interval.

The material here is taken from Witkin and Kass [27]. Their introductory example will be used to illustrate the basic points of the method. Interested readers are urged to refer to that article as well as follow-up articles on space-time constraints (e.g., [3] [16]).

Space-time particle

Consider a particle's position to be a function of time, $x(t)$. A time-varying force function, $f(t)$, is responsible for moving the particle. Its equation of motion is given in Equation 7.93.

$$m\ddot{x}(t) - f(t) - mg = 0 \tag{7.93}$$

Given the function $f(t)$ and values for the particle's position and velocity, (t_0) and $x'(t_0)$, at some initial time, the position function, $x(t)$, can be obtained by integrating Equation 7.93 to solve the initial value problem.

However, the objective here is to determine the force function, $f(t)$. Initial and final positions for the particle are given as well as the times the particle must be at these positions (Eq. 7.94). These equations along with Equation 7.93 are the constraints on the motion.

$$\begin{aligned} x(t_0) &= a \\ x(t_1) &= b \end{aligned} \tag{7.94}$$

The function to be minimized is the fuel consumption, which here, for simplicity, is given as $|f|^2$. For a given time period $t_0 < t < t_1$, this results in Equation 7.95 as the function to be minimized subject to the time-space constraints and the motion equation constraint.

$$R = \int_{t_0}^{t_1} |f|^2 \, dt \tag{7.95}$$

In solving this, discrete representations of the functions $x(t)$ and $f(t)$ are considered. Time derivatives of x are approximated by finite differences (Eqs. 7.96 and 7.97) and substituted into Equation 7.93 to form n physics constraints (Eq. 7.98) and the two boundary constraints (Eq. 7.99).

$$\dot{x}_i = \frac{x_i - x_{i-1}}{h} \tag{7.96}$$

$$\ddot{x}_i = \frac{x_{i+1} - 2x_i + x_{i-1}}{h^2} \tag{7.97}$$

$$p_i = m \frac{x_{i+1} - 2x_i + x_{i-1}}{h^2} - f - mg = 0 \tag{7.98}$$

$$c_a = |x_1 - a| = 0$$
$$c_b = |x_n - b| = 0 \tag{7.99}$$

If one assumes that the force function is constant between samples, the object function, R, becomes a sum of the discrete values of f. The discrete function R is to be minimized subject to the discrete constraint functions, which are expressed in terms of the sample values, x_i and f_i, that are to be solved for.

Numerical solution

The problem as stated fits into the generic form of constrained optimization problems, which is to "find the S_j values that minimize R subject to $C_i(S_j) = 0$." The S_j values are the x_i and f_i. Solution methods can be considered black boxes that request current values for the S_j values, R, and the C_i values as well as the derivatives of R and the C_i with respect to the S_j as the solver iterates toward a solution.

The solution method used by Witkin and Kass [27] is a variant of sequential quadratic programming (SQP) [10]. This method computes a second-order Newton-Raphson step in R, which is taken irrespective of any constraints on the system. A first-order Newton-Raphson step is computed in the C_i to reduce the constraint functions. The step to minimize the objective function, R, is projected onto the null space of the constraint step, that subspace in which the constraints are constant to a first-order approximation. Therefore, as steps are taken to minimize the constraints, a particular direction for the step is chosen that does not affect the constraint minimization and that reduces the objective function.

Because it is first-order in the constraint functions, the first derivative matrix (the Jacobian) of the constraint function must be computed (Eq. 7.100). Because it is second-order in the objective function, the second derivative matrix (the Hessian) of the objective function must be computed (Eq. 7.101). The first derivative vector of the objective function must also be computed, $\partial R/\partial S_j$.

$$J_{ij} = \frac{\partial C_i}{\partial S_j} \tag{7.100}$$

$$H_{ij} = \frac{\partial^2 R}{\partial S_i \partial S_j} \tag{7.101}$$

The second-order step to minimize R, irrespective of the constraints, is taken by solving the linear system of equations shown in Equation 7.102. A first-order step to drive the C_js to zero is taken by solving the linear system of equations shown in Equation 7.103. The final update is $\Delta S_j = \hat{S}_j + \tilde{S}_j$. The algorithm reaches a fixed point when the constraints are satisfied, and any further step that minimizes R would violate one or more of the constraints.

$$-\frac{\partial}{\partial S_j}(R) = \sum_j H_{ij}\hat{S}_j \tag{7.102}$$

$$-C_i = \sum_j J_{ij}(\hat{S}_j + \tilde{S}_j) \tag{7.103}$$

Although one of the more complex methods presented here, space-time constraints are a powerful and useful tool for producing realistic motion while maintaining given constraints. They are particularly effective for creating the illusion of self-propelled objects whose movement is subject to user-supplied time constraints.

7.7 Chapter summary

If complex realistic motion is needed, the models used to control the motion can become complex and mathematically expensive. However, the resulting motion is oftentimes more natural looking and believable than that produced using kinematic (e.g. interpolation) techniques. While modeling physics is a non-trivial task, oftentimes it is worth the investment. It should be noted that there has been much work in this area that has not been included in this chapter. Some other physically based techniques are covered in later chapters that address specific animation tasks such as the animation of clothes and water.

References

[1] Baraff D. Rigid Body Simulation. In: Course Notes, SIGGRAPH 92. Chicago, Ill.; July 1992.

[2] Blinn J. Light Reflection Functions for Simulation of Clouds and Dusty Surfaces. In: Computer Graphics. Proceedings of SIGGRAPH 82, vol. 16(3). Boston, Mass.; July 1982. p. 21–9.

[3] Cohen M. Interactive Spacetime Control for Animation. In: Catmull EE, editor. Computer Graphics. Proceedings of SIGGRAPH 92, vol. 26(2). Chicago, Ill.; July 1992. p. 293–302. ISBN 0-201-51585-7.

[4] Ebert D, Carlson W, Parent R. Solid Spaces and Inverse Particle Systems for Controlling the Animation of Gases and Fluids. Visual Computer March 1994;10(4):179–90.

[5] Featherstone R, Orin D. Robot Dynamics: Equations and Algorithms.

[6] Frautschi S, Olenick R, Apostol T, Goodstein D. The Mechanical Universe: Mechanics and Heat, Advanced Edition. Cambridge: Cambridge University Press; 1986.

[7] Gill P, Hammarling S, Murray W, Saunders M, Wright M. User's Guide for LSSOL: A Fortran Package for Constrained Linear Least-Squares and Convex Quadratic Programming, Technical Report Sol 84-2. Systems Optimization Laboratory, Department of Operations Research, Stanford University; 1986.

[8] Gill P, Murray W, Saunders M, Wright M. User's Guide for NPSOL: A Fortran Package for Nonlinear Programming, Technical Report Sol 84-2. Systems Optimization Laboratory. Department of Operations Research, Stanford University; 1986.

[9] Gill P, Murray W, Saunders M, Wright M. User's Guide for QPSOL: A Fortran Package for Quadratic Programming, Technical Report Sol 84-6. Systems Optimization Laboratory, Department of Operations Research, Stanford University; 1984.

[10] Gill P, Murray W, Wright M. Practical Optimization. New York: Academic Press; 1981.

[11] Gourret J, Magnenat-Thalmann N, Thalmann D. Simulation of Object and Human Skin Deformations in a Grasping Task. In: Lane J, editor. Computer Graphics. Proceedings of SIGGRAPH 89, vol. 23(3). Boston, Mass.; July 1989. p. 21–30.

[12] Hahn J. Introduction to Issues in Motion Control. In: Tutorial 3, SIGGRAPH 88. Atlanta, Ga.; August 1988.

[13] Haumann D. Modeling the Physical Behavior of Flexible Objects. In: SIGGRAPH 87 Course Notes: Topics in Physically Based Modeling. Anaheim, Calif.; July 1987.

[14] Kokkevis E. Practical Physics for Articulated Characters. In: Game Developers Conference. 2004.

[15] Mirtich B. Impulse-Based Dynamic Simulation of Rigid Body Systems. Ph.D. dissertation. University of California at Berkeley, Fall; 1996.

[16] Ngo J, Marks J. Spacetime Constraints Revisited. In: Kajiya JT, editor. Computer Graphics. Proceedings of SIGGRAPH 93, Annual Conference Series. Anaheim, Calif.; August 1993. p. 343–50. ISBN 0-201-58889-7.

[17] O'Brien J, Hodges J. Graphical Modeling and Animation of Brittle Fracture. In: Rockwood A, editor. Computer Graphics. Proceedings of SIGGRAPH 99, Annual Conference Series. Los Angeles, Calif.: Addison-Wesley Longman; August 1999. p. 137–46. ISBN 0-20148-560-5.

[18] van Overveld C, Barenburg B. All You Need is Force: A Constraint-based Approach for Rigid Body Dynamics in Computer Animation. In: Terzopoulos D, Thalmann D, editors. Proceedings of Computer Animation and Simulation '95. Springer Computer Science. 1995. p. 80–94.

[19] Press W, Flannery B, Teukolsky S, Vetterling W. Numerical Recipes: The Art of Scientific Computing. Cambridge: Cambridge University Press; 1986.

[20] Reeves W. Particle Systems: A Technique for Modeling a Class of Fuzzy Objects. In: Computer Graphics. Proceedings of SIGGRAPH 83, vol. 17(3). Detroit, Mich.; July 1983. p. 359–76.

[21] Reeves W, Blau R. Approximate and Probabilistic Algorithms for Shading and Rendering Structured Particle Systems. In: Barsky BA, editor. Computer Graphics. Proceedings of SIGGRAPH 85, vol. 19(3). San Francisco, Calif.; August 1985. p. 31–40.

[22] Sims K. Particle Animation and Rendering Using Data Parallel Computation. In: Baskett F, editor. Computer Graphics. Proceedings of SIGGRAPH 90, vol. 24(4). Dallas, Tex.; August 1990. p. 405–14. ISBN 0-201-50933-4.

[23] Terzopoulos D, Fleischer K. Modeling Inelastic Deformation: Viscoelasticity, Plasticity, Fracture. In: Dill J, editor. Computer Graphics. Proceedings of SIGGRAPH 88, vol. 22(4). Atlanta, Ga.; August 1988. p. 269–78.

[24] Terzopoulos D, Platt J, Barr A, Fleischer K. Elastically Deformable Models. In: Stone MC, editor. Computer Graphics. Proceedings of SIGGRAPH 87, vol. 21(4). Anaheim, Calif.; July 1987. p. 205–14.

[25] Terzopoulos D, Witkin A. Physically Based Models with Rigid and Deformable Components. IEEE Comput Graph Appl November 1988;41–51.

[26] Witkin A, Fleischer K, Barr A. Energy Constraints on Parameterized Models. In: Stone MC, editor. Computer Graphics. Proceedings of SIGGRAPH 87, vol. 21(4). Anaheim, Calif.; July 1987. p. 225–32.

[27] Witkin A, Kass M. Spacetime Constraints. In: Dill J, editor. Computer Graphics. Proceedings of SIGGRAPH 88, vol. 22(4). Atlanta, Ga.; August 1988. p. 159–68.

[28] Weil J. The Synthesis of Cloth Objects. In: Evans DC, Athay RJ, editors. Computer Graphics. Proceedings of SIGGRAPH 86, vol. 20(4). Dallas, Tex.; August 1986. p. 49–54.

[29] DeRose T, Kass M, Truong T. Subdivision Surfaces for Character Animation. In: Cohen M, editor. Computer Graphics. Proceedings of SIGGRAPH 98, Annual Conference Series Orlando, Fla.: Addison-Wesley; July 1998. p. 85–94. ISBN 0-89791-999-8.

[30] Vassilev T, Spanlang B, Chrysanthou Y. Fast Cloth Animation on Walking Avatars. Computer Graphics Forum 2001;20(3):1–8.

[31] Choi K-J, Ko H-S. Stable but Responsive Cloth. Transactions on Graphics 2002;21(3):604–11, SIGGRAPH.

Fluids: Liquids and Gases

Among the most difficult graphical objects to model and animate are those that are not defined by a static, rigid, topologically simple structure. Many of these complex forms are found in nature. They present especially difficult challenges for those intent on controlling their motion. In some cases, special models of a specific phenomenon will suffice. We begin the chapter presenting special models for water, clouds, and fire that approximate their behavior under certain conditions. These models identify salient characteristics of the phenomena and attempt to model those characteristics explicitly. Such approaches are useful for a specific appearance or motion, but for a more robust model, a more rigorous scientific approach is needed. As far as a computational science is concerned, all of the phenomena mentioned earlier fall under the classification of *fluids* and computing their motion is called *computational fluid dynamics* (CFD). The basics of CFD are given at the end of this chapter.

Many of the time-varying models described in this chapter represent work that is state of the art. It is not the objective here to cover all aspects of recent research. The basic models are covered, with only brief reference to the more advanced algorithms.

8.1 Specific fluid models

Fire, smoke, and clouds are gaseous phenomena that have no well-defined surface to model. They are inherently volumetric models, although surface-based techniques have been applied with limited success. For example, water when relatively still has a well-defined surface; however, water changes its shape as it moves. In the case of ocean waves, features on the water's surface move, but the water itself does not travel. The simple surface topology can become arbitrarily complex when the water becomes turbulent. Splashing, foaming, and breaking waves are complex processes best modeled by particle systems and volumetric techniques, but these techniques are inefficient in nonturbulent situations. In addition, water can travel from one place to another, form streams, split into separate pools, and collect again. In modeling these phenomena for purposes of computer graphics, programmers often make simplifying assumptions in order to limit the computational complexity and model only those features of the physical processes that are visually important in the specific situation of interest.

8.1.1 Models of water

Water presents a particular challenge for computer animation because its appearance and motion take various forms [5] [7] [17] [20]. Water can be modeled as a still, rigid-looking surface to which ripples can be added as display attributes by perturbing the surface normal using bump mapping [1].

Alternatively, water can be modeled as a smoothly rolling height field in which time-varying ripples are incorporated into the geometry of the surface [12]. In ocean waves, it is assumed that there is no transport of water even though the waves travel along the surface in forms that vary from sinusoidal to cycloidal [6] [15].[1] Breaking, foaming, and splashing of the waves can be added on top of the base model in a postprocessing step [6] [15]. The transport of water from one location to another adds more computational complexity to the modeling problem [10].

Still waters and small-amplitude waves

The simplest way to model water is merely to assign the color blue to anything below a given height. If the y-axis is "up," then color any pixel blue (with, for example, an illumination model that uses a consistent normal) in which the world space coordinate of the corresponding visible surface has a y-value less than some given constant. This creates the illusion of still water at a consistent "sea level." It is sufficient for placid lakes and puddles of standing water. Equivalently, a flat blue plane perpendicular to the y-axis and at the height of the water can be used to represent the water's surface. These models, of course, do not produce any animation of the water.

Normal vector perturbation (the approach employed in bump mapping) can be used to simulate the appearance of small amplitude waves on an otherwise still body of water. To perturb the normal, one or more simple sinusoidal functions are used to modify the direction of the surface's normal vector. The functions are parameterized in terms of a single variable, usually relating to distance from a source point. It is not necessarily the case that the wave starts with zero amplitude at the source. When standing waves in a large body of water are modeled, each function usually has a constant amplitude. The wave crests can be linear, in which case all the waves generated by a particular function travel in a uniform direction, or the wave crests can radiate from a single user-specified or randomly generated source point. Linear wave crests tend to form self-replicating patterns when viewed from a distance. For a different effect, radially symmetrical functions that help to break up these global patterns can be used. Radial functions also simulate the effect of a thrown pebble or raindrop hitting the water (Figure 8.1). The time-varying height for a point at which the wave begins at time zero is a function of the amplitude and wavelength of the wave (Figure 8.2). Combining the two, Figure 8.3 shows the height of a point at some distance d from the start of the wave. This is a two-dimensional function relative to a point at which the function is zero at time zero. This function can be rotated and translated so that it is positioned and oriented appropriately in world space. Once the height function for a given point is defined, the normal to the point at any instance in time can be determined by computing the tangent vector and forming the vector perpendicular to it. These vectors should then be oriented in world space, so the plane they define contains the direction that the wave is traveling.

Superimposing multiple sinusoidal functions of different amplitude and with various source points (in the radial case) or directions (in the linear case) can generate interesting patterns of overlapping ripples. Typically, the higher the frequency of the wave component, the lower the amplitude. Notice that these do not change the geometry of the surface used to represent the water (e.g., a flat blue plane) but are used only to change the shading properties. Also notice that it must be a time-varying function that propagates the wave along the surface.

The same approach used to calculate wave normals can be used to modify the height of the surface (e.g., [11]). A mesh of points can be used to model the surface of the water and the heights of the

---

[1]A *cycloid* is the curve traced out by a point on the perimeter of a rolling disk.

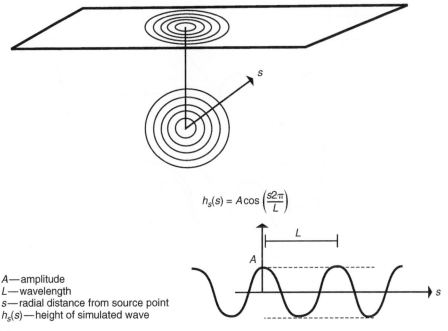

$$h_s(s) = A\cos\left(\frac{s2\pi}{L}\right)$$

A—amplitude
L—wavelength
s—radial distance from source point
$h_s(s)$—height of simulated wave

FIGURE 8.1

Radially symmetric standing wave.

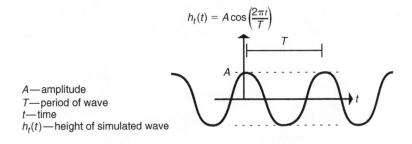

$$h_t(t) = A\cos\left(\frac{2\pi t}{T}\right)$$

A—amplitude
T—period of wave
t—time
$h_t(t)$—height of simulated wave

FIGURE 8.2

Time-varying height of a stationary point.

individual points can be controlled by the overlapping sinusoidal functions. Either a faceted surface with smooth shading can be used or the points can be the control points of a higher order surface such as a B-spline surface. The points must be sufficiently dense to sample the height function accurately enough for rendering. Calculating the normals without changing the actual surface creates the illusion of waves on the surface of the water. However, whenever the water meets a protruding surface, like a rock, the lack of surface displacement will be evident. See Figures 8.4 and 8.5 for a comparison between a water surface modeled by normal vector perturbation and a water surface modeled by a height field. An option to reduce the density of mesh points required is to use only the low-frequency, high-amplitude functions to control the height of the surface points and to include the high-frequency, low-amplitude functions to calculate the normals (Figure 8.6).

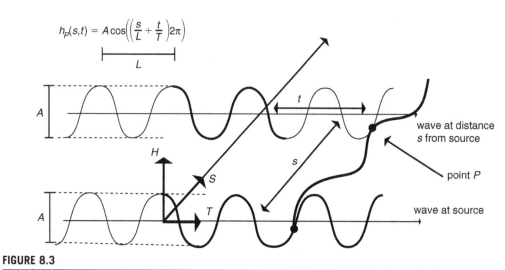

$$h_p(s,t) = A\cos\left(\left(\frac{s}{L} + \frac{t}{T}\right)2\pi\right)$$

FIGURE 8.3

Time-varying function at point P.

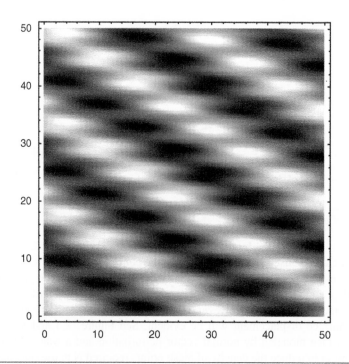

FIGURE 8.4

Superimposed linear waves of various amplitudes and frequencies.

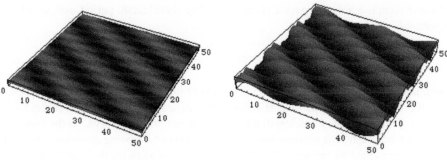

FIGURE 8.5

Normal vector displacement versus height displacement.

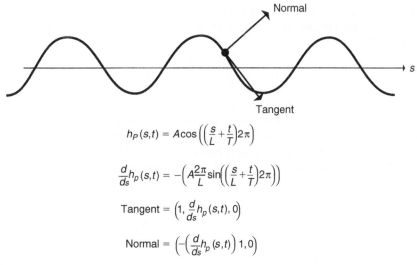

$$h_p(s,t) = A\cos\left(\left(\frac{s}{L} + \frac{t}{T}\right)2\pi\right)$$

$$\frac{d}{ds}h_p(s,t) = -\left(A\frac{2\pi}{L}\sin\left(\left(\frac{s}{L} + \frac{t}{T}\right)2\pi\right)\right)$$

$$\text{Tangent} = \left(1, \frac{d}{ds}h_p(s,t), 0\right)$$

$$\text{Normal} = \left(-\left(\frac{d}{ds}h_p(s,t)\right)1, 0\right)$$

FIGURE 8.6

Normal vector for two-dimensional wave function.

The anatomy of waves

A more sophisticated model must be used to model waves with greater realism, one that incorporates more of the physical effects that produce their appearance and behavior. Waves come in various frequencies, from tidal waves to capillary waves, which are created by wind passing over the surface of the water. The waves collectively called wind waves are those of most interest for visual effects. The sinusoidal form of a simple wave has already been described and is reviewed here in a more appropriate form for the equations that follow. In Equation 8.1, the function $f(s, t)$ describes the amplitude of the wave in which s is the distance from the source point, t is a point in time, A is the maximum amplitude, C is the propagation speed, and L is the wavelength. The period of the wave, T, is the time it takes for one complete wave to pass a given point. The wavelength, period, and speed are related by the equation $C = L/T$.

$$f(s,t) = \cos^{-1}\left(\frac{2\pi(s - Ct)}{L}\right) \tag{8.1}$$

The motion of the wave is different from the motion of the water. The wave travels linearly across the surface of the water, while a particle of water moves in nearly a circular orbit (Figure 8.7). While riding the crest of the wave, the particle will move in the direction of the wave. As the wave passes and the particle drops into the trough between waves, it will travel in the reverse direction. The steepness, S, of the wave is represented by the term H/L where H is defined as twice the amplitude.

Waves with a small steepness value have basically a sinusoidal shape. As the steepness value increases, the shape of the wave gradually changes into a sharply crested peak with flatter troughs. Mathematically, the shape approaches that of a cycloid.

In an idealized wave, there is no net transport of water. The particle of water completes one orbit in the time it takes for one complete cycle of the wave to pass. The average orbital speed of a particle of water is given by the circumference of the orbit, pH, divided by the time it takes to complete the orbit, T (Eq. 8.2).

$$Q_{\text{ave}} = \frac{\pi H}{T} = \frac{\pi HC}{L} = \pi SC \tag{8.2}$$

If the orbital speed, Q, of the water at the crest exceeds the speed of the wave, C, then the water will spill over the wave, resulting in a breaking wave. Because the average speed, Q, increases as the steepness, S, of the wave increases, this limits the steepness of a nonbreaking wave. The observed steepness of ocean waves, as reported by Peachey [15] is between 0.5 and 1.0.

A common simplification of the full CFD simulation of ocean waves is called the Airy model, and it relates the depth of the water, d, the propagation speed, C, and the wavelength of the wave, L (Eq. 8.3).

$$C = \sqrt{\frac{g}{\kappa} \tanh(\kappa d)} = \sqrt{\frac{gL}{2\pi}} \times \tanh\left(\frac{2\pi d}{L}\right)$$

$$L = CT \tag{8.3}$$

In Equation 8.3, g is the acceleration of a body due to gravity at sea level, 9.81 m/sec^2, and $k = 2\pi/L$ is the spatial equivalent of wave frequency. As the depth of the water increases, the function $\tanh(kd)$ tends toward one, so C approaches $gL/2\pi$. As the depth decreases and approaches zero, $\tanh(kd)$ approaches kd, so C approaches $\sqrt{gd}$. Peachey suggests using *deep* to mean $d > L/4$ and *shallow* to mean $d < L/20$.

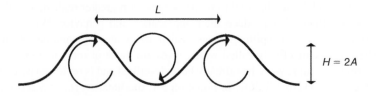

FIGURE 8.7

Circular paths of particles of water subjected to waves.

As a wave approaches the shoreline at an angle, the part of the wave that approaches first will slow down as it encounters a shallower area. The wave will progressively slow down along its length as more of it encounters the shallow area. This will tend to straighten out the wave and is called *wave refraction*.

Interestingly, even though speed (C) and wavelength (L) of the wave are reduced as the wave enters shallow water, the period, T, of the wave remains the same and the amplitude, A (and, equivalently, H), remains the same or increases. As a result, the orbital speed, Q (Eq. 8.2), of the water remains the same. Because orbital speed remains the same as the speed of the wave decreases, waves tend to break as they approach the shoreline because the speed of the water exceeds the speed of the wave. The breaking of a wave means that water particles break off from the wave surface as they are "thrown forward" beyond the front of the wave.

Modeling ocean waves

The description of modeling ocean waves presented here follows Peachey [15]. The ocean surface is represented as a height field, $y=f(x, z, t)$, where (x, z) defines the two-dimensional ground plane, t is time, and y is the height. The wave function f is a sum of various waveforms at different amplitudes (Eq. 8.4).

$$f(x, z, t) = \sum_{i=1}^{n} A_i W_i(x, z, t) \tag{8.4}$$

The wave function, W_i, is formed as the composition of two functions: a wave profile, w_i, and a phase function, $\theta_i(x, z, t)$, according to Equation 8.5. This allows the description of the wave profile and phase function to be addressed separately.

$$W_i(x, z, t) = \omega_i(\text{fraction}[\theta_i(x, z, t)]) \tag{8.5}$$

Each waveform is described by its period, amplitude, source point, and direction. It is convenient to define each waveform, actually each phase function, as a linear rather than radial wave and to orient it so the wave is perpendicular to the x-axis and originates at the source point. The phase function is then a function only of the x-coordinate and can then be rotated and translated into position in world space.

Equation 8.6 gives the time dependence of the phase function. Thus, if the phase function is known for all points x (assuming the alignment of the waveform along the x-axis), then the phase function can be easily computed at any time at any position. If the depth of water is constant, the Airy model states that the wavelength and speed are also constant. In this case, the aligned phase function is given in Equation 8.7.

$$\theta_i(x, y, t) = \theta_i(x, y, t_0) - \frac{t - t_0}{T_i} \tag{8.6}$$

$$\theta_i(x, z, t) = \frac{x_i}{L_i} \tag{8.7}$$

However, if the depth of the water is variable, then L_i is a function of depth and u_i is the integral of the depth-dependent phase-change function from the origin to the point of interest (Eq. 8.8). Numerical integration can be used to produce phase values at predetermined grid points. These grid points can be used to interpolate values within grid cells. Peachey [15] successfully uses bilinear interpolation to accomplish this.

$$\theta_i(x, z, t) = \int_0^x \theta_i'(u, z, t)du \tag{8.8}$$

The wave profile function, wi, is a single-value periodic function of the fraction of the phase function (Eq. 8.5) so that $w_i(u)$ is defined for $0.0 \leq u \leq 1.0$. The values of the wave profile function range over the interval $[-1, 1]$. The wave profile function is designed so that its value is one at both ends of the interval (Eq. 8.9).

$$\omega_i(0,0) = \omega_i(1.0) = 1.0 \tag{8.9}$$

Linear interpolation can be used to model the changing profile of the wave according to steepness. Steepness (H/L) can be used to blend between a sinusoidal function (Eq. 8.1) and a cycloid-like function (Eq. 8.10) designed to resemble a sharp-crested wave profile. In addition, wave asymmetry is introduced as a function of the depth of the water to simulate effects observed in waves as they approach a coastline. The asymmetry interpolant, k, is defined as the ratio between the water depth, d, and deep-water wavelength, L_i (see Eq. 8.11). When k is large, the wave profile is handled with no further modification. When k is small, u is raised to a power in order to shift its value toward the low end of the range between zero and one. This has the effect of stretching out the back of the wave and steepening the front of the wave as it approaches the shore.

$$\omega_i(u) = 8|u - 1/2|^2 - 1 \tag{8.10}$$

$$L_i^{\text{deep}} = \frac{gT_i^2}{2\pi}$$

$$k = \frac{d}{L_i^{\text{deep}}} \tag{8.11}$$

As the wave enters very shallow water, the amplitudes of the various wave components are reduced so the overall amplitude of the wave is kept from exceeding the depth of the water.

Spray and foam resulting from breaking waves and waves hitting obstacles can be simulated using a stochastic but controlled (e.g., Gaussian distribution) particle system. When the speed of the water, Q_{average}, exceeds the speed of the wave, C, then water spray leaves the surface of the wave and is thrown forward. Equation 8.2 indicates that this condition happens when $\pi S > 1.0$ or, equivalently, $S > 1.0/\pi$. Breaking waves are observed with steepness values less than this (around 0.1), which indicates that the water probably does not travel at a uniform orbital speed. Instead, the speed of the water at the top of the orbit is faster than at other points in the orbit. Thus, a user-specified spray-threshold steepness value can be used to trigger the particle system. The number of particles generated is based on the difference between the calculated wave steepness and the spray-threshold steepness.

For a wave hitting an obstacle, a particle system can be used to generate spray in the direction of reflection based on the incoming direction of the wave and the normal of the obstacle surface. A small number of particles are generated just before the moment of impact, are increased to a maximum number at the point of impact, and are then decreased as the wave passes the obstacle. As always, stochastic perturbation should be used to control both speed and direction.

For a more complete treatment of modeling height-field displacement-mapped surface for ocean waves using a fast Fourier transform description, including modeling and rendering underwater environmental effects, the interested reader is directed to the course notes by J. Tessendorf [19].

Finding its way downhill

One of the assumptions used to model ocean waves is that there is no transport of water. However, in many situations, such as a stream of water running downhill, it is useful to model how water travels from one location to another. In situations in which the water can be considered a height field and the motion is assumed to be uniform through a vertical column of water, the vertical component of the velocity can be ignored. In such cases, differential equations can be used to simulate a wide range of convincing motion [10]. The Navier-Stokes equations (which describe flow through a volume) can be simplified to model the flow.

To develop the equations in two dimensions, the user parameterizes functions that are in terms of distance x. Let $z = h(x)$ be the height of the water and $z = b(x)$ be the height of the ground at location x. The height of the water is $d(x) = h(x) - b(x)$. If one assumes that motion is uniform through a vertical column of water and that $v(x)$ is the velocity of a vertical column of water, then the shallow-water equations are as shown in Equations 8.12 and 8.13, where g is the gravitational acceleration (see Figure 8.8). Equation 8.12 considers the change in velocity of the water and relates its acceleration, the difference in adjacent velocities, and the acceleration due to gravity when adjacent columns of water are at different heights. Equation 8.13 considers the transport of water by relating the temporal change in the height of the vertical column of water with the spatial change in the amount of water moving.

$$\frac{\partial v}{\partial t} + v\frac{\partial v}{\partial x} + g\frac{\partial h}{\partial x} = 0 \tag{8.12}$$

$$\frac{\partial d}{\partial t} + \frac{\partial}{\partial x}(vd) = 0 \tag{8.13}$$

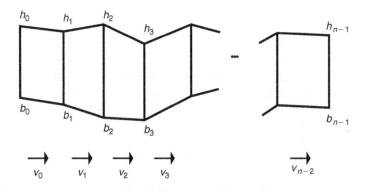

FIGURE 8.8

Discrete two-dimensional representation of height field with water surface h, ground b, and horizontal water velocity v.

These equations can be further simplified if the assumptions of small fluid velocity and slowly varying depth are used. The former assumption eliminates the second term of Equation 8.12, while the latter assumption implies that the term d can be removed from inside the derivative in Equation 8.13. These simplifications result in Equations 8.14 and 8.15.

$$\frac{\partial v}{\partial t} + g\frac{\partial v}{\partial x} = 0 \tag{8.14}$$

$$\frac{\partial d}{\partial t} + d\frac{\partial v}{\partial x} = 0 \tag{8.15}$$

Differentiating Equation 8.14 with respect to x and Equation 8.15 with respect to t and substituting for the cross-derivatives results in Equation 8.16. This is the one-dimensional wave equation with a wave velocity $\sqrt{gd}$. As Kass and Miller [10] note, this degree of simplification is probably not accurate enough for engineering applications.

$$\frac{\partial^2 h}{\partial t^2} + gd\frac{\partial^2 h}{\partial x^2} \tag{8.16}$$

This partial differential equation is solved using finite differences. The discretization, as used by Kass and Miller [10], is set up as in Figure 8.8, with samples of v positioned halfway between the samples of h. The authors report a stable discretization, resulting in Equations 8.17 and 8.18. Putting these two equations together results in Equation 8.19, which is the discrete version of Equation 8.16.

$$\frac{\partial h_i}{\partial t} = \left(\frac{d_{i-1} + d_i}{2\Delta x}\right)\Big|_{v_{i-1}} - \left(\frac{d_i + d_{i+1}}{2\Delta x}\right)v_i \tag{8.17}$$

$$\frac{\partial v_i}{\partial t} = \frac{(-g)(h_{i+1} - h_i)}{\Delta x} \tag{8.18}$$

$$\frac{\partial^2 h_i}{\partial t^2} = -g\left(\frac{d_{i-1} + d_i}{2(\Delta x)^2}\right)(h_i - h_{i-1}) + g\left(\frac{d_i + d_{i+1}}{2(\Delta x)^2}\right)(h_{i+1} - h_i) \tag{8.19}$$

Equation 8.19 states the relationship of the height of the water surface to the height's acceleration in time. This could be solved by using values of h_i to compute the left-hand side and then using this value to update the next time step. As Kass and Miller [10] report, however, this approach diverges quickly because of the sample spacing.

A first-order implicit numerical integration technique is used to provide a stable solution to Equation 8.19. Numerical integration uses current sample values to approximate derivatives. Explicit methods use approximated derivatives to update the current samples to their new values. Implicit integration techniques find the value whose derivative matches the discrete approximation of the current samples. Implicit techniques typically require more computation per step, but, because they are less likely to diverge significantly from the correct values, larger time steps can be taken, thus producing an overall savings.

Kass and Miller [10] find that a first-order implicit method (Eqs. 8.20 and 8.21) is sufficient for this application. Using these equations to solve for $h(n)$ and substituting Equation 8.19 for the second derivative $(\ddot{h}(n))$ results in Equation 8.22.

$$\dot{h}(n) = \frac{h(n) - h(n-1)}{\Delta t} \tag{8.20}$$

$$\ddot{h}(n) = \frac{\dot{h}(n) - \dot{h}(n-1)}{\Delta t} \tag{8.21}$$

$$
\begin{aligned}
h_i(n) = {} & 2h_i(n-1) - h_i(n-2) \\
& -g(\Delta t)^2 \frac{d_{i-1} + d_i}{2 \cdot \Delta x^2} (h_i(n) - h_{i-1}(n)) \\
& +g(\Delta t)^2 \frac{d_{i-1} + d_i}{2 \cdot \Delta x^2} (h_{i+1}(n) - h_i(n))
\end{aligned}
\tag{8.22}
$$

Assuming d is constant during the iteration, the next value of h can be calculated from previous values with the symmetric tridiagonal linear system represented by Equations 8.23–8.25.

$$Ah_i(n) = 2h_i(n-1) + h_i(n-2) \tag{8.23}$$

$$
A = \begin{bmatrix}
e_0 & f_0 & 0 & 0 & 0 & 0 & 0 \\
f_0 & e_0 & f_i & 0 & 0 & 0 & 0 \\
0 & f_i & e_2 & \ldots & 0 & 0 & 0 \\
0 & 0 & \ldots & \ldots & \ldots & 0 & 0 \\
0 & 0 & 0 & \ldots & e_{n-3} & f_{n-3} & 0 \\
0 & 0 & 0 & 0 & f_{n-3} & e_{n-2} & f_{n-2} \\
0 & 0 & 0 & 0 & 0 & f_{n-2} & e_{n-1}
\end{bmatrix}
\tag{8.24}
$$

$$e_0 = 1 + g(\Delta t)^2 \left(\frac{d_0 + d_1}{2(\Delta x)^2} \right)$$

$$e_i = 1 + g(\Delta t)^2 \left(\frac{d_{i-1} + 2d_1 + d_{i+1}}{2(\Delta x)^2} \right) (0 < i < n-1)$$

$$e_{n-1} = 1 + g(\Delta t)^2 \left(\frac{d_{n-2} + d_{n-1}}{2(\Delta x)^2} \right)$$

$$f_i = - \left(g(\Delta t)^2 \left(\frac{d_i + d_{i+1}}{2(\Delta x)^2} \right) \right) \tag{8.25}$$

To simulate a viscous fluid, Equation 8.23 can be modified to incorporate a parameter that controls the viscosity, thus producing Equation 8.26. The parameter τ ranges between 0 and 1. When $\tau = 0$, Equation 8.26 reduces to Equation 8.3.

$$Ah_i(n) = h_i(n-1) + (1-\tau)(h_i(n-1) - h_i(n-2)) \tag{8.26}$$

```
Specify h(t=0), h(t=1), and b
for j=2 . . . n
    to simulate sinks and sources, modify appropriate h values
    calculate d from h(j-1) and b; if hi<bi, then di=0
    use Equation 8.10 (Equation 8.13) to calculate h(j) from h(j-1) and
        h(j-2)
    adjust the values of h to conserve volume (as discussed above)
    if hi<bi, set hi(j) and hi-1(j) to bi-ε
```

FIGURE 8.9

Two-dimensional algorithm for water transport.

Volume preservation can be compromised when $h_i < b_i$. To compensate for this, search for the connected areas of water ($h_j < b_j$ for $j=i, i+1, \ldots, i+n$) at each iteration and compute the volume. If the volume changes from the last iteration. then distribute the difference among the elements of that connected region. The algorithm for the two-dimensional case is shown in Figure 8.9.

Extending the algorithm to the three-dimensional case considers a two-dimensional height field. The computations are done by decoupling the two dimensions of the field. Each iteration is decomposed into two subiterations, one in the x-direction and one in the y-direction.

Summary

Animating all of the aspects of water is a difficult task because of water's ability not only to change shape but also to change its topology over time. Great strides have been made in animating individual aspects of water such as standing waves, ocean waves, spray, and flowing water. But an efficient approach to an integrated model of water remains a challenge.

8.1.2 Modeling and animating clouds

Most any outdoor daytime scene includes some type of clouds as a visual element. Clouds can form a backdrop for activity in the foreground, in which case they may be slow moving or even still. Clouds can be used as an area of activity allowing planes to fly around or through them or, as fog, cars to drive through them. Clouds viewed from afar have certain characteristics that are much different from clouds that are viewed close up. From a distance, they exhibit high level opaque appearance features that transparently disappear upon inspection at a closer distance. Cloud formations can inspire, impart calm, be ominous and foreboding, or produce magnificent sunsets. Distant clouds can have a puffy opaque appearance, often resembling familiar shapes of, for example, various animals or other common objects. Other clouds have a layered or fibrous appearance. While some cloud formations viewed from a distance might appear as a stationary backdrop or slow moving rigid formations, storm clouds can

FIGURE 8.10

An example of cirrus and cirrostratus clouds at sunset. (Copyright 1998 David S. Ebert.)

form relatively quickly and can appear threatening in a manner of minutes. Complex atmospheric processes form clouds. These processes can be modeled and animated to show the formation of storms, hurricanes, and inclement weather (see Figures 8.10 and 8.11).

Clouds are naturally occurring atmospheric phenomena. They are composed of visible water droplets or ice crystals suspended in air, and are formed by moisture droplets adhering to dust particles. When moisture is added to the air and/or the temperature of the air decreases, clouds can form. In either case, moisture in the air condenses into droplets because the air cannot transparently hold that amount of water at that temperature. Clouds occur at a variety of altitudes. Some clouds can be found at 60,000 feet while others can be as low as ground level. Although typically one thinks of clouds as high in the sky, fog phenomena are just very low-level clouds. In addition, a single cloud formation can start near the surface and extend upward as much as 10,000 feet.

Anatomy of clouds and cloud formation

Clouds are categorized according to their altitude, their general shape, and whether they carry moisture (e.g., see [23]). The naming convention combines a standard set of terms that describe the altitude, shape, and moisture of the cloud.

FIGURE 8.11

An example of a cumulus cloud. (Copyright 1997 David S. Ebert.)

Altitude	
Cirrus/cirro	High clouds (by itself, it refers to high level fibrous clouds)
Altus/alto	Middle clouds
Otherwise	Low-level clouds
Shape	
Cumulus/cumulo	Puffy (Latin for "stack")
Stratus/strato	Layer (Latin for "sheet")
Moisture	
Nimubs/nimbo	Water bearing

Certain types of clouds can typically be found at certain elevations. For example, high-level clouds are cirrus, cirrostratus, and cirrocumulus; mid-level clouds are altocumulus, altostratus, and nimbostratus; low-level clouds are cumulonimbus, stratocumulus, stratus, and cumulus. Clouds typified by vertical development are fair weather cumulus and cumulonimbus. Other cloud names less frequently encountered include contrails, billow clouds, mammatus, orographic, and pileus clouds.

Clouds are formed by low-level processes such as particles in the air, turbulence, wind shear, and temperature gradients combined with a significant moisture content. The basic moisture cycle consists of the following: transpiration → evaporation → condensation → precipitation → and back to transpiration.

The processes that form clouds consist of adding moisture to the air or cooling the air. In either case the moisture content becomes higher than what can be handled transparently by the air. As a result, water droplets form and cling to dust particles in the air. The cooling of an air mass is often due to an air mass rising in altitude; air tends to cool as it rises and loses pressure. The cause for this rise is usually due to something forcing the air mass upward: the traveling air mass encounters an incline like a mountain slope; the air mass travels up and over a colder, denser air mass (a warm front); and a cold, dense air mass forces its way underneath the warmer air mass and raises the warm air mass (a cold front). The air mass temperature can also be cooled by the cooling of the earth's surface, often creating fog or other low-level clouds. Adding moisture to the air mass is often accomplished by water evaporating at the earth's surface underneath the air mass.

Cloud models in CG

Although modeling clouds as a single, static background painting can be effective in some animations, limited cloud animation can be done cost effectively. This can just be a texture mapped quadrilateral. Simply moving (i.e., panning) the background provides some degree of animation. Most animation effects can be introduced by using an animated texture map, essentially a movie, of clouds for the background. Additionally, using multiple, semitransparent background images, representing clouds at varying heights, and moving them relative to one another, simulating parallax, can produce the impression of a deep three-dimensional space at minimal cost. In some situations, it is more appropriate to use a spherical shell instead of a planar surface, although care must be taken at the horizon to avoid unrealistic transitions between sky and ground. To display a fog effect, a simple linear distance-based attenuation of object shading can be used. However, the animation of fog swirls requires an approach similar to CFD discussed earlier in this chapter.

For more control and added visual realism, appropriately shaded geometric primitives can be used as cloud building blocks. Two-dimensional primitives can be used for distant clouds while three-dimensional primitives give a greater sense of distance and depth when viewed from various angles and are appropriate for more immersive environments. The primitives can be moved relative to one another and scaled up or down, providing easy control for an animator. While not physically based, this approach can produce effective visuals if the primitives are carefully designed.

The use of such building blocks was pioneered in 1984 by Geoffrey Gardner [8][9] and the general approach is still an attractive way to model clouds. Gardner's technique uses a sky plane that is parallel to the ground plane and some distance above it. Ellipsoids are used as the cloud building blocks and are positioned by the user relative to the sky plane. The ellipsoid surface is textured (as described in the following paragraphs) and then mapped onto a sky plane. Gardner also uses these textured ellipsoids to build full three-dimensional cloud models, which allows for a full three-dimensional effect as the viewing angle changes relative to the ellipsoids. Notice, however, that the texturing function is not defined on the interior of the ellipsoid shells, so the visuals associated with flying into a cloud are not supported. In addition to clouds, Gardner also used this approach, with suitable modification to the texture function, to model other environmental features such as trees and hills as in Figure 8.12.

The surface texturing function is based on the sum of two weighted sums of sine waves, defined to produce a pseudorandom cloud-like texture. The two sets of sine waves are orthogonal to each other. In each set, each frequency is twice the frequency of its predecessor and has an amplitude of one over radical two times its predecessor. In addition, the phase shift of each frequency is based on its position in the other dimension. A user-defined threshold is set so that

FIGURE 8.12

Landscape constructed using multiple partially transparent ellipsoids. (© 1985 ACM, Gardner, G., *Visual Simulation of Clouds*. Computer Graphics. Proceedings of SIGGRAPH 85, 19(3). San Francisco, CA., August 1985. With permission.)

any lesser texture value is considered totally transparent. In order to remove any discontinuities that might arise at the ellipsoid boundary, a monotonically decreasing attenuation function is defined to have a value of one at the middle of the ellipsoid and zero at the boundary. This is used to scale the texture function so that final function values scale down to zero as they approach the limit of the ellipsoid. This creates a cloud primitive that is generally in the shape of an ellipsoid, but whose texture feathers away close to the boundary, thus avoiding any noticeable discontinuities in the shading.

Surface-based textures, such as those used by Gardner, are sufficient for clouds viewed from a distance and represented on a sky plane or viewed from a distance, but to accommodate close viewing and even immersion in the clouds, such as that encountered in current flight simulators and games, a volumetric approach must be taken. Volumetric approaches define a density value for any point in space. Volumes are rendered by considering the entire line of sight through the volume. This makes realistic shading more difficult because it depends on the accumulated shading of each point along the line of sight inside the cloud. Each point along the light of sight has a certain density that controls its opacity/transparency. Each of these points along the line of sight reflects the light that gets to it from a light source. Direct illumination from the light source is light that filters through the cloud on a line from the light source directly to the point in question. Indirect illumination from the light source is light that comes from the light source by way of reflections off of nearby points in the cloud. This in turn means that the illumination of nearby points must be known in order to know how much light they reflect. Usually only one or two levels of such reflections are considered to make shading computations tractable. Mathematically, the shading is an integration along the line of sight and illumination is integration along lines from each light source. Indirect lighting is integration from all directions. For most applications, these integrals can be computed using appropriate discrete approximations or, in some cases, ignored altogether. In any case, the illumination is computationally expensive and not a topic here. The interested reader is directed to the book *Texturing and Modeling: A Procedural Approach*, edited by David Ebert [3].

The same building block approach to cloud formation can be taken with the volumetric approach, as with the ellipsoid shells of Gardner. See Figure 8.13 for an example of work by David Ebert. At the

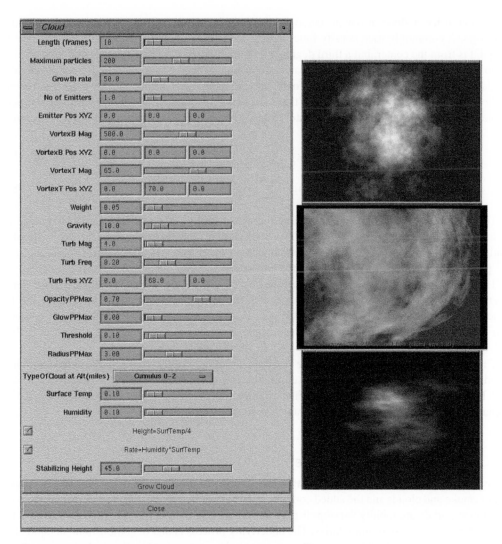

A GUI used to control cloud formation   B Example clouds

FIGURE 8.13

(A) An example of cloud dynamics GUI and (B) example images created in Maya™. (© 1999 Rchi Gartha.)

heart of Ebert's volumetric approach are implicitly defined density functions that provide high-level organization of the cloud, combined with turbulence functions that provide the fine detail.

The implicit function used here is really a sum of individual spherical implicit functions. Each of these spherical functions maps a point in space to a density based on the point's distance from the sphere's center and a density function. One property of the density function is that it is one at its center and monotonically decreases to zero at its outermost edge, which is called its *radius of influence*.

In addition, it has a slope of one at its radius of influence, allowing it to blend smoothly with other primitives. A commonly used density function is a sixth degree polynomial in terms of distance that the point is from the center, but a third degree polynomial in terms of distance-squared, thus avoiding the cost of the square root of the distance calculation (Eq. 8.27). The implicit function for the cloud is a weighted sum of these individual implicit functions (Eq. 8.28). See Chapter 12.1 for more discussion on implicit functions and surfaces. The use of these implicit density functions provides the macrostructure for the cloud formation, allowing an animator to easily form the cloud with simple high-level controls.

$$F(r) = -\frac{4}{9}\frac{r^6}{R^6} + \frac{17}{9}\frac{r^4}{R^4} - \frac{22}{9}\frac{r^2}{R^2} + 1 \qquad (8.27)$$

$$D(p) = \sum_i w_i F(|p - q|) \qquad (8.28)$$

To introduce some pseudorandomness into the density values, the animator can procedurally alter the location of points before evaluating the density function. This perturbation of the points should be continuous with limited high-frequency components in order to avoid aliasing artifacts in the result. The perturbation should also avoid unwanted patterns in the final values.

The microstructure of the clouds is provided by procedural noise and turbulence-based density function (see Appendix B.6). This is blended with the macrostructure density to produce the final values for the cloud density. Various cloud types can be modeled by modifying the primitive shapes, their relative positioning, the density parameters, and the turbulence parameters. See Figures 8.10 and 8.11 for examples.

The building-block approach allows the clouds to be animated either at the macro or the micro level. Cloud migration can be modeled by user-supplied scripts or simplified particle system physics can be used. Internal cloud dynamics can be modeled by animating the texture function parameters, in the case of Gardner-type clouds, or turbulence parameters in the case of Ebert.

8.1.3 Modeling and animating fire

Fire is a particularly difficult and computationally intensive process to model. It has all the complexities of smoke and clouds and the added complexity of very active internal processes that produce light and motion and create rapidly varying display attributes. Fire exhibits temporally and spatially transient features at a variety of granularies. The underlying process is that of combustion—a rapid chemical process that releases heat (i.e., is *exothermic*) and light accompanied by flame. A common example is that of a wood fire in which the hydrocarbon atoms of the wood fuel join with oxygen atoms to form water vapor, carbon monoxide, and carbon dioxide. As a consequence of the heated gases rising quickly, air is sucked into the combustion area, creating turbulence.

Recently, impressive advances have been made in the modeling of fire. At one extreme, the most realistic approaches require sophisticated techniques from CFD and are difficult to control (e.g., [18]). Work has also been performed in simulating the development and spread of fire for purposes of tracking its movement in an environment (e.g., [2] [16]). These models tend to be only global, extrinsic representations of the fire's movement and less concerned with the intrinsic motion of the fire itself. Falling somewhere between these two extremes, procedural image generation and particle systems provide visually effective, yet computationally attractive, approaches to fire.

Procedurally generated image

A particularly simple way to generate the appearance of fire is to procedurally generate a two-dimensional image by coloring pixels suggestive of flames and smoke. The procedure iterates through pixels of an image buffer and sets the palette indices based on the indices of the surrounding pixels [14]. Modifying the pixels top to bottom allows the imagery to progress up the buffer. By using multiple buffers and the alpha (transparency) channel, a limited three-dimensional effect can be achieved. In Figure 8.14 (Color Plate 3), a color palette filled with hues from red to yellow is used to hold RGB values. The bottom row of the image buffer is randomly initialized with color indices.

Particle system approach

One of the first and most popularly viewed examples of computer-generated fire appears in the movie *Star Trek II: The Wrath of Khan* [13]. In the sequence referred to as the *genesis effect*, an expanding wall of fire spreads out over the surface of the planet from a single point of impact. The simulation is not a completely convincing model of fire, although the sequence is effective in the movie. The model uses a two-level hierarchy of particles. The first level of particles is located at the point of impact to simulate the initial blast (Figure 8.15); the second level consists of concentric rings of particles, timed to progress from the central point outward, forming the wall of fire and explosions.

Each ring of second-level hierarchy consists of a number of individual particle systems that are positioned on the ring and overlap to form a continuous ring. The individual particle systems are

FIGURE 8.14

Image representing fire generated using pixel averaging operations.

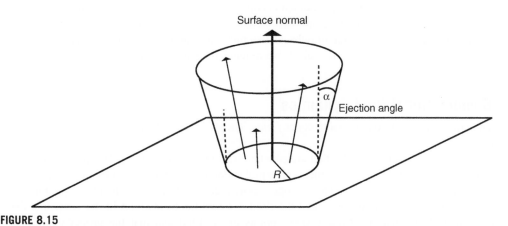

FIGURE 8.15

Explosion-like particle system.

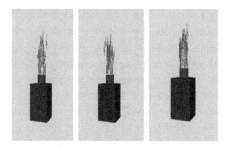

FIGURE 8.16

Images from a particle system simulation of fire.

modeled to look like explosions. The particles in each one of these particle systems are oriented to fly up and away from the surface of the planet. The initial position for a particle is randomly chosen from the circular base of the particle system. The initial direction of travel for each particle is constrained to deviate less than the ejection angle away from the surface normal. Figure 8.16 (Color Plate 4) shows a simple particle system fire.

Other approaches

Various other approaches have been used in animations with varying levels of success. Two-dimensional animated texture maps have been used to create the effect of the upward movement of burning gas, but such models are effective only when viewed from a specific direction. Using a two-dimensional multiple planes approach adds some depth to the fire, but viewing directions are still limited. Stam and Fiume [18] present advection-diffusion equations to evolve both density and temperature fields. The user controls the simulation by specifying a wind field. The results are effective, but the foundation mathematics are complicated and the model is difficult to control.

8.1.4 Summary

Modeling and animating amorphous phenomena is difficult. Gases are constantly changing shape and lack even a definable surface. Volume graphics hold the most promise for modeling and animating gas, but currently they have computational drawbacks that make such approaches of limited use for animation. A useful and visually accurate model of fire remains the subject of research.

8.2 Computational fluid dynamics

Gases and liquids are referred to collectively as *fluids*. The study of methods to compute their behavior is called CFD.

Fluids are composed of molecules that are constantly moving. These molecules are constantly colliding with one another and with other objects. The molecular motion transports measurable amounts of fluid properties throughout the fluid. These properties include such things as density, temperature, and momentum. In CFD, the assumption is that the fluid is a continuous medium in the sense that these properties are well-defined at infinitely small points of the fluid and that the property values are smoothly varying throughout the fluid. This is called the *continuity assumption*.

Modeling gaseous phenomena (smoke, clouds, and fire) is particularly challenging because of their ethereal nature. Gas has no definite geometry, no definite topology. Gas is usually treated as *compressible*, meaning that density is spatially variable and computing the changes in density is part of the computational cost. Liquids are usually treated as *incompressible*, which means the density of the material remains constant. In fact, the equations in Section 8.1.1 were derived from the CFD equations.

In a *steady-state flow*, the motion attributes (e.g., velocity and acceleration) at any point in space are constant. Particles traveling through a steady-state flow can be tracked similarly to how a space curve can be traced out when the derivatives are known. *Vortices*, circular swirls of material, are important features in fluid dynamics. In steady-state flow, vortices are attributes of space and are time invariant. In time-varying flows, particles that carry a non-zero vortex strength can travel through the environment and can be used to modify the acceleration of other particles in the system by incorporating a distance-based force.

8.2.1 General approaches to modeling fluids

There are three approaches to modeling gas: grid-based methods (*Eulerian formulations*), particle-based methods (*Lagrangian formulations*), and hybrid methods. The approaches are illustrated here in two dimensions, but the extension to three dimensions should be obvious.

Grid-based method

The grid-based method decomposes space into individual cells, and the flow of the gas into and out of each cell is calculated (Figure 8.17). In this way, the density of gas in each cell is updated from time step to time step. The density in each cell is used to determine the visibility and illumination of the gas during rendering. Attributes of the gas within a cell, such as velocity, acceleration, and density, can be used to track the gas as it travels from cell to cell.

The flow out of the cell can be computed based on the cell velocity, the size of the cell, and the cell density. The flow into a cell is determined by distributing the densities out of adjacent cells. External forces, such as wind and obstacles, are used to modify the acceleration of the particles within a cell.

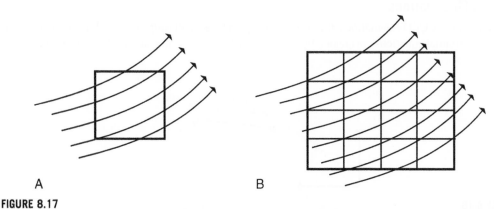

A   B

FIGURE 8.17

Grid-based method. (A) Gas flowing through an individual cell. (B) Grid of cells.

The rendering phase uses standard volumetric graphics techniques to produce an image based on the densities projected onto the image plane. Illumination of a cell from a light source through the volume must also be incorporated into the display procedure.

The disadvantage of this approach is that if a static data structure for the cellular decomposition is used, the extent of the interesting space must be identified before the simulation takes place in order to initialize the cells that will be needed during the simulation of the gas phenomena. Alternatively, a dynamic data structure that adapts to the traversal of the gas through space could be used, but this increases the complexity of the simulation.

Particle-based method

In the particle-based method, particles or globs of gas are tracked as they progress through space, often with a standard particle system approach (Figure 8.18). The particles can be rendered individually, or they can be rendered as spheres of gas with a given density. The advantage of this technique is that it is similar to rigid body dynamics and therefore the equations are relatively simple and familiar. The equations can be simplified if the rotational dynamics are ignored. In addition, there are no restrictions imposed by the simulation setup as to where the gas may travel. Particles are assigned masses, and external forces can be easily incorporated by updating particle accelerations and, subsequently, velocities. The disadvantage of this approach is that a large number of particles are needed to simulate a dense expansive gas and that to display the surface of a liquid, the boundary between and fluid and nonfluid must be reconstructed from the particle samples. Recent techniques generalize on the idea of using rigidly defined particles to include a particle that represents a distribution of fluid mass in the surrounding area.

Hybrid method

Some models of gas trace particles through a spatial grid. Particles are passed from cell to cell as they traverse the interesting space (Figure 8.19). The display attributes of individual cells are determined by the number and type of particles contained in the cell at the time of display. The particles are used to carry and distribute attributes through the grid, and then the grid is used to produce the display.

8.2.2 CFD equations

In developing the CFD calculations, the volume occupied by a small (*differential*) fluid element is analyzed. Differential equations are created by describing what is happening at this element. (Figure 8.20).

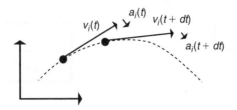

FIGURE 8.18

Particle-based method.

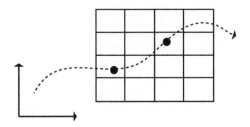

FIGURE 8.19

Hybrid method.

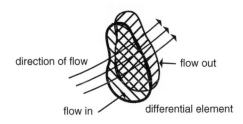

FIGURE 8.20

Differential element used in Navier-Stokes.

The full CFD equations consist of equations that state, in a closed system, the following:

- Mass is conserved.
- Momentum is conserved.
- Energy is conserved.

When expressed as equations, these are the *Navier-Stokes* (NS) equations. The NS equations are nonlinear differential equations. There are various simplifications that can be made in order to make the NS equations more tractable. First of all, for most all graphics applications, energy conservation is ignored. Another common simplification used in graphics is to assume the fluid is nonviscous. These assumptions result in what are called *Euler equations* [12]. By ignoring energy conservation and viscosity, the Euler equations describing the conservation of mass and conservation of momentum can be readily constructed for a given flow field. When dealing with liquids (as opposed to a gas), another useful assumption is that the fluid is incompressible, which means that its density does not change.

Conservation of mass

Conservation of mass is expressed by the *continuity equation* [4]. The underlying assumption of the conservation of mass is that mass is neither created nor destroyed inside of the fluid. The fluid flows from location to location and, if compressible, can become more dense or less dense at various locations. To determine how the total mass inside of an element is changing, all of the sources and sinks in the element have to be added up. The *divergence* of a flow field describes the strength of the source or sink at a location; the divergence for a field, $F=(Fx, Fy, Fz)$, is $div\, F = \nabla \cdot F = \dfrac{\partial F_x}{\partial x} + \dfrac{\partial F_y}{\partial y} + \dfrac{\partial F_z}{\partial z}$. If no mass is created or destroyed inside the element, then it is reasonable that the total divergence within the element must be equal to the flow differences across the element at the x, y, and z boundaries. This is stated by the *Divergence Theorem*: the integral of the flow field's divergence over the volume of the element is the same as the integral of the flow field over the element's boundary. So, instead of worrying about all the sources and sinks inside of the element, all that has to be computed is the flow at the boundaries.

For now, for a given three-dimensional element, consider only flow in the x-direction. To make the discussion a bit more concrete, assume that the fluid is flowing left to right, as x increases, so that the mass flows into a element on the left, at x, and out of the element on the right, at $x + dx$. In the following equations, ρ is density, p is pressure, A is the area of the face perpendicular to the x-direction, V is the volume of the element, and the element dimensions are dx by dy by dz. To compute the mass flowing across a surface, multiply the density of the fluid times area of the surface times the velocity component normal to the surface, v_x in this case. The conservation of mass equation states that the rate at which mass changes is equal to difference between mass flowing into the element and the mass flowing out of the element (Eq. 8.29). The notation for a value at a given location is $<value>_{(location}$ or, when appropriate, $<value>_{(location\ surface}$.

$$-\frac{\partial(\rho V)}{\partial t} = (\rho v_x A)|_{x+dx} - (\rho v_x A)|_x \tag{8.29}$$

Replace the volume, V, and area, A, by their definitions in terms of the element dimensions (Eq. 8.30).

$$-\frac{d(\rho dx dy dz)}{dt} = (\rho v_x dy dz)|_{x+dx} - (\rho v_x dy dz)|_x \tag{8.30}$$

Divide through by $dx dy dz$ (Eq. 8.31).

$$-\frac{\partial \rho}{\partial t} = \frac{\rho v_x}{dx}|_{x+dx} - \frac{\rho v_x}{dx}|_x \tag{8.31}$$

Replace the finite differences by their corresponding differential forms (Eq. 8.32).

$$-\frac{\partial \rho}{\partial t} = \frac{\partial(\rho v_x)}{\partial x} \tag{8.32}$$

Finally, include flow in all three directions (Eq. 8.33).

$$-\frac{\partial \rho}{\partial t} = \frac{\partial(\rho v_x)}{\partial x} + \frac{\partial(\rho v_y)}{\partial y} + \frac{\partial(\rho v_z)}{\partial z} \tag{8.33}$$

If ∇ is used as the gradient operator, $\nabla F = \left(\dfrac{\partial F}{\partial x}, \dfrac{\partial F}{\partial y}, \dfrac{\partial F}{\partial z} \right)$ then a common notation is to define the divergence operator as in Equation 8.34.

$$\nabla \cdot F = \frac{\partial F_x}{\partial x} + \frac{\partial F_y}{\partial y} + \frac{\partial F_z}{\partial z} \tag{8.34}$$

Using the divergence operator in Equation 8.33 produces Equation 8.35.

$$-\frac{\partial \rho}{\partial t} = \nabla \cdot (\rho v) \tag{8.35}$$

If the fluid is incompressible, then the density does not change. This means that $\frac{\partial \rho}{\partial t}$ is zero and the constant density can be divided out, resulting in Equation 8.36.

$$0 = \nabla \cdot v \tag{8.36}$$

Conservation of momentum

The momentum of an object is its mass times its velocity. The momentum of a fluid is its density times the volume of the fluid times the average velocity of the fluid.

A change in momentum $\frac{d(mv)}{dt}$ must be induced by a force (i.e., $f = ma = m\frac{dv}{dt}$). The force in the element of a flow field is either a global force, such as viscosity and gravity, or a change in pressure across the element, $\frac{dp}{dx}$. As noted previously, effects of viscosity are commonly ignored in order to simplify the equations; we will also ignore gravity for now.

For an element, the change in momentum is the change of momentum inside the element plus the difference between the momentum entering the cell and the momentum leaving the element. The mass inside the cell is ρV and, thus, the momentum inside the element is $\rho V v$. The rate of change of momentum inside the element is $\frac{\partial(\rho V v)}{\partial t}$.

Given a three-dimensional element, consider flow only in the x-direction. The mass flowing out one surface is density $\times$ area of the surface times velocity in the x-direction, v_x. The momentum out one surface is the mass flowing out the surface times its velocity, $\rho A v_x v$. The force on the element in the x-direction is the pressure difference across the element from x to $x+dx$. The force is equal to the negative of the change in momentum Equation 8.37.

$$-\left(p\big|_{x+dx}^A - p\big|_x^A\right) = \frac{\partial(\rho V v)}{\partial t} + \left((\rho v_x A)v\big|_{x+dx} - (\rho v_x A)v\big|_x\right) \tag{8.37}$$

Following the steps above in developing the conservation of mass equation by replacing the area, A, with its definition in terms of element dimensions, $dydz$, replacing the volume, V, with $dxdydz$, dividing through by $dxdydz$, and putting everything in differential form gives Equation 8.38.

$$-\frac{\partial p}{\partial x} = \frac{\partial(\rho v_x v)}{\partial x} + \frac{\partial(\rho v)}{\partial t} \tag{8.38}$$

Now, considering flow in all three directions, and separating out momentum only in the x direction, produces Equation 8.39.

$$-\frac{\partial p}{\partial t} = \frac{\partial\left(\rho v_x^2\right)}{\partial x} + \frac{\partial\left(\rho v_x v_y\right)}{\partial y} + \frac{\partial(\rho v_x v_z)}{\partial z} + \frac{\partial(\rho v_x)}{\partial t} \tag{8.39}$$

If desired, viscosity and other forces can be added to the pressure differential force on the left-hand side of the equation. Similar equations can be derived for the y and z directions.

8.2.3 Grid-based approach

The basic process for computing fluid dynamics using a grid-based (Eulerian) approach is to do the following:

- Discretize the fluid continuum by constructing a grid of cells that, collectively, cover the space occupied by the fluid and at each node (cell), the fluid variables (e.g., density and velocity) are approximated and used to initialize the grid cells.
- Create the discrete equations using the approximate values at the element.
- Numerically solve the system of equations to compute new values at the cells using a Newton-like method for the large, sparse system of equations at each time step.

Before solving these equations, boundary conditions must be set up to handle the cells at the limits of the domain. For example, the *Dirichlet boundary condition*, which sets solution values at the boundary cells, is commonly used. The CFD equations, set up for each cell of the grid, produce a sparse set of matrix equations. There are various ways to solve such systems. For example, LU and Conjugate Gradient solvers are often used. Efficient, accurate methods of solving symmetric and asymmetric sparse matrices, such as these, are the topic of ongoing research. Thankfully, in computer animation believability is more of a concern than accuracy, so approximate techniques typically suffice.

Stable fluids

Stam presents a method that solves the full NS equations that is easy to implement and allows real-time user interaction [17]. Although too inaccurate for engineering, it is useful in computer graphics applications. In particular, the procedures are designed to be stable more than accurate while still producing interesting visuals. The procedures operate on a velocity vector field and density field. The velocity field moves the density values around.

Density update

Density values are moved through the density grid assuming a fixed velocity field over the time step. The density equation (Eq. 8.40) states that density (ρ) changes over time according to density sources (s), the diffusion of the density at some specified rate (k), and advection of the density according to the velocity field ($\mathbf{u}$).

$$-\frac{\partial \rho}{\partial t} = s = k\nabla^2\rho - (\mathbf{u} \cdot \nabla)\rho \qquad (8.40)$$

In implementing stable fluids, density is kept in an array of scalars. The first term, s, is an array of density sources set by user interaction or by the environment. These are used to initialize the new density grid at each time step. The second term diffuses the density from each cell to its adjacent cells. The diffusion of the grid can be solved by relaxation. The third term advects the density according to the velocity field.

To diffuse, each cell is updated by a weighted average of its four neighbors added to the original value. Each neighboring value is multiplied by a where $a = dt*diff*N*N$, where *diff* is a user-supplied parameter, and N is the resolution of the grid. This sum is added to the cell's original value and then this sum is divided by $1 + 4*a$.

To advect, the position is interpolated backward according to the velocity field. In the simplest case, the velocity vector at the particle position is accessed and linearly stepping backward according to the selected time step, making sure to clamp within the grid boundaries. The density is then bilinearly interpolated from nearby values and this interpolated value is used as the density of the position.

The velocity update

The change of the velocity over time is given by Equation 8.41 and is due to external forces (f), diffusion of the velocities, and self-advection.

$$-\frac{\partial u}{\partial t} = f + \mu \nabla^2 u - (u \cdot \nabla)u \qquad (8.41)$$

The velocity is kept in an array of vectors. The first term, f, adds in external forces from the user or environment. The second term allows for the diffusion of the velocity, and the third term allows the velocity to be carried along by the velocity field.

These equations are implemented by the sequence: add force, advect, diffuse, and project. The projection step is to adjust the density so that mass is preserved.

The simulation

The simulation is conducted by updating the contributions from density forces and external forces, then stepping the velocity field a step in time, taking the density field a step in time, then drawing the density.

8.2.4 Particle-based approaches including smoothed particle hydrodynamics

As previously noted, particle-based approaches have the advantage of being intuitive, easy to program, and fast to execute—at the expense of accuracy. However, for computer graphics and animation, accuracy is a common sacrifice. The basic idea is to model the fluid by a mass of particles. A simple particle system can be used, but the rigid nature of standard particles means that a massive amount of particles has to be used in order to create the impression of a fluid. In cases of modeling a gas or a spray of liquid this can be effective, but in cases where it is useful to represent a liquid's surface (e.g., for display) this approach is problematic.

In order to give a better impression of a cohesive fluid, spherical implicit surfaces, often referred to as *metaballs* (see Chapter 12.1), can be used in place of rigid particles. Because metaballs allow for the reconstruction of smooth intersections between particles, this approach can produce a gooey-like appearance similar to crunchy peanut butter. Metaballs have the ability to generate organic, blobby shapes but fall short of being able to model liquids effectively. While this is an improvement over rigid particles, is still takes a very large number of particles to produce a smooth surface. In addition, metaballs lack any physical characteristics that make them suitable for effectively representing liquids.

For a more principled approach and one that produces a smoother surface, smooth particle hydrodynamics (SPH), developed for astrophysics, does a better and faster job of rendering a liquid from a collection of particles. SPH can be viewed as an extension of the metaball approach (Figure 8.21). The SPH particle is similar to that in a typical particle system and is represented by its position and velocity. An SPH particle, similar to an implicit surface particle, has a density function that contributes to a field value over its area of influence. The difference is that each SPH particle has a mass and density and each SPH particle also carries and distributes its mass over an area. The SPH particle has a kernel function that plays the role of a metaball's density function. The main difference

FIGURE 8.21

Water simulation using sSPH.

(Image courtesy of Di Cao.)

between an SPH particle and an implicitly defined sphere is that one of the kernel's parameters is its support distance, which is a function of its mass. It should be noted that there are different ways to model particles and compute values in SPH and that the material here follows the work of Muller et al. [21][22]. The interested reader can refer to these papers for more details.

In implementing the fluid dynamics using SPH particles, the main concerns are about conserving mass and updating the fluid particle representation over time due to the forces on the particles.

An additional practical consideration is how to display the fluid—how to reconstruct the fluid from the particle definition at any one point in time for display—and, especially for liquid, how to identify its surface. Similar to a standard particle system, at any one point in time a particle has a position and a velocity and is subject to various forces. The forces considered are those captured in the CFD equations—the pressure gradient and viscosity of the fluid, as well as environmental forces such as gravity. In addition, in the case of liquids, a surface tension force can be modeled.

An SPH particle carries an additional property of mass. This mass is distributed through space according to a smoothing kernel function. In fact, any property of the fluid can be distributed through space using a smoothing function. A property, s, at any given location, r, can be computed using Equation 8.42. In particular, the density can be computed from Equation 8.42, resulting in Equation 8.44. The gradient of a property and the Laplacian of a property, useful in computer pressure gradient and viscosity, can be computed using Equation 8.43.

$$s(r) = \sum_j m_j \frac{s_j}{\rho_j}\, W_h\left(r - r_j\right) \qquad (8.42)$$

$$\nabla s(r) = \sum_j m_j \frac{s_j}{\rho_j} \nabla W_h(r - r_j)$$

$$\nabla^2 s(r) = \sum_j m_j \frac{s_j}{\rho_j} \nabla^2 W_h(r - r_j)$$

(8.43)

$$\rho(r) = \sum_j m_j W_h(r - r_j) \rho_s(r) = \sum_j m_j W_h(r - r_j)$$

(8.44)

The use of particles guarantees that mass is conserved as long as the mass of a particle does not change and as long as particles are not created or destroyed. Updating the fluid according to CFD principles is effected by migration of particles through space due to the forces modeled by the CFD equations. First, the density at a location is computed using Equation 8.43.

Then the pressure is computed using the ideal gas state equation $p = k(\rho - \rho_0)$, where ρ_0 is the rest density for water and k is a constant in the range of 100 to 1000. Forces derived directly by SPH are not guaranteed to be symmetric, so symmetric versions of the force equations are then computed (Eq. 8.45):

$$f_{pressure}(r_i) = -\sum_j M_j \frac{p_i + p_j}{2\rho_j} \nabla W(r_i - r_j, h)$$

$$f_{gravity}(r_i) = \rho_i g$$

$$f_{viscosity}(r_i) = \mu \sum_j m_j \frac{v_j - v_i}{\rho_j} \nabla^2 W(r_i - r_j, h)$$

(8.45)

Quantities in the particle field can be computed at any given location by the use of a radial, symmetric, smoothing kernel. Smoothing kernels with a finite area of support are typically used for computational efficiency. Smoothing kernels should integrate to 1, which is referred to as *normalized*. Different smoothing kernels can be used when computing various values as long as the smoothing kernel has the basic properties of being normalized and symmetric. As examples, the kernels used in the work being followed here are a general purpose kernel (Eq. 8.46), a kernel for pressure to avoid a zero gradient at the center (Eq. 8.47), and a kernel for the viscosity (Eq. 8.48) and its corresponding Laplacian (Eq. 8.49). In this work, a Leap Frog integrator is used to compute the accelerations from the forces.

$$W_{poly6}(r, h) = \frac{315}{64\pi h^9} \begin{cases} (h^2 - r^2) & 0 \leq r \leq hb \\ 0 & otherwise \end{cases}$$

(8.46)

$$W_{spiky}(r, h) = \frac{15}{\pi h^6} \begin{cases} (h - r)^3 & 0 \leq r \leq h \\ 0 & otherwise \end{cases}$$

(8.47)

$$W_{viscosity}(r, h) = \frac{15}{2\pi h^3} \begin{cases} -\frac{r^3}{2h^3} + \frac{r^2}{h^2} + \frac{h}{2r} - 1 & 0 \leq r \leq h \\ 0 & otherwise \end{cases}$$

(8.48)

$$\nabla^2 W_{viscosity}(r, h) = \frac{45}{\pi h^6}(h - r)$$

(8.49)

8.3 Chapter summary

Most of the techniques discussed in this chapter are still the subject of research efforts. Water, smoke, clouds, and fire share an amorphous nature, which makes them difficult to model and animate. Approaches that incorporate a greater amount of physics have been developed recently for these phenomena. As processing power becomes cheaper, techniques such as CFD become more practical (and more desirable) tools for animating water and gas, but convenient controls for such models have yet to be developed.

References

[1] Blinn J. Simulation of Wrinkled Surfaces. In: Computer Graphics. Proceedings of SIGGRAPH 78, 12(3). Atlanta, Ga.; August 1978. p. 286–92.

[2] Bukowski R, Sequin C. Interactive Simulation of Fire in Virtual Building Environments. In: Whitted T, editor. Computer Graphics. Proceedings of SIGGRAPH 97, Annual Conference Series. Los Angeles, Calif.: Addison-Wesley; August 1997. p. 35–44. ISBN 0-89791-896-7.

[3] Ebert D, Musgrave K, Peachey D, Perlin K, Worley S. Texturing and Modeling: A Procedural Approach. Cambridge, Massachusetts: AP Professional; 1998.

[4] Engineers Edge. Continuity Equation—Fluid Flow, http://www.engineersedge.com/fluid_flow/continuity_equation.htm; August 2005.

[5] Foster N, Metaxas D. Realistic Animation of Liquids. Graphical Models and Image-Processing, September 1996;58(5):471–83. Academic Press.

[6] Fournier A, Reeves W. A Simple Model of Ocean Waves. In: Evans DC, Athay RJ, editors. Computer Graphics. Proceedings of SIGGRAPH 86, 20(4). Dallas, Tex.; August 1986. p. 75–84.

[7] Fournier P, Habibi A, Poulin P. Simulating the Flow of Liquid Droplets. In: Booth K, Fournier A, editors. Graphics Interface '98. June 1998. p. 133–42. ISBN 0-9695338-6-1.

[8] Gardner G. Simulation of Natural Scenes Using Textured Quadric Surfaces. In: Computer Graphics. Proceedings of SIGGRAPH 84, 18(3). Minneapolis, Minn.; July 1984. p. 11–20.

[9] Gardner G. Visual Simulation of Clouds. In: Barsky BA, editor. Computer Graphics. Proceedings of SIGGRAPH 85, 19(3). San Francisco, Calif.; August 1985. p. 297–303.

[10] Kass M, Miller G. Rapid, Stable Fluid Dynamics for Computer Graphics. In: Baskett F, editor. Computer Graphics. Proceedings of SIGGRAPH 90, 24(4). Dallas, Tex.; August 1990. p. 49–57. ISBN 0-201-50933-4.

[11] Max N. Vectorized Procedural Models for Natural Terrains: Waves and Islands in the Sunset. In: Computer Graphics. Proceedings of SIGGRAPH 81, 15(3). Dallas, Tex.; August 1981. p. 317–24.

[12] NASA. Euler Equations, http://www.grc.nasa.gov/WWW/K-12/airplane/eulereqs.html; August, 2005.

[13] Paramount. Star Trek II: The Wrath of Khan (film). June 1982.

[14] Parusel A. Simple Fire Effect, GameDev.net; January, 2007. http://www.gamedev.net/reference/articles/article222.asp.

[15] Peachey D. Modeling Waves and Surf. In: Evans DC, Athay RJ, editors. Computer Graphics. Proceedings of SIGGRAPH 86, 20(4). Dallas, Tex.; August 1986. p. 65–74.

[16] Rushmeier H, Hamins A, Choi M. Volume Rendering of Pool Fire Data. IEEE Comput Graph Appl July 1995;15(4):62–7.

[17] Stam J. Stable Fluids. In: Rockwood A, editor. Computer Graphics. Proceedings of SIGGRAPH 99, Annual Conference Series. Los Angeles, Calif.: Addison-Wesley Longman; August 1999. p. 121–8. ISBN 0-20148-560-5.

[18] Stam J, Fiume E. Depicting Fire and Other Gaseous Phenomena Using Diffusion Processes. In: Computer Graphics. Proceedings of SIGGRAPH 95), In Annual Conference Series. Los Angeles, Calif: ACM SIGGRAPH; August 1995. p. 129–36.

[19] Tessendorf J. Simulating Ocean Water. In: SIGGRAPH 2002 Course Notes #9 (Simulating Nature: Realistic and Interactive Techniques). ACM Press; 2002.

[20] Weimer H, Warren J. Subdivision Schemes for Fluid Flow. In: Rockwood A, editor. Computer Graphics. Proceedings of SIGGRAPH 99, Annual Conference Series. Los Angeles, Calif.: Addison-Wesley Longman; August 1999. p. 111–20. ISBN 0-20148-560-5.

[21] Muller M, Charypar D, Gross M. Particle-based Fluid Simulation for Interactive Applications. In: Eurographics/SIGGRAPH Symposium on Computer Animation. 2003.

[22] Muller M, Schirm S, Teschner M. Interactive Blood Simulation for Virtual Surgery Based on Smoothed Particle Hydrodynamics. Technology and Health Care April 19, 2004;12(1/2004):25–31.

[23] WeatherOnline Ltd. Weather Facts: Cloud Types, www.weatheronline.co.uk/reports/wxfacts/Cloud-types. html; Nov. 2011.

Modeling and Animating Human Figures

9

Modeling and animating an articulated figure is one of the most formidable tasks that an animator can be faced with. It is especially challenging when the figure is meant to realistically represent a human. There are several major reasons for this. First, the human figure is a very familiar form. This familiarity makes each person a critical observer. When confronted with an animated figure, a person readily recognizes when its movement does not "feel" or "look" right. Second, the human form is very complex, with more than two hundred bones and six hundred muscles. When fully modeled with linked rigid segments, the human form is endowed with approximately two hundred degrees of freedom (DOFs). The deformable nature of the body's surface and underlying structures further complicates the modeling and animating task. Third, human-like motion is not computationally well defined. Some studies have tried to accurately describe human-like motion, but typically these descriptions apply only to certain constrained situations. Fourth, there is no one definitive motion that is humanlike. Differences resulting from genetics, culture, personality, and emotional state all can affect how a particular motion is carried out. General strategies for motion production have not been described, nor have the nuances of motion that make each of us unique and uniquely recognizable. Although the discussion in this chapter focuses primarily on the human form, many of the techniques apply to any type of articulated figure.

Table 9.1 provides the definitions of the anatomical terms used here. Particularly noteworthy for this discussion are the terms that name planes relative to the human figure: *sagittal*, *coronal*, and *transverse*.

9.1 Overview of virtual human representation

One of the most difficult challenges in computer graphics is the creation of virtual humans. Efforts to establish standard representations of articulated bodies are now emerging [16] [31]. Representing human figures successfully requires solutions to several very different problems. The visible geometry of the skin can be created by a variety of techniques, differing primarily in the skin detail and the degree of representation of underlying internal structures such as bone, muscle, and fatty tissue. Geometry for hair and clothing can be simulated with a clear trade-off between accuracy and computational complexity. The way in which light interacts with skin, clothing, and hair can also be calculated with varying degrees of correctness, depending on visual requirements and available resources.

Techniques for simulating virtual humans have been created that allow for modest visual results that can be computed in real time (i.e., 30 Hz). Other approaches may give extremely realistic results by simulating individual hairs, muscles, wrinkles, or clothing threads. These methods may consequently also require excessive amounts of computation time to generate a single frame of animation. The

Table 9.1 Selected Terms from Anatomy

Sagittal plane	Perpendicular to the ground and divides the body into right and left halves
Coronal plane	Perpendicular to the ground and divides the body into front and back halves
Transverse plane	Parallel to the ground and divides the body into top and bottom halves
Distal	Away from the attachment of the limb
Proximal	Toward the attachment of the limb
Flexion	Movement of the joint that decreases the angle between two bones
Extension	Movement of the joint that increases the angle between two bones

discussion in this section focuses solely on the representation of virtual humans. It addresses animation issues where necessary to discuss how the animation techniques affect the figure's creation.

9.1.1 Representing body geometry

Many methods have been developed for creating and representing the geometry of a virtual human's body. They vary primarily in visual quality and computational complexity. Usually these two measures are inversely proportional.

The vast majority of human figures are modeled using a boundary representation constructed from either polygons (often triangles) or patches (usually nonuniform rational B-splines; NURBS). These boundary shell models are usually modeled manually in one of the common off-the-shelf modeling packages (e.g., [1] [4] [35]). The purpose of the model being produced dictates the technique used to create it. If the figure is constructed for real-time display in a game on a low-end PC or gaming console, usually it will be assembled from a relatively low number of triangular polygons, which, while giving a chunky appearance to the model, can be rendered quickly. If the figure will be used in an animation that will be rendered off-line by a high-end rendering package, it might be modeled with NURBS patch data, to obtain smooth curved contours. Factors such as viewing distance and the importance of the figure to the scene can be used to select from various levels of detail at which to model the figure for a particular sequence of frames.

Polygonal representations

Polygonal models typically consist of a set of vertices and a set of faces. Polygonal human figures can be constructed out of multiple objects (frequently referred to as segments), or they can consist of a single polygonal mesh. When multiple objects are used, they are generally arranged in a hierarchy of joints and rigid segments. Rotating a joint rotates all of that joint's children (e.g., rotating a hip joint rotates all of the child's leg segments and joints around the hip). If a single mesh is used, then rotating a joint must deform the vertices surrounding that joint, as well as rotate the vertices in the affected limb.

Various constraints may be placed on the polygonal mesh's topology depending on the software that will be displaying the human. Many real-time rendering engines require polygonal figures to be constructed from triangles. Some modeling programs require that the object remain closed.

Polygonal representations are primarily used either when rendering speed is of the essence, as is the case in real-time systems such as games, or when topological flexibility is required. The primary problem with using polygons as a modeling primitive is that it takes far too many of them to represent a smoothly curving surface. It might require hundreds or thousands of polygons to achieve the same visual quality as could be obtained with a single NURBS patch.

Patch representations

Virtual humans constructed with an emphasis on visual quality are frequently built from a network of cubic patches, usually NURBS. The control points defining these patches are manipulated to sculpt the surfaces of the figure. Smooth continuity must be maintained at the edges of the patches, which often proves challenging. Complex topologies also can cause difficulties, given the rectangular nature of the patches. While patches can easily provide much smoother surfaces than polygons in general, it is more challenging to add localized detail to a figure without adding a great deal of more global data. Hierarchical splines provide a partial solution to this problem [21].

Other representations

Several other methods have been used for representing virtual human figures. However, they are used more infrequently because of a lack of modeling tools or because of their computational complexity.

Implicit surfaces (Chapter 12.1) can be employed as sculpting material for building virtual humans. Frequently the term "metaballs" is used for spherical implicit surfaces. Metaballs resemble clay in their ability to blend with other nearby primitives. While computationally expensive to render, they provide an excellent organic look that is perfect for representing skin stretched over underlying tissue [35] [41].

Subdivision surfaces (Chapter 12.3) combine the topological flexibility of polygonal objects with the resultant smoothness of patch data. They transform a low-resolution polygonal model into a smooth form by recursively subdividing the polygons as necessary [17] [19].

Probably the most computationally demanding representation method is volumetric modeling. While all of the previously mentioned techniques merely store information about the surface qualities of a virtual human, volumetric models store information about the entire interior space as well. Because of its extreme computational requirements, this technique is limited almost exclusively to the medical research domain, where knowledge of the interior of a virtual human is crucial.

As computers continue to become more powerful, more attempts are being made to more accurately model the interior of humans to get more realistic results on the visible surfaces. There have been several "layered" approaches, where some attempt has been made to model underlying bone and/or muscle and its effect on the skin.

Chen and Zeltzer [14] use a finite element approach to accurately model a human knee, representing each underlying muscle precisely, based on medical data. Several authors have attempted to create visually reasonable muscles attached to bones and then generate skin over the top of the muscle (e.g., [57] [66]). Thalmann's lab takes the interesting hybrid approach of modeling muscles with metaballs, producing cross sections of these metaballs along the body's segments, and then lofting polygons between the cross sections to produce the final surface geometry [10]. Chadwick et al. [13] use free form deformations (FFDs) to produce artist-driven muscle bulging and skin folding, as described in Section 9.1.5.

9.1.2 Geometry data acquisition

Geometric data can be acquired by a number of means. By far the most common method is to have an artist create the figure using interactive software tools. The quality of the data obtained by these means is of course completely dependent on the artist's skills and experience. Another method of obtaining data, digitizing real humans, is becoming more prevalent as the necessary hardware becomes more affordable. Data representing a specific individual are often captured using a laser scanner or by performing image processing on video images [28] [33] [43].

There have also been various parametric approaches to human figure data generation. Virtual human software intended for use in ergonomics simulation tends to use parameters with a strong anthropometric basis [6]. Software with an artistic or entertainment orientation allows for more free-form parametric control of the generated body data [35] [59]. A few efforts use exemplar-based models to allow for data generation by mixing known attributes [9] [43].

9.1.3 Geometry deformation

For a user to animate a virtual human figure, the figure's limbs and other parts must be able to be moved and/or deformed. The method of manipulation used is largely determined by the way the figure is being represented. Very simple figures, usually used for real-time display, are often broken into multiple rigid subparts, such as a forearm, a thigh, or a head. These parts are arranged in a hierarchy of joints and segments such that rotating a joint rotates all of the child segments and joints beneath it in the hierarchy [31]. While this method is quick, it yields suboptimal visual results at the joints, particularly if the body is textured, because the rigid parts merely overlap and occlude each other at the joint.

A single skin is more commonly used for polygonal figures. When a joint is rotated, the appropriate vertices are deformed to simulate rotation around the joint. Several different methods can be used for this, and, as with most of these techniques, they involve trade-offs of realism and speed. The simplest and fastest method is to bind each vertex to exactly one bone. When a bone rotates, the vertices move along with it [7]. Better results can be obtained, at the cost of additional computation, if the vertices surrounding a joint are weighted so that their position is affected by multiple bones [41]. While weighting the effects of bone rotations on vertices results in smoother skin around joints, severe problems can still occur with extreme joint bends. FFDs have been used in this case to simulate the skin-on-skin collision and the accompanying squashing and bulging that occur when a joint such as the elbow is fully bent [13]. The precise placement of joints within the body greatly affects the realism of the surrounding deformations. Joints must be placed strictly according to anthropometric data, or unrealistic bulges will result [27] [53].

Some joints require more complicated methods for realistic deformation. Using only a simple, single three degrees of freedom (DOF) rotation for a shoulder or vertebrae can yield very poor results. A few systems have attempted to construct more complex, anatomically based joints [21] [50]. The hands in particular can require significantly more advanced deformation techniques to animate realistically [23]. Surface deformations that would be caused by changes in underlying material, such as muscle and fat in a real human, can be produced in a virtual human by a number of means. Methods range from those that simply allow an animator to specify muscular deformations [13] to those that require complex dynamics simulations of the various tissue layers and densities [49] [57]. A great deal of muscle simulation research has been conducted for facial animation. See the Chapter 10 for more details. Finally, deformations resulting from interaction with the environment have been simulated both with traditional dynamics systems and with implicit surfaces [60].

9.1.4 Surface detail

After the geometry for a virtual figure has been constructed, its surface properties must also be specified. As with the figure's geometry, surface properties can be produced by an artist, scanned from real life, or procedurally generated. Color, as well as specular, diffuse, bump, and displacement maps may be generated. Accurately positioning the resulting textures requires the generation of texture coordinates [56]. The skin may be simulated using complex physically based simulations [25] [44]. Wrinkles may also be simulated by various means [47] [60].

9.1.5 Layered approach to human figure modeling

A common approach to animating the human figure is to construct the figure in layers consisting of skeleton, muscles, and skin. The skeletal layer is responsible for the articulation of the form. The muscle layer is responsible for deforming the shape as a result of skeletal articulation. The skin is responsible for carrying the appearance of the figure.

Chadwick et al. [13] introduced the layered approach to figure animation by incorporating an articulated skeleton, surface geometry representing the skin, and an intermediate muscle layer that ties the two together. The muscle layer is not anatomically based, and its only function is to deform the surface geometry as a function of joint articulation. The muscle layer implements a system of FFD lattices in which the surface geometry is embedded. The lattice is organized with two planes on each end that are responsible for maintaining continuity with adjoining lattices, and the interior planes are responsible for deforming the skin to simulate muscle bulging (Figure 9.1). As the joint flexes, the interior planes elongate perpendicular to the axis of the link. The elongation is usually not symmetrical about the axis and is designed by the animator. For example, the upper-arm FFD lattice elongates as the elbow flexes. Typically, the FFD for the upper arm is designed to produce the majority of the skin deformation in the region of the biceps. A pair of FFD lattices is used on either side of each joint to isolate the FFDs responsible for muscle bulging from the more rigid area around the joint. In addition, the joint FFDs are designed to maintain continuity on the outside of the joint and create the skin crease on the inside of the joint (see Figure 9.2).

Artistic anatomy can be used to guide analysis of the human form [57][58][66]. Bones, muscles, tendons, and fatty tissue are modeled in order to occupy the appropriate volumes. Scheepers [57] identifies the types of muscles sufficient for modeling the upper torso of the human figure: linear muscles,

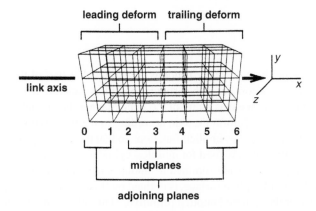

FIGURE 9.1

The basic FFD lattice [13].

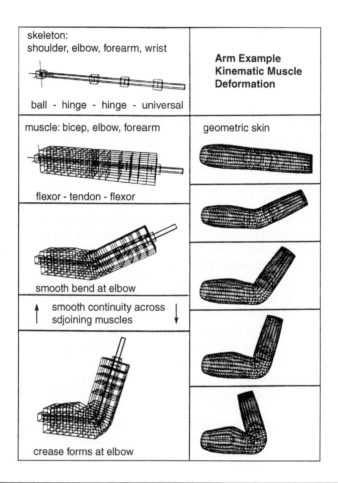

FIGURE 9.2

Deformation induced by FFDs as a result of joint articulation [13].

sheet muscles, and bendable sheet muscles. Tendons are modeled as part of the muscles and attach to the skeleton. Muscles deform according to the articulation of the skeleton. See Figure 9.3 for an example. These muscles populate the skeletal figure in the same manner that actual muscles are arranged in the human body (Figure 9.4). To deform the skin according to the underlying structure (muscles, tendons, and fatty tissue), the user defines implicit functions so that the densities occupy the volume of the corresponding anatomy. Ellipsoids are used for muscles, cylinders for tendons, and flattened ellipsoids for fat pads. The implicit primitives are summed to smooth the surface and further model underlying tissue. The skin, modeled as a B-spline surface, is defined by floating the control points of the B-spline patches to the isosurface of the summed implicit primitives. This allows the skin to distort as the underlying skeletal structure articulates and muscles deform (Figure 9.5).

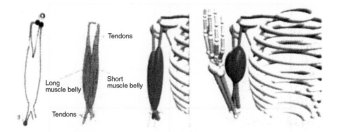

FIGURE 9.3

Linear muscle model [57].

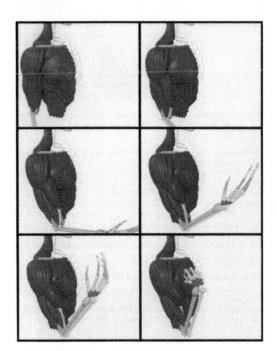

FIGURE 9.4

Muscles of upper torso [57].

Rigging

Rigging, as used here, refers to setting up interactive controls to facilitate animation of a character or other object. In the case of a character, a rig might be set up to control taking a step or waving a hand. Typically, a character rig facilitates the animation of a character by providing a high-level control that manipulates multiple DOFs in a coordinated manner. This not only makes animating the character easier, it also provides consistency for the character's movements.

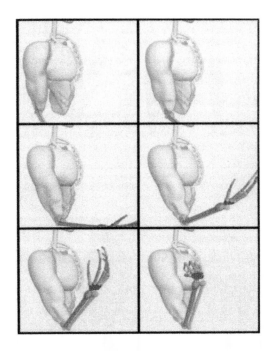

FIGURE 9.5

Skin model over muscles, tendons, and fatty tissue [57].

9.2 Reaching and grasping

One of the most common human figure animation tasks involves movement of the upper limbs. A synthetic figure may be required to reach and operate a control, raise a coffee cup from a table up to his mouth to drink, or turn a complex object over and over to examine it. It is computationally simpler to consider the arm as an appendage that moves independently of the rest of the body. In some cases, this can result in unnatural-looking motion. To produce more realistic motion, the user often involves additional joints of the body in executing the motion. In this section, the arm is considered in isolation. It is assumed that additional joints, if needed, can be added to the reaching motion in a preprocessing step that positions the figure and readies it for independently considered arm motion.

9.2.1 Modeling the arm

The basic model of the human arm, ignoring the joints of the hand for now, can be most simply represented as a seven DOFs manipulator (Figure 9.6): three DOFs are at the shoulder joint, one at the elbow, and three at the wrist. See Chapter 5 for an explanation of joint representation, forward kinematics, and inverse kinematics. A *configuration* or *pose* for the arm is a set of seven joint angles, one for each of the seven DOFs of the model.

Forearm rotation presents a problem. In Figure 9.6, the forearm rotation is associated with the wrist. However, in reality, the forearm rotation is not associated with a localized joint like most of the other

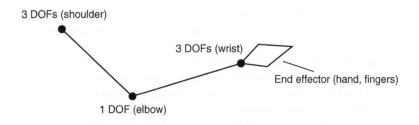

FIGURE 9.6

Basic model of the human arm.

DOFs of the human figure; rather, it is distributed along the forearm itself as the two forearm bones (radius and ulna) rotate around each other. Sometimes this rotation is associated with the elbow instead; other implementations create a "virtual" joint midway along the forearm to handle forearm rotation.

Of course, the joints of the human arm have limits. For example, the elbow can flex to approximately 20 degrees and extend to typically around 180 degrees. Allowing a figure's limbs to exceed the joint limits would certainly contribute to an unnatural look. Most joints are positioned with least strain somewhere in the middle of their range and rarely attain the boundaries of joint rotation unless forced. More subtly, joint limits may vary with the position of other joints, and further limits are imposed on joints to avoid intersection of appendages with other parts of the body. For example, if the arm is moved in a large circle parallel to and to the side of the torso, the muscle strain causes the arm to distort the circle at the back. As another example, in some individuals tendons make it more difficult to fully extend the knee when one is bending at the hip (the motion used to touch one's toes).

If joint limits are enforced, some general motions can be successfully obtained by using forward kinematics. Even if an object is to be carried by the hand, forward kinematics in conjunction with attaching the object to the end effector creates a fairly convincing motion. But if the arm/hand must operate relative to a fixed object, such as a knob, inverse kinematics is necessary. Unfortunately, the normal methods of inverse kinematics, using the pseudo-inverse of the Jacobian, are not guaranteed to give human-like motion. As explained in Chapter 5, in some orientations a singularity may exist in which a DOF is "lost" in Cartesian space. For example, the motion can be hard to control in cases in which the arm is fully extended.

According to the model shown in Figure 9.6, if only the desired end effector position is given, then the solution space is underconstrained. In this case, multiple solutions exist, and inverse kinematic methods may result in configurations that do not look natural. As noted in Chapter 5, there are methods for biasing the solution toward desired joint angles. This helps to avoid violating joint limits and produces more human-like motion but still lacks any anatomical basis for producing human-like configurations.

It is often useful to specify the goal position of the wrist instead of the fingers to better control the configurations produced. But even if the wrist is fixed (i.e., treated as the end effector) at a desired location, and the shoulder is similarly fixed, there are still a large number of positions that might be adopted that satisfy both the constraints and the joint limits. Biasing the joint angles to orientations preferable for certain tasks can reduce or eliminate multiple solutions.

To more precisely control the movement, the user can specify intermediate positions and orientations for the end effector as well as for intermediate joints. Essentially, this establishes key poses for the linkage. Inverse kinematics can then be used to step from one pose to the next so that the arm is still guided along the path. This affords some of the savings of using inverse kinematics while giving the animator more control over the final motion.

The formal inverse Jacobian approach can be replaced with a more procedural approach based on the same principles to produce more human-like motion. In human motion, the joints farther away from the end effector (the hand) have the most effect on it. The joints closer to the hand change angles in order to perform the fine orientation changes necessary for final alignment. This can be implemented procedurally by computing the effect of each DOF on the end effector by taking the cross-product of the axis of rotation, ω_1, with the vector from the joint to the end effector, V_1 (Figure 9.7). In addition, since the arm contains a one DOF angle (elbow), a plane between the shoulder, the elbow, and the wrist is formed, and the arm's preferred positions dictate a relatively limited rotation range for that plane. Once the plane is fixed, the shoulder and elbow angles are easy to calculate and can be easily adjusted on that plane (Figure 9.8). Some animation packages (e.g., Maya^{TM}) allow the animator to specify an inverse kinematic solution based on such a plane and to rotate the plane as desired.

Some neurological studies, notably those by Lacquaniti and Soechting [40] and Soechting and Flanders [61], suggest that the arm's posture is determined from the desired location of the end effector (roughly "fixing the wrist's orientation"), and then the final wrist orientation is tweaked for the nature

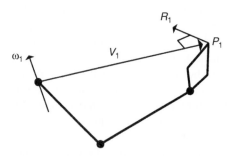

FIGURE 9.7

Effect of the first DOF on the end effector: $R_1 = \omega_1 \times V_1$.

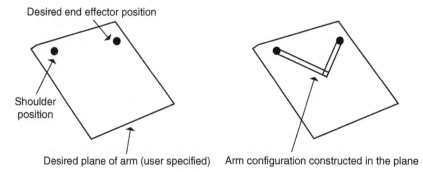

FIGURE 9.8

Constructing the arm in a user-specified plane.

of the object and the task. The model developed by Kondo [38] for this computation makes use of a spherical coordinate system. A set of angles for the shoulder and elbow is calculated from the desired hand and shoulder position and then adjusted if joint limitations are violated. Finally, a wrist orientation is calculated separately. The method is described, along with a manipulation planner for trajectories of cooperating arms, by Koga et al. [36].

9.2.2 The shoulder joint

The shoulder joint requires special consideration. It is commonly modeled as a ball joint with three coincident DOFs. The human shoulder system is actually more complex. Scheepers [57] described a more realistic model of the clavicle and scapula along with a shoulder joint, in which three separate joints with limited range provide very realistic-looking arm and shoulder motion. Scheepers also provided an approach to the forearm rotation problem using a radioulnar (mid-forearm) joint (see Figure 9.9).

9.2.3 The hand

To include a fully articulated hand in the arm model, one must introduce many more joints (and thus DOFs). A simple hand configuration may consist of a palm, four fingers, and a thumb, with joints and DOFs as shown in Figure 9.10.

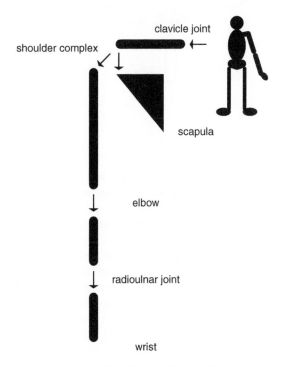

FIGURE 9.9

Conceptual model of the upper limb.

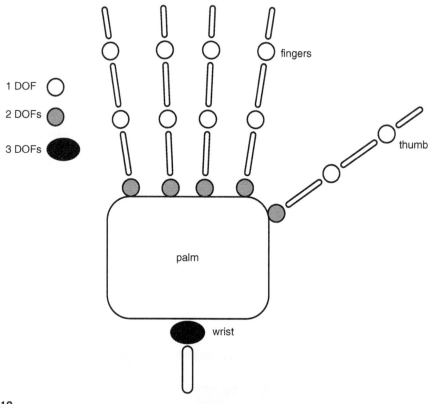

FIGURE 9.10

Simple model of hands and fingers.

A model similar to Figure 9.10 is used by Rijpkema and Girard [54] in their work on grasping. Scheepers [57] uses 27 bones, but only 16 joints are movable. Others use models with subtler joints inside the palm area in order to get human-like action.

If the hand is to be animated in detail, the designer must pay attention to types of grasp and how the grasps are to be used. The opposable thumb provides humans with great manual dexterity: the ability to point, grasp objects of many shapes, and exert force such as that needed to open a large jar of pickles or a small jewelry clasp. This requires carefully designed skeletal systems. Studies of grasping show at least 16 different categories of grasp, most involving the thumb and one or more fingers. For a given task, the problem of choosing which grasp is best (most efficient and/or most credible) adds much more complexity to the mere ability to form the grasp. Approaches to grasping are either procedural (e.g., [5] [8] [30] [46] [54]) or data driven (e.g., [18] [39] [51]).

Simpler models combine the four fingers into one surface and may eliminate the thumb (Figure 9.11). This reduces both the display complexity and the motion control complexity. Display

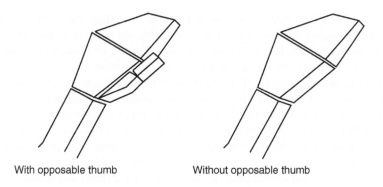

With opposable thumb   Without opposable thumb

FIGURE 9.11

Simplified hands.

complexity, and therefore image quality, can be maintained by using the full-detail hand model but coordinating the movement of all the joints of the four fingers with one "grasping" parameter (Figure 9.12), even though this only approximates real grasping action.

9.2.4 Coordinated movement

Adding to the difficulties of modeling and controlling the various parts of the upper limb is the difficulty of inter-joint cooperation in a movement and assigning a specific motion to a particular joint. It is easy to demonstrate this difficulty. Stretch out your arm to the side and turn the palm of the hand so it is first facing up; then rotate the hand so it is facing down and try to continue rotating it all the way around so that the palm faces up again. Try to do this motion by involving first only the hand/wrist/forearm and

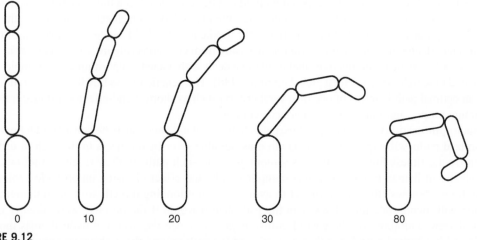

0   10   20   30   80

FIGURE 9.12

Finger flexion controlled by single parameter; the increase in joint angle (degrees) per joint is shown.

then the upper arm/shoulder. Adding motion of the torso including the clavicle and spine, which involves more DOFs, makes this task simpler to accomplish, but also makes the specification of realistic joint angles more complex. It is difficult to determine exactly what rotation should be assigned to which joints at what time in order to realistically model this motion.

Interaction between body parts is a concern beyond the determination of which joints to use in a particular motion. While viewing the arm and hand as a separate independent system simplifies the control strategy, its relation to the rest of the body must be taken into account for a more robust treatment of reaching. Repositioning, twisting, and bending of the torso, reactive motions by the other arm, and even counterbalancing by the legs are often part of movements that only appear to belong to a single arm. It is nearly impossible for a person reaching for an object to keep the rest of the body in a fixed position. Rather than extend joints to the edges of their limits and induce stress, other body parts may cooperate to relieve muscle strain or maintain equilibrium.

Arm manipulation is used in many different full-body movements. Even walking, which is often modeled as an activity of the legs only, involves the torso, the arms, and even the head. The arm often seems like a simple and rewarding place to begin modeling human figure animation, but it is difficult to keep even this task at a simple level.

9.2.5 Reaching around obstacles

To further complicate the specification and control of reaching motion, there may be obstacles in the environment that must be avoided. Of course, it is not enough to merely plan a collision-free path for the end effector. The entire limb sweeps out a volume of space during reach that must be completely devoid of other objects to avoid collisions. For sparse environments, simple reasoning strategies can be used to determine the best way to avoid obstacles.

As more obstacles populate the environment, more complex search strategies might be employed to determine the path. Various path-planning strategies have been proposed. For example, given an environment with obstacles, an artificial potential field can be constructed as a function of the local geometry. Obstacles impart a high potential field that attenuates based on distance. Similarly, the goal position imparts a low potential into the field. The gradient of the field suggests a direction of travel for the end effector and directs the entire linkage away from collisions (Figure 9.13). Such approaches are susceptible to local minima traps that need to be dealt with. Genetic algorithms, for example, have been used to search the space for a global minimum [48]. The genetic fitness function can be tailored to find an optimal path in terms of one of several criteria such as shortest end-effector distance traveled, minimum torque, and minimum angular acceleration.

Such optimizations, however, produce paths that would not necessarily be considered humanlike. Optimized paths will typically come as close as possible to critical objects in the path in order to minimize the fitness function. Humans seldom generate such paths in their reaching motions. The complexity of human motion is further complicated by the effect of vision on obstacle avoidance. If the figure "knows" there is an object to avoid but is not looking directly at it, then the reaching motion will incorporate more leeway in the path than it would if the obstacle were directly in the field of view. Furthermore, the cost of collision can influence the resulting path: it costs more to collide with a barbed-wire fence than a towel, and the path around these obstacles should probably be quite different.

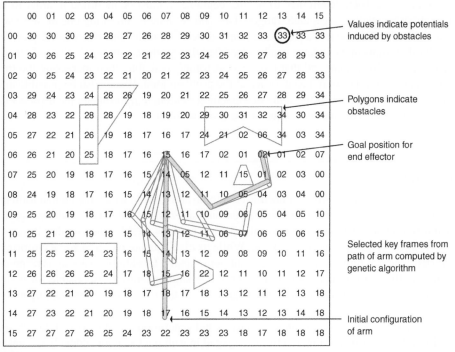

FIGURE 9.13

Path planning result [48].

9.2.6 Strength

As anyone who has ever changed a spark plug in a car knows, avoiding all the obstacles and getting a tool in the correct position is only half the battle. Once in position, the arm and hand must be in a configuration in which there is enough strength available to actually dislodge the plug. To simulate more realistic motions, strength criteria can be incorporated into the task planning [42]. As previously noted, typical reaching motion problems are usually underconstrained, allowing for multiple solutions. The solution space can be searched for a specific motion that is acceptable in terms of the amount of strain it places on the figure.

When a specific motion is proposed by a kinematic planner, it can be evaluated according to the strain it places on the body. The strain is determined by computing the torque necessary at each joint to carry out the motion and rating the torque requirements according to desirability. Given the current pose for the figure, the required joint accelerations, and any external forces, the torque required at each joint can be calculated. For each joint, the maximum possible torque for both flexion and extension is given as a function of the joint's angle as well as that of neighboring joints. A *comfort* metric can be formed as the ratio of currently requested torque and maximum possible torque. The *comfort level* for the figure is computed by finding the maximum torque ratio for the entire body. The most desirable motions are those that minimize the maximum torque ratio over the duration of the motion.

Once a motion has been determined to be unacceptable, it must be modified in order to bring its comfort level back to within acceptable ranges. This can be done by initiating one or more strategies that reduce the strain. Assume that a particular joint has been identified that exceeds the accepted comfort range. If other joints in the linkage can be identified that produce a motion in the end effector similar to that of the problem joint and that have excess torque available, then increasing the torques at these joints can compensate for reduced torque at the problem joint. It may also be possible to include more joints in the linkage, such as the spine in a reaching motion, to reformulate the inverse kinematic problem in the hope of reducing the torque at the problem joint.

9.3 Walking

Walking, along with reaching, is one of the most common activities in which the human form engages. It is a complex activity that, for humans, is learned only after an extended trial-and-error process. An aspect that differentiates walking from typical reaching motions, besides the fact that it uses the legs instead of the arms, is that it is basically cyclic. While its cyclic nature provides some uniformity, acyclic components such as turning and tripping occur periodically. In addition, walking is responsible for transporting the figure from one place to another and is simultaneously responsible for maintaining balance. Thus, dynamics plays a much more integral role in the formation of the walking motion than it does in reaching.

An aspect of walking that complicates its analysis and generation is that it is dynamically stable but not statically stable. This means that if a figure engaged in walking behavior suddenly freezes, the figure is not necessarily in a balanced state and would probably fall to the ground. For animation purposes, this means that the walking motion cannot be frozen in time and statically analyzed to determine the correct forces and torques that produce the motion. As a result, knowledge of the walking motion, in the form of either empirically gathered data [26] [32] or a set of parameters adjustable by the animator, is typically used as the global control mechanism for walking behavior. Attributes such as stride length, hip rotation, and foot placement can be used to specify what a particular walk should look like. A state transition diagram, or its equivalent, is typically used to transition from phase to phase of the gait [11] [12] [22] [29] [52]. Calculation of forces and torques can then be added, if desired, to make the nuances of the motion more physically accurate and more visually satisfying. Kinematics can be used to entirely control the legs, while the forces implied by the movement of the legs are used to affect the motion of the upper body [23] [63]. Alternatively, kinematics can be used to establish constraints on leg motion such as leg swing duration and foot placement. Then the forces and torques necessary to satisfy the constraints can be used to resolve the remaining DOF of the legs [11] [29] [52]. In some cases, forward dynamic control can be used after determining the forces and torques necessary to drive the legs from state to state [47].

9.3.1 The mechanics of locomotion

Understanding the interaction of the various joints involved in locomotion is the first step in understanding and modeling locomotion. The walking and running cycles are presented first. Then the walk cycle is broken down in more detail, showing the complex movements involved. For this discussion, walking is considered to be locomotion characterized by one or both feet touching the ground at any point in time as opposed to running where at most only one foot is on the ground at any point in time.

Walk cycle

The walk cycle can be broken down into various phases [11] based on the relation of the feet to their points of contact with the ground (see Figure 9.14). The *stride* is defined by the sequence of motions between two consecutive repetitions of a body configuration [32]. The *left stance* phase of a stride is initiated with the right foot on the ground and the left heel just starting to strike the ground. During this phase, the body is supported by both feet until the right foot pivots up and the right toe leaves the ground. The left stance phase continues as the right foot leaves the ground and starts swinging forward and as the right heel strikes the ground and both feet are once again on the ground. The left toe leaving the ground terminates the left stance phase. The *right swing phase* is the period in which the right toe leaves the ground, the right leg swings forward, and the right heel strikes the ground. Notice that the right swing phase is a subinterval of the left stance phase. The end of the right swing phase initiates the right stance phase, and analogous phases now proceed with the roles of the left leg and the right leg switched. The walking cycle is characterized by alternating periods of single and double support.

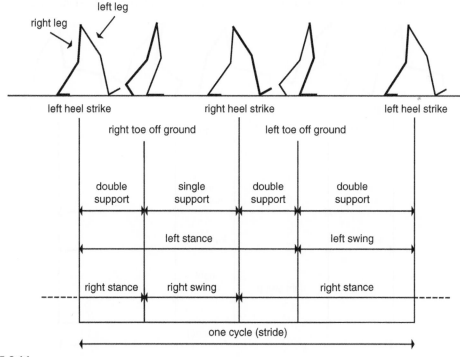

FIGURE 9.14

Walk cycle [32].

Run cycle

The run cycle can also be described as a sequence of phases. It differs from the walk cycle in that both feet are off the ground at one time and at no time are both feet on the ground. As in the walk cycle, the *stance* is the duration that a foot is on the ground. Thus, the *left stance*, defined by the left heel strike and left toe lift, has the right foot off the ground. This is followed by a period of *flight*, during which both feet are off the ground, with the right foot swinging forward. The flight is terminated by the right heel strike, which starts the *right stance* (see Figure 9.15). Notice that the left and right stances do not overlap and are separated by periods of flight.

Pelvic transport

For this discussion, let the pelvis represent the mass of the upper body being transported by the legs. Using a simplified representation for the legs, Figure 9.16 shows how the pelvis is supported by the stance leg at various points during the stance phase of the walk cycle. Figure 9.17 shows these positions superposed during a full stride and illustrates the abutting of two-dimensional circular arcs describing the basic path of the pelvis as it is transported by the legs.

Pelvic rotation

The pelvis represents the connection between the legs and the structure that separates the legs in the third dimension. Figure 9.18 shows the position of the pelvis during various points in the walking cycle, as viewed from above. The pelvis rotates about a vertical axis centered at the stance leg, helping to

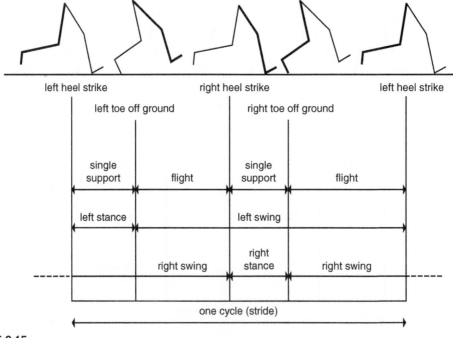

FIGURE 9.15

Run cycle [12].

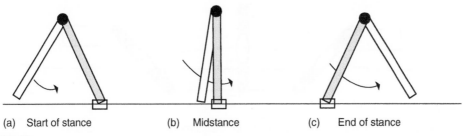

(a)   Start of stance   (b)   Midstance   (c)   End of stance

FIGURE 9.16

Position of pelvis during stance phase (sagittal plane). (a) Start of stance, (b) midstance, and (c) end of stance. Box indicates supporting contact with floor.

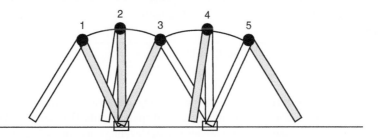

FIGURE 9.17

Transport of pelvis by intersecting circular arcs (sagittal plane).

Start of stance   Midstance   End of stance

FIGURE 9.18

Pelvic orientation during stance phase (transverse plane).

lengthen the stride as the swing leg stretches out for its new foot placement. This rotation of the pelvis above the stance leg means that the center of the pelvis follows a circular arc relative to the top of that leg. The top of the stance leg is rotating above the point of contact with the floor (Figure 9.18), so the path of the center of the pelvis resembles a sinusoidal curve (Figure 9.19).

Pelvic list

The transport of the pelvis requires the legs to lift the weight of the body as the pelvis rotates above the point of contact with the floor (Figure 9.20). To reduce the amount of lift, the pelvis lists by rotating in the coronal plane.

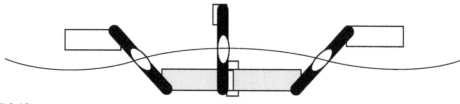

FIGURE 9.19

Path of the pelvic center from above (transverse plane), exaggerated for illustrative purposes.

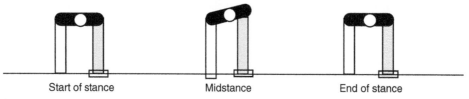

Start of stance   Midstance   End of stance

FIGURE 9.20

Pelvic list to reduce the amount of lift (coronal plane).

Knee flexion

As shown in Figure 9.20, in a pelvic list with one-piece legs, the swing leg would penetrate the floor. Bending at the knee joint (flexion) allows the swing leg to safely pass over the floor and avoid contact (Figure 9.21). Flexion at the knee of the stance leg also produces some leveling out of the pelvic arcs produced by the rotation of the pelvis over the point of contact with the floor. In addition, extension just before contact with the floor followed by flexion of the new stance leg at impact provides a degree of shock absorption.

Ankle and toe joints

The final part of the puzzle to the walking motion is the foot complex, consisting of the ankle, the toes, and the foot itself. This complex comprises several bones and many DOFs and can be simply modeled as two hinge joints per foot (Figure 9.22). The ankle and toe joints serve to further flatten out the rotation of the pelvis above the foot as well as to absorb some shock.

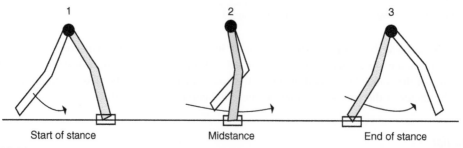

1   2   3

Start of stance   Midstance   End of stance

FIGURE 9.21

Knee flexion allowing for the swing leg to avoid penetrating the floor, leveling the path of the pelvis over the point of contact, and providing some shock absorption (sagittal plane).

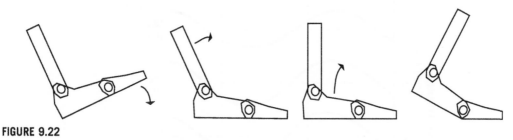

FIGURE 9.22

Rotation due to ankle-toe joints.

9.3.2 **The kinematics of the walk**

Animation of the leg can be performed by appropriate control of the joint angles. As previously mentioned, a leg's walk cycle is composed of a stance phase and a swing phase. The stance phase duration is the time from heel strike to toe lift. The swing phase duration is the time between contact with the ground—from toe lift to heel strike. The most basic approach to generating the walking motion is for the animator to specify a list of joint angle values for each DOF involved in the walk. There are various sources for empirical data describing the kinematics of various walks at various speeds. Figures 9.23 through 9.27, from Inman, Ralson, and Todd [32], graph the angles over time for the various joints involved in the walk cycle, as well as giving values for the lateral displacement of the pelvis.

Specifying all the joint angles, either on a frame-by-frame basis or by interpolation of values between key frames, is an onerous task for the animator. In addition, it takes a skilled artist to design values that create unique walks that deviate in any way from precisely collected clinical data. When creating new walks, the animator can specify kinematic values such as pelvic movement, foot placement, and foot trajectories. Inverse kinematics can be used to determine the angles of the intermediate joints [12]. By constructing the time-space curves traced by the pelvis and each foot, the user can determine the position of each for a given frame of the animation. Each leg can then be positioned by considering the pelvis fixed and the leg a linked appendage whose desired end-effector position is the corresponding position on the foot trajectory curve (Figure 9.28). Sensitivity to segment lengths can cause even clinical data to produce configurations that fail to keep the feet in solid contact with the floor during walking. Inverse kinematics is also useful for forcing clinical data to maintain proper foot placement.

9.3.3 **Using dynamics to help produce realistic motion**

Dynamic simulation can be used to map specified actions and constraints to make the movement more accurate physically. However, as Girard and Maciejewski [22] point out, an animator who wants a particular look for a behavior often wants more control over the motion than a total physical simulation provides (Girard and Maciejewski discussed this in relation to walking, but it obviously applies in many situations where physically reasonable, yet artistically controlled, motion is desired). Dynamics must be intelligently applied so that it aids the animator and does not become an obstacle that the animator must work around. In addition, to make the computations tractable, the animator almost always simplifies the dynamics. There are several common types of simplifications: (1) some dynamic effects are ignored, such as the effect of the swing leg on balance; (2) relatively small temporal variations are

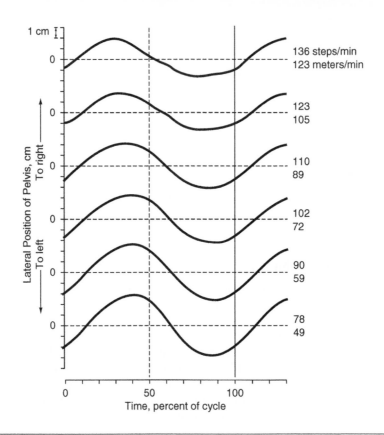

FIGURE 9.23

Lateral displacement of pelvis [32].

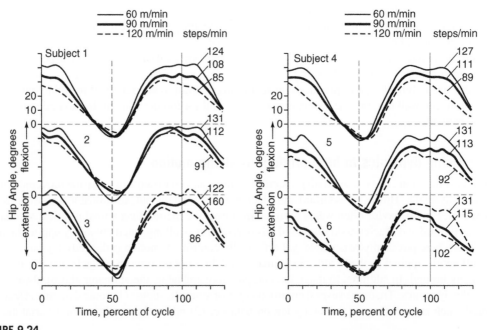

FIGURE 9.24

Hip angles [32].

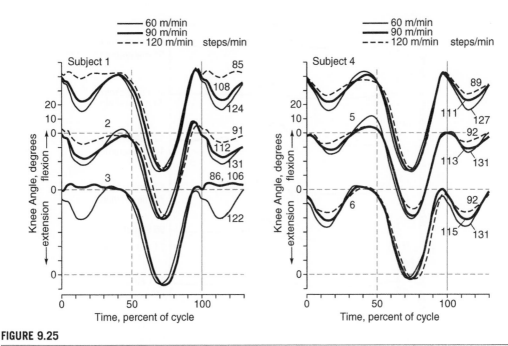

FIGURE 9.25

Knee angles [32].

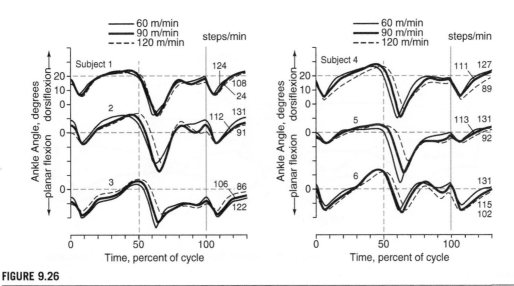

FIGURE 9.26

Ankle angles [32].

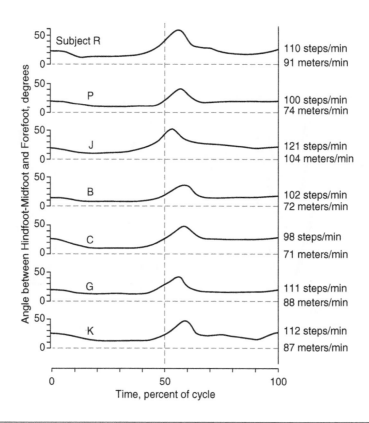

FIGURE 9.27

Toe angles [32].

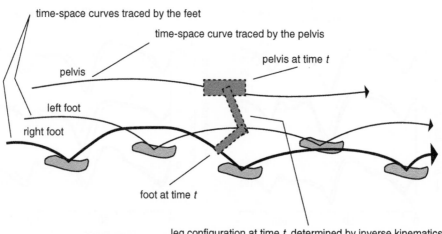

FIGURE 9.28

Pelvis and feet constraints satisfied by inverse kinematics.

ignored and a force is considered constant over some time interval, such as the upward push of the stance leg; (3) a complex structure, such as the seven DOF leg, is replaced for purposes of the dynamic computations by a simplified but somewhat dynamically equivalent structure, such as a one DOF telescoping leg; and (4) computing arbitrarily complex dynamic effects is replaced by computing decoupled dynamic components, such as separate horizontal and vertical components, which are computed independently of each other and then summed.

In achieving the proper motion of the upper body, the legs are used to impart forces to the mass of the upper body as carried by the pelvis. An upward force can be imparted by the stance leg on the mass of the upper body at the point of connection between the legs and the torso, that is, the hips [22] [63]. To achieve support of the upper body, the total upward acceleration due to the support of the legs has to cancel to total downward acceleration due to gravity. As simplifying assumptions, the effect of each leg can be considered independent of the other(s), and the upward force of a leg on the upper body during the leg's stance phase can be considered constant. As additional assumptions, horizontal acceleration can be computed independently for each leg and can be considered constant during the leg's stance phase (Figure 9.29). The horizontal force of the legs can be adjusted automatically to produce the average velocity for the body as specified by the animator, but the fluctuations in instantaneous velocity over time help to create visually more appealing motion than constant velocity alone would. The temporal variations in upward and forward acceleration at each hip due to the alternating stance phases can be used to compute pelvic rotation and pelvic list to produce even more realistic motion.

More physics can be introduced into the lower body by modeling the leg dynamics with a telescoping joint (implemented as a parallel spring-damper system) during the stance phase. The upward force of the leg during the stance phase becomes time varying as the telescoping joint compresses under the weight of the upper body and expands under the restoring forces of the leg complex (upper leg, lower leg, foot, and associated joints). The telescoping mechanism simulates the shock-absorbing effect of the knee-ankle-toe joints. The leg complex is then fit to the length of the telescoping joint by inverse kinematics (Figure 9.30). During the swing phase, the leg is typically controlled kinematically and does not enter into any dynamic considerations.

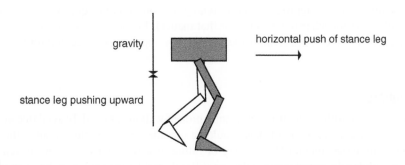

FIGURE 9.29

Horizontal and vertical dynamics of stance leg: gravity and the vertical component must cancel over the duration of the cycle. The horizontal push must account for the average forward motion of the figure over the duration of the cycle.

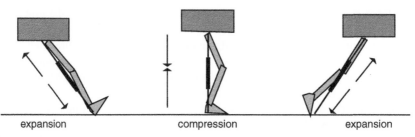

expansion   compression   expansion

FIGURE 9.30

Telescoping joint with kinematically fit leg complex.

Incorporating more physics into the model, the kinematic control information for the figure can be used to guide the motion (as opposed to constraining it). A simple inverse dynamics computation is then used to try to match the behavior of the system with the kinematic control values. Desired joint angles are computed based on high-level animator-supplied parameters such as speed and the number of steps per unit distance. Joint torques are computed based on proportional-derivative servos (Eq. 9.1). The difference between the desired angle, denoted by the underbar, and the current angle at each joint is used to compute the torque to be applied at the next time step. The angular velocities are treated similarly. These torque values are smoothed to prevent abrupt changes in the computed motion. However, choosing good values for the gains (k_s, k_v) can be difficult and usually requires a trial-and-error approach.

$$\tau = k_s(\underline{\theta}_i - \theta_i) + k_v\left(\underline{\dot{\theta}}_i - \dot{\theta}_i\right) \tag{9.1}$$

9.3.4 Forward dynamic control

In some cases, forward dynamic control instead of kinematic control can be effectively used. Kinematics still plays a role. Certain kinematic states, such as maximum forward extension of a leg, trigger activation of forces and torques. These forces and torques move the legs to a new kinematic state, such as maximum backward extension of a leg, which triggers a different sequence of forces and torques [47]. The difficulty with this approach is in designing the forces and torques necessary to produce a reasonable walk cycle (or other movement, for that matter). In some cases it may be possible to use empirical data found in the biomechanics literature as the appropriate force and torque sequence.

9.3.5 Summary

Implementations of algorithms for procedural animation of walking are widely available in commercial graphics packages. However, none of these could be considered to completely solve the locomotion problem, and many issues remain for ongoing or future research. Walking over uneven terrain and walking around arbitrarily complex obstacles are difficult problems to solve in the most general case. The coordinated movement required for climbing is especially difficult. A recurring theme of this chapter is that developing general, robust computational models of human motion is difficult, to say the least.

9.4 Coverings

9.4.1 Clothing

It is the rare application or animation that calls for totally nude figures. Simulating clothing and the corresponding interaction of the clothing with the surfaces of the figure can be one of the most computationally intensive parts of representing virtual humans. The clothes that protect and decorate the body contribute importantly to the appearance of a human figure. For most human figures in most situations, cloth covers the majority of the body. Cloth provides important visual qualities for a figure and imparts certain attributes and characteristics to it. The way in which cloth drapes over and highlights or hides aspects of the figure can make the figure more or less attractive, more or less threatening, and/or more or less approachable. For a figure in motion, clothes can provide important visual cues that indicate the type and speed of a character's motion. For example, the swirling of a skirt or the bouncing of a shirttail indicates the pace or smoothness of a walk.

The simulation of clothing on a virtual human is a complicated task that has not been fully solved, although significant advances are being made. Real-time application, such as computer games, still often use virtual humans that sport rigid body armor that merely rotates along with whatever limb it is attached to. Some applications can get away with simply texture mapping a pattern onto a single-sheet-defined human figure to simulate tight-fitting spandex clothes. High-end off-line animation systems are starting to offer advanced clothing simulation modules that attempt to calculate the effects of gravity as well as cloth–cloth and cloth–body collisions by using mass-spring networks or energy functions. Streamlined versions of these procedures are finding their way into interactive or even real-time applications.

The animation of cloth has been discussed in Section 7.5, but clothing presents special considerations, especially for real-time situations. Collisions and surfaces in extended contact are an almost constant occurrence for a figure wearing clothes. Calculation of impulse forces that result from such collisions is costly for an animated figure. Whenever a vertex is identified that violates a face in the environment, some procedure, such as a transient spring force or enforcing positional constraints, must be invoked so that the geometry is restored to an acceptable position.

To simulate clothing, the user must incorporate the dynamic aspect of cloth into the model to produce the wrinkling and bulging that naturally occur when cloth is worn by an articulated figure. In order for the figure to affect the shape of the cloth, extensive collision detection and response must be calculated as the cloth collides with the figure and with itself almost constantly. The level of detail at which clothes must be modeled to create realistic wrinkles and bulges requires relatively small triangles. Therefore, it takes a large number of geometric elements to clothe a human figure. As a result, one must attend to the efficiency of the methods used to implement the dynamic simulation and collision handling of the clothing.

9.4.4 Hair

One of the most significant hurdles for making virtual humans that are indistinguishable from a real person is the accurate simulation of a full head of hair. Hair is extremely complex. A head of hair consists of approximately 100,000 strands, each strand a flexible cylinder [45]. Because the strands are attached to the scalp and are in close proximity to each other, the motion of a single strand is constantly affected by external forces. Collisions, friction, and even influences such as static electricity can be significant. Surface properties of the hair strands, such as wetness, cleanliness, curliness, oiliness, split ends, and cosmetic applications, affect its motion. In addition, hair can be classified into one of three main types based on the geometry of a strand: Asian, African, and Caucasian. Asian generally has smooth and regular strands with a circular cross section while African strands are irregular and rough with elliptical cross sections. The qualities of Caucasian hair are between these two extremes.

Although the hair's motion is of primary interest for the present discussion, it should also be noted that providing the user with hair style design tools and the complexities of the rendering of hair are also difficult problems. All of these are the subject of ongoing research.

Depending on the requirements of the animation, the modeling and animation of hair can take place on one or more levels of detail. The most common, visually poor but computationally inexpensive technique has been to merely construct a rigid piece of geometry in the rough shape of the volume of hair and attach it to the top of the head, like a helmet, as in Figure 9.31. Texture maps with transparency information are sometimes used to improve the appearance, providing for a wispy boundary. FFDs can be used to animate the global shape of rigidly defined hair.

To add more realism to the hair, a hairstyle feature, such as a ponytail or lock of hair, can be animated independent of the rest of the hair [59]. This provides some dynamics to the hair at minimal computational cost. The animation of cartoon hair has even received some attention in the literature, e.g., [62].

For more realistic hair, the strands, or at least groups of strands, must be able to move relative to the rest of the hair [37]. Clusters of hair strands can be animated as a unit and the hair modeled as a collection of clusters. Sheets of hair strands can be used (Figure 9.32, Color Plate 5) or a generalized cylindrical collection of strands can be modeled, depending on the type of hair modeled. The strand clusters can be animated as flexible sheets and held close together by connecting springs. Alternatively, a few select master strands can be individually animated and then intermediate strands can be interpolated in order to fill out the hair.

For the most realism, but the greatest computational cost, individual strands can be modeled and animated separately (e.g., [55]). Strands are generated using small geometric tubes [3] [47] or particle trails. Individual strands can be animated using the mass-spring-damper system or using the dynamics of rigid body linkages. Often, strand–strand interaction is ignored in order to keep the computational cost within reason.

FIGURE 9.31

Hair modeled as rigid geometry. (Image courtesy of Mike Altman.)

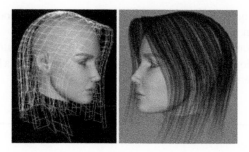

FIGURE 9.32

Hair modeled as strips of strands [37].

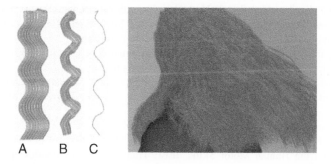

FIGURE 9.33

Hair modeled using multiple levels of detail [65].

To provide both the computational savings of strand clusters as well as the realism of animating individual strands, a multiple level of detail approach can be used (Figure 9.33, Color Plate 6) [65] and, to provide more flexibility, adaptive grouping can be used [64]. Other approaches include modeling hair as a thin shell [34] and as a continuum [24].

9.5 Chapter summary

The human figure is an interesting and complex form. It has a uniform structure but contains infinite variety. As an articulated rigid structure, it contains many DOFs, but its surface is deformable. Modeling and animating hair in any detail is also enormously complex. Moreover, the constant collision and sliding of cloth on the surface of the body represents significant computational complexity.

One of the things that makes human motion so challenging is that no single motion can be identified as correct human motion. Human motion varies from individual to individual for a number of reasons, but it is still recognizable as reasonable human motion, and slight variations can seem odd or awkward. Research in computer animation is just starting to delve into the nuances of human motion. Many of these nuances vary from person to person but, at the same time, are very consistent for a particular person, for a particular ethnic group, for a particular age group, for a particular weight group, for a particular emotional state, and so on. Computer animation is only beginning to analyze, record, and synthesize these important qualities of movement. Most of the work has focused on modeling changes in motion resulting from an individual's emotional state (e.g., [2] [15]).

Human figure animation remains a challenge for computer animators and fertile ground for graphics researchers. Developing a synthetic actor indistinguishable from a human counterpart remains the Holy Grail of researchers in computer animation.

References

[1] Alias/Wavefront. Maya, http://www.aliaswavefront.com; 1998.
[2] Amaya K, Bruderlin A, Calvert T. Emotion from Motion. In: Davis WA, Bartels R, editors. Graphics Interface '96. Canadian Human-Computer Communications Society; May 1996. p. 222–9. ISBN 0-9695338-5-3.
[3] Anjyo K, Usami Y, Kurihara T. A Simple Method for Extracting the Natural Beauty of Hair. In: Dill J, editor. Computer Graphics. Proceedings of SIGGRAPH 88, vol. 22(4). Atlanta, Ga.; August 1988. p. 111–20.
[4] Avid Technology, Inc. SOFTIMAGEI3D, http://www.softimage.com/.
[5] Aydin Y, Kakajiima M. Database Guided Computer Animation of Human Grasping Using Forward and Inverse Kinematics. Computers and Graphics 1999;23:145–54.
[6] Azulola F, Badler N, Hoon TK, Wei S. Sass v.2.1 Anthropometric Spreadsheet and Database for the Iris. Technical Report, Dept. of Computer and Information Science, University of Pennsylvania; 1993. MS-CIS-93-63.
[7] Baca K. Poor Man's Skinning. Game Developer, July 1998;48–51.
[8] Bekey GA, Liu H, Tomovic R, Karplus WJ. Knowledge-Based Control of Grasping in Robot Hands Using Heuristics from Human Motor Skills. IEEE Transactions on Robotics and Automation 1993;9 (6):709–22.
[9] Blanz V, Vetter T. A Morphable Model for the Synthesis of 3D Faces, In: Rockwood A, editor. Computer Graphics. Proceedings of SIGGRAPH 99, Annual Conference Series. Los Angeles, Calif,: Addison-Wesley Longman; August 1999. p. 187–94. ISBN 0-20148-560-5.
[10] Boulic R, Capin T, Huang Z, Kalra P, Linterrmann B, Magnenat-Thalmann N, et al. The HUMANOID Environment for Interactive Animation of Multiple Deformable Human Characters. In: Post F, Göbel M, editors. Computer Graphics Forum, vol. 14(3). Blackwell Publishers; August 1995. p. 337–48. ISSN 1067-7055.
[11] Bruderlin A, Calvert T. Goal-Directed, Dynamic Animation of Human Walking. In: Lane J, editor. Computer Graphics. Proceedings of SIGGRAPH 89, vol. 23(3). Boston, Mass.; July 1989. p. 233–42.
[12] Bruderlin A, Calvert T. Knowledge-Driven, Interactive Animation of Human Running. In: Davis WA, Bartels R, editors. Graphics Interface '96. Canadian Human-Computer Communications Society; May 1996. p. 213–21. ISBN 0-9695338-5-3.
[13] Chadwick J, Haumann D, Parent R. Layered Construction for Deformable Animated Characters. In: Lane J, editor. Computer Graphics. Proceedings of SIGGRAPH 89, vol. 23(3). Boston, Mass.; July 1989. p. 243–52.
[14] Chen D, Zeltzer D. Pump It Up: Computer Animation of a Biomechanically Based Model of Muscle Using the Finite Element Method. In: Catmull EE, editor. Computer Graphics. Computer Graphics. Proceedings of SIGGRAPH 92, vol. 26(2). Chicago, Ill.; July 1992. p. 89–98. ISBN 0-201-51585-7.
[15] Chi D, Costa M, Zhao L, Badler N. The EMOTE Model for Effort and Shape, In: Akeley K, editor. Computer Graphics. Proceedings of SIGGRAPH 2000, Annual Conference Series. ACM Press/ACM SIGGRAPH/ Addison-Wesley Longman; July 2000. p. 173–82. 1-58113-208-5.
[16] COVEN, http://coven.lancs.ac.uk/mpeg4/index.html, January 2.
[17] DeRose T, Kass M, Truong T. Subdivision Surfaces for Character Animation, In: Cohen M, editor. Computer Graphics. Proceedings of SIGGRAPH 98, Annual Conference Series. Orlando, Fla.: Addison-Wesley; July 1998. p. 85–94 ISBN 0-89791-999-8.

[18] ElKoura G, Singh K. Handrix: Animating the Human Hand, In: Symposium on Computer Animation; 2003.

[19] Elson M. Displacement Animation: Development and Application. In: Course #10: Character Animation by Computer, SIGGRAPH 1990. Dallas, Tex.; August 1990.

[20] Engin A, Peindl R. On the Biomechanics of Human Shoulder Complex—I: Kinematics for Determination of the Shoulder Complex Sinus. J Biomech 1987;20(2):103–17.

[21] Forsey D, Bartels R. Hierarchical B-Spline Refinement. In: Dill J, editor. Computer Graphics. Proceedings of SIGGRAPH 88, vol. 22(4). Atlanta, Ga.; August 1988. p. 205–12.

[22] Girard M, Maciejewski A. Computational Modeling for the Computer Animation of Legged Figures. In: Barsky BA, editor. Computer Graphics. Proceedings of SIGGRAPH 85, vol. 19(3). San Francisco, Calif.; August 1985. p. 263–70.

[23] Gourret J-P, Magnenat-Thalmann N, Thalmann D. Simulation of Object and Human Skin Deformations in a Grasping Task. In: Lane J, editor. Computer Graphics. Proceedings of SIGGRAPH 89, vol. 23(3). Boston, Mass.; July 1989. p. 21–30.

[24] Hadap S, Magnenat-Thalmann N. Modeling Dynamic Hair as a Continuum. Computer Graphics Forum 2001;20(3):329–38.

[25] Hanrahan P, Wolfgang K. Reflection from Layered Surfaces due to Subsurface Scattering. In: Kajiya JT, editor. Computer Graphics. Proceedings of SIGGRAPH 93, Annual Conference Series. Anaheim, Calif.; August 1993. p. 165–74. ISBN 0-201-58889-7.

[26] Harris G, Smith P, editors. Human Motion Analysis. Piscataway, New Jersey: IEEE Press; 1996.

[27] Henry-Biskup S. Anatomically Correct Character Modeling. Gamasutra November 13 1998;2(45). http://www.gamasutra.com/features/visual_arts/19981113/charmod_01.htm.

[28] Hilton A, Beresford D, Gentils T, Smith R, Sun W. Virtual People: Capturing Human Models to Populate Virtual Worlds, http://www.ee.surrey.ac.uk/Research/VSSP/3DVision/VirtualPeople/.

[29] Hodgins J, Wooten W, Brogan D, O'Brien J. Animating Human Athletics, In: Cook R, editor. Computer Graphics. Proceedings of SIGGRAPH 95, Annual Conference Series. Los Angeles, Calif: Addison-Wesley; August 1995. p. 71–8. ISBN 0-201-84776-0.

[30] Huang Z, Boulic R, Magnenat-Thalmann N, Thalmann D. A Multi-Sensor Approach Fograsping and 3D Interaction. Proc Computer Graphics International 1995.

[31] Humanoid Animation Working Group . Specification for a Standard Humanoid, http://ece.uwaterloo.ca/~h-anim/spec1.1/; 1999.

[32] Inman V, Ralston H, Todd F. Human Walking. Baltimore, Maryland: Williams & Wilkins; 1981.

[33] Kakadiaris I, Metaxas D. 3D Human Body Model Acquisition from Multiple Views. In: Proceedings of the Fifth International Conference on Computer Vision. Boston, Mass.; June 20–23, 1995.

[34] Kim T-Y, Neumann U. A Thin Shell Volume for Modeling Human Hair. In: Computer Animation- 2000. IEEE Computer Society; 2000. p. 121–8.

[35] Kinetix. 3D Studio MAX, http://www.ktx.com; February 2007.

[36] Koga Y, Kondo K, Kuffner J, Latombe J-C. Planning Motions with Intentions, In: Glassner A, editor. Computer Graphics. Proceedings of SIGGRAPH 94, Annual Conference Series. Orlando, Fla.: ACM Press; July 1994. p. 395–408. ISBN 0-89791-667-0.

[37] Koh C, Huang Z. Real-Time Animation of Human Hair Modeled in Strips. In: Computer Animation and Simulation '00. Sept. 2000. p. 101–12.

[38] Kondo K. Inverse Kinematics of a Human Arm. Technical Report STAN-CS-TR-94-1508, Stanford University; 1994.

[39] Kry P, Pai D. Interaction Capture and Synthesis. Transactions on Graphics 2006;25(3).

[40] Lacquaniti F, Soechting JF. Coordination of Arm and Wrist Motion during a Reaching Task. Journal of Neuroscience 1982;2(2):399–408.

[41] Lander J. On Creating Cool Real-Time 3D, Gamasutra October 17, 1997;1(8). http://www.gamasutra.com/features/visual_arts/101797/rt3d_01.htm.

[42] Lee P, Wei S, Zhao J, Badler N. Strength Guided Motion. In: Baskett F, editor. Computer Graphics. Proceedings of SIGGRAPH 90, vol. 24(4). Dallas, Tex.; August 1990. p. 253–62. ISBN 0-201-50933-4.

[43] Lee W-S, Magnenat-Thalmann N. From Real Faces to Virtual Faces: Problems and Solutions. In: Proc. 3IA'98. Limoges (France); 1998.

[44] Lewis JP, Cordner M, Fong N. Pose Space Deformations: A Unified Approach to Shape Interpolation and Skeleton-Driven Deformation, In: Akeley K, editor. Computer Graphics. Proceedings of SIGGRAPH 2000, Annual Conference Series. ACM Press/ACM SIGGRAPH/Addison-Wesley Longman; July 2000. p. 165–72. ISBN 1-58113-208-5.

[45] Magnenat-Thalmann N, Hadap S. State of the Art in Hair Simulation. In: International Workshop on Human Modeling and Animation. Korea Computer Graphics Society; June 2000. p. 3–9.

[46] Mas Sanso R, Thalmann D. A Hand Control and Automatic Grasping System for Synthetic Actors. In: Eurographics '94. 1994.

[47] McKenna M, Zeltzer D. Dynamic Simulation of Autonomous Legged Locomotion. In: Baskett F, editor. Computer Graphics. Proceedings of SIGGRAPH 90, vol. 24(4). Dallas, Tex.; August 1990. p. 29–38 ISBN: 0-201-50933-4.

[48] Miller D. The Generation of Human-Like Reaching Motion for an Arm in an Obstacle-Filled 3-D Static Environment. Ph.D. dissertation, Ohio State University; 1993.

[49] Nedel L, Thalmann D. Modeling and Deformation of the Human Body Using an Anatomically-Based Approach. In: Computer Animation '98. Philadelphia, Pa: IEEE Computer Society; June 1998.

[50] Pandzic I, Capin T, Magnenat-Thalmann N, Thalmann D. Developing Simulation Techniques for an Interactive Clothing System. In: Proceedings of VSMM '97. Geneva, Switzerland; 1997. p. 109–18.

[51] Pollard NS, Zordan VB. Physically Based Grasping Control from Example. In: Symposium on Computer Animation; 2005.

[52] Raibert M, Hodgins J. Animation of Dynamic Legged Locomotion. In: Sederberg TW, editor. Computer Graphics. Proceedings of SIGGRAPH 91, vol. 25(4). Las Vegas, Nev.; July 1991. p. 349–58. ISBN 0-201-56291-X.

[53] REM Infografica. MetaReyes and ClothReyes, http://www.infografica.com/.

[54] Rijpkema H, Girard M. Computer Animation of Knowledge-Based Human Grasping. In: Sederberg TW, editor. Computer Graphics, Proceedings of SIGGRAPH 91, vol. 25(4). Las Vegas, Nev.; July 1991. p. 339–48. ISBN 0-201-56291-X.

[55] Rosenblum R, Carlson W, Tripp E. Simulating the Structure and Dynamics of Human Hair: Modeling, Rendering, and Animation. The Journal of Visualization and Computer Animation 1991;2(4):141–8.

[56] Sannier G, Magnenat-Thalmann N. A User-Friendly Texture-Fitting Methodology for Virtual Humans. In: Computer Graphics International 1997. Hasselt/Diepenbeek, Belgium: IEEE Computer Society; June 1997.

[57] Scheepers F. Anatomy-Based Surface Generation for Articulated Models of Human Figures. Ph.D. dissertation, Ohio State University; 1996.

[58] Scheepers F, Parent R, Carlson W, May S. Anatomy-Based Modeling of the Human Musculature. In: Whitted T, editor. Computer Graphics. Proceedings of SIGGRAPH 97, Annual Conference Series. p. 163–72. ISBN 0-89791-896-7.

[59] Sega. Virtual Fighter 3, http://www.sega.com/games/games_vf3.html.

[60] Singh K. Realistic Human Figure Synthesis and Animation for VR Applications. Ph.D. dissertation, Ohio State University; 1995.

[61] Soechting J, Flanders M. Errors in Pointing Are Due to Approximations in Sensorimotor Transformations. J Neurophysiol August 1989;62(2):595–608.

[62] Sugisaki E, Yu Y, Anjyo K, Morishima S. Simulation-Based Cartoon Hair Animation, In: 13th International Conference in Central Europe on Computer Graphics, Visualization and Computer Vision (WSCG). Plzen, Czech Republic; January, 2005.

[63] Torkos N, van de Panne M. Footprint-Based Quadruped Motion Synthesis. In: Booth K, Fournier A, editors. Graphics Interface '98. June 1998. p. 151–60. ISBN 0-9695338-6-1.

[64] Ward K, Lin M. Adaptive Grouping and Subdivision for Simulating Hair Dynamics. In: Pacific Graphics Conference on Computer Graphics and Applications. October 2003. p. 234–43.

[65] Ward K, Lin M, Lee J, Fisher S, Macri D. Modeling Hair Using Level-of-Detail Representations. In: International Conference on Computer Animation and Social Agents. May 2003. p. 41–7.

[66] Wilhelms J, van Gelder A. Anatomically Based Modeling, In: Whitted T, editor. Computer Graphics. Proceedings of SIGGRAPH 97, Annual Conference SeriesLos Angeles, Calif.: Addison-Wesley; August 1997. p. 173–80. ISBN 0-89791-896-7.

Facial Animation

Realistic facial animation is one of the most difficult tasks that a computer animator can be asked to do. Human faces are familiar to us all. Facial motion adheres to an underlying, well-defined structure but is idiosyncratic. A face has a single mechanical articulator but has a flexible covering capable of subtle expression and rapid, complex lip movements during speech.

The face is an important component in modeling a figure, because it is the main instrument for communication and for defining a figure's mood and personality. Animation of speech is especially demanding because of the requirement for audio synchronization (and, therefore, is often referred to as *lip-sync animation*). In addition, a good facial model should be capable of geometrically representing a specific person (called *conformation* by Parke [24], *static* by others, e.g., [28]).

Facial models can be used for cartooning, for realistic character animation, for telecommunications to reduce bandwidth, and for human–computer interaction (HCI). In cartooning, facial animation has to be able to convey expression and personality that is often exaggerated. In realistic character animation, the geometry and movement of the face must adhere to the constraints of realistic human anatomy. Telecommunications and HCI have the added requirement that the facial model and motion must be computationally efficient. In some applications, the model must correspond closely to a specific target individual.

In addition to the issues addressed by other animation tasks, facial animation often has the constraint of precise timing with respect to an audio track during lip-synching and some expressions. Despite its name, lip-synching involves more than just the lips; the rigid articulation of the jaw and the muscle deformation of the tongue must also be considered.

10.1 The human face

10.1.1 Anatomic structure

The human face has an underlying skeletal structure and one main skeletal articulatory component—the jaw. In addition to the jaw, the other rigid articulatory components are the eyes.

The surface of the skull is covered with muscles, most of which (at least indirectly) connect areas of the skin to positions on the skull. These muscles tug on the skin to create movement, often recognizable as expressions. Features of the skin include eyelids, mouth, and eyebrows. The skin has various expressive wrinkles such as on the forehead and around the mouth.

As with the rest of the body, muscles provide the driving force for the face (Figure 10.1, Color Plate 7). However, unlike the rest of the body, the muscles mainly move the skin into interesting

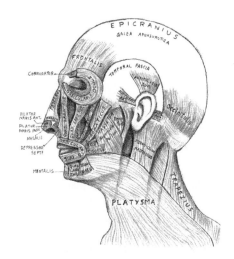

FIGURE 10.1

Muscles of the face and head [30].

(Image courtesy of Arun Somasundaram.)

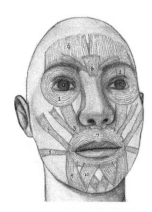

FIGURE 10.2

Muscles of the face that are significant in speech and linguistic facial expressions [30].

(Image courtesy of Arun Somasundaram.)

positions, producing recognizable lip and skin configurations and, in speech, modify the sounds emanating from the labial opening. In particular, the muscles around the mouth are extremely flexible and capable of producing a variety of shapes and subtle movements. A muscle of particular importance is the orbicularis oris (Figure 10.2, muscle number 7; Color Plate 8), which wraps around the mouth and is connected to several muscles that are connected to the skull.

10.1.2 **The facial action coding system**

The facial action coding system (FACS) is the result of research conducted by the psychologists Ekman and Friesen [13] with the objective of deconstructing all facial expressions into a set of basic facial movements. These movements, called action units, or AUs, when considered in combinations, can be used to describe all facial expressions.

Forty-six AUs are identified in the study, and they provide a clinical basis from which to build a facial animation system. Examples of AUs are brow lowerer, inner brow raiser, wink, cheek raiser, upper lip raiser, and jaw drop. See Figure 10.3 for an example of diagrammed AUs. Given the AUs, an animator can build a facial model that is parameterized according to the motions of the AUs. A facial animation system can be built by giving a user access to a set of variables that are in one-to-one correspondence with the AUs. A parametric value controls the amount of the facial motion that results from the associated AU. By setting a variable for each AU, the user can generate all of the facial expressions analyzed by Ekman and Friesen. By using the value of the variables to interpolate the degree to which the motion is realized and by interpolating their value over time, the user can then animate the facial model. By combining the AUs in nonstandard ways, the user can also generate many truly strange expressions.

While this work is impressive and is certainly relevant to facial animation, two of its characteristics should be noted before it is used as a basis for a facial animation system. First, the FACS is meant to be descriptive of a static expression, not generative. Second, the FACS is not time based, and facial movements are analyzed only relative to a neutral pose. This means that the AUs were not designed to animate a facial model in all the ways that an animator may want to control a face. In addition, the FACS describes facial expressions, not speech. The movements for forming individual phonemes, the basic units of speech, were not specifically incorporated into the system. While the AUs provide a good starting point for describing the basic motions that must be in a facial animation system, they were never intended for this purpose.

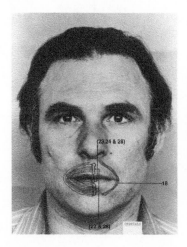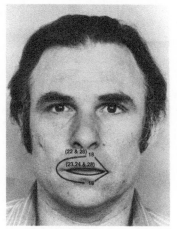

FIGURE 10.3

Three AUs of the lower face [13].

10.2 **Facial models**

Depending on the objective of the animation, there are various ways to model the human face. As with all animation, trade-offs include realism and computational complexity.

If a cartoon type of animation is desired, a simple geometric shape for the head (such as a sphere) coupled with the use of animated texture maps often suffices for facial animation. The eyes and mouth can be animated using a series of texture maps applied to a simple head shape (see Figure 10.4). The nose and ears may be part of the head geometry, or, simpler still, they may be incorporated into the texture map.

Stylized models of the head with only a pivot jaw and rotating spheres for eyeballs that mimic the basic mechanical motions of the human face may also be used. Eyelids can be skin-colored hemispheres that rotate to enclose the visible portion of the eyeball. The mouth can be a separate geometry positioned on the surface of the face geometry, and it can be animated independently or sequentially replaced in its entirety by a series of mouth shapes to simulate motion of deformable lips (see Figure 10.5, Color Plate 9). These approaches are analogous to techniques used in conventional hand-drawn and stop-motion animation.

For more realistic facial animation, more complex facial models are used whose surface geometry more closely corresponds to that of a real face, and the animation of these models is correspondingly more complex. For an excellent in-depth presentation of facial animation see the book by Parke and Waters [25]. An overview is given here.

FIGURE 10.4

Texture-mapped facial animation from *Getting into Art*. (©1990 David S. Ebert [10].)

FIGURE 10.5

Simple facial model using rigid components for animation.

(Image courtesy of John Parent.)

The first problem confronting an animator in facial animation is creating the geometry of the facial model to make it suitable for animation. This in itself can be very difficult. Facial animation models vary widely from simple geometry to anatomy based. Generally, the complexity is dictated by the intended use. When deciding on the construction of the model, important factors are geometry data acquisition method, motion control and corresponding data acquisition method, rendering quality, and motion quality. The first factor concerns the method by which the actual geometry of the head of the subject or character is obtained. The second factor concerns the method by which the data describing changes to the geometry are obtained. The quality of the rendered image with respect to smoothness and surface attributes is the third concern. The final concern is the corresponding quality of the computed motion.

The model can be discussed in terms of its static properties and its dynamic properties. The statics deal with the geometry of the model in its neutral form, while the dynamics deal with the deformation of the geometry of the model during animation. Facial models are either polygonal or higher order. Polygonal models are used most often for their simplicity (e.g., [24] [25] [32] [33]); splines are chosen when a smooth surface is desired.

Polygonal models are relatively easy to create and can be deformed easily. However, the smoothness of the surface is directly related to the complexity of the model, and polygonal models are visually inferior to other methods of modeling the facial surface. Currently, data acquisition methods only sample the surface, producing discrete data, and surface fitting techniques are subsequently applied.

Spline models typically use bicubic, quadrilateral surface patches, such as Bezier or B-spline, to represent the face. While surface patches offer the advantage of low data complexity in comparison to polygonal techniques when generating smooth surfaces, they have several disadvantages when it comes to modeling an object such as the human face. With standard surface patch technology, a rectangular grid of control points is used to model the entire object. As a result it is difficult to maintain low data complexity while incorporating small details and sharp localized features, because entire rows and/or entire columns of control information must be added. Thus, a small addition to one local area of the surface to better represent a facial feature means that information has to be added across the entire surface.

Hierarchical B-splines, introduced by Forsey and Bartels [16], are a mechanism by which local detail can be added to a B-spline surface while avoiding the global modifications required by standard B-splines. Finer resolution control points are carefully laid over the coarser surface while continuity is carefully maintained. In this way, local detail can be added to a surface. The organization is hierarchical, so finer and finer detail can be added. The detail is defined relative to the coarser surface so that editing can take place at any level.

10.2.1 Creating a continuous surface model

Creating a model of a human head from scratch is not easy. Not only must the correct shape be generated, but when facial animation is the objective, the geometric elements (vertices, edges) must be placed appropriately so that the motion of the surface can be controlled precisely. If the model is dense in the number of geometric elements used, then the placement becomes less of a concern, but in relatively low resolution models it can be an issue. Of course, one approach is to use an interactive system and let the user construct the model. This is useful when the model to be constructed is a fanciful creature or a caricature or must meet some aesthetic design criteria. While this approach gives an artist the most freedom in creating a model, it requires the most skill. There are three approaches used to construct a model: refining from a low-resolution model, modifying a high-resolution simple shape, and designing the surface out from high-resolution areas.

Subdivision surfaces (e.g., [9]) use a polygonal control mesh that is refined, in the limit, to a smooth surface. The refinement can be terminated at an intermediate resolution, rendered as a polygonal mesh, vertices adjusted, and then refined again. In this manner, the designer can make changes to the general shape of the head at relatively low resolution. Figure 10.6 shows the initial stages of a subdivision-based facial design [27]. Subdivision surfaces have the advantage of being able to create local complexity without global complexity. They provide an easy-to-use, intuitive interface for developing new models, and provisions for discontinuity of arbitrary order can be accommodated [9]. However, they are difficult to interpolate to a specific dataset, which makes modeling a specific face problematic.

Alternatively, the designer may start with a relatively high-resolution simple shape such as a sphere. The designer pushes and pulls on the surface to form the general features of the face and refines areas as necessary to form the face [17] (see Figure 10.7).

Another alternative is to build the surface out from the expressive regions of the face—the mouth and eyes. This approach can be used to ensure that these regions are flexible enough to support the intended animation. See Figure 10.8 for an example of an initial surface design [29].

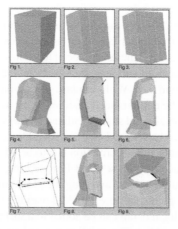

FIGURE 10.6

Early stages in facial modeling using subdivision surfaces [27].

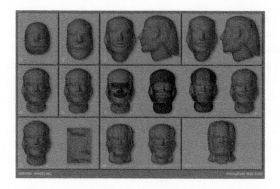

FIGURE 10.7

Facial design by pushing and pulling surface of high-resolution sphere-like shape [17].

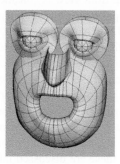

FIGURE 10.8

Facial design by initially defining the eyes and mouth regions [29].

Besides the interactive design approach, there are two main methods for creating facial models: digitization using some physical reference and modification of an existing model. The former is useful when the model of a particular person is desired; the latter is useful when animation control is already built into a generic model.

As with any model, a physical sculpture of the desired object can be generated with clay, wood, or plaster and then digitized, most often using a mechanical or magnetic digitizing device. A two-dimensional, surface-based coordinate grid can be drawn on the physical model, and the polygons can be digitized on a polygon-by-polygon basis. Post-processing can identify unique vertices, and a polygonal mesh can be easily generated. The digitization process can be fairly labor intensive when large numbers of polygons are involved, which makes using an actual person's face a bit problematic. If a model is used, this approach still requires some artistic talent to generate the physical model, but it is easy to implement at a relatively low cost if small mechanical digitizers are used.

Laser scanners use a laser to calculate distance to a model surface and can create very accurate models. They have the advantage of being able to directly digitize a person's face. The scanners sample the surface at regular intervals to create an unorganized set of surface points. The facial model can be constructed in a variety of ways. Polygonal models are most commonly generated. Scanners have the added advantage of being able to capture color information that can be used to generate a texture map. This is particularly important with facial animation: a texture map can often cover flaws in the model and motion. Laser scanners also have drawbacks; they are expensive, bulky, and require a physical model.

Muraki [22] presents a method for fitting a blobby model (implicitly defined surface formed by summed, spherical density functions) to range data by minimizing an energy function that measures the difference between the isosurface and the range data. By splitting primitives and modifying parameters, the user can refine the isosurface to improve the fit.

Models can also be generated from photographs. This has the advantage of not requiring the presence of the physical model once the photograph has been taken, and it has applications for video conferencing and compression. While most of the photographic approaches modify an existing model by locating feature points, a common method of generating a model from scratch is to take front and side images of a face on which grid lines have been drawn (Figure 10.9). Point correspondences can be established between the two images either interactively or by locating common features automatically, and the grid in three-space can be reconstructed. Symmetry is usually assumed for the face, so only one side view is needed and only half of the front view is considered.

Modifying an existing model is a popular technique for generating a face model. Of course, someone had to first generate a generic model. But once this is done, if it is created as a parameterized model and the parameters were well designed, the model can be used to try to match a particular face, to design a face, or to generate a family of faces. In addition, the animation controls can be built into the model so that they require little or no modification of the generic model for particular instances.

One of the most often used approaches to facial animation employs a parameterized model originally created by Parke [23] [24]. The parameters for his model of the human face are divided into two categories: *conformational* and *expressive*. The conformational parameters are those that distinguish one individual's head and face from another's. The expressive parameters are those concerned with animation of an individual's face; these are discussed later. Symmetry between the sides of the face is assumed. Conformal parameters control the shape of the forehead, cheekbone, cheek hollow,

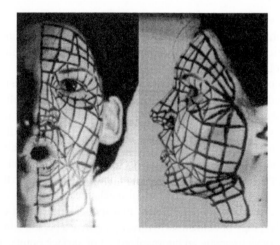

FIGURE 10.9

Photographs from which a face may be digitized [25].

chin, and neck. There are several scale distances between facial features:[1] head x, y, z; chin to mouth and chin to eye; eye to forehead; eye x and y; and widths of the jaw, cheeks, nose bridge, and nostril. Other conformal parameters translate features of the face: chin in x and z; end of nose x and z; eyebrow z. Even these are not enough to generate all possible faces, although they can be used to generate a wide variety.

Parke's model was not developed based on any anatomical principles but from the intuitions from artistic renderings of the human face. Facial anthropometric statistics and proportions can be used to constrain the facial surface to generate realistic geometries of a human head [8]. Variational techniques can then be used to create realistic facial geometry from a deformed prototype that fits the constraints. This approach is useful for generating heads for a crowd scene or a background character. It may also be useful as a possible starting point for some other character; however, the result will be influenced heavily by the prototype used.

The MPEG-4 standard proposes tools for efficient encoding of multimedia scenes. It includes a set of *facial definition parameters* (FDPs) [15] that are devoted mainly to facial animation for purposes of video teleconferencing. Figure 10.10 shows the feature points defined by the standard. Once the model is defined in this way, it can be animated by an associated set of *facial animation parameters* (FAPs) [14], also defined in the MPEG-4 standard. MPEG-4 defines 68 FAPs. The FAPs control rigid rotation of the head, eyeballs, eyelids, and mandible. Other low-level parameters indicate the translation of a corresponding feature point, with respect to its position in the neutral face, along one of the coordinate axes [7].

One other interesting approach to generating a model of a face from a generic model is fitting it to images in a video sequence [8]. While not a technique developed for animation applications, it is useful for generating a model of a face of a specific individual. A parameterized model of a face is set up in a three-dimensional viewing configuration closely matching that of the camera that produced the video images. Feature points are located on the image of the face in the video and are also located on the three-dimensional synthetic model. Camera parameters and face model parameters are then modified to more closely match the video by using the pseudoinverse of the Jacobian. (The Jacobian is the matrix of partial derivatives that relates changes in parameters to changes in measurements.) By computing the difference in the measurements between the feature points in the image and the projected feature points from the synthetic setup, the pseudoinverse of the Jacobian indicates how to change the parametric values to reduce the measurement differences.

10.2.2 Textures

Texture maps are very important in facial animation. Most objects created by computer graphics techniques have a plastic or metallic look, which, in the case of facial animation, seriously detracts from the believability of the image. Texture maps can give a facial model a much more organic look and can give the observer more visual cues when interacting with the images. The texture map can be taken directly from a person's head; however, it must be registered with the geometry. The lighting situation during digitization of the texture must also be considered.

---

[1]In Parke's model, the z-axis is up, the x-axis is oriented from the back of the head toward the front, and the y-axis is from the middle of the head out to the left side.

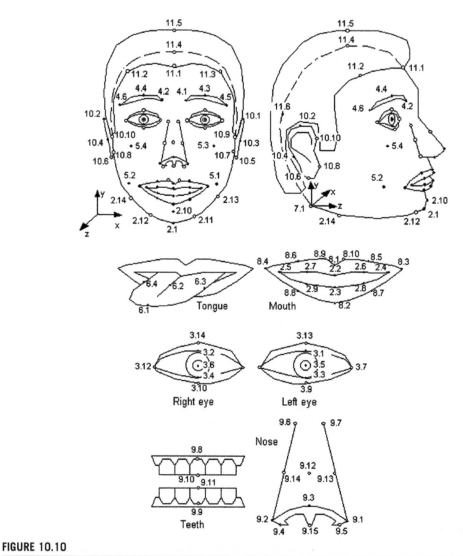

FIGURE 10.10

Feature points corresponding to the MPEG-4 FDPs [15].

Laser scanners are capable of collecting information on intensity as well as depth, resulting in a high-resolution surface with a matching high-resolution texture. However, once the face deforms, the texture no longer matches exactly. Since the scanner revolves around the model, the texture resolution is evenly spread over the head. However, places are missed where the head is self-occluding (at the ears and maybe the chin) and at the top of the head.

Texture maps can also be created from photographs by simply combining top and side views using pixel blending where the textures overlap [1]. Lighting effects must be taken into consideration, and because the model is not captured in the same process as the texture map, registration with a model is an issue. Using a sequence of images from a video can improve the process.

10.3 Animating the face

Attempts to animate the face raise the questions: What are the primitive motions of the face? And how many degrees of freedom are there in the face?

10.3.1 Parameterized models

As introduced in the discussion of FACS, parameterizing the facial model according to primitive actions and then controlling the values of the parameters over time is one of the most common ways to implement facial animation. Abstractly, any possible or imaginable facial contortion can be considered as a point in an *n*-dimensional space of all possible facial poses. Any parameterization of a space should have complete coverage and be easy to use. Complete coverage means that the space reachable by (linear) combinations of the parameters includes all (at least most) of the interesting points in that space. Of course, the definition of the word *interesting* may vary from application to application, so a generally useful parameterization includes as much of the space as possible. For a parameterization to be easy to use, the set of parameters should be as small as possible, the effect of each parameter should be independent of the effect of any other parameter, and the effect of each parameter should be intuitive. Of course, in something as complex as facial animation, attaining all of these objectives is probably not possible, so determining appropriate trade-offs is an important activity in designing a parameterization. Animation brings an additional requirement to the table: the animator should be able to generate common, important, or interesting motions through the space by manipulating one or just a few parameters.

The most popular parameterized facial model is credited to Parke [23] [24] [25] and has already been discussed in terms of creating facial models based on the so-called conformational parameters of a generic facial model. In addition to the conformational parameters, there are *expression parameters*. Examples of expression parameters are upper-lip position, eye gaze, jaw rotation, and eyebrow separation. Figure 10.11 shows a diagram of the parameter set with the (interpolated) expression parameters identified. Most of the parameters are concerned with the eyes and the mouth, where most facial expression takes place. With something as complex as the face, it is usually not possible to animate interesting expressions with a single parameter. Experience with the parameter set is necessary for understanding the relationship between a parameter and the facial model. Higher level abstractions can be used to aid in animating common motions.

10.3.2 Blend shapes

The simplest approach to facial animation is to define a set of key poses, also called *blend shapes*. For a set of example blend shapes, see Figure 10.12 (Color Plate 10). Facial animation is produced by selecting two of the key poses and interpolating between the positions of their corresponding vertices in the two poses.

This restricts the available motions to be the interpolation from one key pose to another. To generalize this a bit more, a weighted sum of two or more key poses can be used in which the weights sum

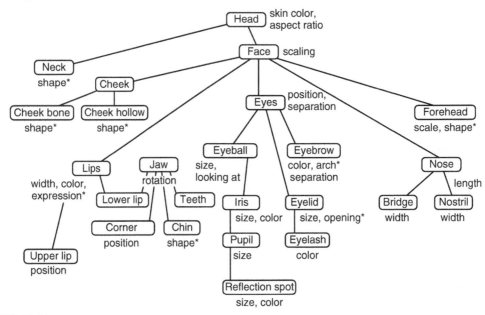

FIGURE 10.11

Parke model. *Indicates interpolated parameters [25].

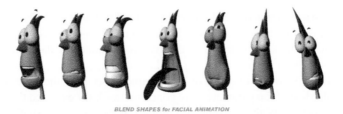

FIGURE 10.12

Blend shapes of a character. (Character design by Chris Oatley; rig design by Cara Christeson; character modeling and posing by Daniel Guinn.)

to one. Each vertex position is then computed as a linear combination of its corresponding position in each of the poses whose weight is non-zero. This can be used to produce facial poses not directly represented by the keys. However, this is still fairly restrictive because the various parts of the facial model are not individually controllable by the animator. The animation is still restricted to those poses represented as a linear combination of the keys. If the animator allows for a wide variety of facial motions, the key poses quickly increase to an unmanageable number.

10.3.3 **Muscle models**

Parametric models encode geometric displacement of the skin in terms of an arbitrary parametric value. Muscle-based models (e.g., Figure 10.13, Color Plate 11) are more sophisticated, although there is wide variation in the reliance on a physical basis for the models. There are typically three types of muscles that need to be modeled for the face: linear, sheet, and sphincter. The *linear muscle* is a muscle that contracts and pulls one point (the *point of insertion*) toward another (the *point of attachment*). The *sheet muscle* acts as a parallel array of muscles and has a line of attachment at each of its two ends rather than a single point of attachment as in the linear model. The *sphincter muscle* contracts a loop of muscle. It can be thought of as contracting radially toward an imaginary center. The user, either directly or indirectly, specifies muscle activity to which the facial model reacts. Three aspects differentiate one muscle-based model from another: the geometry of the muscle-skin arrangement, the skin model used, and the muscle model used.

The main distinguishing feature in the geometric arrangement of the muscles is whether they are modeled on the surface of the face or whether they are attached to a structural layer beneath the skin (e.g., bone or tissue [27],[33]). The former case is simpler in that only the surface model of the face is needed for the animation system (Figure 10.14). The latter case is more anatomically correct and thus promises more accurate results, but it requires much more geometric structure in the model and is therefore much more difficult to construct (Figure 10.15).

The model used for the skin will dictate how the area around the point of insertion of a (linear) muscle reacts when that muscle is activated; the point of insertion will move an amount determined

FIGURE 10.13

Example muscles for facial animation [30].

(Image courtesy of Arun Somasundaram.)

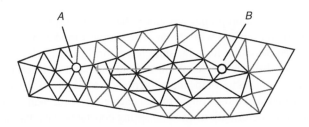

FIGURE 10.14

Part of the surface geometry of the face showing the point of attachment (A) and the point of insertion (B) of a linear muscle; point B is pulled toward point A.

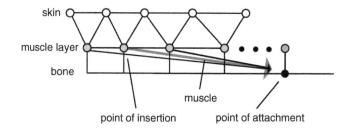

FIGURE 10.15

Cross section of the tri-layer muscle as presented by Parke and Waters [27]; the muscle only directly affects nodes in the muscle layer.

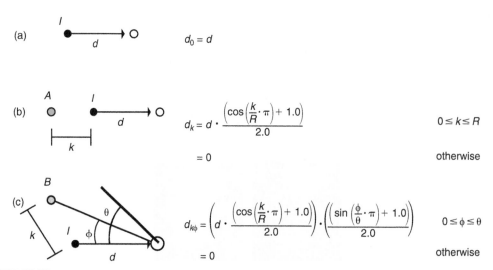

(a) $d_0 = d$

(b)
$$d_k = d \cdot \frac{\left(\cos\left(\frac{k}{R} \cdot \pi \right) + 1.0 \right)}{2.0} \qquad 0 \le k \le R$$
$$= 0 \qquad\qquad \text{otherwise}$$

(c)
$$d_{k\phi} = \left(d \cdot \frac{\left(\cos\left(\frac{k}{R} \cdot \pi \right) + 1.0 \right)}{2.0} \right) \cdot \left(\frac{\left(\sin\left(\frac{\phi}{\theta} \cdot \pi \right) + 1.0 \right)}{2.0} \right) \qquad 0 \le \phi \le \theta$$
$$= 0 \qquad\qquad \text{otherwise}$$

FIGURE 10.16

Sample attenuation: (a) insertion point I is moved d by muscle; (b) point A is moved d_k based on linear distance from the insertion point; and (c) point B is moved $d_{k\phi}$ based on the linear distance and the deviation from the insertion point displacement vector.

by the muscle. How the deformation propagates along the skin as a result of that muscle determines how rubbery or how plastic the surface will appear. The simplest model to use is based on geometric distance from the point and deviation from the muscle vector. For example, the effect of the muscle may attenuate based on the distance a given point is from the point of insertion and on the angle of deviation from the displacement vector of the insertion point. See Figure 10.16 for sample calculations. A slightly more sophisticated skin model might model each edge of the skin geometry as a spring and control the propagation of the deformation based on spring constants. The insertion point is moved by the action of the muscle, and this displacement creates restoring forces in the springs attached to the insertion point, which moves the adjacent vertices, which in turn moves the vertices attached to them, and so on (see Figure 10.17). The more complicated Voight model treats the skin as a viscoelastic

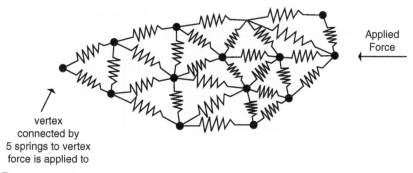

vertex
connected by
5 springs to vertex
force is applied to

FIGURE 10.17

Spring mesh as skin model; the displacement of the insertion point propagates through the mesh according to the forces imparted by the springs.

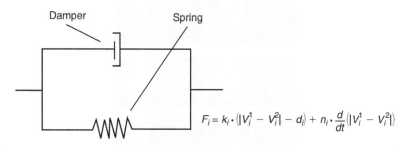

Damper   Spring

$$F_i = k_i \cdot \left(|V_i^1 - V_i^2| - d_i \right) + n_i \cdot \frac{d}{dt} \left(|V_i^1 - V_i^2| \right)$$

FIGURE 10.18

Voight viscoelastic model; the motion induced by the spring forces is damped. Variables k and n are spring and damper constants, respectively; and d_i is the rest length for the spring.

element by combining a spring and a damper in parallel (Figure 10.18). The movement induced by the spring is damped as a function of the change in length of the edge.

The muscle model determines the function used to compute the contraction of the muscle. The alternatives for the muscle model are similar to those for the skin, with the distinction that the muscles are active elements, whereas the skin is composed of passive elements. Using a linear muscle as an example, the displacement of the insertion point is produced as a result of muscle activation. Simple models for the muscle will simply specify a displacement of the insertion point based on activation amount. Because, in the case of the orbicularis oris, muscles are attached to other muscles, care must be taken in determining the skin displacement when both muscles are activated.

More physically accurate muscle models will compute the effect of muscular forces. The simplest dynamic model uses a spring to represent the muscle. Activating the muscle results in a change of its rest length so as to induce a force at the point of insertion. More sophisticated muscle models include damping effects. A muscle model developed by clinical observation is shown in Figure 10.19. However, spring-based facial muscles often result in a computationally expensive approach, and jiggling of the skin can be difficult to control.

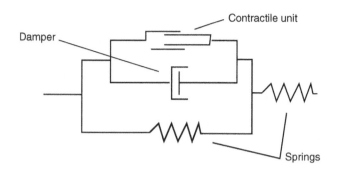

FIGURE 10.19

Hill's model for the muscle.

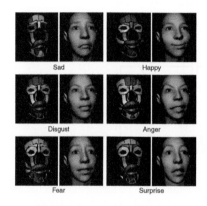

FIGURE 10.20

A basic set of facial expressions [30].

(Image courtesy of Arun Somasundaram.)

10.3.4 Expressions

Facial expressions are a powerful means of communication and are the basis for most facial animation. Any facial animation system will provide for a basic set of expressions. A commonly used set is: happy, angry, sad, fear, disgust, and surprise (Figure 10.20, Color Plate 12).

10.3.5 Summary

A wide range of approaches can be used to model and animate the face. Which approach to use depends greatly on how realistic the result is meant to be and what kind of control the animator is provided. Results vary from cartoon faces to parameterized surface models to skull-muscle-skin simulations. Realistic facial animation remains one of the interesting challenges in computer animation.

10.4 **Lip-sync animation**
10.4.1 **Articulators of speech**

Speech is a complex process and not even completely understood by linguists. Involved in the production of speech are the lungs, the vocal folds, the velum, lips, teeth, and tongue. These articulators of speech constitute the *vocal tract* (see Figure 10.21). The lungs produce the primary air flow necessary for sound production. Sound is a sensation produced by vibrations transmitted through air. The various sounds of speech are produced primarily by controlling the frequency, amplitude, and duration of these vibrations. Vibrations are generated by either the vocal folds in the throat or by the tongue in the oral cavity, or by lips. When vibration is produced by the vocal folds, the sound is called *voiced*. For example, the "th" sound in "then" is voiced whereas the "th" of "thin" is not voiced. The sound travels up into either nasal cavity (e.g., "n") or oral cavity depending on the configuration of the velum. In the oral cavity, the sound is modified by the configuration of the lips, jaw, and tongue with certain frequencies resonating in the cavity. In addition, the tongue and lips can vibrate by either partially obstructing the air flow or by completely stopping and then releasing it. If the air flow is relatively unrestricted, then the sound is considered a vowel, otherwise it is a consonant. If the air flow is completely stopped and then released (e.g., "t," "d") then it is referred to as a *stop*. If the air flow is partially restricted creating a vibration (e.g., "f," "th") then it is referred to as a *fricative*.

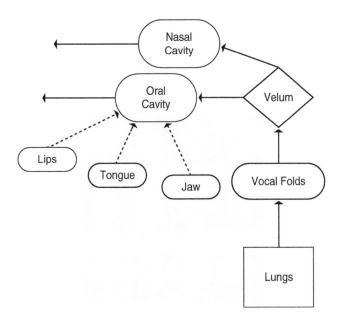

FIGURE 10.21

Schematic of air flow through the vocal tract. Lungs produce air flow. Vocal folds may vibrate. Velum deflects air flow to either nasal cavity or oral cavity. Harmonics of vibrations initiated by vocal folds may be reinforced in oral cavity depending on its configuration; vibrations may be initiated in oral cavity by lips or tongue. Dashed lines show agents that modify oral cavity configuration.

The fundamental frequency of a sound is referred to as F0 and is present in voiced sounds [21]. Frequencies induced by the speech cavities, called *formants*, are referred to as F1, F2, ... in order of their amplitude. The fundamental frequency and formants are arguably the most important concepts in processing speech.

While most of this activity is interior and therefore not directly observable, it can produce motion in the skin that may be important in some animation. Certainly it is important to correctly animate the lips and the surrounding area of the face. Animating the tongue is also usually of some importance.

Some information can be gleaned from a time-amplitude graph of the sound, but more informative is a time-frequency graph with amplitude encoded using color. Using these *spectrographs*, trained professionals can determine the basic sounds of speech.

10.4.2 Phonemes

In trying to understand speech and how a person produces it, a common approach is to break it down into a simple set of constituent, atomic sound segments. The most commonly used segments are called *phonemes*. Although the specific number of phonemes varies from source to source, there are generally considered to be around 42 phonemes.

The corresponding facial poses that produce these sounds are referred to as *visemes*. Visemes that are similar enough can be combined into a single unique viseme and the resulting set of facial poses can be used (Figure 10.22), for example, as blend shapes for a simple type of lip-sync animation.

However, the sounds and associated lip movements are much more complex than can be represented by simply interpolating between static poses. Within the context of speech, a phoneme is

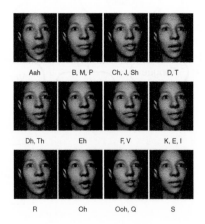

FIGURE 10.22

Viseme set [30].

manifested as variations of the basic sound. These variations of a phoneme are called *allophones* and the mechanism producing the variations is referred to as *coarticulation*. While the precise mechanism of coarticulation is a subject for debate, the basic reason for it is that the ideal sounds are slurred together as phonemes are strung one after the other. Similarly, the visemes of visual speech are modified by the context of the speech in which they occur. The speed at which speech is produced and the physical limitations of the speech articulators result in an inability to attain ideal pronunciation. This slurs the visemes together and, in turn, slurs the phonemes. The computation of this slurring is the subject of research.

10.4.3 Coarticulation

One of the complicating factors in automatically producing realistic (both audial and visual) lip-sync animation is the effect that one phoneme has on adjacent phonemes. Adjacent phonemes affect the motion of the speech articulators as they form the sound for a phoneme. The resulting subtle change in sound of the phoneme produces what is referred to as an allophone of the phoneme. This effect is known as *coarticulation*. Lack of coarticulation is one of the main reasons that lip-sync animation using blend shapes appears unrealistic. While there have been various strategies proposed in the literature to compute the effects of coarticulation, Cohen and Massaro [6] have used weighting functions, called *dominance functions*, to perform a priority-based blend of adjacent phonemes. King and Parent [18] have modified and extended the idea to animation song. Pelachaud et al. [26] cluster phonemes based on deformability and use a look-ahead procedure that applies forward and backward coarticulation rules. Other approaches include the use of constraints [12], physics [2] [30], rules [5], and syllables [19]. None have proven to be a completely satisfying solution for automatically producing realistic audiovisual speech.

10.4.4 Prosody

Another complicating factor to realistic lip-sync animation is changing neutral speech to reflect emotional stress. Such stress is referred to as *prosody*. Affects of prosody include changing the duration, pitch, and amplitude of words or phrases of an utterance. This is an active area of research (e.g., [3] [4] [5] [11] [20]).

10.5 Chapter summary

Facial animation presents interesting challenges. As opposed to most other areas of computer animation, the foundational science is largely incomplete. This, coupled with the complexity inherent in the facial structure of muscles, bones, fatty tissue, and other anatomic elements, makes facial animation one area that has not been conquered by the computer.

References

[1] Akimoto T, Suenaga Y. Three-Dimensional Facial Model Creation Using Generic Model and Front and Side View of Face. IEICE Transactions on Information and Systems March 1992;E75-D(2):191–7.

[2] Albrecht I, Haber J, Seidel H-P. Speech Synchronization for Physics-Based Facial Animation. In: Proceedings of WSCG. Feb 2002. p. 9–16.

[3] Byun M, Badler N. Facemote: Qualitative Parametric Modifiers for Facial Animations. In: Proceedings of the ACM SIGGRAPH Symposium on Computer Animation. 2002. p. 65–71.

[4] Cao Y, Faloutsos P, Pighin F. Unsupervised Learning for Speech Motion Editing. Proceedings of ACM SIGGRAPH/Eurographics Symposium on Computer Animation. 2003. p. 225–31.

[5] Cassell J, Pelachaud C, Badler N, Steedman M, Achorn B, Becket W, et al. Animated Conversation: Rule-Based Generation of Facial Expression, Gesture and Spoken Intonation for Multiple Conversational Agents. In: Proceedings of ACM SIGGRAPH 1994. p. 413–20.

[6] Cohen M, Massaro D. Modeling Coarticulation in Synthetic Visual Speech. In: Models and Techniques in Computer Animation. Tokyo: Springer-Verlag; 1993.

[7] COVEN. http://coven.lancs.ac.uk/mpeg4/index.html; January 2001.

[8] DeCarlo D, Metaxas D, Stone M. An Anthropometric Face Model Using Variational Techniques. In: Cohen M, editor. Computer Graphics. Proceedings of SIGGRAPH 98, Annual Conference Series. Orlando, Fla. Addison-Wesley; July 1998. p. 67–74. ISBN 0-89791-999-8.

[9] DeRose T, Kass M, Truong T. Subdivision Surfaces for Character Animation. In: Cohen M, editor. Computer Graphics. Proceedings of SIGGRAPH 98, Annual Conference Series. Orlando, Fla. Addison-Wesley; July 1998. p. 85–94. ISBN 0-89791-999-8.

[10] Ebert D, Ebert J, Boyer K. "Getting into Art" (animation). CIS Department, Ohio State University; May 1990.

[11] Edge J, Maddock S. Expressive Visual Speech Using Geometric Muscle Functions. In: Proc. of Eurographics UK Chapter Annual Conference (EGUK). 2001. p. 11–8.

[12] Edge J, Maddock S. Constraint-Based Synthesis of Visual Speech. In: Conference Abstracts and Applications of SIGGRAPH. 2004.

[13] Ekman P, Friesen W. Facial Action Coding System. Palo Alto, California: Consulting Psychologists Press; 1978.

[14] FAP Specifications. http://www-dsp.com.dist.unige.it/~pok/RESEARCH/MPEGfapspec.htm; January 2001.

[15] FDP Specifications. http://www-dsp.com.dist.unige.it/~pok/RESEARCH/MPEGfdpspec.htm; January 2001.

[16] Forsey D, Bartels R. Hierarchical B-Spline Refinement. In: Dill J, editor. Computer Graphics. Proceedings of SIGGRAPH 88 vol. 22(4). Atlanta, Ga.; August 1988. p. 205–12.

[17] International Computer. Three-Dimensional Tutorials: Pixologic ZBrush: Head Modeling Part 1, www.3dlinks.com/oldsite/tutorials_ZOldHead.cfm; August 2006.

[18] King S, Parent R. Animating Song. Computer Animation and Virtual Worlds March 2004;15(1):53–61.

[19] Kshirsagar S, Magnenat-Thalmann N. Visyllable Based Speech Animation. Proceedings of Eurographics 2003;22(3):631–9.

[20] Kshirsagar S, Molet T, Magnenat-Thalmann N. Principle Components of Expressive Speech Animation. In: Computer Graphics International. IEEE Computer Society; 2001. p. 38–44.

[21] Lemmetty S. Review of Speech Synthesis Technology. M.S. Thesis, Helsinki University; 1999. http://www.acoustics.hut.fi/publications/files/theses/lemmetty_mst/.

[22] Muraki S. Volumetric Shape Description of Range Data Using Blobby Model. In: Sederberg TW, editor. Computer Graphics. Proceedings of SIGGRAPH 91 vol. 25(4). Las Vegas, Nev.; July 1991. p. 227–35. ISBN 0-201-56291-X.

[23] Parke F. Computer-Generated Animation of Faces. In: Proceedings of the ACM Annual Conference; August 1972.

[24] Parke F. A Parametric Model for Human Faces. Ph.D. dissertation, University of Utah; 1974.

[25] Parke F, Waters K. Computer Facial Animation. Wellesley, Massachusetts: A. K. Peters; 1996. ISBN 1-56881-014-8.

[26] Pelachaud C, Badler N, Steedman M. Generating Facial Expressions for Speech. Cognitive Science 1996;20 (1):1–46.

[27] Ratner P. Subdivision Modeling of a Human, www.highend3d.com/maya/tutorials/-modeling/polygon/189. html; August 2006.

[28] Rydfalk M. CANDIDE: A Parameterized Face. Technical Report LiTH-ISY-I-0866, Sweden: Linköping University; 1987.

[29] Silas F. FrankSilas.com Face Tutorial, www.franksilas.com/FaceTutorial.htm; August 2006.

[30] Somasundaram A. A Facial Animation Model for Expressive Audio-Visual Speech. Ph.D. Dissertation, Ohio State University; 2006.

[31] Waters K. A Muscle Model for Animating Three-Dimensional Facial Expressions. In: Proceedings of SIGGRAPH, vol. 21(4); 1987. p. 17–24.

[32] Waters K. A Physical Model of Facial Tissue and Muscle Articulation Derived from Computer Tomography Data. SPIE Visualization in Biomedical Computing 1992;1808:574–83.

[33] Waters K, Terzopoulos D. Modeling and Animating Faces Using Scanned Data. Journal of Visualization and Computer Animation 1991;2(4):123–8.

Behavioral Animation

11

As discussed in the previous chapters, there are a variety of ways to specify motion. Some are very direct. In key-frame animations, for example, the animator precisely specifies values for animation parameters as specific frames and selects an interpolation procedure to fill in the remaining values for the in-between frames. There is not much mystery as to what the parametric values will be or what the motion will look like. In physically based animation, such as rigid body dynamics, the animator might specify starting values for some of physical simulation parameters, for example, initial position and velocity. The animator generally knows what the animation will look like but does not have a good idea of the intermediate parametric values as an object falls, collides, and bounces. This is an example of model-based animation where there is an underlying model controlling the motion. In this case the model is physics—or at least some approximation to it.

Behavioral animation is another type of model-based animation in which it is the cognitive processes that are being modeled. The term *cognitive process* refers to any behavior that can be attributed to mental activity, voluntary or involuntary, that is reacting to the environment—basically non-physically based procedural motion. The cognitive processes can be as simple as moving toward a light or as complex as balancing self-preservation, mood swings, and personality disorders.

Whereas modeling complex cognitive processes might be necessary for creating a believable autonomous character in a game, simple cognitive processes are often sufficient for modeling crowd behavior. Development of behavioral animation has been motivated by applications of virtual reality (VR), computer games, military simulations, and the film industry. Behavioral animation can be used to evaluate environmental design such as escape facilities in a burning building. It also can be used to create massive crowds for films, to supply background animation, or to provide stimulus for virtual environments.

Modeling behavior can be useful for several reasons:

- To relieve the animator from dealing with details when background animation is not of primary importance
- To generate animation on the fly, as in a game or VR application
- To inspect the results of a simulation, that is, to evaluate the environment relative to a certain behavior
- To populate virtual environments

Synthetic characters with built-in rules of behavior are often referred to as *actors* or *intelligence agents*. When a figure is meant to be the embodiment of a user in a virtual environment, the term *avatar*[1] is often used.

---

[1] In Hindu mythology, an *avatar* is an incarnation of a deity (especially Vishnu) in human or animal form.

At the highest levels of abstraction, the animator takes on the role of a director. At his command are intelligent characters who know how to get the job done, they only need to be told what to do. It is the character's job to find a way to do it. The animator ceases to animate and only directs the action in a very general way. This chapter is about behavior and how it is modeled in computer animation applications. The implied objective is that the behavior be realistic, or at least believable, so that character appears to be *autonomous*.

There are two general aspects of behavioral animation: cognitive modeling and aggregate behavior. Cognitive modeling can range from a simple cause-and-effect-based system to an attempt to model psychological changes in an agent's mental make-up under environmental influences. This aspect of computer animation is in the realm of artificial intelligence (AI). Aggregate behavior refers to how the individuals making up a group are modeled and how their behavior contributes to the quality of the group's motion as a whole.

Cognitive modeling

In general, the task of modeling intelligence and cognition is the problem domain of AI. Aspects of behavior modeling are addressed in computer animation because of the spatial aspects and the imagery produced, and not by focusing on the accuracy of the mental processes being modeled. At all levels of cognitive modeling, the character senses the environment, processes the sensory information possibly with respect to his/her internal state and traits, and then responds with actions that interact with the synthetic environment and possibly with changes to the internal state. At the basic level, a simple reasoning mechanism, such as use of a decision tree or case-based reasoning, suffices. More sophisticated approaches might model personality and mood of the character. For example, a rule-based approach is often used and can prove effective, although there are problems of scale. A common approach is to use a database of simple actions (e.g., simple SIMS games) as a library of behaviors and have appropriate behavior kick in when conditions are satisfied. Simple rules of the form ($<$cond$>$,$<$behavior$>$) can determine activity with the implication of executing the activity when the condition is met: if ($<$cond_i$>$ is TRUE) then do activity_i. Of course, the condition can be arbitrarily complex as can the activity.

When more than one precondition is met, some arbitration strategy must be in place. A simple method is to say that the first condition satisfied is activated, so order of rules sets the priority of rules. For a simple example, consider the behaviors of walking, running, standing, and turning. One of the major issues with this approach is transitioning from one activity to another. The simple solution is to have a neutral behavior to start and end each behavior, for example, standing.

$$\text{stand} \rightarrow \text{walk} \rightarrow \text{turn} \rightarrow \text{run} \rightarrow \text{stand}$$

A more sophisticated approach is to blend from one activity to another with either precomputed transitions or transitions computed on the fly. Adding more activities requires that more rules are added and more transitions are needed. Meta-rules can be added to reason about the rules such as when to apply, which are compatible, and so forth.

Higher levels of cognitive modeling often refer to themselves as modeling intelligence and individuality. Instead of directly reacting to environmental conditions, a cognitive structure is created that can more realistically sense the environment, alter the mental state, and reason about conditions using internal representations while filtering everything through mental states.

Aggregate behavior

Managing complexity is one of the most important uses of the computer in animation, and nothing exemplifies that better than particle systems. A *particle system* is a large collection of individual elements, which, taken together, represent a conglomerate fuzzy object. Both the behavior and the appearance of each individual particle are very simple. The individual particles typically behave according to simple physical principles with respect to their environment but not with respect to other particles of the system. When viewed together, the particles create the impression of a single, dynamic complex object. This illusion of a greater whole is referred to as *emergent behavior* and is an identifying characteristic of particle systems, flocking, and, to a lesser extent, crowds.

Often behavioral animation is concerned with a large number of characters. The primitive behaviors of flocking and prey–predator activity, as well as crowd modeling, are common examples. One of the problems when dealing with a large group of behavioral characters is that knowledge about nearby characters is often required. Particle systems do not exhibit this complexity because there is (typically) no particle–particle interaction. There is only simple physical interaction with the environment. But when knowledge of nearby characters is required, the processing complexity is n-squared where n is the number of characters. Even when interactions are limited to some k nearest neighbors, it is still necessary to find those k nearest neighbors out of the total population of n.

One way to find the nearest neighbors efficiently is to perform a three-dimensional bucket sort and then check adjacent buckets for neighbors. Such a bucket sort can be updated incrementally by adjusting bucket positions of any members that deviate too much from the bucket center as the buckets are transformed along with the flock. There is, of course, a time-space trade-off involved in bucket size—the smaller the buckets, the more buckets needed but the fewer members per bucket on average. This does not completely eliminate the n-squared problem because of worst-case distributions, but it is effective in practice.

The members of a flock, typically fewer in number than particles in a particle system, usually behave according to more sophisticated physics (e.g., flight) and a bit of intelligence (e.g., collision avoidance). Simple cognitive processes that control the movement of a member are modeled and might include such behavior as goal-directed motion and the steering to maintain separation from neighbors.

Adding more intelligence to the members in a group results in more interesting individual behaviors, which is sometimes referred to as *autonomous behavior*. Modeling autonomous behavior tends to involve fewer participants, less physics, and more intelligence. Particle systems, flocking, and crowd behavior are examples of independently behaving members of groups with varying levels of autonomy, physical characteristics, and simulated motions (Table 11.1). If a larger number of autonomous agents are modeled, crowds are created that share some of the same emergent qualities as flocks and particle systems.

Table 11.1 Aggregate Behavior

Type	Number of Elements	Incorporated Physics	Intelligence
Particles	10^2–10^4	Much—with environment	None
Flocks	10^1–10^3	Some—with environment and other elements	Limited
Crowds	10^1–10^2	Usually little physics, but depends on interaction with environment	Varies from little to much

11.1 Primitive behaviors

Primitive behavior, for purposes of this discussion, is characterized by being immediately reactive to the sensed environment as opposed to any temporally extended reasoning process. The environment is sensed and an immediate reactive behavior is made to the conditions extracted from the sensory data. There is little reasoning, no memory, and no planning. Usually, a simple rule-based system is involved. While a limited number of internal states might be modeled, they control basic urges such as hunger and flight-from-danger. Flocking and prey–predator behavior are the primary examples of such primitive behavior that is of a direct cause-and-effect form.

11.1.1 Flocking behavior

Flocking can be characterized as having a moderate number of members (relative to particle systems and autonomous behavior), each of which is controlled by a relatively simple set of rules that operate locally. The members exhibit limited intelligence and are governed by basic physics. The physics-based modeling typically includes collision response, gravity, and drag. As compared to particle systems, there are fewer elements and some interactions with nearby neighbors are modeled. In addition, members' behavior usually models some limited intelligence as opposed to being strictly physics based.

Flocking is one of the lowest forms of behavioral modeling. Members only have the most primitive intelligence that tells them how to be a member of a flock. From these local rules of the individual members, a global flocking behavior can emerge. While flocks typically consist of uniformly modeled members, prey–predator behavior can result from mixing two competing types of mobile agents.

The flocking behavior manifests itself as a goal-directed body, able to split into sections and reform, creating organized patterns of flock members that can perform coordinated maneuvers. For purposes of this discussion, the birds-in-flight analogy will be used, although, in general, any collection of participants that exhibits this kind of group behavior falls under the label of "flocking." For example, flocking behavior is often used to control herds of animals moving over terrain. Of course, in this case, their motion is limited to the surface of a two-dimensional manifold. To use the flock analogy but acknowledge that it refers to a more general concept, Reynolds uses the term *boid* to refer to a member of the generalized flock. Much of this discussion is taken from his seminal paper [34].

There are two main urges at work in the members of a flock: *avoiding collision* and *staying part of the flock*. These are competing tendencies and must be balanced, with collision avoidance taking precedence.

Avoiding collisions is relative to other members of the flock as well as other obstacles in the environment. Avoiding collision with other members in the flock means that some spacing must be maintained between the members even though all members are usually in motion and that motion usually has some element of randomness associated with it to break up unnatural-looking regularity. However, because the objects, as members of a flock, are typically heading in the same direction, the relative motion between flock members is usually small. This facilitates maintaining spacing between flock members and, therefore, collision avoidance among members. The objective of collision avoidance should be that resulting adjustments bring about small and smooth movements.

Staying part of the flock has to do with each member trying to be just that—a member of the flock. In order to stay part of the flock, a member has an urge to be close to other members of the flock, thus working in direct opposition to the collision avoidance urge. In Reynolds' flock model, a boid stays part of the flock by a *flock centering force* [34]. But as he points out, a global flock centering force does not work well in practice because it prohibits flock splitting, such as that often observed when a flock passes around an obstacle. Flock centering should be a localized tendency so that members in the middle of a flock will stay that way and members on the border of a flock will have a tendency to veer toward their neighbors on one side. Localizing the flocking behavior also reduces the order of complexity of the controlling procedure.

Another force that is useful in controlling the member's reaction to movements of the flock is *velocity matching*, whereby a flock member has an urge to match its own velocity with that of its immediate neighbors. By keeping its motion relative to nearby members small, velocity matching helps a flock member avoid collision with other members while keeping it close to other members of the flock.

Local control
Controlling the behavior of a flock member with strictly local behavior rules is not only computationally desirable, it also seems to be intuitively the way that flocks operate in the real world. Thus, the objective is for control to be as local as possible, with no reference to global conditions of the flock or environment. There are three processes that might be modeled: *physics*, *perception*, and *reasoning and reaction*. The physics modeled is similar to that described in particle systems: gravity, collision detection, and collision response. Perception concerns the information about the environment to which the flock member has access. Reasoning and reaction are incorporated into the module that negotiates among the various demands produced as a result of the perception.

Perception
The main distinction between flocking and particle systems is the modeling of perception and the subsequent use of the resulting information to drive the reaction and reasoning processes. When one localizes the control of the members, a localized area of perception is modeled. Usually the "sight" of the member is restricted to just those members around it, or further to just those members generally in front of it. The position of the member can be made more precise to generate better defined arrangements of the flock members. For example, if a flock member always stays at a 45° angle, to the side and behind, to adjacent members, then a tendency to form a diamond pattern results. If a flock member always stays behind and to the side of one member with no members on the other side, then a **V** pattern can be formed. Speed can also affect perception; the field of view (fov) can extend forward and be slimmer in width as speed increases. To affect a localized fov, a boid should do the following:

Be aware of itself and two or three of its neighbors
Be aware of what is in front of it and have a limited fov
Have a distance-limited fov
Be influenced by objects within the line of sight
Be influenced by objects based on distance and size (angle subtended in the fov)
Be affected by things using an inverse distance-squared or distance-cubed weighting function
Have a general migratory urge but no global objective
Not follow a designated leader
Not have knowledge about a global flock center

Interacting with other members

A member interacts with other members of the flock to maintain separation without collision while trying to maintain membership in the flock. There is an attractive force toward other members of the flock while a stronger, but shorter range, repulsion from individual members of the flock exists. In analyzing the behavior of actual flocks, Potts [33] observed a *chorus line effect* in which a wave motion travels faster than any rate chained reaction time could produce. This may be due to perception extending beyond a simple closest-neighbor relationship.

Interacting with the environment

The main interaction between a flock member and the environment is collision avoidance, for which various approaches can be used. Force fields are the simplest to implement and give good results in simple cases. However, in more demanding situations, force fields can give undesirable results. The trade-offs of various strategies are discussed later, under the section Collision Avoidance.

Global control

There is usually a global goal that is used to control and direct the flock. This can be used to influence all members of the flock or to influence just the leader. The animation of the current leader of the flock is often scripted to follow a specific path or is given a specific global objective. Members can have a migratory urge, follow the leader, stay with the pack, or exhibit some combination of these urges.

Flock leader

To simulate the behavior of actual flocks, the animator can have the leader change periodically. Presumably, actual flocks change leaders because the wind resistance is strongest for the leader and rotating the job allows the birds to conserve energy. However, unless changing the flock leader adds something substantive to the resulting animation, it is easier to have one designated leader whose motion is scripted along a path to control the flock's general behavior.

Negotiating the motion

In producing the motion, three low-level controllers are commonly used. They are, in order of priority, collision avoidance, velocity matching, and flock centering. Each of these controllers produces a directive that indicates desired speed and direction (a velocity vector). The task is to negotiate a resultant velocity vector given the various desires.

As previously mentioned, control can be enforced with repulsion from other members and environmental obstacles and attraction to flock center. However, this has major problems as forces can cancel each other out. Reynolds refers to the programmatic entity that resolves competing urges as the *navigation module*. As Reynolds points out, averaging the requests is usually a bad idea in that requests can cancel each other out and result in nonintuitive motion. He suggests a prioritized acceleration allocation strategy in which there is a finite amount of control available, for example, one unit. A control value is generated by the low-level controllers in addition to the velocity vector. A fraction of control is allocated according to priority order of controllers. If the amount of control runs out, then one or more of the controllers receives less than what they requested. If less than the amount of total possible control is allocated, then the values are normalized (e.g., to sum to the value of one). A weighted average is then used to compute the final velocity vector. Governors may be used to dampen the resulting motion by clamping the maximum velocity or clamping the maximum acceleration.

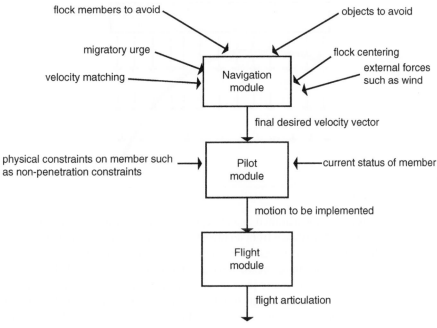

FIGURE 11.1

Negotiating the motion.

In addition to prioritized behaviors, the physical constraints of the flock member being modeled need to be incorporated into the control strategy. Reynolds suggests a three-stage process consisting of navigation, piloting, and flying (see Figure 11.1). Navigation, as discussed earlier, negotiates among competing desires and resolves them into a final desire. The pilot module incorporates this desire into something the flock member model is capable of doing at the time, and the flight module is responsible for the final specification of commands from the pilot module.

The navigation module arbitrates among the urges and produces a resultant directive to the member. This information is passed to the pilot model, which instructs the flock member to react in a certain way in order to satisfy the directive. The pilot model is responsible for incorporating the directive into the constraints imposed on the flock member. For example, the weight, current speed and acceleration, and internal state of the member can be taken into account at this stage. Common constraints include clamping acceleration, clamping velocity, and clamping velocity from below. The result from the pilot module is the specific action that is to be put into effect by the flock member model. The flight module is responsible for producing the parameters that will animate that action.

Collision avoidance

Several strategies can be used to avoid collisions. The ones mentioned here are from Reynolds's paper [34] and from his course notes [35]. These strategies, in one way or another, model the flock member's fov and visual processing. A trade-off must be made between the complexity of computation involved and how effective the technique is in producing realistic and intuitive motion.

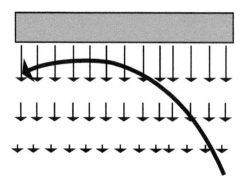

FIGURE 11.2

Force field collision avoidance.

The simple strategy is to position a limited-extent, repelling force field around every object. As long as a flock member maintains a safe distance from an object, there is no force imparted by the object to the flock member. Whether this occurs can easily be determined by calculating the distance[2] between the center point of the flock member and the center of the object. Once this distance gets below a certain threshold, the distance-based force starts to gently push the flock member away from the object. As the flock member gets closer, the force grows accordingly (see Figure 11.2). The advantages of this are that in many cases it effectively directs a flock member away from collisions, it is easy to implement, and its effect smoothly decreases the farther away the flock member is from the object, but it also has its drawbacks.

In some cases, the force field approach fails to produce the motion desired (or at least expected). It prevents a flock member from traveling close and parallel to the surface of an object. The repelling force is as strong for a flock member moving parallel to a surface as it is for a flock member moving directly toward the surface. A typical behavior for a flock member attempting to fly parallel to a surface is to veer away and then toward the surface. The collision avoidance force is initially strong enough to repel the member so that it drifts away from the surface. When the force weakens because of increasing distance, the member heads back toward the surface. This cycle of veering away and then toward the surface keeps repeating. Another problem with simple collision avoidance forces occurs when they result in a force vector that points directly toward the flock member. In this case, there is no direction indicated in which the member should veer; it has to stop and back up, which is an unnatural behavior. Aggregate forces can also prevent a flock member from finding and moving through an opening that may be more than big enough for passage but for which the forces generated by surrounding objects are enough to repel the member (see Figure 11.3).

The problem with a simple repelling force field approach is that there is no reasoned strategy for avoiding the potential collision. Various path planning heuristics that can be viewed as attempts to model simple cognitive processes in the flock member are useful. For example, a bounding sphere can be used to divert the flock member's path around objects by steering away from the center toward the rim of the sphere (Figure 11.4).

Once the flock member is inside the sphere of influence of the object, its direction vector can be tested to see if it indicates a potential intersection with the bounding sphere of the object. This calculation is the same as that used in ray tracing to see if a ray intersects a sphere (Figure 11.5).

---

[2]As in many cases in which the calculation of distances is required, distance-squared can be used thus avoiding the square root required in the calculation of distance.

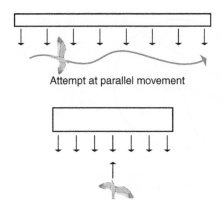

Attempt at parallel movement

Attempt to fly directly toward a surface

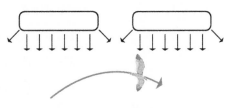

Attempt at finding a passageway

FIGURE 11.3

Problems with force field collision avoidance.

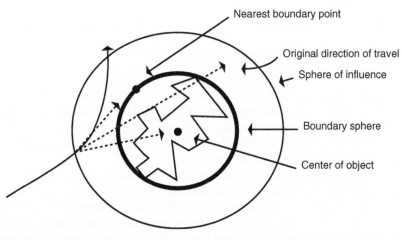

FIGURE 11.4

Steering to avoid a bounding sphere.

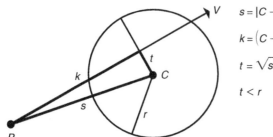

$$s = |C - P|$$

$$k = (C - P) \cdot \frac{V}{|V|}$$

$$t = \sqrt{s^2 - s^2}$$

$$t < r \qquad \qquad \text{indicates penetration} \\ \text{with bounding sphere}$$

FIGURE 11.5

Testing for potential collision with a bounding sphere.

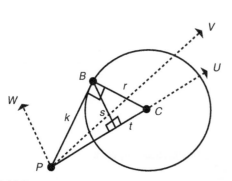

$$k = \sqrt{|C - P|^2 - r^2}$$

$$r^2 = s^2 + t^2$$

$$k^2 = s^2 + (|C - P| - t)^2$$

$$k^2 = r^2 - t^2 + (|C - P| - t)^2$$

$$k^2 = r^2 - t^2 + (|C - P|^2 - 2|C - P|t + t^2$$

$$t = \frac{k^2 - r^2 - |C - P|^2}{-2|C - P|}$$

$$s = \sqrt{r^2 - t^2}$$

$$U = \frac{C - P}{|C - P|}$$

$$W = \frac{(U \times V) \times U}{|(U \times V) \times U|}$$

$$B = P + (|C - P| - t) U + sW$$

FIGURE 11.6

Calculation of point B on the boundary of a sphere.

When a potential collision has been detected, the steer-to-avoid procedure can be invoked. It is useful to calculate the point, B, on the boundary of the bounding sphere that is in the plane defined by the direction vector, V, the location of the flock member, P, and the center of the sphere, C. The point B is located so that the vector from P to B is in the plane defined earlier, is tangent to the sphere, and is on the same side of C as the vector from P to V (Figure 11.6).

Steering to the boundary point on the bounding sphere is a useful strategy if a close approach to the object's edge is not required. For a more accurate and more computationally expensive strategy, the flock member can steer to the silhouette edges of the object that share a back face and a front face with respect to the flock member's position and, collectively, define a nonplanar manifold. Faces can be tagged as front or back by taking the dot product of the normal with a vector from the flock member to the face. The closest distance between the semi-infinite direction of the travel vector to a silhouette edge can be calculated. If there is

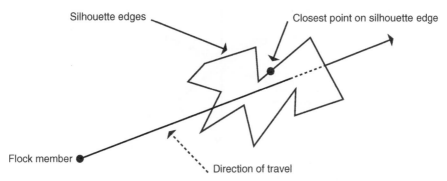

FIGURE 11.7

Determining steer-to-point from silhouette edges.

space for the flock member's body and a comfort zone in which to fly around this point, then this point can be steered toward that closest point on the silhouette (Figure 11.7).

An alternative strategy is to sample the environment by emitting virtual feelers to detect potential collision surfaces. The feelers can be emitted in a progressively divergent pattern, such as a spiral, until a clear path is found.

Another strategy is to model vision by projecting the environment to an image plane from the point of view of the flock member. By generating a binary image, the user can search for an uncovered group of pixels. Depth information can be included to allow searching for discontinuities and to estimate the size of the opening in three-space. The image can be smoothed until a gradient image is attained, and the gradient can be followed to the closest edge.

Splitting and rejoining

When a flock is navigating among obstacles in the environment, one of their more interesting behaviors is the splitting and rejoining that result as members veer in different directions and break into groups as they attempt to avoid collisions. If the groups stay relatively close, then flock membership urges can bring the groups back together to re-form the original single flock. Unfortunately, this behavior is difficult to produce because a balance must be created between collision avoidance and the flock membership urge, with collision avoidance taking precedence in critical situations. Without this precise balance, a flock faction will split and never return to the flock or flock members will not split off as a separate group; instead they will only individually avoid the obstacle and disrupt the flock formation.

Modeling flight

Since flocking is often used to model the behavior of flying objects, it is useful to review the principles of flight. A local coordinate system of *roll*, *pitch*, and *yaw* is commonly used to discuss the orientation of the flying object. The roll of an object is the amount that it rotates to one side from an initial orientation. The pitch is the amount that its nose rotates up or down, and the yaw is the amount it rotates about its up vector (see Figure 11.8).

Specific to modeling flight, but also of more general use, is a description of the forces involved in aerodynamics. *Geometric flight*, as defined by Reynolds [34], is the "dynamic, incremental, rigid transformation of an object moving along and tangent to a three-dimensional curve." Such flight is controlled by *thrust*, *drag*, *gravity*, and *lift* (Figure 11.9). Thrust is the force used to propel the object

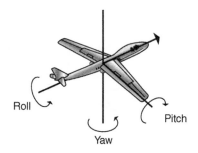

FIGURE 11.8

Roll, pitch, and yaw of local coordinate system.

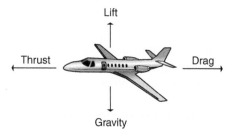

FIGURE 11.9

Forces of flight.

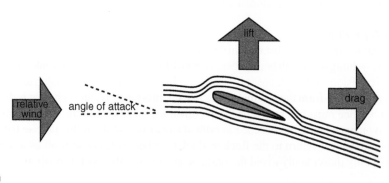

FIGURE 11.10

Lift produced by an airfoil.

forward and is produced by an airplane's engine or the flapping of a bird's wings. Drag is the force induced by an object traveling through a medium such as air (i.e., wind resistance) and works in the direction directly against that in which the object is traveling relative to the air. Gravity is the force that attracts all objects to the earth and is modeled by a constant downward acceleration. Lift is the upward force created by the wing diverting air downward. During flight, the airfoil's shape and orientation is such that air traveling over it is bent. A reaction to the change of momentum of the air is the lift (Figure 11.10). At higher angles of attack the downward velocity of the diverted air is greater. For straight and level flight, lift cancels gravity and thrust equals drag.

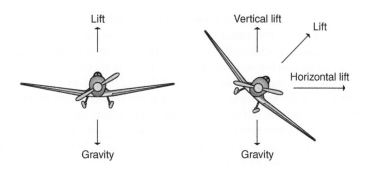

FIGURE 11.11

Lifting forces.

In flight, a turn is induced by a *roll*, in which the up vector of the object rotates to one side. Because the up vector is rotated to one side, the lift vector is not directly opposite to gravity. Such a lift vector can be decomposed into a vertical component, the component that is directly opposite to gravity, and the horizontal component, the component to the side of the object. A flying object turns by being lifted sideways by the horizontal component of the lift; this is why planes tilt into a turn. For the object to be directed into the turn, there must be some yaw induced. If a plane is flying level so that lift cancels gravity, tilting the lift vector to one side will result in a decrease in the vertical component of lift. Thus, to maintain level flight during a turn, one must maintain the vertical component of lift by increasing the thrust (see Figure 11.11).

Increasing the pitch increases the angle the wing makes with the direction of travel. This results in increased lift and drag. However, if thrust is not increased when pitch is increased, the increased drag will result in a decrease in velocity, which, in turn, results in decreased lift. So, to fly up, both thrust and pitch need to be increased.

These same principles are applicable to a soaring bird. The major difference between the flight of a bird and that of an airplane is in the generation of thrust. In a plane, thrust is produced by the propeller. In a bird, thrust is produced by the flapping of the wings.

Some of the important points to notice in modeling flight are as follows:

Turning is effected by horizontal lift
Turning reduces lift
Increasing pitch increases lift and drag
Increasing speed increases lift and drag

11.1.2 Prey–predator behavior

The previous discussion on flocking discussed modeling the very basics of intelligent behavior (not much beyond that of a college freshman): stay with your friends and avoid bumping into things. While flocking behavior can be interesting, especially when interacting with obstacles in the environment, the objective is to produce a single uniform motion—the emergent behavior of the flock. In order to create more interesting interaction among the members, two sets of characters can be created with competing goals and discriminating physical abilities.

The classic example of competing goals is that of a prey–predator model in which one set of characters is trying to eat the other. Very simple prey–predator models can create interesting animation if the rules of behavior are carefully chosen. Of course, if the two character groups have the same capabilities, then the resulting interaction may result in some pretty boring interaction. In general, interesting behavior might result when there are two or more classes of objects that have different qualities of motion so that, depending on the geometric arrangement of the characters and the environment, a member of one class or the other may win in a given situation. For example, one set of agents could be slower but have better acceleration, a better turning radius, a better turning velocity, and better reasoning capabilities. There is no absolute model to use when modeling character behavior because balancing the models is dependent on the desired motion.

At a minimum, the reasoning component for prey–predator models could be based on simple attraction/repulsion. One character class is attracted to the other (e.g., trying to eat it) while the other class is repulsed by the first (e.g., trying to avoid being eaten). While a simple force field model can produce interesting motion, incorporating some predictive reasoning ability in one or both character classes can make situational behavior more realistic. For example, the ability to compute the trajectory of a character from its current position and speed can produce more interesting and realistic behavior.

11.2 Knowledge of the environment

Behavioral animation is all about cognitive interaction with the environment combined with the limitations imposed by simulation of physical constraints. One of the important issues when modeling behavior is the information a character has access to about its environment.

At the simplest level, a character has direct access to the environment database. It has perfect and complete knowledge about its own attributes as well as the attributes of all the other objects in the environment. A character can "see" behind walls and know about changes to the environment well-removed from its locale. While this might be a handy computational shortcut, such knowledge can produce behaviors that are unrealistic. Characters reacting to events that are obviously hidden from view can destroy believability in the character.

More realistic behavior can be modeled if the character is enabled with locally acquiring knowledge about the environment. Such local acquisition is affected by simulating character-centric sensors, most commonly vision. Modeling vision involves determining what can be viewed from the position and orientation of the character. Modeling other senses might also be useful in some applications. Touch, for example, can be modeled by detecting collisions and might be useful for navigating through dark environments. Modeling sounds might be useful in a forest setting. However, since vision is by far the most common sense modeled in computer animation, the following discussion will be restricted to incorporating a character's sight into behavioral animation.

For a character that senses the environment, added realism can be created by providing the capability of remembering what has been sensed. Simulated memory is a way for senses to be recorded and accumulated. Current positions, motions, and display attributes (e.g., color) of other objects in the environment can be captured and used later when reasoning about the environment.

11.2.1 Vision

Vision systems limit knowledge about the environment by (1) modeling a limited fov and (2) computing visual occlusions. A simple form of vision only models the field of view and ignores occlusions. Using the character's position and (head) orientation, an fov can be easily calculated. Any object within that fov (e.g., by testing if the object's bounding box overlaps the fov), is marked as "visible" and, therefore, is known to the character.

While such fov is fast and easy to compute, it does not provide for appropriate behavior when significant objects or events are within the fov but occluded from view. Occlusions provide the possibility of hide-and-seek behaviors. Occlusion is provided by computing the geometric elements that are visible from the point of view of the characters and is modeled by rendering a scene from the character position. The vision system can deal with occlusions by incorporating (1) ray casting that samples the environment in order to determine the closest object along the ray and the associated distance or (2) a low-resolution z-buffer that pseudocolors objects with identification tags to indicate what object is visible at each of the pixels and the associated distance (z-depth).

For the most part in computer animation applications, the issues associated with image processing and perception are avoided by allowing the actor to access the object database directly once it has been determined that an object is visible (e.g., [13]). Thus, the actor immediately knows what the object is and what its properties are, such as its orientation and speed. The object's position in space can be retrieved from the ray casting depth or z-buffer. Alternatively, the object's position in space as well as other properties can be determined from the object database directly once it is determined that it is visible.

Vision does not help unless it tells the character something about its environment and lets the character modify its behavior accordingly. The simplest information available to the character is spatial occupancy—vision tells the character where things are. In a prey–predator model, enabling better vision can tip the scales in favor of a slower prey.

As an example of vision, Tu and Terzopoulos [41] have developed behavioral animation of artificial fishes with synthetic vision. A 300 degree fov with a radius consistent with translucent water is used. Any object within that field and not fully occluded is "seen." The artificial fish is given partial access to the object database in order to retrieve relevant information about the sighted object. In work by Noser et al. [24], as well as Kuffner and Latombe [18], a z-buffer vision system is used to determine visible objects for navigating through an obstacle-filled environment. The three-dimensional position of a given pixel is computed from the depth information in the z-buffer in Noser's work, whereas in Kuffner's work the object database is accessed to locate objects.

Obviously, reasoning plays a large part in any kind of intelligent behavior. Information about the environment is only as useful as what the agent can do with it. Prediction and evaluation are the primary building blocks of reasoning agents. Prediction allows the agent to guess what a current state will be. Evaluation allows the agent to estimate how good a particular situation is or to estimate how good a given situation will be.

11.2.2 Memory

Vision provides knowledge that is immediate but fleeting. If the only information available to the character is provided by the vision system, what is known is what is seen unless a mechanism exists to record what is seen for later use—memory. The simplest type of memory is accumulating a description of the static world. For example, a vision system that has depth information can record the world space coordinates of observed objects. A three-dimensional representation of the character's observed world can be built as the character traverses the environment. In work by Noser et al. [24], a quadtree data structure is used to partition space and construct an occupancy map from the vision information. As more of the environment is seen, more knowledge is accumulated.

Moveable objects create a problem for an accumulation-only occupancy map. If an object is only inserted and never removed from the occupancy map, then it will pollute the map if it moves throughout the scene and is observed in several places. Thus, dynamic objects demand that an update mechanism be included with the memory module. If an object is seen at a certain location, other references to its location must be removed from the occupancy map.

This raises the issue of the validity of the recorded location of a moving object that has been out of view for awhile. Chances are that if an object is seen moving at one location and then is out of view for a while, the object is no longer at that location. It seems like a good idea to remove memories of locations of moving objects if they have not been seen for awhile. Time-stamping memories of moving objects allows this to be done. As time progresses, the memories that are older than some (possibly object-specific) threshold are to be removed (e.g., [18] [24]).

11.3 Modeling intelligent behavior

Modeling behavior, especially intelligent behavior, is an open-ended task. Even simple flocking and prey–predator models can be difficult to control. Humans are even more complex and modeling their behavior is a task that AI has been addressing for decades. This type of AI falls under the rubric of computer graphics because of the spatial (geometric) reasoning component. In computer animation, the animator usually wants the visuals of character motion to mimic (or at least be a caricature of) how humans are seen behaving in real life. While motion capture is a relatively quick and easy way to model human behavior, developing algorithms to produce human behavior, if done well, are more flexible and more general than motion capture (mocap)-based animation. In addition, such models often contribute to a greater understanding of human behavior. Human and animal behavior have been modeled since the early days of computer animation but only recently have more sophisticated methods been employed for higher level, autonomous cognitive behavior.

For example, the Improv system of Ken Perlin and Athomas Goldberg uses layered scripts with decision rules to create novel animation [29]. The AlphaWolf project at Massachusetts Institute of Technology has modeled not only expressiveness, but also learning and development [40]. Attempts have also been made to apply neural net technology to learning behavior (e.g., [7]). Recently, there has been a more formal approach by referring to the literature in psychology (e.g., [36]). It has been applied to expressions and gestures, personality, and crowd behavior.

11.3.1 Autonomous behavior

Autonomous behavior, as used here, refers to the motion of an object that results from modeling its cognitive processes. Usually, such behavior is applied to relatively few objects in the environment. As a result, emergent behavior is not typically associated with autonomous behavior, as it is with particle systems and flocking behavior, simply because the numbers are not large enough to support the sensation of an aggregate body. To the extent that cognitive processes are modeled in flocking, autonomous behavior shares many of the same issues, most of which result from the necessity to balance competing urges.

Autonomous behavior models an object that knows about the local environment in which it exists, reasons about the state it is in, plans a reaction to its circumstances, and carries out actions that affect its environment. The environment, as it is sensed and perceived by the object, constitutes the external state. In addition, there is an internal state associated with the object made up of time-varying urges, desires, and emotions, as well as (usually static) rules of behavior.

It should not be hard to appreciate that modeling behavior can become arbitrarily complex. Autonomous behavior can be described as various levels of complexity, depending on how many and what type of cognitive processes are modeled. Is the objective simply to produce some interesting behavior, or is it to simulate how the actual object would operate in a given environment? How much and what kinds of control are to be exercised by the user over the autonomous agent? Possible aspects to include in the simulation are sensors, especially vision and touch; perception; memory; causal knowledge; commonsense reasoning; emotions; and predispositions. Many of these issues are more the domain of AI than of computer graphics. However, besides its obvious role in rendering, computer graphics is also a relevant domain for which to discuss this work because the objective of the cognitive simulation is motion control. In addition, spatial reasoning is usually a major component of cognitive modeling and therefore draws heavily on algorithms associated with computer graphics. Applications include military simulations (e.g., [21]), pedestrian traffic (e.g., [37]), and, of course, computer games (e.g., [15]).

Autonomous behavior is usually associated with animal-like articulated objects. However, it can be used with any type of object, especially if that object is typically controlled by a reasoning agent. Obvious examples are cars on a highway, planes in the air, or tanks on a battlefield. Autonomous behavior can also be imparted to inanimate objects whose behavior can be humanized, such as kites, falling leaves, or clouds. The current discussion focuses on the fundamentals of modeling human behavior.

Internal state

Internal state is modeled partially by intentions. Intentions take on varied importance depending on the urges they are meant to satisfy. The instinct to survive is perhaps the strongest urge and, as such, takes precedence over, say, the desire to scratch an itch. Internal state also includes such things as inhibitions, identification of areas of interest, and emotional state. These are the internal state variables that are inputs to the rest of the behavioral model. While the internal state variables may actually represent a continuum of importance, Blumberg and Galyean [1] group them into three precedence classes: imperatives, things that must be done; desires, things that should be done, if they can be accommodated by the reasoning system; and suggestions, ways to do something should the reasoning system decide to do that something.

Levels of behavior

There are various levels at which an object's motion can be modeled. In addition to that discussed by Blumberg and Galyean [1], Zeltzer [44] and Korein and Badler [17] discuss the importance of decomposing high-level goals into object-specific manipulation of the available degrees of freedom (DOFs) afforded by the geometry of the object. The levels of behavior are differentiated according to the level of abstraction at which the motion is conceptualized. They provide a convenient hierarchy in which the implementation of behaviors can be discussed. The number of levels used is somewhat arbitrary. Those presented here are used to emphasize the motion control aspects of autonomous behavior as opposed to the actual articulation of that motion (refer to Figure 11.12).

Internal state and *knowledge of the external world* are inputs to the *reasoning unit*, which produces a strategy intended to satisfy some objective. A *strategy* is meant to define the *what* that needs to be done. The *planner* turns the strategy into a *sequence of actions* (the *how*), which is passed to the *movement coordinator*. The movement coordinator selects the appropriate *motor activities* at the appropriate time. The motor activities control specific DOFs of the object.

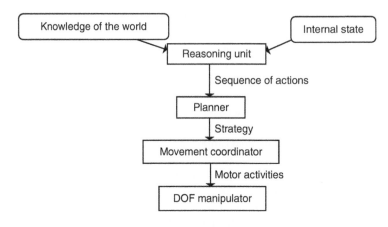

FIGURE 11.12

Levels of behavior.

Keeping behavior under control

One of the fundamental concerns with autonomous behavior, as with all high-level motion control, is how to model the behaviors so that they are generated automatically but are still under the control of the animator. One of the motivations for modeling behaviors is to relieve the animator from specifying the time-varying values for each of the DOFs of the underlying geometry associated with the object. Pure autonomy is often not the ultimate goal in designing autonomous agents; often behavior needs to be controlled in order to be relevant to the objectives of the animator. Also, the control needs to occur at various levels of specificity.

The various levels of behavior provide hooks with which control over behavior can be exercised. The animator can insert external control over the behavior by generating imperatives at any of the various levels: strategies, action sequences, and/or activity invocation. In addition, more general control can be exercised by setting internal state variables.

Arbitration between competing intentions

As with arbitration among competing forces in flocking, behaviors cannot, in general, be averaged and expected to produce reasonable results. The highest precedent behavior must be accommodated at all costs, and other behaviors may be accommodated if they do not contradict the effect of the dominant behavior. One approach is to group behaviors into sets in which one behavior is selected. Selected behaviors may then be merged to form the final behavior.

11.3.2 Expressions and gestures

Nothing is more boring than a speaker who stands up and dryly repeats a script without expression, without a personality. As we have seen in Chapter 10, there has been a fair amount of work on so-called talking heads. These are animations where a synthetic face lip-syncs a script. Most of these are cartoon-like in their lip movement, but they use facial expressions to add interest to the animation.

Several "talking heads" have been created that include facial expressions coordinated with synthetic speech. *Greta* [31] includes wrinkles and furrows to enhance realism. *Ruth*, the Rutgers

University Talking Head [5] [6], animates nonverbal signals in synchrony with speech and lip movements. The intonation is specified using the tones and break indices (ToBI) standard [27]. Brow expressions are categorized in terms of Ekman's facial action units [11]. The expressions are generated by rules inferred from lexical structure along with observed behavior of live performances. Additionally, in face-to-face communication the eyes alone can convey information about the participants, beliefs, goals, and emotions [30].

Upper body gestures (i.e., arm and torso) allow more opportunity for expressive speech. *BEAT*, a toolkit by Justine Cassell et al. [2], allows animators to type in text they wish spoken by a synthetic figure. Using linguistic and contextual information contained in the text, the movements of the hands, arms, and face and intonation of the speech can be controlled. A knowledge base relating gestures to emotions within the context of grammatical constructions is used to suggest gestural behaviors. These are passed through an activity filter that produces a schedule of behaviors that is converted to word timing and associated gestural animation. Laban movement analysis (LMA) (e.g., [19]) is used as a theoretical framework for generating the qualitative aspects of linguistic gestures by Chi et al. in the *EMOTE* system [4]. The shape and effort components of LMA are used to control the timing of the gesture, the articulation of the arms during the gesture, and animation of the torso. As an alternative to parameterized models of behavior, data-driven techniques use captured human motion to animate a conversational character [3] [38]. Performance data are segmented and annotated in order to provide a database of motion and speech samples. These samples are recombined and blended together to create extended utterances.

11.4.3 Modeling individuality: personality and emotions

In the previous section, the ability to generate gestures along with speech makes a character more believable. In the EMOTE system, the gestures can be modified by shape and effort parameters [4]. While in EMOTE these values are externally supplied, it demonstrates the ability to vary behavior based on parameters. By making these parameters associated with the internal state of a character, individuals within a population can be differentiated by the quality of their behavior by their *personality* [43]. For the current discussion, personality refers to the time-invariant traits that control the idiosyncratic actions of an individual. *Emotions* are considered time-varying attributes of an individual. *Mood* is sometimes considered a third attribute of individuality, and is a personal trait that is time varying but on a longer timescale than emotion.

Graphics research into modeling personalities often borrows from research in psychology, where there are already models proposed for characterizing personalities. These models include OCEAN and PEN. The OCEAN model of personality, also called the Five-Factor Model, or the "Big 5," consists of openness, conscientiousness, extroversion, agreeableness, and neuroticism [32]. The PEN model of Hans Eysenck has three dimensions of personality: extraversion, neuroticism, and psychoticism [14]. The Eysenck Personality Profiler is widely used in business. It is a questionnaire that measures 21 personality traits along the three dimensions.

As with personality, graphics research in emotions has borrowed heavily from psychology. Psychology, however, is not definitive when defining basic emotions and what those basic emotions are. As identified by Ekman [10], there are six basic emotions: happy, sad, fear, disgust, surprise, and anger. Emotional expressions and gestures contribute to communication (vis-á-vis linguists). Five basic emotions are suggested by Oatley and Johnson-Laird [25]: happiness, anxiety, sadness, anger, and disgust. Ekman [12] has also suggested there are 15 emotions: amusement, anger, contempt, contentment, disgust, embarrassment, excitement, fear, guilt, pride in achievement, relief, sadness/distress, satisfaction, sensory pleasure, and shame. The Ortony Clore Collins (OCC) model describes how the perceptions of people dictate how they experience emotions [28].

The Improv system [29] provides tools for an animator to script the behavior of characters that respond to users and to other characters in real time. It consists of a behavior engine that is responsible for executing, in parallel, scripts that are applicable to the current situation. The scripts allow for layered behavior so that the animator can build high-level behaviors on top of low-level ones. It provides for nondeterministic behavior by including probabilistic weights to be associated with alternative choices. Underneath the behavior engine is the animation engine, which is responsible for procedurally generating motion segments and blending between them. Personality and emotions are not explicitly modeled but must be encoded in the choices and probabilities supplied with the scripted behavior.

The *AlphaWolf* project of Tomlinson and Blumberg [40] models three parameters of emotion: pleasure, arousal, and dominance. These are updated based on past emotion, an emotion's rate of drift, and environmental factors. In addition, there is context-specific emotional memory that can bring a character back to a previously experienced emotion.

Egges et al. described a system based on the OCC model [8] [9]. The big five personality traits are used to modify emotional states. An intelligent three-dimensional talking head with lip-sync speech and emotional states creates behavior. Emotions are used to control the conversational behavior of an agent. A method is employed to update a character's emotional state based on personality, emotions, mood, and emotional influences.

11.4 Crowds

A crowd is a large group of people physically grouped (crowded) together. But the term "crowd" has been applied to a wide variety of scenarios, such as a collection of individuals reacting to the environment (e.g., [42]), a mass of hundreds of thousands of armed warriors in a feature length film where individual identification is practically impossible, and dots on a screen showing movement from room to room during a fire simulation. For this discussion, a crowd is considered to be the animation of representations of a large number of individuals where the individuals are not the primary focus of the animation and detailed inspection of individuals by the viewer is not expected. This distinguishes a crowd from an environment containing multiple agents in which the individual activity of the agents is important. The use of multiple agents is just that—some number of behaviorally animated characters where the actions of individuals are discernible and important and the viewer would be inspecting the activity of individuals in the environment. The emphasis here is on animation of multiple figures where the global nature of the activity is the primary consideration.

There are two main applications of graphical crowd modeling. The first is as a visual effect. Perhaps the obvious one is the use in feature length film to fill in the movement or activity of a mass of people in battle scenes or populating an environment with activity or creating an audience in the bleachers of a stadium. In battle scenes, for example, it is common to have the first row or two of characters to be live actors while all the background activity is synthetic characters. Managing the complexity of such crowds has become an important issue in the visual effects industry and is being addressed by software products such as Massive [20] [23]. Computer games are another obvious example of this application.

In this type of crowd, the appearance and motion of the individuals must be organized to be believable, not distracting, and computationally efficient.

The second application of graphical crowd modeling is as a simulation of activity in a space in order to better understand that activity. Here, the visuals are not the focus. The distribution and spatial relationship of crowd members with respect to each other and with respect to the environment are the main concern. For example, crowd animation might be used to model the translocation of a mass of people, such as a crowd entering an amusement park or exiting a stadium, in order to evaluate the various pathways through an environment for traffic flow. A related use that has received a fair amount of attention recently is evacuation simulation. In these cases, the appearance of the individuals has no value, it is only the location of the members and how they traverse the environment while trying to escape some threat, such as fire, usually in a building. Here, the cognitive modeling and subsequent data gathering are the important tasks. The visuals are either not important at all or are of secondary importance. Often the crowd members are represented merely as dots in a bird's-eye view of a building floor plan.

11.4.1 Crowd behaviors

There are certain behaviors that individuals of a crowd assume because they are members of the crowd. These behaviors are similar to that found in members of a flock: collision avoidance and maintaining crowd membership. Simulating crowd behavior is both an interesting and challenging research area. Using Reynolds' flock of boids as a reference, crowd members engage in collision avoidance, match the velocity of neighbors, and stay close to other crowd members. For example, in finding seats in a theater or evacuating a building, the flow of the crowd can be studied [16]. However, as opposed to typical models of flocking behavior, crowd flow can be multidirectional. Using psychology literature as a source for how people behave in crowds (e.g., [36]), one of the main activities found in crowds is the avoidance of oncoming people when traversing an environment. Various rules can be derived by looking at actual behavioral studies, as seen in the following:

- In high density, start avoiding at 5 feet or nearer; in low density, avoidance can start at 100 feet
- In low density, change paths; in high density, rotate body and side step (*step and slide*)
- Open (front to other person, men) versus closed (back to other person, women)
- In changing path, move to open side else tend to move to right
- Avoid people moving in the same direction by slowing down or overtaking them and resuming the original course after the pass has been completed

11.4.2 Internal structure

There are higher cognitive behaviors that create substructures within the group. While a crowd can exhibit an aggregate behavior (e.g., so-called *mob mentality*) that is similar to the flocking behavior of birds and emerges from localized motion of the flock members, it is usually made up of smaller groups of people in which the group members interact closely. Group membership is dynamic, and group interaction often occurs [22].

Groups are formed by common urges (e.g., going for lunch), belief systems (e.g., political allies), and emotional state (e.g., soccer fans) as well as by spatial proximity. Individuals of a crowd may or may not have a group membership. Qualities of an individual relevant to group membership include the following:

- A changeable belief system: From experiences, senses (e.g., vision) understanding the nature of things as a result of learning, and causes and effects (from seeing or experiencing a sequence of events)
- Time-varying biases, propensities, disposition, inclination, tendency, and emotional state: Things that will be satisfied if possible and if there are enough extra DOFs in attaining a goal or desire to satisfy secondary desires
- Goals, desires: Definite objectives that the agent is trying to satisfy

11.4.3 Crowd control

Similar to flocking behavior, individual crowd members are typically represented as being aware of nearby crowd members and must be managed to avoid collisions. As opposed to flocking behavior, there is less concern about the effect of physics on the crowd members, because crowds are often land-based animals (e.g., humans) where collisions are avoided rather than detected and movement occurs in two-dimensional space. However, because of the large number of members typically found in a crowd, strategies for controlling the motion of each member must be very efficient for many applications. Three main strategies have been used to control crowds: rules, forces, and flows. Rules have been discussed under the topic of intelligent behavior. For crowds, usually simple rules are used in order to keep the computation cost low. Forces can be used that are similar to those found in particle systems but must be organized so as to prevent collisions. The work by Helbing (e.g., [45]) concerning what he calls *social forces* is basically an extension of Reynolds' work on flocking behavior that is applied to social interactions in crowds. In the third approach to crowd control, the movement of crowd members has been compared to flow fields. This is usually effective for crowds that are massive in the number of members and only positional information of an individual is needed.

11.4.4 Managing *n*-squared complexity

Because of the typically large number of members in a crowd, the task of comparing each crowd member to every other crowd member for discovering the spatial proximity relationship is a major concern. As a consequence, some type of spatial decomposition approach has often been used in cases where the domain of interest is known ahead of time, for example, in building evacuation simulations. In such cases, a spatial sort can be used to sort the crowd members into a data structure conducive for identifying neighbors within a certain distance. For example, a grid-based approach can be used in which all of the interesting space is broken down into a rectangular matrix of cells. With a small enough cell size, each cell holds one group member so that movement into a cell is restricted to empty cells. This can be useful in maintaining the "personal space" of a crowd member, for example. Of course this also results in a large number of cells, so updating the occupancy of each cell can start to incur an undesirable overhead. Increasing the cell size reduces the number of cells, but requires maintaining a list of crowd members inside of each cell and increases the complexity of finding the nearest neighbor in a given direction. Other structures can also be used such as binary space partitioning, a hierarchical quadtree, or dynamic Voronoi diagrams, but these require modification as the crowd members change location over time.

11.4.5 Appearance

In cases where the appearance of individuals is of interest (e.g., feature films), the objective is to produce a population with enough variability so the viewer is not distracted. If every individual in the crowd had exactly the same appearance, it would noticeable to, and therefore distracting for, the viewer. The same holds for detectable patterns in appearance. The objective here is to not have individual features, or relationships among features, of the crowd noticeable to the casual viewer. A common strategy is to create mutually exclusive, yet compatible, sets with variations. For appearance, the sets might be hair color, hair style, shirt color, shirt type, pants color, and pants type. By combinatorial arithmetic, it is easy to see that even just 2 variations per set results in 64 different individual appearances. This same approach can be taken for the movements of the individuals. For example, walking behavior can be divided into the following sets: gait length, walking speed, arm swing distance, frequency of head swivel, and frequency of head nods. Pseudorandomness can be an important part of breaking noticeable patterns in the crowd. Even with the same gait, introducing slight phase shifts in the walk can improve the visual impression of the crowd.

Because crowds can be viewed at a variety of distances, they lend themselves to geometric level-of-detail modeling. The term *impostor* has been used to mean an alternate representation of a graphical element that is of reduced complexity and is used in place of the original geometry when viewed from a distance. Impostors can be used to reduce the display complexity of crowd members at large viewing distances. Crowd members also lend themselves to level-of-detail behavioral modeling [26]. At large viewing distances, the motion of the member can be simplified as well as the appearance. For example, walking can be represented by a series of sprites (two-dimensional graphical objects overlaid on the background) that have simple representations of leg and arm swing behavior. Because individual behaviors are arbitrarily complex, there are opportunities to abstract out "typical crowd behavior" in order to simplify modeling of the individuals (if the viewing conditions permit) when the behavior of the individual member is only statistically relevant (e.g., [39]). Actions can be selected on a statistical basis and strung together to create the visual effect of a crowd.

11.6 Chapter summary

We are only beginning to understand the complexities of spatial reasoning, idiosyncratic behavior, and generic behavior. There is no firm theoretical footing on which to stand when modeling personalities, emotions, and memory like when modeling physical interactions of rigid objects. This represents both a challenge and an opportunity for researchers and developers in computer animation.

References

[1] Blumberg J, Galyean T. Multi-Level Direction of Autonomous Creatures for Real-Time Virtual Environments. In: Cook R, editor. Computer Graphics. Proceedings of SIGGRAPH 95, Annual Conference Series. Los Angeles, Calif: Addison-Wesley; August 1995. p. 47–54. ISBN 0-201-84776-0.

[2] Cassell J, Vilhjalmsson H, Bickmore T. BEAT: the Behavior Expression Animation Toolkit. In: SIGGRAPH.

[3] Cassell J, Pelachaud C, Badler N, Steedman M, Achorn B, Becket T, et al. ANIMATED CONVERSATION: Rule-Based Generation of Facial Expression, Gesture & Spoken Intonation for Multiple Conversational Agents. In: SIGGRAPH 94.

[4] Chi D, Costa M, Zhao L, Badler N. The EMOTE Model for Effort and Shape. In: Computer Graphics. Proceedings of SIGGRAPH 2000. New Orleans, La.; 23–28 July; 2000. p. 173–82.

[5] DeCarlo D, Stone M. The Rutgers University Talking Head: RUTH. In: Rutgers Manual. www.cs.rutgers.edu/~village/ruth/ruthmanual10.pdf.

[6] DeCarlo D, Revilla C, Stone M. Making Discourse Visible: Coding and Animating Conversational Facial Displays. In: Proceedings of Computer Animation; 2002. p. 11–6.

[7] Dinerstein J, Egbert P, de Garis H, Dinerstein N. Fast and Learnable Behavioral and Cognitive Modeling for Virtual Character Animation: Research Articles. Computer Animation and Virtual Worlds May 2004;15(2).

[8] Egges A, Kshirsagar S, Magnenat-Thalmann N. A Model for Personality and Emotion Simulation. In: Knowledge-Based Intelligent Information & Engineering Systems (KES2003). 2003. p. 453–61.

[9] Egges A, Kshirsagar S, Magnenat-Thalmann N. Generic Personality and Emotion Simulation for Conversational Agents. Computer Animation and Virtual Worlds January 2004;15(1):1–13.

[10] Ekman P. Universals and Cultural Differences in Facial Expressions of Emotion. In: Cole J, editor. Nebraska Symposium on Motivation 1971, vol. 19. Lincoln, NE: University of Nebraska Press; 1972. p. 207–83.

[11] Ekman P. About Brows: Emotional and Conversational Signals. In: von Cranach V, Foppa K, Lepenies W, Ploog D, editors. Human Ethology: Claims and Limits of a New Discipline: Contributions to the Colloquium. Cambridge: Cambridge University Press; 1979. p. 169–202.

[12] Ekman P. Basic Emotions. In: Dalgleish T, Power M, editors. Handbook of Cognition and Emotion. Sussex, U.K.: John Wiley & Sons, Ltd.; 1999.

[13] Enrique S, Watt A, Maddock S, Policarpo F. Using Synthetic Vision for Autonomous Non-Player Characters in Computer Games. In: 4th Argentine Symposium on Artificial Intelligence Santa Fe, Argentina; 2002.

[14] Eysenck H. Biological Dimensions of Personality. In: Pervin L, editor. Handbook of Personality: Theory and Research. New York: Guilford; 1990. p. 244–76.

[15] Isla D, Blumberg B. New Challenges for Character-Based AI in Games. In: Artificial Intelligence and Interactive Entertainment: Papers from the 2002 AAAI Spring Symposium. AAAI Press; 2002.

[16] Johnson C. The Glasgow-Hospital Evacuation Simulator: Using Computer Simulations to Support a Risk-Based Approach to Hospital Evacuation, http://www.dcs.gla.ac.uk/~johnson/papers/G-HES.PDF; Sept. 2005.

[17] Korein J, Badler N. Techniques for Generating the Goal-Directed Motion of Articulated Structures. IEEE Computer Graph Appl November 1982.

[18] Kuffner J, Latombe JC. Fast Synthetic Vision, Memory, and Learning Models for Virtual Humans. In: Proceedings of Computer Animation; 1999. p. 118–27.

[19] Maletic V. Body, Space, Expression: The Development of Rudolf Laban's Movement and Dance Concepts. New York: Mouton de Gruyte; 1987.

[20] Massive Software. Massive Software, http://www.massivesoftware.com/massive.php; Santa Fe, Argentina 2007.

[21] McKenzie F, Petty M, Kruszewski R, Gaskins R, Nguyen Q-A, Seevinck J, et al. Crowd Modeling for Military Simulations Using Game Technology. In: SCS International Journal of Intelligent Games & Simulation (IJIGS). 2004.

[22] Musse S, Thalmann D. A Model of Human Crowd Behavior: Group Inter-Relationship and Collision Detection Analysis. In: Computer Animations and Simulations '97. Proc. Eurographics Workshop. Budapest: Springer Verlag, Wien; 1997. p. 39–51.

[23] New Zealand Trade & Enterprise. Crowd Control, http://www.nzte.govt.nz/common/files/la2-massive.pdf; February 2007.

[24] Noser H, Renault O, Thalmann D, Magnenat-Thalmann N. Navigation for Digital Actors Based on Synthetic Vision, Memory and Learning. Computer and Graphics 1995;19(1):7–19.

[25] Oatley K, Johnson-Laird P. Towards a Cognitive Theory of the Emotions. Cognition and Emotion 1:29–50.

[26] O'Sullivan C, Cassell J, Vilhjalmsson H, Dingliana J, Dobbyn S, McNamee B, et al. Levels of Detail for Crowds and Groups. Computer Graphics Forum 21(4):733–42.

[27] The Ohio State University Department of Linguistics. ToBI, http://ling.ohio-state.edu/~tobi/, Sept. 2005.

[28] Ortony A, Clore G, Collins A. The Cognitive Structure of Emotions. Cambridge, U.K.: Cambridge University Press; 1988.

[29] Perlin K, Goldberg A. Improv: A System for Scripting Interactive Actors in Virtual Worlds. In: Rushmeier H, editor. Proc. SIGGRAPH 96. New York: ACM Press; 1996. p. 205–16.

[30] Poggi I, Pelachaud C. Signals and Meanings of Gaze in Animated Agents. In: Prevost S, Cassell J, Sullivan J, Churchill E, editors. Embodied Conversational Characters. Cambridge, MA: MIT Press; 2000.

[31] Pelachaud C, Poggi I. Subtleties of Facial Expressions in Embodied Agents. Journal of Visualization and Computer Animation 2002;13(5):301–12.

[32] Popkins N. The Five-Factor Model: Emergence of a Taxonomic Model for Personality Psychology, www.personalityresearch.org/papers/popkins.html; Sept. 2005.

[33] Potts W. The Chorus Line Hypothesis of Maneuver Coordination in Avian Flocks. Nature May 1984;309:344–5.

[34] Reynolds C. Flocks, Herds, and Schools: A Distributed Behavioral Model. In: Stone MC, editor. Computer Graphics. Proceedings of SIGGRAPH 87, vol. 21(4). Anaheim, Calif.; July 1987. p. 25–34.

[35] Reynolds C. Not Bumping into Things. In: Physically Based Modeling Course Notes. SIGGRAPH 88. Atlanta, Ga.; August 1988. p. G1–G13.

[36] Rymill S, Dodgson N. A Psychologically-Based Simulation of Human Behavior. In: Theory and Practice of Computer Graphics. 2005. p. 25–42.

[37] Shao W, Terzopoulos D. Autonomous Pedestrians. In: Symposium on Computer Animations, Proceedings of the 2005 ACM SIGGRAPH/Eurographics Symposium on Computer Animation. Los Angeles; 2005. p. 19–28.

[38] Stone M, DeCarlo D, Oh I, Rodriguez C, Stere A, Lees A, et al. Speaking with Hands: Creating Animated Conversational Characters from Recordings of Human Performance. In: Proceedings of SIGGRAPH. 2004.

[39] Sung M, Gleicher M, Chenney S. Scalable Behaviors for Crowd Simulation. Computer Graphics Forum 2004;23(3), Eurographics '04.

[40] Tomlinson B, Blumberg B. Social Behavior, Emotion and Learning in a Pack of Virtual Wolves. In: AAAI Fall Symposium "Emotional and Intelligent II: the Tangled Knot of Social Cognition". North Falmouth, MA; November 2–3, 2001.

[41] Tu X, Terzopoulos D. Artificial Fishes: Physics, Locomotion, Perception, Behavior. In: Computer Graphics. Proc. of SIGGRAPH 94. p. 43–50.

[42] Ulicny B, Thalmann D. Crowd Simulation for Interactive Virtual Environments and VR training Systems. In: Proceedings of the Eurographic workshop on Computer Animation and Simulation Table of Contents. Manchester, UK; 2001. p. 163–70.

[43] Wong K. Personality Model of a Believable and Socially Intelligent Character. M.Sc Thesis. University of Sheffield; 2004.

[44] Zeltzer D. Task Level Graphical Simulation: Abstraction, Representation and Control. In: Badler N, Barsky B, Zeltzer D, editors. Making Them Move. San Francisco: Morgan Kaufmann; 1991.

[45] Helbing D, Molnár P. Social Force Model for Pedestrian Dynamics. Physical Review E 1995;51(5):4282–6.

Special Models for Animation

12

Often, how an object is defined is intimately woven with how it is animated. We have already seen examples of this, such as hierarchical linkages, special models for fluids, and so forth. This chapter contains a collection of such special models that often have a role to play in animation. These models are implicit surfaces, L-systems, and subdivision surfaces. Some of these modeling techniques have already been mentioned in previous chapters (e.g., implicit surfaces in Chapter 8 and subdivision surfaces in the Chapter 10). This chapter is intended to give the interested reader some additional information on these modeling techniques.

12.1 Implicit surfaces

Implicitly defined surfaces are surfaces defined by all those points that satisfy an equation of the form, $f(P) = 0$; $f(P)$ is called the *implicit function*. A common approach to using implicit surfaces to define objects useful for animation is to construct a compound implicit function as a summation of implicitly defined primitive functions. Interesting animations can be produced by animating the relative position and orientation of the primitive functions or by animating parameters used to define the functions themselves. Implicit surfaces lend themselves to shapes that are smooth and organic looking. Animated implicit surfaces are useful for shapes that change their topology over time. They are often used for modeling liquids, clouds, and animals.

An extensive presentation of implicit surface formulations is not appropriate material for this book, but it can be found in several sources, in particular in Bloomenthal's edited volume [3]. A brief overview of commonly used implicit surfaces and their use in animation is presented here.

12.1.1 Basic implicit surface formulation

Typically, an *implicit surface* is defined by the collection of points that are the zero points of some implicit function, $f(P) = 0$. The surface is referred to as implicit because it is only implicitly defined, not explicitly represented. As a result, when an explicit definition of the surface is needed, as in a graphics display procedure, the surface has to be searched for by inspecting points in space in some organized way.

Implicit surfaces can be directly ray traced. Rays are constructed according to the camera parameters and display resolution, as is typical ray tracers. Points along the ray are then sampled to locate a surface point within some error tolerance.

An explicit polygonal representation of an implicitly defined surface can be constructed by sampling the implicit function at the vertices of a three-dimensional mesh that is constructed so that its extent contains the non-zero extent of the implicit function. The implicit function values at the mesh vertices are then interpolated along mesh edges to estimate the location of points that lie on the implicit surface. Polygonal fragments are then constructed in each cell of the mesh by using any surface points located on the edges of that mesh cell [10].

The best known implicit primitive is often referred to as the *metaball* and is defined by a central point (C), a radius of influence (R), a density function (f), and a threshold value (T). All points in space that are within a distance R from C are said to have a density of $f(distance(P,C)/R)$ with respect to the metaball (where $distance(P,C)$ computes the distance between P and C and where $distance(P,C)/R$ is the *normalized distance*). The set of points for which $f(distance(P,C)/R) - T = 0$ (implicitly) defines the surface, S.

In Figure 12.1, r is the distance at which the surface is defined for the isolated metaball shown because that is the distance at which the function, f, evaluates to the threshold value. Desirable attributes of the function, f, are $f(0.0) = 1.0$, $f(0.5) = 0.5$, $f(1.0) = 0.0$, and $f'(0.0) = f'(1.0) = 0.0$. A common definition for f, as proposed by Wyvill [20], is Equation 12.1.

$$f(s) = 1 - \left(\frac{4}{9}\right)s^6 - \frac{17}{9}s^4 - \frac{22}{9}s^2 \tag{12.1}$$

Two generalizations of this formulation are useful. The first uses the weighted sum of a number of implicit surface primitives so that the surface-defining implicit function becomes a smooth blend of the individual surfaces (Eq. 12.2).

$$F(P) = \Sigma w_i f_i(P) - T \tag{12.2}$$

The weights are arbitrarily specified by the user to construct some desired surface. If all of the weights, w_i, are one, then the implicit primitives are summed. Because the standard density function has zero slope at one, the primitives will blend together smoothly. Using negative weights will create smooth concavities. Several concavities can be created by making the weight more negative. Integer weights are usually sufficient for most applications, but noninteger weights can also be used (see Figure 12.2).

The second generalization provides more primitives with which the user can build interesting objects. Most primitives are distance based, and most of those are offset surfaces. Typical primitives use the same formulation as the metaball but allow a wider range of central elements. Besides a single

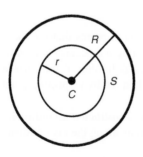
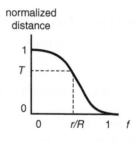

FIGURE 12.1

The metaball and a sample density function.

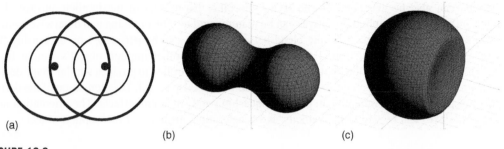

FIGURE 12.2

Overlapping metaball density functions. (a) Schematic showing overlapping density functions. (b) Surface constructed when positive weight are associated with density functions. (c) Surface constructed when one positive weight and one negative weight are associated with density functions.

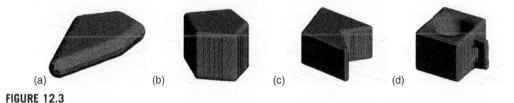

FIGURE 12.3

Various surfaces extracted from implicit distance-based functions. (a) Distance-based implicit primitive based on a single polygon. (b) Distance-based implicit primitive based on a single convex polyhedron. (c) Distance-based implicit primitive based on a single concave polyhedron. (d) Compound implicitly defined object.

point, a primitive can be defined that uses a line segment, a triangle, a convex polyhedron, or even a concave polyhedron. Any central element for which there is an associated well-defined distance function can be used. The drawback to using more complex central elements, such as a concave polyhedron, is that the distance function is more computationally expensive. Other primitives, which are not strictly offset surfaces but which are still distance based, are the cone-sphere and the ellipse. See Figure 12.3a–c for examples of distance-based primitives. As with metaballs, these implicit primitives can be combined to create interesting compound density functions whose isosurfaces can be extracted and displayed (Figure 12.3d).

12.1.2 Animation using implicitly defined objects

In Bloomenthal's book [3], Wyvill discussed several animation effects that can be produced by modifying the shape of implicit surfaces. The most obvious way to achieve these modifications is to control the movement of the underlying central elements. Topological changes are handled nicely by the implicit surface formulation. The points that define a collection of metaballs can be controlled by a simple particle system, and the resulting implicit surface can be used to model water, taffy, or clouds, depending on the display attributes and the number of particles. Central elements consisting of lines can be articulated as a jointed hierarchy.

Motion can also be generated by modifying the other parameters of the implicit surface. The weights of the different implicit primitives can be manipulated to effect bulging and otherwise control the size of an implicit object. A simple blend between objects can be performed by decreasing the weight of one implicit object from 1 to 0 while increasing another implicit object's weight from 0 to 1. The density function of the implicit objects can also be controlled to produce similar shape deformation effects. Modifying the central element of the implicit is also useful. The elongation of the defining axes of the ellipsoidal implicit can produce squashing and stretching effects. The orientation and length of the axes can be tied to the velocity and acceleration attributes of the implicit object.

12.1.3 Collision detection

Implicitly defined objects, because they are implemented using an implicit function, lend themselves to collision detection. Sample points on the surface of one object can be tested for penetration with an implicit object by merely evaluating the implicit function at those points in space (Figure 12.4). Of course, the effectiveness of this method is dependent on the density and distribution of the sample points on the first object. If a polyhedral object can be satisfactorily approximated by one or more implicit surface primitives, then these can be used to detect collisions of other polyhedral objects. Bounding spheres are a simple example of this, but more complex implicits can be used for more precise fitting.

Because implicit functions can be used effectively in testing for collisions, they can be used to test for collisions between polyhedral objects if they can fit reasonably well on the surface of the polyhedra (Figure 12.5). Of course, the effectiveness of the technique is dependent on the accuracy with which the implicit surfaces approximate the polyhedral surface.

12.1.4 Deforming the implicit surface as a result of collision

Cani has developed a technique to compute the deformation of colliding implicit surfaces [3] [6] [7] (published under the name of Gascuel). This technique first detects the collision of two implicit surfaces by testing sample points on the surface of one object against the implicit function of the other, as previously described. The overlap of the areas of influence of the two implicit objects is called the

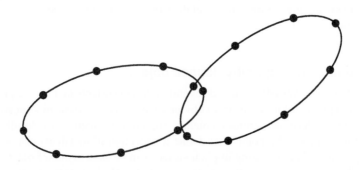

FIGURE 12.4

Point samples used to test for collisions.

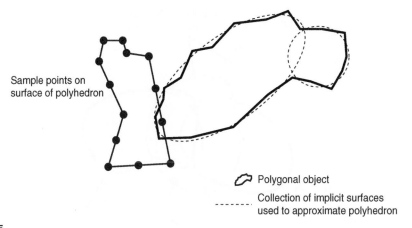

Sample points on
surface of polyhedron

⬭ Polygonal object

- - - - - - - Collection of implicit surfaces
used to approximate polyhedron

FIGURE 12.5

Using implicit surfaces for detecting collisions between polyhedral objects.

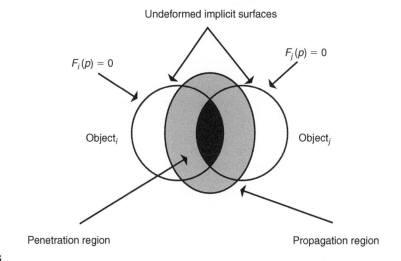

Undeformed implicit surfaces

$F_i(p) = 0$

$F_j(p) = 0$

Object$_i$

Object$_j$

Penetration region

Propagation region

FIGURE 12.6

Penetrating implicit surfaces.

penetration region. An additional region just outside the penetration region is called the *propagation region* (Figure 12.6).

The density function of each object is modified by the overlapping density function of the other object to deform the implicitly defined surface of both objects so that they coincide in the region of overlap, thus creating a contact surface (Figure 12.7). As shown in the example in Figure 12.7, a deformation term is added to F_i as a function of Object$_j$'s overlap with Object$_i$, G_{ij}, to form the contact surface. Similarly, a deformation term is added to F_j as a function of Object$_i$'s overlap with Object$_j$, G_{ji}. The deformation functions are defined so that the isosurface of the modified density functions,

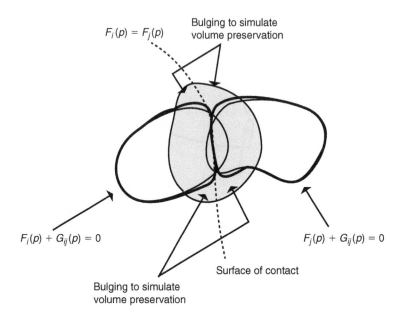

$F_i(p) = F_j(p)$
Bulging to simulate
volume preservation

$F_i(p) + G_{ij}(p) = 0$ $F_j(p) + G_{ij}(p) = 0$

Surface of contact

Bulging to simulate
volume preservation

FIGURE 12.7

Implicit surfaces after deformation due to collision.

$F_i(p) + G_{ij}(p) = 0$ and $F_j(p) + G_{ji}(p) = 0$, coincide with the surface defined by $F_i(p) = F_j(p)$. Thus, the implicit functions, after they have been modified for deformation and evaluated for points p on the contact surface, are merely $F_i(p) - F_j(p) = 0$ and $F_j(p) - F_i(p) = 0$, respectively. Thus, G_{ij} evaluates to F_j at the surface of contact.

In the penetration region, the Gs are negative in order to compress the respective implicit surfaces as a result of the collision. They are defined so that their effect smoothly fades away for points at the boundary of the penetration region. Consequently, the Gs are merely the negative of the Fs for points in the penetration region (Eq. 12.3).

$$G_{ij}(p) = -F_j(p)$$
$$G_{ji}(p) = -F_i(p)$$

(12.3)

To simulate volume preservation, an additional term is added to the Gs so that they evaluate to a positive value in the propagation region immediately adjacent to, but just outside, the penetration region. The effect of the positive evaluation will be to bulge out the implicit surface around the newly formed surface of contact (Figure 12.7).

In the propagation region, $G(p)$ is defined as a function, h, of the distance to the border of the interpenetration region. To ensure C^1 continuity between the interpenetration region and the propagation region, $h'(0)$ must be equal to the directional derivative of G along the gradient at point p (k in Figure 12.8). See Gascuel [6] and Gascuel and Gascuel [7] for more details.

Restoring forces, which arise as a result of the deformed surfaces, are computed and added to any other external forces acting on the two objects. The magnitude of the force is simply the deformation term, G; it is in the direction of the normal to the deformed surface at point p.

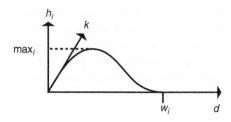

FIGURE 12.8

Deformation function in the propagation region.

12.1.5 Level set methods

"Level set methods add dynamics to implicit surfaces" [13]. These methods operate on signed distance functions using a grid representation for an object. Each grid cell contains the value of the signed distance function at that location in space. An isosurface, φ, is computed on the grid as the interface between positive and negative values. The isosurface is updated according to a velocity field defined over the interface. For example, in two-dimensional space, $d\varphi/dt=V(x,y,t)$. The function V can be given as an externally generated velocity field. The isosurface advects (i.e., moves) in the direction of the velocity function. A commonly used function uses the direction of the (positive or negative) gradient of φ and a constant magnitude. Thus, the isosurface advects in the direction of its normal at a constant speed. Alternatively, the speed may be based on the magnitude of the curvature, $d^2\varphi/dt^2$. A Euler update of $\varphi(t+Dt) = \varphi(t) + dt*V$ can be used to actually update the grid values. This is called *solving the level set equations.*

Every step taken in time corrupts the signed distance function of the grid as values are modified. Thus, the grid values need to be updated, called *renormalized*, in order to construct a signed distance field again.

The level set equations for advection of a surface by a vector field are represented by Equations 12.4 and 12.5.

$$H = V \times \nabla\phi \tag{12.4}$$

$$\nabla\phi = \left(\frac{d\phi}{dx}, \frac{d\phi}{dy}\right) \tag{12.5}$$

where V is a vector field.

Solving the level set equations moves the interface in the direction of the vector field. For each coordinate, calculate $V \cdot \nabla\varphi$ and see where it takes the interface. $\nabla\varphi$ is calculated from the grid representation.

To approximate $\nabla\varphi$, use the upwind scheme such that, in the horizontal case,

$$\text{if, } v_x < 0, \frac{d\phi}{dx} = \phi(x+1,y) - \phi(x,y)$$

$$\text{if, } v_x > 0, \frac{d\phi}{dx} = \phi(x,y) - \phi(x-1,y)$$

To solve the level set equations advecting in the normal direction, at a given point, get the point's normal and scale it to form vector field. Common functions used to control the speed are curvature at the point and a constant function. The interface can be updated by taking a Euler step in the direction of its velocity. Once updated, the distance function needs to be updated as well. For purposes of efficiency, the entire grid is not usually processed. Instead, just a band of grid values on either side of the interface is updated [13].

12.1.6 Summary

Although their display may be problematic for some graphic systems, implicit surfaces provide unique and interesting opportunities for modeling and animating unusual shapes. They produce very organic-looking shapes and, because of their indifference to changes in genus of the implicitly defined surface, lend themselves to the modeling and animating of fluids and elastic material.

12.2 Plants

The modeling and animation of plants represent an interesting and challenging area for computer animation. Plants seem to exhibit arbitrary complexity while possessing a constrained branching structure. They grow from a single source point, developing a branching structure over time while the individual structural elements elongate. Plants have been modeled using particle systems, fractals, and L-systems. There has been much work on modeling the static representations of various plants (e.g., [1] [2] [8] [12] [15] [17] [18]). The intent here is not to delve into the botanically correct modeling of particular plants but rather to explain those aspects of modeling plants that make the animation of the growth process challenging. The representational issues of synthetic plants are discussed in just enough detail to uncover these aspects. Prusinkiewicz and Lindenmayer [14] [16] provide more information on all aspects of modeling and animating plants.

The topology[1] of a plant is characterized by a recursive branching structure. To this extent, plants share with fractals the characteristics of self-similarity under scale. The two-dimensional branching structures typically of interest are shown in Figure 12.9. The three-dimensional branching structures are analogous.

An encoding of the branching structure of a given plant is one of the objectives of plant modeling. Plants are immensely varied, yet most share many common characteristics. These shared characteristics allow efficient representations to be formed by abstracting out the features that are common to plants of interest. But the representation of the static structure of a mature plant is only part of the story. Because a plant is a living thing, it is subject to changes due to growth. The modeling and animation of the growth process is the subject of this section.

12.2.1 A little bit of botany

Botany is, of course, useful when trying to model and animate realistic-looking plants. For computer graphics and animation, it is only useful to the extent that it addresses the visual characteristics of the plant. Thus, the structural components and surface elements of plants are briefly reviewed here.

---

[1]The term *topology*, as applied to describing the form of plants, refers to the number and arrangement of convex regions of the plant delineated by concave areas of attachment to other convex regions.

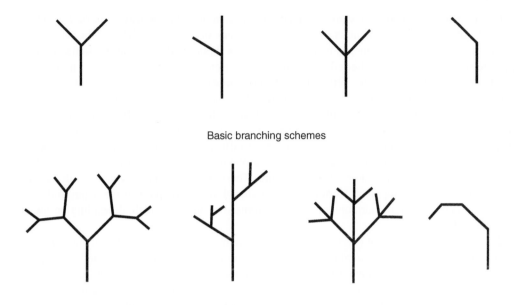

Basic branching schemes

Structures resulting from repeated application of a single branching scheme

FIGURE 12.9

Branching structures of interest in two dimensions.

Simplifications are made to highlight the information most relevant for computer graphics modeling and animation.

The structural components of plants are *stems*, *roots*, *buds*, *leaves*, and *flowers*. Roots are typically not of interest when modeling the visual aspects of plants and have not been incorporated into these plant models. Most plants of interest in visualization have a definite branching structure. Such plants are either *herbaceous* or *woody*. The latter are larger plants whose branches are heavier and more structurally independent. The branches of woody plants tend to interfere and compete with one another. They are also more subject to the effects of wind, gravity, and sunlight. Herbaceous plants are smaller, lighter plants, such as ferns and mosses, whose branching patterns tend to be more regular and less subject to environmental effects.

Stems are usually above ground, grow upward, and bear leaves. The leaves are attached in a regular pattern at nodes along the stem. The portions of the stem between the nodes are called *internodes*. *Branching* is the production of subordinate stems from a main stem (the *axis*). Branches can be formed by the main stem bifurcating into two equally growing stems (*dichotomous*), or they can be formed when a stem grows laterally from the main axis while the main axis continues to grow in its original direction (*monopodial*).

Buds are the embryonic state of stems, leaves, and flowers; they are classified as either *vegetative* or *flower buds*. Flower buds develop into flowers, whereas vegetative buds develop into stems or leaves. A bud at the end of a stem is called a *terminal* bud; a bud that grows along a stem is called a *lateral* bud. Not all buds develop; nondeveloping buds are referred to as *dormant*. Sometimes in woody plants, dormant buds will suddenly become active, producing young growth in an old-growth area of a tree.

Leaves grow from buds. They are arranged on a stem in one of three ways: *alternate, opposite,* or *whorled.* Alternate means that the leaves shoot off one side of a stem and then off the other side in an alternating pattern. Opposite means that a pair of leaves shoot off the stem at the same point, but on opposite sides, of the stem. Whorled means that three or more leaves radiate from a node.

Growth of a cell in a plant has four main influences: *lineage, cellular descent, tropisms,* and *obstacles.* Lineage refers to growth controlled by the age of the cell. *Cellular descent* refers to the passing of nutrients and hormones from adjacent cells. Growth controlled exclusively by lineage would result in older cells always being larger than younger cells. Cellular descent can be bidirectional, depending on the growth processes and reactions to environmental influences occurring in the plant. Thus, cellular descent is responsible for the ends of some plants growing more than the interior sections. Plant hormones are specialized chemicals produced by plants. These are the main internal factors that control the plant's growth and development. Hormones are produced in specific parts of the plants and are transported to others. The same hormone may either promote or inhibit growth, depending on the cells it interacts with.

Tropisms are an important class of responses to external influences that change the direction of a plant's growth. They include *phototropism,* the bending of a stem toward light, and *geotropism,* the response of a stem or root to gravity. Physical obstacles also affect the shape and growth of plants. Collision detection and response can be calculated for temporary changes in the plant's shape. Permanent changes in the growth patterns can occur when such forces are present for extended periods.

12.2.2 L-systems

DOL-systems

L-systems are parallel rewriting systems. They were conceived as mathematical models of plant development by the biologist Aristid Lindenmayer (the "L" in L-systems). The simplest class of L-system is deterministic and context free (as in Figure 12.10); it is called a D0L-system.[2] A D0L-system is a set of production rules of the form $\alpha_i \rightarrow \beta_i$, in which α_i, the *predecessor,* is a single symbol and β_i, the *successor,* is a sequence of symbols. In deterministic L-systems, α_i occurs only once on the left-hand side of a production rule. A sequence of one or more symbols is given as the initial string, or *axiom.* A production rule can be applied to the string if its left-hand side occurs in the string. The effect of applying a production rule to a string means that the occurrence of α_i in the string is rewritten as β_i. Production rules are applied in parallel to the initial string. This replacement happens in parallel for all

```
S -> ABA          S    <--------- axiom
A -> XX           ABA
B -> TT           XXTTXX

Production rules   String sequence
```

FIGURE 12.10

Simple D0L-system and the sequence of strings it generates.

---

[2]The D in D0L clearly stands for *deterministic;* the 0 indicates, as is more fully explained later, that the productions are context free.

occurrences of any left-hand side of a production in the string. Symbols of the string that are not on the left-hand side of any production rule are assumed to be operated on by the identity production rule, $\alpha_i \rightarrow \beta_i$. The parallel application of the production rules produces a new string. The production rules are then applied again to the new string. This happens iteratively until no occurrences of a left-hand side of a production rule occur in the string. Sample production rules and the string they generate are shown in Figure 12.10.

Geometric interpretation of L-systems

The strings produced by L-systems are just that—strings. To produce images from such strings, one must interpret them geometrically. Two common ways of doing this are *geometric replacement* and *turtle graphics*. In geometric replacement, each symbol of a string is replaced by a geometric element. For example, the string XXTTXX can be interpreted by replacing each occurrence of X with a straight line segment and each occurrence of T with a V shape so that the top of the V aligns with the endpoints of the geometric elements on either side of it (see Figure 12.11).

In turtle graphics, a geometry is produced from the string by interpreting the symbols of the string as drawing commands given to a simple cursor called a turtle. The basic idea of turtle graphics interpretation, taken from Prusinkiewicz and Lindenmayer [16], uses the symbols from Table 12.1 to control

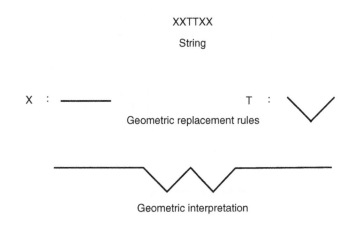

FIGURE 12.11

Geometric interpretation of a simple string.

Table 12.1 Turtle Graphics Commands	
Symbol	**Turtle Graphic Interpretation**
F	Move forward a distance d while drawing a line. Its state will change from (x, y, α) to $(x + d \cos \alpha, y + d \sin \alpha, \alpha)$.
f	Move forward a distance d without drawing a line. Its state will change as above.
+	Turn left by an angle δ. Its state will change from (x, y, α) to $(x, y, \alpha+\delta)$.
−	Turn right by an angle δ. Its state will change from (x, y, α) to $(x, y, \alpha-\delta)$.

the turtle. The state of the turtle at any given time is expressed as a triple (x, y, α), where x and y give its coordinates in a two-dimensional space and α gives the direction it is pointing relative to some given reference direction (here, a positive angle is measured counterclockwise from the reference direction). The values d and δ are user-specified system parameters and are the linear and rotational step sizes, respectively.

Given the reference direction, the initial state of the turtle (x_0, y_0, α_0), and the parameters d and δ, the user can generate the turtle interpretation of a string containing the symbols of Table 12.1 (Figure 12.12).

Bracketed L-systems

The D0L-systems described previously are inherently linear, and the graphical interpretations reflect this. To represent the branching structure of plants, one introduces a mechanism to represent multiple segments attached at the end of a single segment. In *bracketed L-systems*, brackets are used to mark the beginning and the end of additional offshoots from the main lineage. The turtle graphics interpretation of the brackets is given in Table 12.2. A stack of turtle graphic states is used, and the brackets push and pop states onto and off this stack. The state is defined by the current position and orientation of the turtle. This allows branching from a stem to be represented. Further, because a stack is used, it allows an arbitrarily deep branching structure.

Figure 12.13 shows some production rules. The production rules are context free and *nondeterministic*. They are context free because the left-hand side of the production rule does not contain any

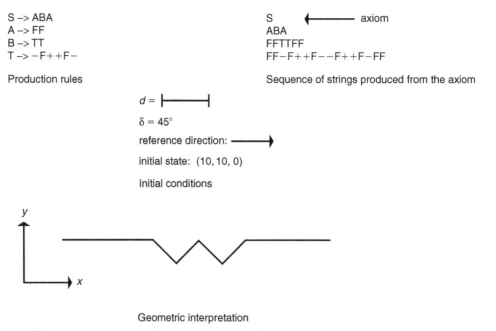

Production rules	Sequence of strings produced from the axiom
S –> ABA	S ⟵ axiom
A –> FF	ABA
B –> TT	FFTTFF
T –> −F++F−	FF−F++F−−F++F−FF

$d = \longmapsto$

$\delta = 45°$

reference direction: $\longrightarrow$

initial state: $(10, 10, 0)$

Initial conditions

Geometric interpretation

FIGURE 12.12

Turtle graphic interpretation of a string generated by an L-system.

Table 12.2 Turtle Graphic Interpretation of Brackets

Symbol	Turtle Graphic Interpretation
[	Push the current state of the turtle onto the stack
]	Pop the top of the state stack and make it the current state

$$S \Rightarrow FAF$$
$$A \Rightarrow [+FBF]$$
$$A \Rightarrow F$$
$$B \Rightarrow [-FBF]$$
$$B \Rightarrow F$$

FIGURE 12.13

Nondeterministic, context-free production rules.

context for the predecessor symbol. They are nondeterministic because there are multiple rules with identical left-hand sides; one is chosen at random from the set whenever the left-hand side is to be replaced. In this set of rules, S is the start symbol, and A and B represent locations of possible branching; A branches to the left and B to the right. The production stops when all symbols have changed into ones that have a turtle graphic interpretation.

Figure 12.14 shows some possible terminal strings and the corresponding graphics produced by the turtle interpretation. An added feature of this turtle interpretation of the bracketed L-system is the reduction of the size of the drawing step by one-half for each branching level, where *branching level* is defined by the current number of states on the stack. L-systems can be expanded to include attribute symbols that explicitly control line length, line width, color, and so on [16] and that are considered part of the state.

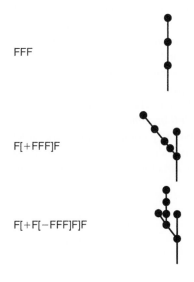

FFF

F[+FFF]F

F[+F[−FFF]F]F

FIGURE 12.14

Some possible terminal strings.

FAF

F[+FBF]F

F[+F[−FFF]F]F

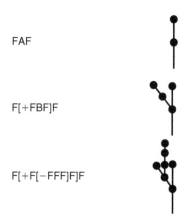

FIGURE 12.15

Sequence of strings produced by bracketed L-system.

Not only does this representation admit *database amplification* [19], but the expansion of the start symbol into a terminal string parallels the topological growth process of the plants. In some sense, the sequence of strings that progress to the final string of all turtle graphic symbols represents the growth of the plant at discrete events in its evolution (see Figure 12.15). This addresses one of the animation issues with respect to plants—that of animating the development of the branching structure. However, the gradual appearance and subsequent elongation of elements must also be addressed if a growing structure is to be animated in a reasonable manner.

Stochastic L-systems

The previous section introduced nondeterminism into the concept of L-systems, but the method used to select the possible applicable productions for a given symbol was not addressed. *Stochastic L-systems* assign a user-specified probability to each production so that the probabilities assigned to productions with the same left-hand side sum to one. These probabilities indicate how likely it is that the production will be applied to the symbol on a symbol-by-symbol basis.

Consider the productions of Figure 12.13 being assigned the probabilities shown in Figure 12.16. These probabilities will control how likely a production will be to form a branch at each possible branching point. In this example, left branches are very likely to form, while right branches are somewhat unlikely. However, any arbitrarily complex branching structure has a non-zero probability of occurring. Using such stochastic (nondeterministic) L-systems, one can set up an L-system that produces a wide variety of branching structures that still exhibit some family-like similarity [16].

$$S_{1.0} \Rightarrow FAF$$
$$A_{0.8} \Rightarrow [+FBF]$$
$$A_{0.2} \Rightarrow F$$
$$B_{0.4} \Rightarrow [-FBF]$$
$$B_{0.6} \Rightarrow F$$

FIGURE 12.16

Stochastic L-system.

$$S \Rightarrow FAT$$
$$A > T \Rightarrow [+FBF]$$
$$A > F \Rightarrow F$$
$$B \Rightarrow [-FAF]$$
$$T \Rightarrow F$$

Production rules

S
FAT
F[+FBF]F
F[+F[−FAF]F]F

String sequence

FIGURE 12.17

Context-sensitive L-system production rules.

Context free versus context sensitive

So far, only context-free L-systems have been presented. *Context-sensitive L-systems* add the ability to specify a context, in which the left-hand side (the predecessor symbol) must appear in order for the production rule to be applicable. For example, in the deterministic productions of Figure 12.17,[3] the symbol A has different productions depending on the context in which it appears. The context-sensitive productions shown in the figure have a single right-side context symbol in two of the productions. This concept can be extended to n left-side context symbols and m right-side context symbols in the productions, called (n, m) *L-systems*, and, of course, they are compatible with nondeterministic L-systems. In (n, m) L-systems, productions with fewer than n context symbols on the left and m on the right are allowable. Productions with longer contexts are usually given precedence over productions with shorter contexts when they are both applicable to the same symbol. If the context is one-sided, then L-systems are referred to as nL-systems, where n is the number of context symbols and the side of the context is specified independently.

12.2.3 Animating plant growth

There are three types of animation in plants. One type is the flexible movement of an otherwise static structure, for example, a plant being subjected to a high wind. Such motion is an example of a flexible body reacting to external forces and is not dealt with in this section. The other types of animation are particular to plants and involve the modeling of the growth process.

The two aspects of the growth process are (1) changes in topology that occur during growth and (2) the elongation of existing structures. The topological changes are captured by the L-systems already described. They occur as discrete events in time and are modeled by the application of a production that encapsulates a branching structure, as in $A \Rightarrow F[+F]B$.

Elongation can be modeled by productions of the form $F \Rightarrow FF$. The problem with this approach is that growth is chunked into units equal to the length of the drawing primitive represented by F. If F represents the smallest unit of growth, then an internode segment can be made to grow arbitrarily long. But the production rule $F \Rightarrow FF$ lacks termination criteria for the growth process. Additional drawing symbols can be introduced to represent successive steps in the elongation process, resulting in a series of productions $F_0 \Rightarrow F_1, F_1 \Rightarrow F_2, F_2 \Rightarrow F_3, F_3 \Rightarrow F_4$, and so on. Each symbol would represent a drawing operation of a different length. However, if the elongation process is to be modeled in,

---

[3]In the notation used here, as it is in the book by Prusinkiewicz and Lindenmayer [16], the predecessor is the symbol on the "greater than" side of the inequality symbols. This is used so that the antecedent can be visually identified when using either left or right contexts.

say, one hundred time steps, then approximately one hundred symbols and productions are required. To avoid this proliferation of symbols and productions, the user can represent the length of the drawing operation parametrically with the drawing symbol in *parametric L-systems*.

Parametric L-systems

In parametric L-systems, symbols can have one or more parameters associated with them. These parameters can be set and modified by productions of the L-system. In addition, optional conditional terms can be associated with productions. The conditional expressions are in terms of parametric values. The production is applicable only if its associated condition is met. In the simple example of Figure 12.18, the symbol A has a parameter t associated with it. The productions create the symbol A with a parameter value of 0.0 and then increase the parametric value in increments of 0.01 until it reaches 1.0. At this point the symbol turns into an F.

Context-sensitive productions can be combined with parametric systems to model the passing of information along a system of symbols. Consider the production of Figure 12.19. In this production, there is a single context symbol on both sides of the left-hand symbol that is to be changed. These productions allow for the relatively easy representation of such processes as passing nutrients along the stem of a plant.

Timed L-systems

Timed L-systems add two more concepts to L-systems: a *global time variable*, which is accessible to all productions and which helps control the evolution of the string; and a *local age value*, τ_i, associated with each letter μ_i. Timed L-system productions are of the form shown in Equation 12.6. By this production, the letter μ_0 has a *terminal age* of β_0 assigned to it. The terminal age must be uniquely assigned to a symbol. Also by this production, each symbol μ_i has an *initial age* of α_i assigned to it. The terminal age assigned to a symbol μ_i must be larger than its initial age so that its *lifetime*, $\beta_i - \alpha_i$, is positive.

$$(\mu_0, \beta_0) \Rightarrow ((\mu_1, \alpha_1), (\mu_2, \alpha_2), \ldots, (\mu_n, \alpha_n)) \tag{12.6}$$

A timed production can be applied to a matching symbol when that symbol's terminal age is reached. When a new symbol is generated by a production, it is commonly initialized with an age of zero. As global time progresses from that point, the local age of the variable increases until its terminal age is reached, at which point a production is applied to it and it is replaced by new symbols.

A string can be derived from an axiom by jumping from terminal age to terminal age. At any point in time, the production to be applied first is the one whose predecessor symbol has the smallest

$$
\begin{aligned}
S \qquad\qquad &=> A(0) \\
A(t) \qquad\qquad &=> A(t + 0.01) \\
A(t) : t>=1.0 \ &=> F
\end{aligned}
$$

FIGURE 12.18

Simple parametric L-system.

$$A(t0)<A(t1)>A(t2):\ t2>t1 \ \& \ t1>t0 => A(t1+0.01)$$

FIGURE 12.19

Parametric, context-sensitive L-system production.

axiom: (*A*,0)

(*A*,3) => (*S*,0) [+ (*B*,0)] (*S*,0)
(*B*,2) => (*S*,0)

FIGURE 12.20

Simple timed L-system.

difference between its terminal age and its local age. Each symbol appearing in a string has a local
age less than its terminal age. The geometric interpretation of each symbol is potentially based on
the local age of that symbol. Thus, the appearance of buds and stems can be modeled according to their
local age.

In the simple example of Figure 12.20, the symbol *A* can be thought of as a plant seed; *S* can be
thought of as an internode stem segment; and *B* can be thought of as a bud that turns into the stem
of a branch. After three units of time the seed becomes a stem segment, a lateral bud, and another stem
segment. After two more time units, the bud develops into a branching stem segment.

Interacting with the environment

The environment can influence plant growth in many ways. There are local influences such as physical
obstacles, including other plants and parts of the plant itself. There are global influences such as amount
of sunlight, length of day, gravity, and wind. But even these global influences are felt locally. The wind
can be blocked from part of the plant, as can the sun. And even gravity can have more of an effect on an
unsupported limb than on a supported part of a vine. The nutrients and moisture in the soil affect
growth. These are transported throughout the plant, more or less effectively depending on local damage
to the plant structure.

Mech and Prusinkiewicz [11] describe a framework for the modeling and animation of plants that
bidirectionally interacts with the environment. They describe *open L-systems*, in which communication
terms of the form ? $E(x_1, x_2, \ldots, x_m)$ are used to transmit information as well as request information
from the environment. In turtle graphic interpretation of an L-system string, the string is scanned left to
right. As communication terms are encountered, information is transmitted between the environment
and the plant model. The exact form of the communication is defined in an auxiliary specification file
so that only relevant information is transmitted. Information about the environment relevant to the plant
model includes distribution of nutrients, direction of sunlight, and length of day. The state of the plant
model can be influenced as a result of this information and can be used to change the rate of elongation
as well as to control the creation of new offshoots. Information from the plant useful to the environmental model includes use of nutrients and shade formation, which, in turn, can influence other plant
models in the environment.

12.2.4 Summary

L-systems, in all the variations, are an interesting and powerful modeling tool. Originally intended only
as a static modeling tool, L-systems can be used to model the time-varying behavior of plantlike growth
and motion. See Figure 12.21 (Color Plate 13) for an example from a video that used L-systems to
animate plantlike creatures. Because of the iterative nature of string development, the topological

FIGURE 12.21

Example from video using L-systems to animate plantlike figures.

(Image courtesy of Vita Berezine-Blackburn, ACCAD.)

changes of plant growth can be successfully captured by L-systems. By adding parameters, time variables, and communication modules, one can model other aspects of plant growth. Most recently, Deussen et al. [5] have used open L-systems to model plant ecosystems.

12.3 Subdivision surfaces

Subdivision surfaces are useful in animation for designing objects in a top-down fashion [4]. Starting from a coarse polyhedron, the geometry is refined a step at a time. Each step introduces more complexity to the object, usually by rounding corners and edges. There are various methods proposed in the literature for conducting the subdivision and much research has been performed in determining what the limit surfaces look like and what their mathematical properties are.

As a simple subdivision example, consider the sequence in Figure 12.22 in which each vertex is replaced by a face made from vertices one-third along the way down each edge emanating from the vertex (Figure 12.23) and each face is redefined to include the new vertices on the original edges (Figure 12.24).

One of the more popular subdivision schemes is due to Charles Loop [9]. For a closed, two-dimensional, triangulated manifold, the Loop subdivision method creates a new vertex at the midpoint of each edge and is repositioned. In addition, each original vertex is repositioned, and each triangle is divided into four new triangles (Figure 12.25). The new edge midpoints are positioned according to $V_M = (3 V_1 + 3 V_2 + V_A + V_B)/8$ where V_1 and V_2 are the end vertices of the original edge and

FIGURE 12.22

Sequence of subdivision steps.

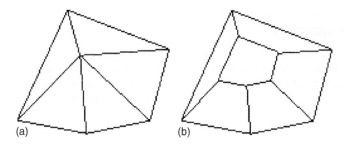

FIGURE 12.23

During subdivision, each vertex is replaced by a face. (a) Original vertex of object to be subdivided. (b) A face replaces the vertex by using new vertices defined on connecting edges.

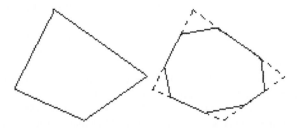

FIGURE 12.24

During subdivision, each face is replaced by a face using the newly defined vertices on its edges. Original face of object to be subdivided.

V_A and V_B are the other vertices of the faces that share the edge. Each original vertex is repositioned by a weighted average of its original position and those of all the vertices connected to it by a single edge. There have been many schemes proposed for repositioning the original vertices. A particularly simple one is $V' = (1 - s)V + s\sum_{i=1}^{n}W_i$ where V is the original vertex position and the n W_is are the connected vertices, $s = \frac{3}{4}n$.

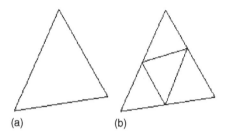

(a)   (b)

FIGURE 12.25

Loop subdivision. (a) Original triangle. (b) After subdivision.

12.4 Chapter summary

Implicit surfaces are useful in a variety of situations, most notably when trying to capture the qualities of organic shapes. Plants exhibit enormous complexity, a well-defined structure, and almost endless variety. Subdivision surfaces have proven to be a powerful design tool in animation applications. These qualities make capturing their essence both intriguing and fascinating.

References

[1] Aono M, Kunii TL. Botanical Tree Image Generation. IEEE Comput Graph Appl May 1984;4(5):10–34.
[2] Bloomenthal J. Modeling the Mighty Maple. In: Barsky BA, editor. Computer Graphics. Proceedings of SIGGRAPH 85, vol. 19(3). San Francisco, Calif.; August 1985. p. 305–11.
[3] Bloomenthal J, Bajaj C, Blinn J, Cani-Gascuel M-P, Rockwood A, Wyvill B, et al. Introduction to Implicit Surfaces. San Francisco: Morgan Kaufmann; 1997.
[4] DeRose T, Kass M, Truong T. Subdivision Surfaces for Character Animation. In: Cohen M, editor. Computer Graphics. Proceedings of SIGGRAPH 98, Annual Conference Series. Orlando, Fla.: Addison-Wesley; July 1998. p. 85–94. ISBN 0-89791-999-8.
[5] Deussen O, Hanrahan P, Lintermann B, Mech R, Pharr M, Prusinkiewicz P. Realistic Modeling and Rendering of Plant Ecosystems. In: Cohen M, editor. Computer Graphics. Proceedings of SIGGRAPH 98, Annual Conference Series. Orlando, Fla: Addison-Wesley; July 1998. p. 275–86. ISBN 0-89791-999-8.
[6] Gascuel M-P. An Implicit Formulation for Precise Contact Modeling between Flexible Solids. In: Kajiya JT, editor. Computer Graphics. Proceedings of SIGGRAPH 93, Annual Conference Series. Anaheim, Calif.; August 1993. p. 313–20. ISBN 0-201-58889-7.
[7] Gascuel J-D, Gascuel M-P. Displacement Constraints for Interactive Modeling and Animation of Articulated Structures. Visual Computer March 1994;10(4):191–204 ISSN 0-178-2789.
[8] Greene N. Voxel Space Automata: Modeling with Stochastic Growth Processes in Voxel Space. In: Lane J, editor. Computer Graphics. Proceedings of SIGGRAPH 89, vol. 23(3). Boston, Mass.; July 1989. p. 175–84.
[9] Loop C. Smooth Subdivision Surfaces Based on Triangles. M.S. Thesis. University of Utah; 1987.
[10] Lorensen W, Cline H. Marching Cubes: A High Resolution 3D Surface Construction Algorithm. In: Proceedings of SIGGRAPH 87, vol. 21(4). p. 163–9.
[11] Mech R, Prusinkiewicz P. Visual Models of Plants Interacting with Their Environment. In: Rushmeier H, editor. Computer Graphics. Proceedings of SIGGRAPH 96, Annual Conference Series. New Orleans, La.: Addison-Wesley; August 1996. p. 397–410. ISBN 0-201-94800-1.

[12] Oppenheimer P. Real-Time Design and Animation of Fractal Plants and Trees. In: Evans DC, Athay RJ, editors. Computer Graphics. Proceedings of SIGGRAPH 86, vol. 20(4). Dallas, Tex.; August 1986. p. 55–64.

[13] Osher S, Fedkiw R. Level Set Methods and Dynamic Implicit Surfaces. New York: Springer-Verlag; 2003.

[14] Prusinkiewicz P, Hammel M, Mjolsness E. Animation of Plant Development. In: Kajiya JT, editor. Computer Graphics. Proceedings of SIGGRAPH 93, Annual Conference Series. Anaheim, Calif.; August 1993. p. 351–60. ISBN 0-201-58889-7.

[15] Prusinkiewicz P, James M, Mech MR. Synthetic Topiary, In: Glassner A, editor. Computer Graphics. Proceedings of SIGGRAPH 94, Annual Conference Series. Orlando, Fla.: ACM Press; July 1994. p. 351–8. ISBN 0-89791-667-0.

[16] Prusinkiewicz P, Lindenmayer A. The Algorithmic Beauty of Plants. New York: Springer-Verlag; 1990.

[17] Prusinkiewicz P, Lindenmayer A, Hanan J. Developmental Models of Herbaceous Plants for Computer Imagery Purposes. In: Dill J, editor. Computer Graphics. Proceedings of SIGGRAPH 88, vol. 22(4). Atlanta, Ga.; August 1988. p. 141–50.

[18] Reffye P, Edelin C, Francon J, Jaeger M, Puech C. Plant Models Faithful to Botanical Structure and Development. In: Dill J, editor. Computer Graphics. Proceedings of SIGGRAPH 88, vol. 22(4). Atlanta, Ga.; August 1988. p. 151–8.

[19] Smith AR. Plants, Fractals, and Formal Languages, In: Computer Graphics. Proceedings of SIGGRAPH 84, vol. 18(3). Minneapolis, Minn.; July 1984. p. 1–10.

[20] Wyvill B, McPheeters C, Wyvill G. Data Structure for Soft Objects. Visual Computer 1986;2(4):227–34.

Rendering Issues

This appendix presents rendering techniques for computing a series of images that are to be played back as an animation sequence. It is assumed that the reader has a solid background in rendering techniques and issues, namely, the use of frame buffers, the z-buffer display algorithm, and aliasing. The techniques presented here concern smoothly displaying a sequence of images on a computer monitor (*double buffering*), efficiently computing images of an animated sequence (*compositing, drop shadows, billboarding*), and effectively rendering moving objects (*motion blur*). An understanding of this material is not necessary for understanding the techniques and algorithms covered in the rest of the book, but computer animators should be familiar with these techniques when considering the trade-offs involved in rendering images for animation.

A.1 Double buffering

Not all computer animation is first recorded onto film or video for later viewing. In many cases, image sequences are displayed in real time on the computer monitor. Computer games are a prime example of this, as is Web animation. Real-time display also occurs in simulators and for previewing animation for later high-quality recording. In some of these cases, the motion is computed and images are rendered as the display is updated; sometimes pre-calculated images are read from the disk and loaded into the frame buffer. In either case, the time it takes to paint an image on a computer screen can be significant (for various reasons). To avoid waiting for the image to update, animators often paint the new image into an off-screen buffer. Then a quick operation is performed to change the off-screen buffer to on-screen (often with hardware display support); the previous on-screen buffer becomes off-screen. This is called *double buffering*.

In double buffering, two (or more) buffers are used. One buffer is used to refresh the computer display, while another is used to assemble the next image. When the next image is complete, the two buffers switch roles. The second buffer is used to refresh the display, while the first buffer is used to assemble the next image. The buffers continue exchanging roles to ensure that the buffer used for display is not actively involved in the update operations (see Figure A.1).

Double buffering is often supported by the hardware of a computer graphics system. For example, the display system may have two built-in frame buffers, both of which are accessible to the user's program and are program selectable for driving the display. Alternatively, the graphics system may be designed so that the screen can be refreshed from an arbitrary section of main memory and the buffers are identified by pointers into the memory. Double buffering is also effective when

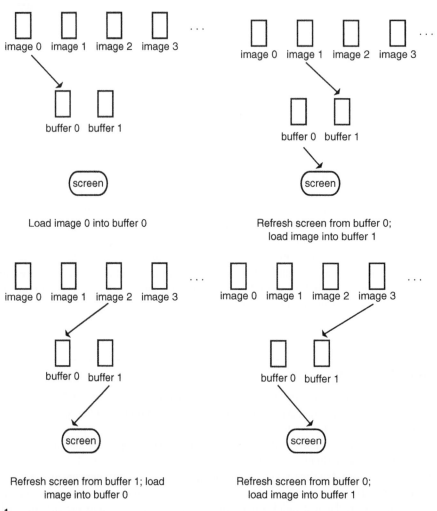

FIGURE A.1

Double buffering.

implemented completely in software, as is popular in JAVA applets. Pseudocode of simple double buffering using two frame buffers is shown in Figure A.2.

A.2 Compositing

Compositing is the act of combining separate image layers into a single picture. It allows the animator, or compositor, to combine elements obtained from different sources to form a single image. It allows artistic decisions to be deferred until all elements are brought together. Individual elements can be

```
open_window (w)
i=0;                    // start using buffer 0
j=0;                    // start with image 0
do{
  load_image(buffer[i],image[j]);/*load the jth image into the ith frame
                          buffer*/
  display_in_window(w,buffer[i]);//display the ith buffer
  i=1-i;                // swap which buffer to load image into
  j=j+1;                  // advance to the next image
}until (done);
```

FIGURE A.2

Double buffering pseudocode.

selectively manipulated without the user having to regenerate the scene as a whole. As frames of computer animation become more complex, the technique allows layers to be calculated separately and then composited after each has been finalized. Compositing provides great flexibility and many advantages to the animation process. Before digital imagery, compositing was performed optically in much the same way that a multi-plane camera is used in conventional animation. With digital technology, compositing operates on digital image representations.

One common and extremely important use of compositing is that of computing the foreground animation separately from the static background image. If the background remains static it need only be computed and rendered once. Then, as each frame of the foreground image is computed, it is composited with the background to form the final image. The foreground can be further segmented into multiple layers so that each object, or even each object part, has its own layer. This technique allows the user to modify just one of the object parts, re-render it, and then recompose it with the previously rendered layers, thus potentially saving a great deal of rendering time and expense. This is very similar to the two-and-a-half dimensional approach taken in conventional hand-drawn cell animation.

Compositing offers a more interesting advantage when digital frame buffers are used to store depth information in the form of z-values, along with the color information at each pixel. To a certain extent, this approach allows the animator to composite layers that actually interleave each other in depth. The initial discussion focuses on compositing without pixel depth information, which effectively mimics the approach of the multi-plane camera. Following this, compositing that takes the z-values into account is explored.

A.2.1 Compositing without pixel depth information

Digital compositing attempts to combine multiple two-dimensional images of three-dimensional scenes so as to approximate the visibility of the combined scenes. Compositing will combine two images at a time with the desire of maintaining the equilibrium shown in Equation A.1.

$$composite(render(scene1), \; render(scene2)) = render(merge(scene1, scene2)) \qquad (A.1)$$

The *composite* operation on the left side of the equation refers to compositing the two rendered images; the *merge* operation on the right side of the equation refers to combining the geometries of the two scenes into one. The *render* function represents the process of rendering an image based on the input scene geometries. In the case in which the composite operator is used, the render function must tag pixels not covered by the scene as transparent. The equality will hold if the scenes are disjoint in depth from the observer and the composite operator gives precedence to the image closer to the observer. The visibility between elements from the different scenes can be accurately represented in two-and-a-half dimensional compositing. The visibility between disjoint planes can be accurately resolved by assigning a single visibility priority to all elements within a single plane.

The image-based **over** operator places one image on top of the other. In order for **over** to operate, there must be some assumption or some additional information indicating which part of the closer image (also referred to as the *overlay plane* or *foreground image*) occludes the image behind it (the *background image*). In the simplest case, all of the foreground image occludes the background image. This is useful for the restricted situation in which the foreground image is smaller than the background image. In this case, the smaller foreground image is often referred to as a *sprite*. There are usually two-dimensional coordinates associated with the sprite that locate it relative to the background image (see Figure A.3).

However, for most cases, additional information in the form of an *occlusion mask* (also referred to as a *matte* or *key*) is provided along with the overlay image. A one-bit matte can be used to indicate which pixels of the foreground image should occlude the background during the compositing process (see Figure A.4). In frame buffer displays, this technique is often used to overlay text or a cursor on top of an image.

Compositing is a binary operation, combining two images into a single image. However, any number of images can be composited to form a final image. The images must be ordered by depth and are

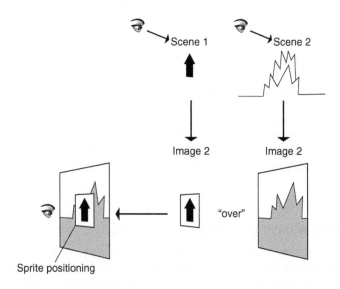

FIGURE A.3

Two-and-a-half dimensional compositing without transparency.

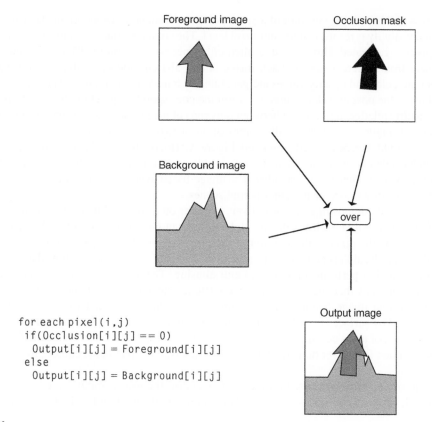

```
for each pixel(i,j)
  if(Occlusion[i][j] == 0)
    Output[i][j] = Foreground[i][j]
  else
    Output[i][j] = Background[i][j]
```

FIGURE A.4

Compositing using a one-bit occlusion mask.

operated on two at a time. Each operation replaces the two input images with the output image. Images can be composited in any order as long as the two composited during any one step are adjacent in their depth ordering.

Compositing the images using a one-bit occlusion mask, that is, using the color of a foreground image pixel in the output image, is an all-or-nothing decision. However, if the foreground image is calculated by considering semitransparent surfaces or partial pixel coverage, then fractional occlusion information must be available and anti-aliasing during the pixel-merging process must be taken into account. Instead of using a one-bit mask for an all-or-nothing decision, using more bits allows a partial decision. This gray-scale mask is called *alpha* and is commonly maintained as a fourth *alpha channel* in addition to the three (red, green, and blue) color channels. The alpha channel is used to hold an opacity factor for each pixel. Even this is a shortcut; to be more accurate, an alpha channel for each of the three colors R, G, and B is required. A typical size for the alpha channel is eight bits.

The fractional occlusion information available in the alpha channel is an approximation used in lieu of detailed knowledge about the three-dimensional geometry of the two scenes to be combined. Ideally, in the process of determining the color of a pixel, polygons[1] from both scenes are made available to the renderer and visibility is resolved at the sub-pixel level. The combined image is anti-aliased, and a color for each pixel is generated. However, it is often either necessary or more efficient to composite the images made from different scenes. Each image is anti-aliased independently, and, for each pixel, the appropriate color and opacity values are generated in rendering the images. These pixels are then combined using the opacity values (alpha) to form the corresponding pixel in the output image. Note that all geometric relationships are lost between polygons of the two scenes once the image pixels have been generated. Figure A.5(a) shows the combination of two scenes to create complex overlapping geometry that could then be correctly rendered. Figure A.5(b), on the other hand, shows the two scenes rendered independently at which point it's impossible to insert the geometric elements of the one between the geometric elements of the other. The image-based **over** operator, as described above, must give visibility priority to one or the other partial scenes.

The pixel-based compositing operator, **over**, operates on a color value, RGB, and an alpha value between 0 and 1 stored at each pixel. The alpha value can be considered either the opacity of the surface that covers the pixel or the fraction of the pixel covered by an opaque surface, or a combination of the two. The alpha channel can be generated by visible surface algorithms that handle transparent surfaces and/or perform some type of anti-aliasing. The alpha value for a given image pixel represents the amount that the pixel's color contributes to the color of the output image when composited with an image behind the given image. To characterize this value as the fraction of the pixel covered by surfaces from the corresponding scene is not entirely accurate. It actually needs to be the coverage of areas contributing to the pixel color weighted by the anti-aliasing filter kernel.[2] In the case of a box filter over non-overlapping pixel areas, the alpha value equates to the fractional coverage.

To composite pixel colors based on the **over** operator, the user computes the new alpha value for the pixel: $\alpha = \alpha_F + (1 - \alpha_F)\alpha_B$. The composited pixel color is then computed by Equation A.2.

$$(\alpha_F RGB_F + (1 - \alpha_F)\alpha_B RGB_B)/\alpha \qquad (A.2)$$

where RGB_F is the color of the foreground, RGB_B is the color of the background, a_F is the alpha channel of the foreground pixel, and a_B is the alpha channel of the background pixel. The **over** operator is not commutative but is associative (see Equation A.3).

$$F \textbf{ over } B = \begin{cases} \alpha_{F \text{ over } B} = \alpha_F + (1 - \alpha_F)\alpha_B & \text{over operator} \\ RGB_{F \text{ over } B} = (\alpha_F RGB_F + (1 - \alpha_F)\alpha_B RGB_B)/\alpha_{F \text{ over } B} \end{cases}$$

$$A \text{ over } B \neq B \text{ over } A \qquad\qquad \text{not commutative}$$

$$(A \text{ over } B) \text{over } C = A \text{ over}(B \text{ over } C) \qquad \text{associative}$$

$$(A.3)$$

The compositing operator, **over**, assumes that the fragments in the two input images are uncorrelated. The assumption is that the color in the images comes from randomly distributed fragments. For example, if the alpha of the foreground image is 0.5, then the color fragments of what is behind the

---

[1] To simplify the discussion and diagrams, one must assume that the scene geometry is defined by a collection of polygons. However, any geometric element can be accommodated provided that coverage, occlusion, color, and opacity can be determined on a subpixel basis.

[2] The *filter kernel* is the weighting function used to blend color fragments that partially cover a pixel's area.

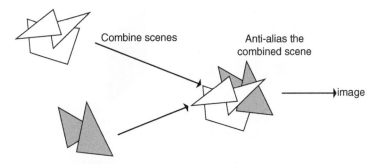

A   Without compositing: anti-alias the combined scenes to produce image

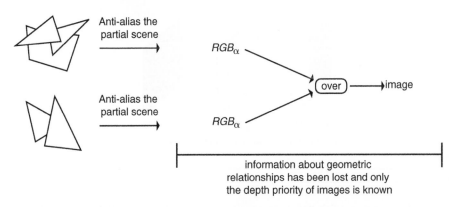

B   Without compositing: anti-alias the partial scenes and then combine to produce image

FIGURE A.5

Anti-aliasing combined scenes versus alpha channel compositing.

foreground image will, on average, show through 50 percent of the time. Consider the case of a foreground pixel and middle ground pixel, both with partial coverage in front of a background pixel with full coverage (Figure A.6).

The result of the compositing operation is shown in Equation A.4.

$$
\begin{aligned}
RGB_{FMB} &= (\alpha_F RGB_F + (1 - \alpha_F)\alpha_M RGB_M) + (1 - \alpha_{FM})\alpha_B RGB_B \\
\text{where} \quad \alpha_{FM} &= (\alpha_F + (1 - \alpha_F)\alpha_M) \text{and} \; \alpha_{FMB} = 1.0 \\
RGB &= 0.5RGB_F + 0.25RGB_M + 0.25RGB_B
\end{aligned}
\tag{A.4}
$$

If the color fragments are correlated, for example, if they share an edge in the image plane, then the result of the compositing operation is incorrect. The computations are the same, but the result does not accurately represent the configuration of the colors in the combined scene. In the example of Figure A.7, none of the background should show through. A similarly erroneous result occurs if the middle ground image has its color completely on the other side of the edge, in which case none of the middle ground color should appear in the composited pixel. Because geometric information has been discarded, compositing fails to handle these cases correctly.

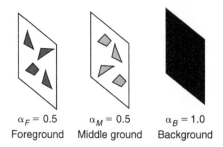

$$\alpha_F = 0.5 \qquad \alpha_M = 0.5 \qquad \alpha_B = 1.0$$

Foreground   Middle ground   Background

FIGURE A.6

Compositing randomly distributed color fragments for a pixel.

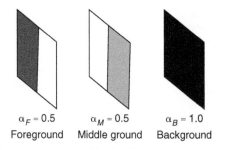

$$\alpha_F = 0.5 \qquad \alpha_M = 0.5 \qquad \alpha_B = 1.0$$

Foreground   Middle ground   Background

FIGURE A.7

Compositing correlated colors for a pixel.

When α_F and α_B represent full coverage opacities or uncorrelated partial coverage fractions, the **over** operator computes a valid result. However, if the alphas represent partial coverages that share an edge, then the compositing **over** operator does not have enough information to tell whether, for example, the partial coverages overlap in the pixel area or whether the areas of coverage they represent are partially or completely disjoint. Resolution of this ambiguity requires that additional information be stored at each pixel indicating which part of the pixel is covered by a surface fragment. For example, the A-buffer[3] algorithm [1] provides this information.

Alpha channel is a term that represents a combination of the partial coverage and the transparency of the surface or surfaces whose color is represented at the pixel. Notice that when compositing colors, a color always appears in the equation multiplied by its alpha value. It is therefore expedient to store the color value already scaled by its alpha value. In the following discussion, lowercase *rgb* refers to a color value that has already been scaled by alpha. Pixels and images whose colors have been scaled by alpha are called *premultiplied*. Uppercase *RGB* refers to a color value that has not been scaled by alpha. In a premultiplied image, the color at a pixel is considered to already be scaled down by its alpha factor, so that if a surface is white with an *RGB* value of $(1, 1, 1)$ and it covers half a pixel as indicated by an alpha value of 0.5, then the *rgb* stored at that pixel will be $(0.5, 0.5, 0.5)$. It is important to recognize that storing premultiplied images is very useful.

---

[3]The A-buffer is a Z-buffer in which information recorded at each pixel includes the relative depth and coverage of all fragments, in z-sorted order, which contribute to the pixel's final color.

A.2.2 **Compositing with pixel depth information**

In compositing, independently generated images may sometimes not be disjoint in depth. In such cases, it is necessary to interleave the images in the compositing process. Duff [5] presents a method for compositing three-dimensional rendered images in which depth separation between images is not assumed. An ***rgb*αz** (premultiplied) representation is used for each pixel that is simply a combination of an *rgb* value, the alpha channel, and the *z*-, or depth, value. The *z*-value associated with a pixel is the depth of the surface visible at that pixel; this value is produced by most rendering algorithms.

Binary operators are defined to operate on a pair of images *f* and *b* on a pixel-by-pixel basis to generate a resultant image (Equation A.5). Applying the operators to a sequence of images in an appropriate order will produce a final image.

$$c = f \textbf{ op } b \tag{A.5}$$

The first operator to define is the **over** operator. Here, it is defined using colors that have been premultiplied by their corresponding alpha values. The **over** operator blends together the color and alpha values of an ordered pair of images on a pixel-by-pixel basis. The first image is assumed to be "over" or "in front of" the second image. The color of the resultant image is the color of the first image plus the product of the color of the second image and the transparency (one minus opacity) of the first image. The alpha value of the resultant image is computed as the alpha value of the first image plus the product of the transparency of the first and the opacity of the second. Values stored at each pixel of the image, resulting from $c = f$ **over** b, are defined as shown in Equation A.6.

$$rgb_c = rgb_f + (1 - \alpha_f)rgb_b$$
$$\alpha_c = \alpha_f + (1 - \alpha_f)\alpha_b \tag{A.6}$$

For a given foreground image with corresponding alpha values, the foreground *rgb*s will be unattenuated during compositing with the **over** operator and the background will show through more as α_f decreases. Notice that when $\alpha_f = 1$, then $rgb_c = rgb_f$ and $\alpha_c = \alpha_f = 1$; when $\alpha_f = 0$ (and therefore $rgb_f = 0, 0, 0$), then $rgb_c = rgb_b$ and $\alpha_c = \alpha_b$. Using **over** with more than two layers requires that their ordering in *z* be taken into account when compositing. The **over** operator can be successfully used when compositing planes adjacent in *z*. If non-adjacent planes are composited, a plane lying between these two cannot be accurately composited; the opacity of the closest surface is not separately represented in the composited image. **Over** is not commutative, although it is associative.

The second operator to define is the *z*-depth operator, **zmin**, which operates on the *rgb*, alpha, and *z*-values stored at each pixel. The **zmin** operator simply selects the ***rgb*αz** values of the closer pixel (the one with the minimum *z*). Values stored at each pixel of the image resulting from $c = f$ **zmin** b are defined by Equation A.7.

$$rgb\alpha_c = \text{if}(z_f < z_b) \quad \text{then}(rgb\alpha_f) \quad \text{else}(rgb\alpha_b)$$
$$z_c = \min(z_f, z_b) \tag{A.7}$$

The order in which the surfaces are processed by **zmin** is irrelevant; it is commutative and associative and can be successfully used on non-adjacent layers.

Comp is an operator that combines the action of **zmin** and **over**. As before, each pixel contains an *rgb* value and an α value. However, for an estimate of relative coverage, each pixel has *z*-values at each of its four corners. Because each *z*-value is shared by four pixels, the upper left *z*-value can be stored at each pixel location. This requires that an extra row and extra column of pixels be kept in order to provide the z-values for the pixels in the rightmost row and bottommost column of the image; the rgb and α values for these pixels in the extra row and column are never used.

To compute $c = f \mathbf{comp}\ b$ at a pixel, one must first compute the z-values at the corners to see which is larger. There are $2^4 = 16$ possible outcomes of the four corner comparisons. If the comparisons are not the same at all four corners, the pixel is referred to as *confused*. This means that within this single pixel, the layers cross each other in z. For any edge of the pixel whose endpoints compare differently, the z-values are interpolated to estimate where along the edge the surfaces actually meet in z. Figure A.8 illustrates the implied division of a pixel into areas of coverage based on the relative z-values at the corners of the foreground and background pixels.

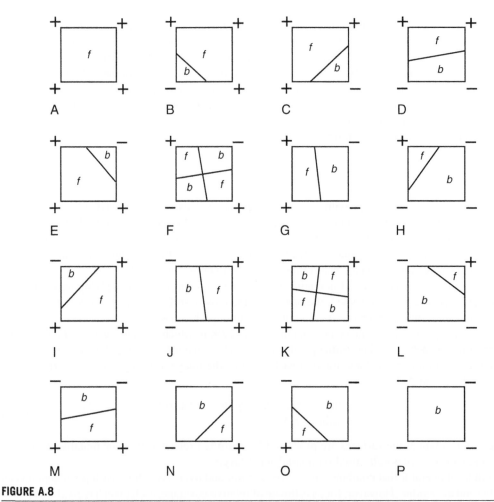

FIGURE A.8

Categories of pixels based on z comparisons at the corners; the label at each corner is sign($z_f - z_b$).

Considering symmetry, there are really only four cases to consider in computing β, the coverage fraction for the surface f: whole, corner, split, and two-opposite corners. *Whole* refers to the simple cases in which the entire pixel is covered by one surface (cases a and p in Figure A.8); in this case, β equals either 1 or 0. *Corner* refers to the cases in which one surface covers the entire pixel except for one corner (cases b, c, e, h, i, l, n, and o in Figure A.8). If f is the corner surface, then the coverage is $\beta = \dfrac{s \times t}{2}$, where s and t represent the fraction of the edge indicated by the z-value interpolation as measured from the corner vertex. If b is the corner surface, then $\beta = 1.0 - \dfrac{s \times t}{2}$. *Split* refers to the situation in which the vertices of opposite edges are tagged differently (cases d, g, j, and m in Figure A.8); the coverage of the surface is $\beta = \dfrac{s+t}{2}$, where s and t represent the fraction of the edge indicated by the z-value interpolation as measured from the vertices tagged "1" toward the other vertices. The fraction for the other surface would be $\beta = 1.0 - \dfrac{s+t}{2}$. If each vertex is tagged the same as that diagonally opposite of it, but the diagonally opposite pairs are tagged differently (cases f and k in Figure A.8), the equation for β can be approximated by computing the fractional distance along each pixel edge according to the z-value interpolation from the corners—for opposite pairs, use the same direction to determine the fraction. For the opposite pairs of edges of the pixel, average the two fractional distances. Use these distances as an axis-aligned subdivision of the pixel into four quadrants and compute the areas of f and b.

Once β is computed, then that fraction of the pixel is considered as having the surface f as the front surface and the rest of the pixel is considered as having the surface b as the front surface. The **comp** operator is defined as the linear blend, according to the interpolant β, of two applications of the **over** operator—one with surface f in front and one with surface b in front (Equation A.8).

$$
\begin{aligned}
rgb_c &= \beta(rgb_f + (1 - \alpha_f)rgb_b) + (1 - \beta)(rgb_b + (1 - \alpha_b)rgb_f) \\
\alpha_c &= \beta(\alpha_f + (1 - \alpha_f)\alpha_b) + (1 - \beta)(\alpha_b + (1 - \alpha_b)\alpha_f) \\
z_c &= \min(z_f, z_b)
\end{aligned}
\qquad \text{(A.8)}
$$

Comp decides on a pixel-by-pixel basis which image represents a surface in front of the other, including the situation in which the surfaces change relative depth within a single pixel. Thus, images that are not disjoint in depth can be successfully composited using the **comp** operator, which is commutative (with β becoming $(1 - \beta)$) but not associative. See Duff's paper [5] for a discussion.

A.3 Displaying moving objects: motion blur

If it is assumed that objects are moving in time, it is worth addressing the issue of effectively displaying these objects in a frame of animation [6] [8]. In the same way that aliasing is a sampling issue in the spatial domain, it is also a sampling issue in the temporal domain. As frames of animation are calculated, the positions of objects change in the image, and this movement changes the color of a pixel as a function of time. In an animation, the color of a pixel is sampled in the time domain. If the temporal frequency of a pixel's color function is too high, then the temporal sampling can miss important information.

Consider an object moving back and forth in space in front of an observer. Assume the animation is calculated at the rate of thirty frames per second. Now assume that the object starts on the left side of the screen and moves to the right side in one-sixtieth of a second and moves back to the left side in another one-sixtieth of a second. This means that every thirtieth of a second (the rate at which frames are calculated and therefore the rate at which positions of objects are sampled) the object is on the left side of the screen. The entire motion of the object is missed because the sampling rate is too low to capture the high-frequency motion of the object, resulting in temporal aliasing. Even if the motion is not this contrived, displaying a rapidly moving object by a single instantaneous sample can result in motions that appear jerky and unnatural. As mentioned in Chapter 1, Section 1.1, images of a fast-moving object can appear to be disjointed, resulting in jerky motion similar to that of live action under a strobe light (and this is often called *strobing*).

Conventional animation has developed its own techniques for representing fast motion. Speed lines can be added to moving objects, objects can be stretched in the direction of travel, or both speed lines and object stretching can be used [9] (see Figure A.9).

There is an upper limit on the amount of detail the human eye can resolve when viewing a moving object. If the object moves too fast, mechanical limitations of muscles and joints that control head and eye movement will fail to maintain an accurate track. This will result in an integration of various samples from the environment as the eye tries to keep up. If the eye is not tracking the object and the object moves across the field of view, receptors will again receive various samples from the environment integrated together, forming a cumulative effect in the eye-brain. Similarly, a movie camera will open its shutter for an interval of time, and an object moving across the field of view will create a blurred image on that frame of the film. This will smooth out the apparent motion of the object. In much the same way, the synthetic camera can (and should) consider a frame to be an interval of time instead of an instance in time. Unfortunately, accurately calculating the effect of moving objects in an image requires a nontrivial amount of computation.

To fully consider the effect that moving objects have on an image pixel, one must take into account the area of the image represented by the pixel, the time interval for which a given object is visible in that pixel, the area of the pixel in which the object is visible, and the color variation of the object over that time interval in that area, as well as such dynamic effects as rotating textured objects, shadows, and specular highlights [8].

There are two analytic approaches to calculating motion blur: continuous and discrete. Continuous approaches attempt to be more accurate but are only tractable in limited situations. Discrete approaches, while less accurate, are more generally applicable and more robust; only discrete approaches are considered here. Ray tracing is probably the easiest domain in which to understand the discrete process. It is common in ray tracing to generate more than one ray per pixel in order to

Speed line   Speed lines and object stretching

FIGURE A.9

Methods used in conventional animation for displaying speed.

FIGURE A.10

An example of synthetic (calculated) motion blur.

spatially anti-alias the image. These rays can be distributed in a regular pattern or stochastically [2] [3] [10]. To incorporate temporal anti-aliasing, one need only take the rays for a pixel and distribute them in time as well as in space. In this case, the frame is not considered to be an instant in time but rather an interval. The interval can be broken down into subintervals and the rays distributed into these subintervals. The various samples are then filtered to form the final image. See Figure A.10 for an example.

One of the extra costs associated with temporal anti-aliasing is that the motion of the objects must be computed at a higher rate than the frame rate. For example, if a 4×4 grid of subsamples is used for anti-aliasing in ray tracing and these are distributed over separate time subintervals, then the motion of the objects must be calculated at sixteen times the frame rate. If the subframe motion is computed using complex motion control algorithms, this may be a significant computational cost. Linear interpolation is often used to estimate the positions of objects at subframe intervals.

Although discussed above in terms of distributed ray tracing, this same strategy can be used with any display algorithm, as Korein and Badler [6] note. Multiple frame buffers can be used to hold rendered images at subframe intervals. These can then be filtered to form the final output image. The rendering algorithm can be a standard z-buffer, a ray tracer, a scanline algorithm, or any other technique. Because of the discrete sampling, this is still susceptible to temporal aliasing artifacts if the object is moving too fast relative to the size of its features in the direction of travel; instead of motion blur, multiple images of the object may result because intermediate pixels are effectively "jumped over" by the object during the sampling process.

In addition to analytic methods, hand manipulation of the object shape by the animator can reduce the amount of strobing. For example, the animator can stretch the object in the direction of travel. This will tend to reduce or eliminate the amount of separation between images of the object in adjacent frames.

A.4 Drop shadows

The shadow cast by an object onto a surface is an important visual cue in establishing the distance between the two. Contact shadows, or shadows produced by an object contacting the ground, are especially important. Without them, objects appear to float just above the ground plane. In high-quality animation, shadow decisions are prompted by lighting, cinematography, and visual understanding considerations. However, for most other animations, computational expense is an important concern and computing all of the shadows cast by all objects in the scene onto all other objects in the scene is overkill. Much computation can be saved if the display system supports the user specification of which objects in a scene cast shadows onto which other objects in a scene. For example, the self-shadowing[4] of an object is often not important to understanding the scene visually. Shadows cast by moving objects onto other moving objects are also often not of great importance. These principles can be observed in traditional hand-drawn animation in which only a select set of shadows is drawn.

By far the most useful type of shadow in animated sequences is the drop shadow. The drop shadow is the shadow that an object projects to the ground plane. The drop shadow lets the viewer know how far above the ground plane an object is as well as the object's relative depth and therefore relative size (see Figures A.11 and A.12).

Drop shadows can be produced in several different ways. When an object is perspectively projected from the light source to a flat ground plane, an image of the object can be formed (see Figure A.13). If this image is colored dark and displayed on the ground plane (as a texture map, for example), then it is an effective shortcut.

Another inexpensive method for creating a drop shadow is to make a copy of the object, scale it flat vertically, color it black (or make it dark and transparent), and position it just on top of the ground plane

FIGURE A.11

Scene without drop shadows; without shadows, it is nearly impossible to estimate relative heights and distances if the sizes of the objects are not known.

---

[4]*Self-shadowing* refers to an object casting shadows onto its own surface.

FIGURE A.12

Scene with drop shadows indicating relative depth and, therefore, relative height and size.

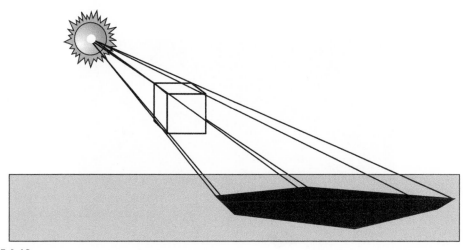

FIGURE A.13

Computing the drop shadow by perspective projection.

(see Figures A.14 and A.15). The drop shadow has the correct silhouette for a light source directly overhead but without the computational expense of the perspective projection method.

The drop shadow can be effective even if it is only an approximation of the real shadow that would be produced by a light source in the environment. The height of the object can be effectively indicated by merely controlling the relative size and softness of a drop shadow, which only suggests the shape of the object that casts it. For simple drop shadows, circles can be used, as in Figure A.16.

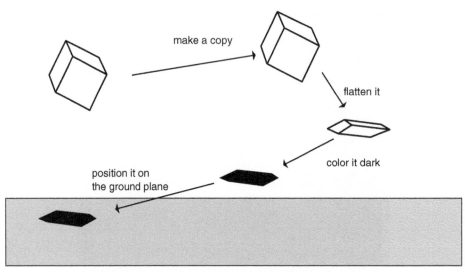

FIGURE A.14

Computing the drop shadow by flattening a copy of the object.

FIGURE A.15

Drop shadow using a flattened copy of the object: right side shows flattened copy with original coloring; left side shows flattened copy as black shadow.

FIGURE A.16

Circular drop shadows.

Globular shapes that may more closely indicate the shapes of the original objects can also be used as shadows. The drop shadow, however it is produced, can be represented on the ground plane in several different ways. It can be colored black and placed just above the ground plane. It can be made into a darkly colored transparent object so that any detail of the ground plane shows through. It can be colored darker in the middle with more transparency toward the edges to simulate a shadow's penumbra[5] (Figure A.17).

When placing a shadow over the ground plane, one must take care to keep it close enough to the ground so that it does not appear as a separate object and at the same time does not conflict with the ground geometry. To avoid the problems of using a separate geometric element to represent the drop shadow, the user can incorporate the shadow directly into the texture map of the ground plane.

A.5 Billboarding and impostors

Billboarding and the use of impostors are a rendering shortcut that reduces the computational complexity of a scene while maintaining its visual complexity (e.g., [4] [7]). These related techniques do this by using a two-dimensional substitute (such as a planar projection) for a complex three-dimensional object that is seamlessly incorporated into a three-dimensional scene. Typically, a billboard is a partially transparent, textured quadrilateral. The texture map is an image of the object represented. The quadrilateral is partially transparent in cases when the object's image does entirely cover the quadrilateral;

---

[5]The *penumbra* is the area partially shadowed by an opaque body; it receives partial illumination from a light source. The *umbra* is the area completely shadowed by an opaque body; it receives no illumination from the light source.

FIGURE A.17

Various ways to render a drop shadow: black, transparent, and black with transparent edges.

transparent sections allow the camera to see objects behind the quadrilateral but not obscured by the billboard image. An impostor usually implies that it is a temporary substitute for a three-dimensional object, whereas billboards usually refer to independent graphical elements.

Billboards are useful when a three-dimensional graphical element is too complex to model and render for its perceptual effect. Billboards are suitable for objects that are to be viewed at a distance, that are spherically or cylindrically perceptually symmetric, and that are just as effectively modeled in two dimensions as in three dimensions. Common uses of billboarding include the representation of trees, clouds, mountains, explosions, fire, and lens flares. See Figure A.18 for an example. Billboard representations can also be incorporated as one level of an object's level of detail representation.

FIGURE A.18

A scene containing several billboards of trees (Image courtesy of Ysheng Chen.)

Background and ground planes could be considered types of billboards, although usually the term is reserved for substitutes of objects that would otherwise be modeled with three-dimensional geometry. However, by using multiple billboards for distance background objects at various depths, a parallax effect, typical of conventional two-dimensional animation, can be created. As an example, consider a passenger looking out the window of a moving train. The passenger may see trees, houses, and mountains moving across the window view at different speeds due to their relative depth. For objects not close enough to the train for their sides to be seen, these objects can be modeled effectively by using two-dimensional elements—billboards.

The two-dimensional background elements discussed above are examples of static billboards. Animation of a billboard's orientation is also useful. When a camera has the ability to travel in and around complex elements, billboard representations can be used that change their orientation depending on the position and orientation of the camera. Tree billboards are good examples. A tree billboard can be procedurally rotated so that its plane is always perpendicular to the direction to the camera. As the camera travels through a terrain environment, for example, the tree billboards are smoothly rotated to always face the camera. In this way, the billboards effectively hide their two dimensionality and always present their visual complexity to the camera. Orienting a billboard toward the camera can create perspective distortion when the billboard is to the side of the center of interest. As an alternative to directly facing the camera, billboards can also be oriented so that they are parallel to the view plane. This produces less perspective distortion when projected to the side of the image. Tree billboards are constrained to only rotate around the y-axis because the tree up vector must be aligned with the world up vector (y-axis). Other billboards, such as representations of clouds, explosions, or snowflakes, can be oriented to always face a certain direction as well but can be free to rotate in their plane about their own center.

Some billboards may contain animated images. These are similar to sprites used in two-and-a-half dimensional graphics common in many early digital arcade games. Animated billboards can be used for fire, explosions, and smoke and typically loop through a sequence of pre-rendered images to create the impression of a dynamic process. A billboard animation loop can be used to create a walk cycle for a figure.

A.6 Summary

Although these techniques are not concerned with specifying or controlling the motion of graphical objects, they are important to the process of generating images for computer animation. Double buffering helps to smoothly update a display of images. Compositing helps to conserve resources and combine elements from different sources. Motion blur prevents fast-moving objects from appearing jerky and distracting to the viewer. Shadows help to locate objects relative to surfaces, but only those shadows that effectively serve such a purpose need to be generated. Billboarding and the use of impostors reduce the computational complexity of the modeling and rendering while maintaining the visual complexity of the scene.

References

[1] Carpenter L. The A-Buffer Hidden Surface Method. In: Computer Graphics. Proceedings of SIGGRAPH 84, vol. 18(3). Minneapolis, Minn.; July 1984. p. 103–8.
[2] Cook R. Stochastic Sampling in Computer Graphics. ACM Transactions on Graphics January 1986;5(1):51–71.

[3] Cook R, Porter T, Carpenter L. Distributed Ray Tracing. In: Computer Graphics. Proceedings of SIGGRAPH 84, vol. 18(3). Minneapolis, Minn.; July 1984. p. 137–46.

[4] Decoret X, Durand F, Sillion F, Dorsey J. Billboard Clouds for Extreme Model Simplification. Transactions on Graphics July 2003;22(3), Special Issue: Proceedings of ACM SIGGRAPH 2003. pp. 689–96.

[5] Duff T. Compositing 3-D Rendered Images. In: Barsky BA., editor. Computer Graphics. Proceedings of SIGGRAPH 85, vol. 19(3). San Francisco, Calif.; August 1985. p. 41–4.

[6] Korein J, Badler N. Temporal Anti-Aliasing in Computer Generated Animation. In: Computer Graphics. Proceedings of SIGGRAPH 83, vol. 17(3). Detroit, Mich.; July 1983. p. 377–88.

[7] Maciel P, Shirley P. Visual Navigation of Large Environments Using Textured Clusters. In: Proc 1995 Symposium on Interactive 3D Graphics. Monterey, California; April 9–12, 1995. p. 95–102.

[8] Potmesil M, Chadkravarty I. Modeling Motion Blur in Computer Generated Images. In: Computer Graphics. Proceedings of SIGGRAPH 83, vol. 17(3). Detroit, Mich.; July 1983. p. 389–400.

[9] Thomas F, Johnson O. The Illusion of Life. New York: Hyperion; 1981.

[10] Whitted T. An Improved Illumination Model for Shaded Display. Communication of the ACM June 1980;23(6):343–9.

Background Information and Techniques

B.1 Vectors and matrices

A *vector* is a one-dimensional list of values. This list can be shown as a row vector or a column vector (e.g., Eq. B.1). In general, a matrix is an *n*-dimensional array of values. For purposes of this book, a matrix is two-dimensional (e.g., Eq. B.2).

$$[a \ b \ c] \quad \begin{bmatrix} a \\ b \\ c \end{bmatrix} \tag{B.1}$$

$$\begin{bmatrix} a & b & c \\ d & e & f \\ g & h & i \end{bmatrix} \tag{B.2}$$

Matrices are multiplied together by taking the ith row of the first matrix and multiplying each element by the corresponding element of the jth column of the second matrix and summing all the products to produce the i,jth element. When computing $C = AB$, where A has v elements in each row and B has v elements in each column, an element C_{ij} is computed according to Equation B.3.

$$C_{ij} = A_{i1}B_{1j} + A_{i2}B_{2j} + A_{i3}B_{3j} + \ldots + A_{iv}B_{vj}$$
$$= \sum_{k=1}^{v} A_{ik}B_{kj} \tag{B.3}$$

The "inside" dimension of the matrices must match in order for the matrices to be multiplied together. That is, if A and B are multiplied and A is a matrix with U rows and V columns (a $U \times V$ matrix), then B must be a $V \times W$ matrix; the result will be a $U \times W$ matrix. In other words, the number of columns (the number of elements in a row) of A must be equal to the number of rows (the number of elements in a column) of B. As a more concrete example, consider multiplying two 3×3 matrices. Equation B.4 shows the computation for the first element.

$$AB = \begin{bmatrix} A_{11} & A_{12} & A_{13} \\ A_{21} & A_{22} & A_{23} \\ A_{31} & A_{32} & A_{33} \end{bmatrix} \begin{bmatrix} B_{11} & B_{12} & B_{13} \\ B_{21} & B_{22} & B_{23} \\ B_{31} & B_{32} & B_{33} \end{bmatrix}$$

$$= \begin{bmatrix} A_{11}B_{11} + A_{12}B_{21} + A_{13}B_{31} & \cdots & \cdots \\ & \cdots & & \cdots & \cdots \\ & \cdots & & \cdots & \cdots \end{bmatrix} \tag{B.4}$$

The *transpose* of a vector or matrix is the original vector or matrix with its rows and columns exchanged (e.g., Eq. B.5). The *identity matrix* is a square matrix with ones along its diagonal and zeros elsewhere (e.g., Eq. B.6). The *inverse* of a square matrix when multiplied by the original matrix produces the identity matrix (e.g., Eq. B.7). The *determinant* of a 3×3 matrix is formed as shown in Equation B.8. The determinant of matrices greater than 3×3 can be defined recursively. First, define an element's *submatrix* as the matrix formed when removing the element's row and column from the original matrix. The determinant is formed by considering any row, element by element. The determinant is the first element of the row times the determinant of its submatrix, minus the second element of the row times the determinant of its submatrix, plus the third element of the row times the determinant of its submatrix, and so on. The sum is formed for the entire row, alternating additions and subtractions.

$$\begin{bmatrix} a & b & c \\ d & e & f \end{bmatrix}^T = \begin{bmatrix} a & d \\ b & e \\ c & f \end{bmatrix} \tag{B.5}$$

$$I = \begin{bmatrix} 1 & 0 & 0 \\ 0 & 1 & 0 \\ 0 & 0 & 1 \end{bmatrix} \tag{B.6}$$

$$MM^{-1} = M^{-1}M = I \tag{B.7}$$

$$\begin{vmatrix} a & b & c \\ d & e & f \\ g & h & i \end{vmatrix} = a(ei - fh) - b(di - fg) + c(dh - eg) \tag{B.8}$$

B.1.1 Inverse matrix and solving linear systems

The inverse of a matrix is useful in computer graphics to represent the inverse of a transformation and in computer animation to solve a set of linear equations. There are various ways to compute the inverse. One common method, which is also useful for solving sets of linear equations, is *LU* decomposition. The basic idea is that a square matrix, for example, a 4×4, can be decomposed into a lower triangular matrix times an upper triangular matrix. How this is done is discussed later in this section. For now, it is assumed that the *LU* decomposition is available (Eq. B.9).

$$A = \begin{bmatrix} A_{11} & A_{12} & A_{13} & A_{14} \\ A_{21} & A_{22} & A_{23} & A_{24} \\ A_{31} & A_{32} & A_{33} & A_{34} \\ A_{41} & A_{42} & A_{43} & A_{44} \end{bmatrix}$$

$$= LU = \begin{bmatrix} L_{11} & 0 & 0 & 0 \\ L_{21} & L_{22} & 0 & 0 \\ L_{31} & L_{32} & L_{33} & 0 \\ L_{41} & L_{42} & L_{43} & L_{44} \end{bmatrix} \begin{bmatrix} U_{11} & U_{12} & U_{13} & U_{14} \\ 0 & U_{22} & U_{23} & U_{24} \\ 0 & 0 & U_{33} & U_{34} \\ 0 & 0 & 0 & U_{44} \end{bmatrix}$$

(B.9)

The decomposition of a matrix A into the L and U matrices can be used to easily solve a system of linear equations. For example, consider the case of four unknowns (x) and four equations shown in Equation B.10. Use of the decomposition permits the system of equations to be solved by forming two systems of equations using triangular matrices (Eq. B.11).

$$A_{11}x_1 + A_{12}x_2 + A_{13}x_3 + A_{14}x_4 = b_1$$

$$A_{21}x_1 + A_{22}x_2 + A_{23}x_3 + A_{24}x_4 = b_2$$

$$A_{31}x_1 + A_{32}x_2 + A_{33}x_3 + A_{34}x_4 = b_3$$

$$A_{41}x_1 + A_{42}x_2 + A_{43}x_3 + A_{44}x_4 = b_4$$

(B.10)

$$\begin{bmatrix} A_{11} & A_{12} & A_{13} & A_{14} \\ A_{21} & A_{22} & A_{23} & A_{24} \\ A_{31} & A_{32} & A_{33} & A_{34} \\ A_{41} & A_{42} & A_{43} & A_{44} \end{bmatrix} \begin{bmatrix} x_1 \\ x_2 \\ x_3 \\ x_4 \end{bmatrix} = \begin{bmatrix} b_1 \\ b_2 \\ b_3 \\ b_4 \end{bmatrix}$$

$$Ax = b$$
$$Ax = b$$
$$(LU)x = b$$
$$L(Ux) = b$$
$$Ux = y$$
$$Ly = b$$

(B.11)

This solves the original set of equations. The advantage of doing it this way is that both of the last two equations resulting from the decomposition involve triangular matrices and therefore can be solved trivially with simple substitution methods. For example, Equation B.12 shows the solution to $Ly = b$. Notice that by solving the equations in a top to bottom fashion, the results from the equations of previous rows are used so that there is only one unknown in any equation being considered. Once the solution for y has been determined, it can be used to solve for x in $Ux = y$ using a similar approach. Once the LU decomposition of A is formed, it can be used repeatedly to solve sets of linear equations that differ only in right-hand sides, such as those for computing the inverse of a matrix. This is one of the advantages of LU decomposition over methods such as Gauss-Jordan elimination.

$$Ly = b$$

$$
\begin{bmatrix}
L_{11} & 0 & 0 & 0 \\
L_{21} & L_{22} & 0 & 0 \\
L_{31} & L_{32} & L_{33} & 0 \\
L_{41} & L_{42} & L_{43} & L_{44}
\end{bmatrix}
\begin{bmatrix}
y_1 \\ y_2 \\ y_3 \\ y_4
\end{bmatrix}
=
\begin{bmatrix}
b_1 \\ b_2 \\ b_3 \\ b_4
\end{bmatrix}
$$

First row: $L_{11}y_1 = b_1$

$$y_1 = b_1/L_{11} \tag{B.12}$$

Second row: $L_{21}y_1 + L_{22}y_2 = b_2$

$$y_2 = (b2 - (L_{21}y_1))/L_{22}$$

Third row: $L_{31}y_1 + L_{32}y_2 + L_{33}y_3 = b_3$

$$y_3 = (b_3 - (L_{31}y_1 - L_{32}y_2))/L_{33}$$

Fourth row: $L_{41}y_1 + L_{42}y_2 + L_{43}y_3 + L_{44}y_4 = b_4$

$$y_4 = (b_4 - (L_{41}y_1) - (L_{42}y_2) - (L_{43}y_3))/L_{44}$$

The decomposition procedure sets up equations and orders them so that each is solved simply. Given the matrix equation for the decomposition relationship, one can construct equations on a term-by-term basis for the A matrix. This results in N^2 equations with $N^2 + N$ unknowns. As there are more unknowns than equations, N elements are set to some arbitrary value. A particularly useful set of values is $L_{ii} = 1.0$. Once this is done, the simplest equations (for A_{11}, A_{12}, etc.) are used to establish values for some of the L and U elements. These values are then used in the more complicated equations. In this way the equations can be ordered so there is only one unknown in any single equation by the time it is evaluated. Consider the case of a 4×4 matrix. Equation B.13 repeats the original matrix equation for reference and Equation B.14 shows the resulting sequence of equations in which the underlined variable is the only unknown.

$$
\begin{bmatrix}
1 & 0 & 0 & 0 \\
L_{21} & 1 & 0 & 0 \\
L_{31} & L_{32} & 1 & 0 \\
L_{41} & L_{42} & L_{43} & 1
\end{bmatrix}
\begin{bmatrix}
U_{11} & U_{12} & U_{13} & U_{14} \\
0 & U_{22} & U_{23} & U_{24} \\
0 & 0 & U_{33} & U_{34} \\
0 & 0 & 0 & U_{44}
\end{bmatrix}
=
\begin{bmatrix}
A_{11} & A_{12} & A_{13} & A_{14} \\
A_{21} & A_{22} & A_{23} & A_{24} \\
A_{31} & A_{32} & A_{33} & A_{34} \\
A_{41} & A_{42} & A_{43} & A_{44}
\end{bmatrix}
\tag{B.13}
$$

For the first column of A

$$U_{11} = A_{11}$$
$$\underline{L_{21}}U_{11} = A_{21}$$
$$\underline{L_{31}}U_{11} = A_{31}$$
$$\underline{L_{41}}U_{11} = A_{41}$$

$$\tag{B.14}$$

For the second column of A

$$\underline{U_{12}} = A_{12}$$
$$L_{21}U_{12} + \underline{U_{22}} = A_{22}$$
$$L_{31}U_{12} + \underline{L_{32}}U_{22} = A_{32}$$
$$L_{41}U_{12} + \underline{L_{42}}U_{22} = A_{42}$$

For the third column of A

$$\underline{U_{13}} = A_{13}$$
$$L_{21}U_{13} + \underline{U_{23}} = A_{23}$$
$$L_{31}U_{13} + L_{32}U_{23} + \underline{U_{33}} = A_{33}$$
$$L_{41}U_{13} + L_{42}U_{23} + \underline{L_{43}}U_{33} = A_{43}$$

(B.14)

For the fourth column of A

$$\underline{U_{14}} = A_{14}$$
$$L_{21}U_{14} + \underline{U_{24}} = A_{24}$$
$$L_{31}U_{14} + L_{32}U_{24} + \underline{U_{34}} = A_{34}$$
$$L_{41}U_{14} + L_{42}U_{24} + \underline{L_{43}}U_{34} + \underline{U_{44}} = A_{44}$$

Notice a two-phase pattern in each column in which terms from the U matrix from the first row to the diagonal are determined first, followed by terms from the L matrix below the diagonal. This pattern can be generalized easily to matrices of arbitrary size [16], as shown in Equation B.15. The computations for each column j must be completed before proceeding to the next column.

$$U_{ij} = A_{ij} - \sum_{K=1}L_{ik}U_{kj} \qquad \text{for } i = 1,\ldots,j$$

$$L_{ij} = \frac{1}{U_{ij}}\left(Aij - \sum_{k=1}^{j-1}L_{ik}U_{kj}\right) \qquad \text{for } i = j+1,\ldots,n$$

(B.15)

So far this is fairly simple and easy to follow. Now comes the complication—*partial pivoting*. Notice that some of the equations require a division to compute the value for the unknown. For this computation to be numerically stable (i.e., the numerical error less sensitive to particular input values), this division should be by a relatively large number. By reordering the rows, one can exert some control over what divisor is used. Reordering rows does not affect the computation if the matrices are viewed as a system of linear equations; reordering obviously matters if the inverse of a matrix is being computed. However, as long as the reordering is recorded, it can easily be undone when the inverse matrix is formed from the L and U matrices.

Consider the first column of the 4×4 matrix. The divisor used in the last three equations is U_{11}, which is equal to A_{11}. However, if the rows are reordered, then a different value might end up as A_{11}. So the objective is to swap rows so that the largest value (in the absolute value sense) of $A_{11}, A_{21}, A_{31}, A_{41}$ ends up at A_{11}, which makes the computation more stable. Similarly, in the second column, the divisor is U_{22}, which is equal to $A_{22} - (L_{21}U_{12})$. As in the case of the first column, the rows below this are checked to see if a row swap might make this value larger. The row above is not checked because that row was needed to calculate U_{12}, which is needed in the computations of the rows below it. For each successive column, there are fewer choices because the only rows that are checked are the ones below the topmost row that requires the divisor.

There is one other modification to partial pivoting. Because any linear equation has the same solution under a scale factor, an arbitrary scale factor applied to a linear equation could bias the comparisons made in the partial pivoting. To factor out the effect of an arbitrary scale factor when comparing values for the partial pivoting, one scales the coefficients for that row, just for the purposes of the comparison, so that the largest coefficient of a particular row is equal to one. This is referred to as *implicit pivoting*.

$$\begin{bmatrix} A_{11} & A_{12} & A_{13} & A_{14} \\ A_{21} & A_{22} & A_{23} & A_{24} \\ A_{31} & A_{32} & A_{33} & A_{34} \\ A_{41} & A_{42} & A_{43} & A_{44} \end{bmatrix} \Rightarrow \begin{bmatrix} U_{11} & U_{12} & U_{13} & U_{14} \\ L_{21} & U_{22} & U_{23} & U_{24} \\ L_{31} & L_{32} & U_{33} & U_{34} \\ L_{41} & L_{42} & L_{43} & U_{44} \end{bmatrix}$$

FIGURE B.1

In-place computation of the L and U values assuming no row exchanges.

When performing the decomposition, the values of the input matrix can be replaced with the values of L and U (Figure B.1). Notice that the diagonal elements of the L matrix do not have to be stored because they are routinely set to one. If row exchanges take place, then the rows of the matrices in Figure B.1 will be mixed up. For solving the linear system of equations, the row exchanges have no effect on the computed answers. However, the row exchanges are recorded in a separate array so that they can be undone for the formation of the inverse matrix. In the code that follows (Figure B.2), the LU decomposition approach is used to solve a system of linear equations and is broken down into several procedures. These procedures follow those found in *Numerical Recipes* [16].

After the execution of **LUdecomp**, the A matrix contains the elements of the L and U matrices. This matrix can then be used either for solving a linear system of equations or for computing the inverse of a matrix. The previously discussed simple substitution methods can be used to solve the equations that involve triangular matrices. In the code that follows, the subroutine **LUsubstitute** is called with the newly computed A matrix, the dimension of A, the vector of row swaps, and the right-hand vector (i.e., the b in $Ax = b$).

One of the advantages of the LU decomposition is that the decomposition matrices can be reused if the only change in a system of linear equations is the right-hand vector of values. In such a case, the routine **LUdecomp** only needs to be called once to form the LU matrices. The substitution routine, **LUsubstitute**, needs to be called with each new vector of values (remember to reuse the A matrix that holds the LU decomposition).

To perform matrix inversion, use the L and U matrices repeatedly with the b matrix holding column-by-column values of the identity matrix. In this way the inverse matrix is built up column by column.

B.1.2 Singular value decomposition

Singular value decomposition (SVD) is a popular method used for solving linear least-squares problems ($Ax = b$, where the number of rows of A is greater than the number of columns). It gives the user information about how ill conditioned the set of equations is and allows the user to step in and remove sources of numerical inaccuracy.

As with LU decomposition, the first step decomposes the A matrix into more than one matrix (Eq. B.16). In this case, an $M \times N$ A matrix is decomposed into a column-orthogonal $M \times N$ U matrix, a diagonal $N \times M$ W matrix, and an $N \times N$ orthogonal V matrix.

$$A = UWV^T \tag{B.16}$$

The magnitude of the elements in the W matrix indicates the potential for numerical problems. Zeros on the diagonal indicate singularities. Small values (where *small* is user defined) indicate the

```
/*LU Decomposition
 * with partial implicit pivoting
 * partial means that the pivoting only happens by row
 * implicit means that the pivots are scaled by the maximum value in the row
 */

/*================================================================*/
/*LUdecomp
 * inputs: A matrix of coefficients
 *   n - dimension of A
 * outputs: A matrix replaced with L and U diagonal matrices
   (diagonal   values of L = 1)
 *   Rowswaps - vector to keep track of row swaps
 *   Val - indicator of odd/even number of row swaps
 */
int LUdecomp(float **A,int n,int *rowswaps,int *val)
{
  float epsilon,*rowscale, temp;
  float sum;
  float pvt;
  int   ipvt;
  int   i,j,k;
  rowscale = (float*)malloc(sizeof(float)*n);

  epsilon = 0.00000000001;  /* small value to avoid division by zero */
  *val = 1;                 /* even/odd indicator (valence) */

  /* initialize the rowswap vector to indicate no swaps */
  for (i=0; i<n; i++) rowswaps[i] = i;

  /* for each row, find largest (in absolute value sense) element and
     record in rowscale */
  for (i=0; i<n; i++) {
    temp = fabs(A[i][0]);
    for (j=1; j<n; j++)
     if (fabs(A[i][j]) > temp) temp = fabs(A[i][j]);
    if (temp == 0) return(-1); /* got a row of all zeros - can't deal
                                   with that */
    rowscale[i] = 1/temp; /* later we need to divide by largest
                             element */
  }

  /* loop through the columns of A (and U) */
  for (j=0; j<n; j++) {
```

FIGURE B.2

LU decomposition code.

Continued

```
    /* do the rows down to the diagonal – these don't need a
       division       so no swap */
    for (i=0; i<j; i++) {
      sum = A[i][j];
      for (k=0; k<i; k++) sum = sum - A[i][k]*A[k][j];
      A[i][j] = sum;
    }
    /* do the rows from the diagonal down */
    pvt = 0.0;
    ipvt = -1;
    for (i=j; i<n; i++) {
      sum = A[i][j];
      for (k=0; k<j; k++) sum = sum - A[i][k]*A[k][j];
      A[i][j] = sum;
      /* calculate the scaled value for pivoting consideration */
        temp = rowscale[i]*fabs(sum);
        if (temp >= pvt) {ipvt = i; pvt = temp;}
    }

    /* if a better pivot value is found, interchange the rows */
      if (j != ipvt) {
        for (k=0; k<n; k++) {
          temp = A[ipvt][k];
          A[ipvt][k] = A[j][k];
          A[j][k] = temp;
        }
        *val = -(*val);   /* keep track of even/odd number
                              interchanges */
        rowscale[ipvt] = rowscale[j];   /* and record which
                                           was      swapped */
      }
      rowswaps[j] = ipvt;

      if (A[j][j] == 0.0) A[j][j] = epsilon; /* to guard against
                                                divisions by zero */
      /* now the row is ready for division */
      for (i=j+1; i<n; i++) A[i][j] = A[i][j]/A[j][j];
    }
    return 1;
  }

  /* ================================================================ */
  /* LUsubstitute
```

FIGURE B.2—CONT'D

```
 * inputs: A — matrix holding the L and U matrix values as a result
   of    LUdecomp
 *      n — dimension of A
 *      Rowswaps — vector holding a record of the row swaps
        performed        in LUdecomp
 *      b — vector of right-hand values as in Ax = b
 */
void LUsubstitute(float **A,int n,int *rowswaps,float *b)
{
  int   i,j,ib;
  float  sum;
  int   m;

   /* row swap version */
   ib = -1;
   for (i=0; i<n; i++) {
     m = rowswaps[i];
     sum = b[m];
     b[m] = b[i];
     if (ib != -1) {
         for (j=ib; j<i; j++) sum = sum-A[i][j]*b[j];
     }
     else {
        if (sum != 0.0) ib = i;
     }
     b[i] = sum;
   }

   for (i=n-1; i>=0; i--) {
     sum = b[i];
     for (j=i+1; j<n; j++) sum = sum - A[i][j]*b[j];
        b[i] = sum/A[i][i];
     }
   return;
}
```

FIGURE B.2—CONT'D

potential of numerical instability. It is the user's responsibility to inspect the W matrix and zero out values that are small enough to present numerical problems. Once this is done, the matrices can be used to solve the least-squares problem using a back substitution method similar to that used in LU decomposition. The code for SVD is available in various software packages and can be found in *Numerical Recipes* [16].

B.2 Geometric computations

A vector (a one-dimensional list of numbers) is often used to represent a point in space or a direction and magnitude in space (e.g., Figure B.3). A slight complication in terminology results because a direction and magnitude in space is also referred to as a *vector*. As a practical matter, this distinction is usually not important. A vector in space has no position, only magnitude and direction. For geometric computations, a matrix usually represents a transformation (e.g., Figure B.4).

B.2.1 Components of a vector

A vector, A, with coordinates (A_x, A_y, A_z) can be written as a sum of vectors, as shown in Equation B.17, in which i, j, k are unit vectors along the principal axes, x, y, and z, respectively.

$$A = A_x i + A_y j + A_z k \tag{B.17}$$

B.2.2 Length of a vector

The *length* of a vector is computed as in Equation B.18. If a vector is of unit length, then $|A| = 1.0$, and it is said to be *normalized*. Dividing a vector by its length, forming a unit-length vector, is said to be *normalizing* the vector.

$$|A| = \sqrt{A_x^2 + A_y^2 + A_z^2} \tag{B.18}$$

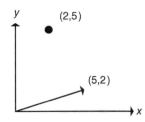

FIGURE B.3

A point and vector in two-space.

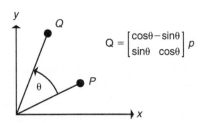

FIGURE B.4

A matrix representing a rotation.

B.2.3 **Dot product of two vectors**

The *dot product*, or *inner product*, of two vectors is computed as in Equation B.19. The computation is commutative (Eq. B.20) and associative (Eq. B.21). The dot product of a vector with itself results in the square of its length (Eq. B.22). The dot product of two vectors, A and B, is equal to the lengths of the vectors times the cosine of the angle between them (Eq. B.23, Figure B.5). As a result, the angle between two vectors can be determined by taking the arccosine of the dot product of the two normalized vectors (or, if they are not normalized, by taking the arccosine of the dot product divided by the lengths of the two vectors). The dot product of two vectors is equal to zero if the vectors are perpendicular to each other, as in Figure B.6 (or if one or both of the vectors are zero vectors). The dot product can also be used to compute the projection of one vector onto another vector (Figure B.7). This is useful in cases in which the coordinates of a vector are needed in an auxiliary coordinate system (Figure B.8).

$$A \bullet B = A_x B_x + A_y B_y + A_z B_z \quad \text{(B.19)}$$

$$A \bullet B = B \bullet A \quad \text{(B.20)}$$

$$(A \bullet B) \bullet C = A \bullet (B \bullet C) \quad \text{(B.21)}$$

$$A \bullet A = |A|^2 \quad \text{(B.22)}$$

$$A \bullet B = |A||B|cos\theta \quad \text{(B.23)}$$

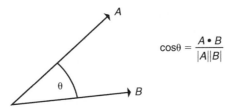

$$cos\theta = \frac{A \bullet B}{|A||B|}$$

FIGURE B.5

Using the dot product to compute the cosine of the angle between two vectors.

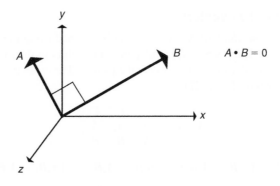

$$A \bullet B = 0$$

FIGURE B.6

The dot product of perpendicular vectors.

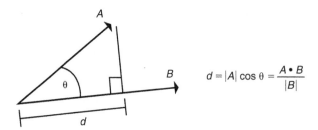

FIGURE B.7

Computing the length of the projection of vector A onto vector B.

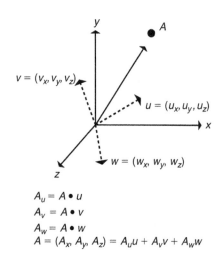

FIGURE B.8

Computing the coordinates of a vector in an auxiliary coordinate system.

B.2.4 Cross-product of two vectors

The *cross-product*, or *outer product*, of two vectors can be defined using the determinant of a 3×3 matrix as shown in Equation B.24, where i, j, and k are unit vectors in the directions of the principal axes. Equation B.25 shows the definition as an explicit equation. The cross-product is not commutative (Eq. B.26), but it is associative (Eq. B.27).

$$A \times B = \begin{vmatrix} i & j & k \\ A_x & A_y & A_z \\ B_x & B_y & B_z \end{vmatrix} \tag{B.24}$$

$$A \times B = (A_y B_z - A_z B_y)i + (A_z B_x - A_x B_z)j + (A_x B_y - A_y B_x)k \tag{B.25}$$

$$A \times B = -(B \times A) = (-A) \times (-B) \tag{B.26}$$

$$A \times (B \times C) = (A \times B) \times C \tag{B.27}$$

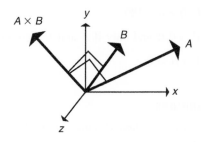

FIGURE B.9

Vector formed by the cross product of two vectors.

The direction of $A \times B$ is perpendicular to both A and B (Figure B.9), and the direction is determined by the right-hand rule (if A and B are in right-hand space). If the thumb of the right hand is put in the direction of the first vector (A) and the index finger is put in the direction of the second vector (B), the cross-product of the two vectors will be in the direction of the middle finger when it is held perpendicular to the first two fingers.

The magnitude of the cross-product is the length of one vector times the length of the other vector times the sine of the angle between them (Eq. B.28). A zero vector will result if the two vectors are colinear or if either vector is a zero vector (Eq. B.29). This relationship is useful for determining the sine of the angle between two vectors (Figure B.10) and for computing the perpendicular distance from a point to a line (Figure B.11).

$$|A \times B| = |A||B|\sin\theta \tag{B.28}$$

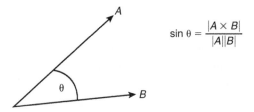

$$\sin\theta = \frac{|A \times B|}{|A||B|}$$

FIGURE B.10

Using the cross-product to compute the sine of the angle between two vectors.

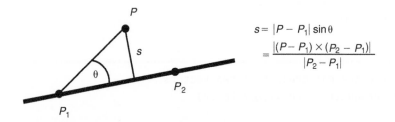

$$s = |P - P_1|\sin\theta$$
$$= \frac{|(P - P_1) \times (P_2 - P_1)|}{|P_2 - P_1|}$$

FIGURE B.11

Computing the perpendicular distance from a point to a line.

where θ is the angle from A to B, $0 < \theta < 180$

$$|A \times B| = 0 \text{ if and only if } A \text{ and } B \text{ are colinear,} \qquad (B.29)$$

i.e., $\sin \theta = \sin 0 = 0$, or $A = 0$ or $B = 0$.

B.2.5 Vector and matrix routines

Simple vector and matrix routines are given here (Figure B.12). These are used in some of the routines presented elsewhere in this appendix.

Vector routines

Some vector routines can be implemented just as easily as in-line code using C's #*define* construction.

B.2.6 Closest point between two lines in three-space

The intersection of two lines in three-space often needs to be calculated. Because of numerical imprecision, the lines rarely, if ever, actually intersect in three-space. As a result, the computation that needs
to be performed is to find the two points, one from each line, at which the lines are closest to each other.

```
/* Vector.c */
typedef   struct xyz_struct {
float     x,y,z;
xyz_td;

}
xyz_td    p;

/* ============================================================ */
/* compute the cross product of two vectors */
xyz_td crossProduct(xyz_td v1,xyz_td v2)
{
  xyz_td    p;

  p.x = v1.y*v2.z - v1.z*v2.y;
  p.y = v1.z*v2.x - v1.x*v2.z;
  p.z = v1.x*v2.y - v1.y*v2.x;
  return    p;
}

/* ============================================================ */
/* compute the dot product of two vectors */
float dotProduct(xyz_td v1,xyz_td v2)
```

FIGURE B.12

Vector routines.

Continued

```
{
  return v1.x*v2.x=v1.y*v2.y+v1.z*v2.z;
}

/* ================================================================= */
/* normalize a vector */
void normalizeVector(xyz_td *v)
{
  float  len;

  len = sqrt(v->x*v->x + v->y*v->y + v->z*v->z);
  v->x /= len;
  v->y /= len;
  v->z /= len;
}

/* ================================================================= */
/* form the vector from the first point to the second */
xyz_td formVector(xyz_td p1, xyz_td p2)
{
  xyz_td      p;

  p.x = p2.x-p1.x;
  p.y = p2.y-p1.y;
  p.z = p2.z-p1.z;
  return p;
}

/* ================================================================= */
/* compute the length of a vector */
float length(xyz_td v)
{
  return sqrt(v.x*v.x+v.y*v.y+v.z*v.z);
}
```

Matrix Routines

```
/* Matrix.c */

/* ================================================================= */
/* Matrix multiplication */
/* C x B = A */
void Matrix4x4MatrixMult (float **C,float **B,float **A)
{
  int  i,j;
```

FIGURE B.12—CONT'D

```
      for (i=0; i<4; i++) {
        for (j=0; j<4; j++) {
          A[i][j] = C[i][0]*B[0][j]+ C[i][1]*B[1][j]+
                  C[i][2]*B[2][j]+ C[i][3]*B[3][j];
        }
      }
    }
    /* ================================================================ */
    /* matrix-vector multiplication */
    /* N = M x V */
    void Matrix4 x 4Vector4Mult (float **M,float *V,float *N)
    {
      N[0] = M[0][0]*V[0]+M[0][1]*V[1]+M[0][2]*V[2]+M[0][3]*V[3];
      N[1] = M[1][0]*V[0]+M[1][1]*V[1]+M[1][2]*V[2]+M[1][3]*V[3];
      N[2] = M[2][0]*V[0]+M[2][1]*V[1]+M[2][2]*V[2]+M[2][3]*V[3];
      N[3] = M[3][0]*V[0]+M[3][1]*V[1]+M[3][2]*V[2]+M[3][3]*V[3];
    }

    /* ================================================================ */
    /* vector-matrix multiplication */
    /* N = V x M */
    void Vector4Matrix4 x 4Mult (float *V,float **M,float *N)
    {
      N[0] = M[0][0]*V[0]+M[1][0]*V[1]+M[2][0]*V[2]+M[3][0]*V[3];
      N[1] = M[0][1]*V[0]+M[1][1]*V[1]+M[2][1]*V[2]+M[3][1]*V[3];
      N[2] = M[0][2]*V[0]+M[1][2]*V[1]+M[2][2]*V[2]+M[3][2]*V[3];
      N[3] = M[0][3]*V[0]+M[1][3]*V[1]+M[2][3]*V[2]+M[3][3]*V[3];
    }

    /* ================================================================ */
    /* compute the inverse of a matrix */
    void ComputeInverse4 x 4(float **M,float **Minv)
    {
      int       rowswaps[4];
      int       val;
      float     b[4];
      int       i,j;
      float**A;

      A = (float **)malloc(sizeof(float *)*4);
      for (i=0; i<4; i++) {
```

FIGURE B.12—CONT'D

```
        A[i] = (float *)malloc(sizeof(float)*4);
      }
      for (i=0; i<4; i++) {
        for (j=0; j<4; j++) {
          A[i][j] = M[i][j];
        }
      }

      LUdecomp(A,4,rowswaps,&val);

      for (i=0; i<4; i++) {
        for (j=0; j<4; j++) b[j] = (i==j) ? 1:0;
        LUsubstitute(A,4,rowswaps,b);
        for (j=0; j<4; j++) Minv[j][i] = b[j];
      }

    }
```

FIGURE B.12—CONT'D

The points P_1 and P_2 at which the lines are closest form a line segment perpendicular to both lines (Figure B.13). They can be represented parametrically as points along the lines, and then the parametric interpolants can be solved for by satisfying the equations that state the requirement for perpendicularity (Eq. B.30).

$$(P_2 - P_1) \bullet V = 0$$

$$(P_2 - P_1) \bullet W = 0$$

$$(B + tW - (A + sV)) \bullet V = 0$$

$$(B + tW - (A + sV)) \bullet W = 0$$

$$t = \frac{-B \bullet V + A \bullet V + sV \bullet V}{W \bullet V}$$

$$t = \frac{-B \bullet W + A \bullet W + sV \bullet W}{W \bullet W}$$

$$\frac{-B \bullet V + A \bullet V + sV \bullet V}{W \bullet V} = \frac{-B \bullet W + A \bullet W + sV \bullet W}{W \bullet W}$$

$$(-B \bullet V + A \bullet V + sV \bullet V)(W \bullet W) = (-B \bullet W + A \bullet W + sV \bullet W) \cdot (W \bullet V)$$

$$s = \frac{(A \bullet W - B \bullet W)(W \bullet V) + (B \bullet V - A \bullet V) \cdot (W \bullet W)}{(V \bullet V)(W \bullet W) - (V \bullet W)^2}$$

$$t = \frac{(B \bullet V - A \bullet V)(W \bullet V) + (A \bullet W - B \bullet W)(V \bullet V)}{(W \bullet W)(V \bullet V) - (V \bullet W)^2}$$

$$(B.30)$$

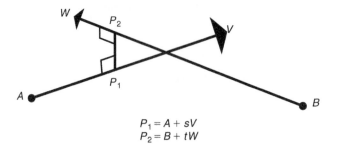

$$P_1 = A + sV$$
$$P_2 = B + tW$$

FIGURE B.13

Two lines are closest to each other at points P_1 and P_2.

B.2.7 Area calculations

Area of a triangle

The area of a triangle consisting of vertices V_1, V_2, V_3 is one-half times the length of one edge times the perpendicular distance from that edge to the other vertex. The perpendicular distance from the edge to a vertex can be computed using the cross-product (see Figure B.14). For triangles in two dimensions, the z-coordinates are essentially considered zero and the cross-product computation is simplified accordingly (only the z-coordinate of the cross-product is non-zero).

The signed area of the triangle is required for computing the area of a polygon. In the two-dimensional case, this is done simply by not taking the absolute value of the z-coordinate of the cross-product. In the three-dimensional case, a vector normal to the polygon can be used to indicate the positive direction. The direction of the vector produced by the cross-product can be compared to the normal (using the dot product) to determine whether it is in the positive or negative direction. The length of the cross-product vector can then be computed and the appropriate sign applied to it.

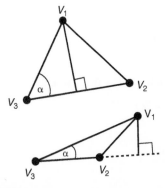

$$\text{Area}(V_1, V_2, V_3) = \tfrac{1}{2}|V_2 - V_3|\,|V_1 - V_3|\sin\alpha$$
$$= \tfrac{1}{2}|V_2 - V_3|\frac{|(V_1 - V_3) \times (V_2 - V_3)|}{|V_2 - V_3|}$$
$$= \tfrac{1}{2}|(V_1 - V_3) \times (V_2 - V_3)|$$

FIGURE B.14

Area of a triangle.

Area of a polygon

The area of a polygon can be computed as a sum of the signed areas of simple elements. In the two-dimensional case, the signed area under each edge of the polygon can be summed to form the area (Figure B.15). The area under an edge is the average height of the edge times its width (Eq. B.31, where subscripts are computed modulo $n + 1$).

$$
\begin{aligned}
\text{Area} &= \sum_{i=1}^{n} \frac{(y_i + y_{i+1})}{2}(x_{i+1} - x_i) \\
&= \frac{1}{2}\sum_{i=1}^{n}(y_i x_{i+1} - y_{i+1} x_i)
\end{aligned}
\tag{B.31}
$$

The area of a polygon can also be computed by using each edge of the polygon to construct a triangle with the origin (Figure B.16). The signed area of the triangle must be used so that edges directed

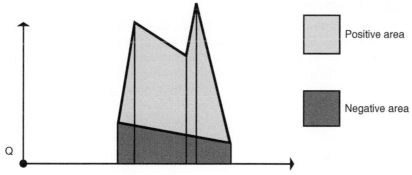

Positive area

Negative area

FIGURE B.15

Computing the area of a polygon.

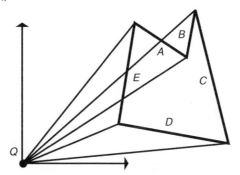

Area of polygon (A, B, C, D, E) = Area of Triangle (Q, A)
+ Area of Triangle(Q, B)
+ Area of Triangle(Q, C)
+ Area of Triangle(Q, D)
+ Area of Triangle(Q, E)

FIGURE B.16

The area of a two-dimensional polygon: the edges are labeled with letters, triangles are constructed from each edge to the origin, and the areas of the triangles are signed according to the direction of the edge with respect to the origin. Area of polygon (A, B, C, D, E) = Area of Triangle (Q, A)

clockwise with respect to the origin cancel out edges directed counterclockwise with respect to the origin. Although this is more computationally expensive than summing the areas under the edges, it suggests a way to compute the area of a polygon in three-space. In the three-dimensional case, one of the vertices of the polygon can be used to construct a triangle with each edge, and the three-dimensional version of the vector equations of Figure B.14 can be used.

B.2.8 The cosine rule

The cosine rule states the relationship between the lengths of the edges of a triangle and the cosine of an interior angle (Figure B.17). It is useful for determining the interior angle of a triangle when the locations of the vertices are known.

B.2.9 Barycentric coordinates

Barycentric coordinates are the coordinates of a point in terms of weights associated with other points. Most commonly used are the barycentric coordinates of a point with respect to vertices of a triangle. The barycentric coordinates (u_1, u_2, u_3) of a point, P, with respect to a triangle with vertices V_1, V_2, V_3 are shown in Figure B.18. Notice that for a point inside the triangle, the coordinates always sum to one. This can be extended easily to any convex polygon by a direct generalization of the equations. However, it cannot be extended to concave polygons.

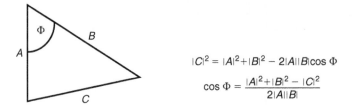

$$|C|^2 = |A|^2 + |B|^2 - 2|A||B|\cos \Phi$$

$$\cos \Phi = \frac{|A|^2 + |B|^2 - |C|^2}{2|A||B|}$$

FIGURE B.17

The cosine rule.

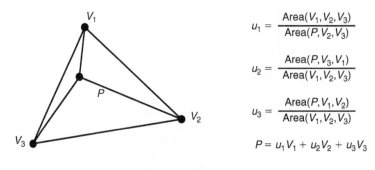

$$u_1 = \frac{\text{Area}(V_1, V_2, V_3)}{\text{Area}(P, V_2, V_3)}$$

$$u_2 = \frac{\text{Area}(P, V_3, V_1)}{\text{Area}(V_1, V_2, V_3)}$$

$$u_3 = \frac{\text{Area}(P, V_1, V_2)}{\text{Area}(V_1, V_2, V_3)}$$

$$P = u_1 V_1 + u_2 V_2 + u_3 V_3$$

FIGURE B.18

The barycentric coordinates of a point with respect to vertices of a triangle.

B.2.10 **Computing bounding shapes**

Bounding volumes are useful as approximate extents of more complex objects. Often, simpler tests can be used to determine the general position of an object by using bounding volumes, thus saving computation. In computer animation, the most obvious example occurs in testing for object collisions. If the bounding volumes of objects are not overlapping, then the objects themselves must not be penetrating each other. Planar polyhedra are considered here because the bounding volumes can be determined by inspecting the vertices of the polyhedra. Nonplanar objects require more sophisticated techniques. *Axis-aligned bounding boxes* and bounding spheres are relatively easy to calculate but can be poor approximations of the object's shape. Slabs and *oriented bounding boxes* (OBBs) can provide a much better fit. OBBs are rectangular bounding boxes at an arbitrary orientation [7]. For collision detection, bounding shapes are often hierarchically organized into tighter and tighter approximations of the object's space. A *convex hull*, the smallest convex shape bounding the object, provides an even tighter approximation but requires more computation.

Bounding boxes

Bounding box typically refers to a boundary cuboid (or rectangular solid) whose sides are aligned with the principal axes. A bounding box for a collection of points is easily computed by searching for minimum and maximum values for the x-, y-, and z-coordinates. A point is inside the bounding box if its coordinates are between min/max values for each of the three coordinate pairs. While the bounding box may be a good fit for some objects, it may not be a good fit for others (Figure B.19). How well the bounding box approximates the shape of the object is rotationally variant, as shown in Figures B.19b and B.19c.

Bounding slabs

Bounding slabs are a generalization of bounding boxes. A pair of arbitrarily oriented planes are used to bound the object. The orientation of the pair of planes is specified by a user-supplied normal vector.

The normal defines a family of planes that vary according to perpendicular distance to the origin. The planar equation $a x + b y + c z = d$, (a, b, c) represents a vector normal to the plane. If this vector has unit length, then d is the perpendicular distance to the plane. If the length of (a, b, c) is not one, then d is the perpendicular distance scaled by the vector's length. Notice that d is equal to the dot product of the vector (a, b, c) and a point on the plane.

Given a user-supplied normal vector, the user computes the dot product of that vector and each vertex of the object and records the minimum and maximum values (Eq. B.32). The normal vector

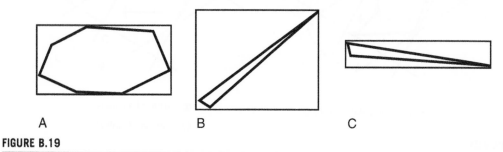

A   B   C

FIGURE B.19

Sample objects and their bounding boxes (in two-dimensional).

and these min/max values define the bounding slab. See Figure B.20 for a two-dimensional diagram illustrating the idea. Multiple slabs can be used to form an arbitrarily tight bounding volume of the convex hull of the polyhedron. A point is inside this bounding volume if the result of the dot product of it and the normal vector is between the corresponding min/max values for each slab (see Figure B.21 for a two-dimensional example).

$$
\begin{aligned}
p &= (x, y, z) & &\text{vertex} \\
N &= (a, b, c) & &\text{normal} \\
P g N &= d & &\text{computing the planar equation constant}
\end{aligned}
\tag{B.32}
$$

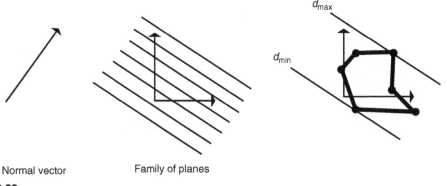

Normal vector   Family of planes

FIGURE B.20

Computing a boundary slab for a polyhedron.

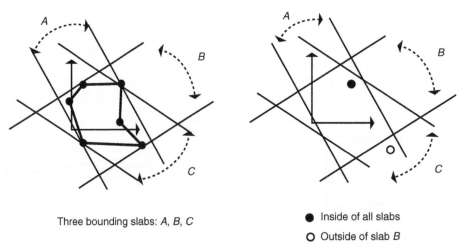

Three bounding slabs: *A, B, C*   ● Inside of all slabs
                                         ○ Outside of slab *B*

FIGURE B.21

Multiple bounding slabs.

Bounding spheres

Computing the optimal bounding sphere for a set of points can be expensive. However, more tractable approximate methods exist. A quick and fairly accurate method of computing an approximate bounding sphere for a collection of points is to make an initial guess at the bounding sphere and then incrementally enlarge the sphere as necessary by inspecting each point in the set. The description here follows Ritter [17].

The first step is to loop through the points and record the minimum and maximum points in each of the three principal directions. The second step is to use the maximally separated pair of points from the three pairs of recorded points and create an initial approximation of the bounding sphere. The third step is, for each point in the original set, to adjust the bounding sphere as necessary to include the point. Once all the points have been processed, a near-optimal bounding sphere has been computed (Figure B.22). This method is fast and it is easy to implement (Figure B.23).

Convex hull

The convex hull of a set of points is the smallest convex polyhedron that contains all the points. A simple algorithm for computing the complex hull is given here, although more efficient techniques exist. Since this is usually a one-time code for an object (unless the object is deforming), this is a case where efficiency can be traded for ease of implementation. Refer to Figure B.24.

1. Find a point on the convex hull by finding the point with the largest y-coordinate. Refer to the point found in this step as P_1.
2. Construct an edge on the convex hull by using the following method. Find the point that, when connected with P_1, makes the smallest angle with the horizontal plane passing through P_1. Use L to refer to the line from P_1 to the point. Finding the smallest sine of the angle is equivalent to finding the smallest angle. The sine of the angle between L and the horizontal plane passing through P_1 is equal to the cosine of the angle between L and the vector $(0, -1, 0)$. The dot product of these two vectors is used to compute the cosine of the angle between them. Refer to the point found in this step as P_2, and refer to the line from P_1 to P_2 as L.
3. Construct a triangle on the convex hull by the following method. First, construct the plane defined by L and a horizontal line perpendicular to L at P_1. The horizontal line is constructed according to $(L \times (0, -1, 0))$. All of the points are below this plane. Find the point that, when connected with P_1, makes the smallest angle with this plane. Use K to refer to the line from P_1 to the point. The sine of the angle between K and the plane is equal to the cosine of the angle between K and the

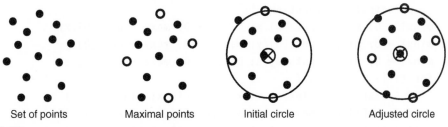

| Set of points | Maximal points | Initial circle | Adjusted circle |

FIGURE B.22

Computing a bounding circle for a set of points.

downward-pointing normal vector of the plane. This normal can be computed as $N = (L \times (0, -1, 0)) \times L$ (in right-hand space). Refer to the point found in this step as P_3. The triangle on the convex hull is defined by these three points. A consistent ordering of the points should be used so that they, for example, compute an outward-pointing normal vector ($N = (P_3 - P_1) \times (P_2 - P_1)$, $N_y > 0.0$) in right-hand space. Initialize the list of convex hull triangles with this triangle and its outward-pointing normal vector. Mark each of its three edges as *unmatched* to indicate that the triangle that shares the edge has not been found yet.

4. Search the current list of convex hull triangles and find an *unmatched* edge. Construct the triangle of the convex hull that shares this edge by the following method. Find the point that, when connected by a line from a point on the edge, makes the smallest angle with the plane of the triangle while creating a dihedral angle (interior angle between two faces measured at a shared edge) greater than 90 degrees. The dihedral angle can be computed using the angle between the normals of the two triangles. When the point has been found, add the triangle defined by this point and the marked edge to the list of convex hull triangles and the unmarked edge. The rest of the unmarked edges in the list

```
/*
** Bounding Sphere Computation
*/
void boundingSphere(xyz_td *pnts,int n, xyz_td *cntr, float *radius)
{
    int    i,minxi,maxxi,minyi,maxyi,minzi,maxzi,p1i,p2i;
    float  minx,maxx,miny,maxy,minz,maxz;
    float  diam2,diam2x,diam2y,diam2z,rad,rad2;
    float  dx,dy,dz;
    float  cntrx,cntry,cntrz;
    float  delta;
    float  dist,dist2;
    float  newrad,newrad2;
    float  newcntrx,newcntry,newcntrz;

    /* step one: find minimal and maximal points in each of 3
        principal    directions */
    minxi = 0; minx = pnts[0].x; maxxi = 0; maxx = pnts[0].x;
    minyi = 0; miny = pnts[0].y; maxyi = 0; maxy = pnts[0].y;
    minzi = 0; minz = pnts[0].z; maxzi = 0; maxz = pnts[0].z;
    for (i=1; i<n; i++) {
        if (pnts[i].x < minx) { minx = pnts[i].x; minxi = i; }
        if (pnts[i].x > maxx) { maxx = pnts[i].x; maxxi = i; }
        if (pnts[i].y < miny) { miny = pnts[i].y; minyi = i; }
```

FIGURE B.23

Bounding sphere code.

Continued

```
     if (pnts[i].y > maxy) { maxy = pnts[i].y; maxyi = i; }
     if (pnts[i].z < minz) { minz = pnts[i].z; minzi = i; }
     if (pnts[i].z > maxz) { maxz = pnts[i].z; maxzi = i; }
   }

   /* step two: find maximally separated points from the 3 pairs; use
      to initialize sphere */
   /* find maximally separated points by comparing the distance squared
      between points */
   dx = pnts[minxi].x - pnts[maxxi].x;
   dy = pnts[minxi].y - pnts[maxxi].y;
   dz = pnts[minxi].z - pnts[maxxi].z;
   diam2x = dx*dx + dy*dy + dz*dz;
   dx = pnts[minyi].x - pnts[maxyi].x;
   dy = pnts[minyi].y - pnts[maxyi].y;
   dz = pnts[minyi].z - pnts[maxyi].z;
   diam2y = dx*dx + dy*dy + dz*dz;
   dx = pnts[minzi].x - pnts[maxzi].x;
   dy = pnts[minzi].y - pnts[maxzi].y;
   dz = pnts[minzi].z - pnts[maxzi].z;
   diam2z = dx*dx + dy*dy + dz*dz;
   diam2 = diam2x; p1i = minxi; p2i = maxxi;
   if (diam2y>diam2) { diam2 = diam2y; p1i=minyi; p2i=maxyi; }
   if (diam2z>diam2) { diam2 = diam2z; p1i=minzi; p2i=maxzi;}
   /* center of initial sphere is average of two points */
   cntrx = (pnts[p1i].x+pnts[p2i].x)/2;
   cntry = (pnts[p1i].y+pnts[p2i].y)/2;
   cntrz = (pnts[p1i].z+pnts[p2i].z)/2;
   /* calculate radius and radius squared of initial sphere - from
      diameter squared*/
   rad2 = diam2/4;
   rad = sqrt(rad2);
   printf("maximally separated pair: (%f,%f,%f):(%f,%f,%f),%f\n",
     pnts[p1i].x,pnts[p1i].y,pnts[p1i].z,
     pnts[p2i].x,pnts[p2i].y,pnts[p2i].z,diam2);
   printf("initial center: (%f,%f,%f)\n",cntrx,cntry,cntrz);
   printf("initial diam2: %f\n",diam2);
   printf("initial radius, radius2 = %f,%f\n",rad,rad2);

   /* third step: now step through the set of points and adjust
      bounding     sphere as necessary */
   for (i=0; i<n; i++) {
     dx = pnts[i].x - cntrx;
```

FIGURE B.23—CONT'D

```
        dy = pnts[i].y - cntry;
        dz = pnts[i].z - cntrz;
        dist2 = dx*dx + dy*dy + dz*dz;      /* distance squared of old
                                               center to pnt */
        if (dist2 > rad2) {                 /* need to update sphere if this
                                               point is outside old radius*/
            dist = sqrt(dist2);
            /* new radius is average of current radius and distance from
               center to pnt */
            newrad = (rad + dist)/2;
            newrad2 = newrad*newrad;
            printf("new radius = %f\n",newrad);
            delta = dist - newrad;        /* distance from old center to new
                                             center */
            /* delta/dist and rad/dist are weights of pnt and old center to
               compute new center */
            newcntrx = (newrad*cntrx+delta*pnts[i].x)/dist;
            newcntry = (newrad*cntry+delta*pnts[i].y)/dist;
            newcntrz = (newrad*cntrz+delta*pnts[i].z)/dist;

            /* test to see if new radius and center contain the point */
            /* this test should only fail by an epsilon due to numeric
               imprecision */
            dx = pnts[i].x - newcntrx;
            dy = pnts[i].y - newcntry;
            dz = pnts[i].z - newcntrz;
            dist2 = dx*dx + dy*dy + dz*dz;
            if (dist2 > newrad2) {
                printf("ERROR by %lf\n",((double)(dist2))-newrad2);
                printf("  center - radius: (%f,%f,%f) -  %f\n",cntrx,cntry,
                        cntrz,rad);
                printf(" New center - radius: (%f,%f,%f) - %f\n",
                  newcntrx,newcntry,newcntrz,newrad);

            }
            cntrx = newcntrx;
            cntry = newcntry;
            cntrz = newcntrz;
            rad = newrad;
            rad2 = rad*rad;

        }
    }
```

FIGURE B.23—CONT'D

```
*radius = rad;
cntr->x = cntrx;
cntr->y = cntry;
cntr->z = cntrz;
return;
}
```

FIGURE B.23—CONT'D

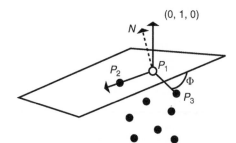

Step 1: Find the highest point

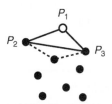

Step 2: Find an edge of the convex hull

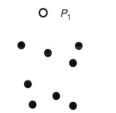

Step 3: Find an initial triangle

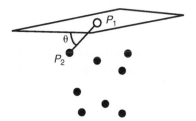

Step 4: Construct each triangle of the convex hull using one or two existing edges from previously formed convex hull triangles

FIGURE B.24

Computing the convex hull.

must be searched to see if the two other edges of the new triangle already occur in the list. If either of the new edges does not have a match in the list, then it should be marked as unmatched. Otherwise, mark the edge as *matched*. Now, go back through the list of convex hull triangles and look for another *unmatched* edge and repeat the procedure to construct a new triangle that shares that edge. When there are no more *unmatched* edges in the list of convex hull triangles, the hull has been constructed.

Step 4 in the previous algorithm does not handle the case in which there are more than three coplanar vertices. To handle these cases, instead of forming a triangle, form a convex hull polygon for the coplanar points. This is done similarly in two and three dimensions. First, collect all of the coplanar points (all of the points within some epsilon of being coplanar) into a set. Second, initialize the current edge to be the unmatched edge of step 4 and add the first point of the current edge to the set of coplanar points. Third, iteratively find the point in the set of coplanar points that makes the smallest angle with the current edge as measured from the first point of the edge to the second point of the edge to the candidate point. When the point is found, remove it from the set of coplanar points and make the new current edge the edge from the second point of the old current edge to the newly found point. Continue iterating until the first point of the original unmatched edge is found. This will complete the convex hull polygon, which is then added to the list of convex hull polygons; its newly formed edges are processed as in step 4 (see Figure B.25).

```
/* ConvexHull.c
 * This code uses a brute force algorithm to construct the convex hull
 * and does not handle more than three coplanar points
 */

#include "Vector.h"

typedef struct conHullTri_struct {
    int    pi[3];
    int    matched[3];
    xyz_td  normal;
    struct conHullTri_struct *next;
} conHullTri_td;

conHullTri_td        *chtList;

/* ================================================================ */
/* CONVEX HULL */
int ConvexHull(xyz_td *pntList,int num,int **triangleList,int *numTriangles)
{
    int       i;
    int       p1i,p2i,p3i,pi;
    xyz_td    yaxis;
    xyz_td    v,n,nn,nnn,v1,v2;
    float     t,t1,t2;
    int       count;
    conHullTri_td *chtPtr,*chtPtrTail,*chtPtrNew,*chtPtrA;
    int          done;
    int          *triList;
    int       dummy,notError;
```

FIGURE B.25

Convex hull code.

Continued

```
yaxis.x = 0; yaxis.y = 1; yaxis.z = 0;

/* find the highest point */
p1i = 0; t = pntList[0].y;
for (i=1; i<num; i++) {
    if (pntList[i].y > t) { p1i=i; t=pntList[i].y;}
}

/* find point that makes minimum angle with horizontal plane */
p2i = (p1i==0) ? 1:0;
v = formVector(pntList[p1i],pntList[p2i]);
normalizeVector(&v);
t = v.y;
if (t>0.0) {
    printf(" ERROR - found higher point\n");
    scanf("%d",&dummy);
    return 1;
}
for (i=p2i+1; i<num; i++) {
    if (i!=p1i) {
        v = formVector(pntList[p1i],pntList[i]);
        normalizeVector(&v);
        t1 = v.y;
        if (t1 > t) {p2i = i; t = t1;}
    }
}

/* find point that makes triangle with minimum angle with horizontal
   plane through edge */
v1 = formVector(pntList[p1i],pntList[p2i]);

if ((p1i!=0) && (p2i!=0)) p3i=0;
else if ((p1i!=1) && (p2i!=1)) p3i=1;
else p3i=2;
v2 = formVector(pntList[p2i],pntList[i]);
n = crossProduct(v2,v1);
normalizeVector(&n);
if (n.y < 0) {
    n.x = -n.x; n.y = -n.y; n.z = -n.z;
}
for (i=p3i+1; i<num; i++) {
 if ((i!=p1i)&&(i!=p2i)) {
     v = formVector(pntList[p2i],pntList[i]);
```

FIGURE B.25—CONT'D

```
      nn = crossProduct(v1,v);
      normalizeVector(&nn);
      if (nn.y < 0) { nn.x = -nn.x; nn.y = -nn.y; nn.z = -nn.z; }
      if (nn.y>n.y) {
      p3i=i;
      n.x = nn.x; n.y = nn.y; n.z = n.z;
    }
  }
}

/* compute outward-pointing normal vector in right-hand space for
   clockwise triangle */
/* recalculate the normal vector */
v1 = formVector(pntList[p1i],pntList[p2i]);
v2 = formVector(pntList[p2i],pntList[p3i]);
n = crossProduct(v2,v1);
normalizeVector(&n);
if (n.y<0) {
   n.x = -n.x; n.y = -n.y; n.z = -n.z;
   pi = p1i; p1i = p2i; p2i = pi;
}

/* make a convex hull entry */
count = 1;
chtPtr = (conHullTri_td *)malloc(sizeof(conHullTri_td));
if (chtPtr == NULL) {
   printf("unsuccessful memory allocation 1\n");
   scanf("%d",&dummy);
   return 1;
}
chtPtr->pi[0] = p1i;
chtPtr->pi[1] = p2i;
chtPtr->pi[2] = p3i;
chtPtr->matched[0] = FALSE;
chtPtr->matched[1] = FALSE;
chtPtr->matched[2] = FALSE;
chtPtr->normal = n;

/* initialize the convex hull triangle list with the triangle */
chtList = chtPtr;
chtPtr->next = NULL;
chtPtrTail = chtPtr;
```

FIGURE B.25—CONT'D

```
/* check and make sure all vertices are 'underneath' the initial
   triangle */
for (i=0; i<num; i++) {
  if ((i!=chtPtr->pi[0]) &&
      (i!=chtPtr->pi[1]) &&
      (i!=chtPtr->pi[2]) ) {
      v = formVector(pntList[chtPtr->pi[0]],pntList[i]);
      t = dotProduct(v,n);
   if (t>0.0) {
      /* ERROR - point above initial triangle */
      printf("ERROR - found a point above initial triangle  (%d)\n",i);
      return 1;
   }
 }
}

/* now loop through the convex hull triangle list and process
   unmatched edges */
done = FALSE;
chtPtr = chtList;
while (chtPtr!=NULL) {
 /* look for first unmatched edge */
 if ((!(chtPtr->matched[0])) ||
     (!(chtPtr->matched[1])) ||
     (!(chtPtr->matched[2])) ) {

    /* set it now as matched, and record 3 points with unmatched as
       first two */
    if (!(chtPtr->matched[0])) {
       p1i=chtPtr->pi[0]; p2i=chtPtr->pi[1]; p3i=chtPtr->pi[2];
       chtPtr->matched[0] = TRUE;
    }
    else if (!(chtPtr->matched[1])) {
     p1i=chtPtr->pi[1]; p2i=chtPtr->pi[2]; p3i=chtPtr->pi[0];
     chtPtr->matched[1] = TRUE;
    }
    else if (!(chtPtr->matched[2])) {
     p1i=chtPtr->pi[2]; p2i=chtPtr->pi[0]; p3i=chtPtr->pi[1];
     chtPtr->matched[2] = TRUE;
    }

    /* get info of triangle of unmatched edge */
    n.x = chtPtr->normal.x;
```

FIGURE B.25—CONT'D

```
    n.y = chtPtr->normal.y;
    n.z = chtPtr->normal.z;
    v1=formVector(pntList[p2i],pntList[p1i]);

    /* find new vertex which, with unmatched edge, makes    triangle */
    /* whose normal is closest to normal of triangle of unmatched
        edge */
    pi = -1;
    for (i=0; i<num; i++) {
      if ((i!=p1i)&&(i!=p2i)&&(i!=p3i)) {
         v=formVector(pntList[p1i],pntList[i]);
    /* test to see if point is above triangle */
    t1 = dotProduct(v,n);
    if (t1>0) {
       /* ERROR - point above initial triangle */
       printf("ERROR - found a point above initial triangle   (%d)\n",i);
       return 1;
    }
    /* compute normal of proposed new triangle */
    nn = crossProduct(v,v1);
    normalizeVector(&nn);
    /* test for concave corner */
    nnn = crossProduct(n,nn);
    t2 = dotProduct(nnn,v1);
    if (t2<0.0) {
       printf("ERROR - concave corner found\n");
       return 1;
    }
    /* compute angle made by faces (=angle made by normals)    */
    t1 = dotProduct(n,nn);
    /* printf(" %d: dot product of normals: %f\n",i,t1); */
    /* printf(" normal for comparison: %f %f    %f\n",n.x,n.y,n.z); */
    /* printf(" normal: %f %f %f\n",nn.x,nn.y,nn.z); */
    /* save smallest angle (largest cosine) */
    if (pi==-1) {pi=i; t=t1;}
    else if (t1>t) {pi=i; t=t1;}
   }
}

/* check and make sure all vertices are 'underneath' this    triangle */
v=formVector(pntList[p1i],pntList[pi]);
nn = crossProduct(v,v1);
normalizeVector(&nn);
```

FIGURE B.25—CONT'D

```
        for (i=0; i<num; i++) {
         if ((i!=p2i) &&
             (i!=p1i) &&
             (i!=pi) ) {
             v = formVector(pntList[p1i],pntList[i]);
             t = dotProduct(v,nn);
             if (t>0.0) {
              /* ERROR - point above new triangle */
              printf("ERROR - found a point above new triangle    (%d)\n",i);
              return 1;
          }
        }
      }

 /* search for p2i-pi or pi-pi1 already in database - error     condition */
  chtPtrA = chtList; notError = TRUE;
  while ((chtPtrA!=NULL)&&notError) {
   if((chtPtrA->pi[0]==p1i)&&(chtPtrA->pi[1]==pi)) notError =    FALSE;
   else if((chtPtrA->pi[1]==p1i)&&(chtPtrA->pi[2]==pi))  notError = FALSE;
   else if((chtPtrA->pi[2]==p1i)&&(chtPtrA->pi[0]==pi))  notError = FALSE;
   else if((chtPtrA->pi[0]==pi)&&(chtPtrA->pi[1]==p2i))  notError = FALSE;
   else if((chtPtrA->pi[1]==pi)&&(chtPtrA->pi[2]==p2i))  notError = FALSE;
   else if((chtPtrA->pi[2]==pi)&&(chtPtrA->pi[0]==p2i))  notError = FALSE;
   else chtPtrA = chtPtrA->next;
   /* end while */
  if (!notError) {
    printf("ERROR - duplicating edge      (%d,%d,%d)\n",p1i,p2i,pi);
    return 1;
  }

  /* add p1i, p2i, pi */
  count++;
  chtPtrNew = (conHullTri_td *)malloc(sizeof(conHullTri_td));
  if (chtPtrNew == NULL) {
    printf(" unsuccessful memory allocation 2\n");
    return 1;
  }

  chtPtrTail->next = chtPtrNew;
  chtPtrNew->pi[0] = p2i;
  chtPtrNew->pi[1] = p1i;
  chtPtrNew->pi[2] = pi;
```

FIGURE B.25—CONT'D

```
chtPtrNew->matched[0] = TRUE;
chtPtrNew->matched[1] = FALSE;
chtPtrNew->matched[2] = FALSE;
chtPtrNew->normal.x = nn.x;
chtPtrNew->normal.y = nn.y;
chtPtrNew->normal.z = nn.z;
chtPtrNew->next = NULL;
chtPtrTail = chtPtrNew;

/* search for p2i-pi or pi-p1i already in database in reverse   order */
chtPtrA = chtList;
while (chtPtrA!=NULL) {
 if (!chtPtrA->matched[0]&&(chtPtrA-  >pi[0]==pi)&&(chtPtrA->
    pi[1]==p1i)) {
    chtPtrA->matched[0] = TRUE;
    chtPtrNew->matched[1] = TRUE;
 }
 else if (!chtPtrA->matched[1]&&(chtPtrA-  > pi[1]==pi)&&
 (chtPtrA->pi[2]==p1i)) {
   chtPtrA->matched[1] = TRUE;
   chtPtrNew->matched[1] = TRUE;
 }
 else if (!chtPtrA->matched[2]&&(chtPtrA-  > pi[2]==pi)&&
 (chtPtrA->pi[0]==p1i)) {
   chtPtrA->matched[2] = TRUE;
   chtPtrNew->matched[1] = TRUE;
  }
  else if (!chtPtrA->matched[0]&&(chtPtrA-  > pi[0]==p2i)&&
  (chtPtrA->pi[1]==pi)) {
   chtPtrA->matched[0] = TRUE;
   chtPtrNew->matched[2] = TRUE;
  }
  else if (!chtPtrA->matched[1]&&(chtPtrA-  > pi[1]==p2i)&&
  (chtPtrA->pi[2]==pi)) {
   chtPtrA->matched[1] = TRUE;
   chtPtrNew->matched[2] = TRUE;
  }
  else if (!chtPtrA->matched[2]&&(chtPtrA-  >pi[2]==p2i)&&
  (chtPtrA->pi[0]==pi)) {
   chtPtrA->matched[2] = TRUE;
```

FIGURE B.25—CONT'D

```
        vchtPtrNew->matched[2] = TRUE;
      }
      else {
        chtPtrA = chtPtrA->next;
        }
    } /* end while */

  } /* end endif */
  else {
    chtPtr = chtPtr->next;
   }
  }

  triList = (int *)malloc(sizeof(int)*count*3);
  chtPtr = chtList;
  for (i=0; i<count; i++) {
   if (chtPtr==NULL) {
      printf("ERROR: count %d doesn't match data structure\n",count);
      return 1;
     }
     triList[3*i] = chtPtr->pi[0];
     triList[3*i+1] = chtPtr->pi[1];
     triList[3*i+2] = chtPtr->pi[2];
     chtPtr=chtPtr->next;
    }
   *numTriangles = count;
   *triangleList = triList;
  return 0;
}

void printCHTlist(conHullTri_td *chtPtr)
{
printf("CHT list\n");
while (chtPtr!=NULL) {
 printf("%d:%d:%d ; %d:%d:%d ; %f,%f,%f\n",
   chtPtr->pi[0],chtPtr->pi[1],chtPtr->pi[2],
   chtPtr->matched[0],chtPtr->matched[1],chtPtr->matched[2],
   chtPtr->normal.x,chtPtr->normal.y,chtPtr->normal.z);
   chtPtr = chtPtr->next;
  }
 }
```

FIGURE B.25—CONT'D

B.3 Transformations

B.3.1 Transforming a point using vector-matrix multiplication

Vector-matrix multiplication is usually how the transformation of a point is represented. Because a vector is just an $N \times 1$ matrix, vector-matrix multiplication is actually a special case of matrix-matrix multiplication. Vector-matrix multiplication is usually performed by premultiplying a column vector by a matrix. This is equivalent to post-multiplying a row vector by the transpose of that same matrix. Both notations are encountered in the graphics literature, but use of the column vector is more common. The examples in Equations B.33 and B.34 use a 4×4 matrix and a point in three-space using homogeneous coordinates, consistent with what is typically encountered in graphics applications.

$$
\begin{bmatrix} Q_x \\ Q_y \\ Q_z \\ Q_w \end{bmatrix} = Q = MP = \begin{bmatrix} M_{11} & M_{12} & M_{13} & M_{14} \\ M_{21} & M_{22} & M_{23} & M_{24} \\ M_{31} & M_{32} & M_{33} & M_{34} \\ M_{41} & M_{42} & M_{43} & M_{44} \end{bmatrix} \begin{bmatrix} P_x \\ P_y \\ P_z \\ 1 \end{bmatrix} \tag{B.33}
$$

$$
\begin{bmatrix} Q_x & Q_y & Q_z & Q_w \end{bmatrix} = Q^T = P^T M^T
$$

$$
= \begin{bmatrix} P_x & P_y & P_z & 1 \end{bmatrix} \begin{bmatrix} M_{11} & M_{21} & M_{31} & M_{41} \\ M_{12} & M_{22} & M_{32} & M_{42} \\ M_{13} & M_{23} & M_{33} & M_{43} \\ M_{14} & M_{24} & M_{34} & M_{44} \end{bmatrix} \tag{B.34}
$$

B.3.2 Transforming a vector using vector-matrix multiplication

In addition to transforming points, it is also often useful to transform vectors, such as normal vectors, from one space to another. However, the computations used to transform a vector are different from those used to transform a point. Vectors have direction and magnitude but do not have a position in space. Thus, for example, a pure translation has no effect on a vector. If the transformation of one space to another is a pure rotation and uniform scale, then those transformations can be applied directly to the vector. However, it is not so obvious how to apply transformations that incorporate nonuniform scale.

The transformation of a vector can be demonstrated by considering a point, P, which satisfies a planar equation (Eq. B.35). Note that (a, b, c) represents a vector normal to the plane. Showing how to transform a planar equation will, in effect, show how to transform a vector. The point is transformed by a matrix, M (Eq. B.36). Because the transformations of rotation, translation, and scale preserve planarity, the transformed point, P', will satisfy some new planar equation, N', in the transformed space (Eq. B.37). Substituting the definition of the transformed point, Equation B.36, into Equation B.37 produces Equation B.38. If the transformed planar equation is equal to the original normal postmultiplied by the inverse of the transformation matrix (Eq. B.39), then Equation B.37 is satisfied, as shown by Equation B.40. The transformed normal vector is, therefore, (a', b', c').

$$
ax + by + cz + d = 0
$$

$$
\begin{bmatrix} a & b & c & d \end{bmatrix} \begin{bmatrix} x \\ y \\ z \\ 1 \end{bmatrix} = 0 \tag{B.35}
$$

$$N^T P = 0$$

$$P' = MP \tag{B.36}$$

$$N'^T P' = 0 \tag{B.37}$$

$$N'^T MP = 0 \tag{B.38}$$

$$N'^T = N^T M^{-1} \tag{B.39}$$

$$N^T M^{-1} MP = N^T P = 0 \tag{B.40}$$

In order to transform a vector (a, b, c), treat it as a normal vector for a plane passing through the origin $[a, b, c, 0]$ and post-multiply it by the inverse of the transformation matrix (Eq. B.39). If it is desirable to keep all vectors as column vectors, then Equation B.41 can be used.

$$N' = \left(N'^T \right)^T = \left(N^T M^{-1} \right)^T = \left(M^{-1} \right)^T N \tag{B.41}$$

B.3.3 Axis-angle rotations

Given an axis of rotation $A = [\, a_x \; a_y \; a_z]$ of unit length and an angle θ to rotate by (Figure B.26), the rotation matrix M can be formed by Equation B.42. This is a more direct way to rotate a point around an axis, as opposed to implementing the rotation as a series of rotations about the global axes.

$$\hat{A} = \begin{bmatrix} a_x a_x & a_x a_y & a_x a_z \\ a_y a_x & a_y a_y & a_y a_z \\ a_z a_x & a_z a_x & a_z a_z \end{bmatrix}$$

$$A^* = \begin{bmatrix} 0 & -a_z & a_y \\ a_z & 0 & -a_x \\ -a_z & a_x & 0 \end{bmatrix} \tag{B.42}$$

$$M = \hat{A} + \cos\theta \left(I - \hat{A} \right) + \sin\theta A^*$$
$$p' = MP$$

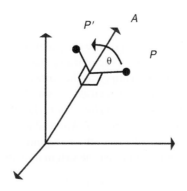

FIGURE B.26

Axis-angle rotation.

B.3.4 Quaternions

Quaternions are discussed in Chapter 2.4 and Chapter 3.1. The equations from those chapters, along with additional equations, are collected here to facilitate the discussion.

Quaternion arithmetic

Quaternions are four-tuples and can be considered as a scalar combined with a vector (Eq. B.43). Addition and multiplication are defined for quaternions by Equations B.44 and B.45, respectively. Quaternion multiplication is associative (Eq. B.46), but it is not commutative (Eq. B.47). The magnitude of a quaternion is computed as the square root of the sum of the squares of its four components (Eq. B.48). Quaternion multiplication has an identity (Eq. B.49) and an inverse (Eq. B.50). The inverse distributes over quaternion multiplication similarly to how the inverse distributes over matrix multiplication (Eq. B.51). A quaternion is normalized by dividing it by its magnitude (Eq. B.52). Equation B.52a is used to compute the quaternion that represents the rotation r such that q represents the half-way rotation between p and r.

$$q = [s, x, y, z] = [s, v] \tag{B.43}$$

$$[s_1, v_1] + [s_2, v_2] = [s_1 + s_2, v_1 + v_2] \tag{B.44}$$

$$[s_1, v_1][s_2, v_2] = [s_1 s_2 - v_1 \cdot v_2, s_1 v_2 + s_2 v_1 + v_1 \times v_2] \tag{B.45}$$

$$(q_1 q_2)q_3 = q_1(q_2 q_3) \tag{B.46}$$

$$q_1 q_2 \neq q_2 q_1 \tag{B.47}$$

$$\|q\| = \sqrt{s^2 + x^2 + y^2 + z^2} \tag{B.48}$$

$$[s, v][1, (0, 0, 0)] = [s, v] \tag{B.49}$$

$$q^{-1} = (1/\|q\|)^2 [s, -v]$$
$$q^{-1}q = qq^{-1} = [1, (0, 0, 0)] \tag{B.50}$$

$$(pq)^{-1} = q^{-1}p^{-1} \tag{B.51}$$

$$qunit = q/(\|q\|) \tag{B.52}$$

$$r = 2(p \cdot q)q - p \tag{B.52a}$$

Rotations by quaternions

A point in space is represented by a vector quantity in quaternion form by using a zero scalar value (Eq. B.53). A quaternion can be used to rotate a vector using quaternion multiplication (Eq. B.54). Compound rotations can be implemented by premultiplying the corresponding quaternions (Eq. B.55), similar to what is routinely done when rotation matrices are used. As should be expected, compounding a rotation with its inverse produces the identity transformation for vectors (Eq. B.56). An axis-angle rotation is represented by a unit quaternion, as shown in Equation B.57. Any scalar multiple of a quaternion represents the same rotation. In particular, the negation of a quaternion (negating each of its four components, $-q = [-s, -x, y, -z]$) represents the same rotation that the original quaternion represents (Eq. B.58).

$$v = [0, x, y, z] \tag{B.53}$$

$$v' = Rot(v) = qvq^{-1} \tag{B.54}$$

$$\begin{aligned}
\text{Rot}_q\left(\text{Rot}_p(v)\right) &= q(pvp^{-1})q^{-1} \\
&= ((qp)v(p^{-1}q^{-1})) \\
&= ((qp)v(qp^{-1})) \\
&= \text{Rot}_{qp}(v)
\end{aligned} \tag{B.55}$$

$$\begin{aligned}
\text{Rot}^{-1}(\text{Rot}(v)) &= q^{-1}(qvq^{-1})q \\
&= (q^{-1}q)v(q^{-1}q) = v
\end{aligned} \tag{B.56}$$

$$\text{Rot}_{[\theta,x,y,z]} \equiv [\cos(\theta/2),\ \sin(\theta/2)(x,y,z)] \tag{B.57}$$

$$\begin{aligned}
-q &\equiv \text{Rot}_{[-\theta,-x,-y,-z]} \\
&= [\cos(-\theta/2),\ \sin(-\theta/2)(-(x,y,z))] \\
&= [\cos(\theta/2),\ -\sin(\theta/2)(-(x,y,z))] \\
&= [\cos(\theta/2),\ \sin(\theta/2)(x,y,z)] \\
&\equiv \text{Rot}_{[\theta,x,y,z]} \\
&\equiv q
\end{aligned} \tag{B.58}$$

Conversions

It is often useful to convert back and forth between rotation matrices and quaternions. Often, quaternions are used to interpolate between orientations, and the result is converted to a rotation matrix so as to combine it with other matrices in the display pipeline.

Given a unit quaternion ($q = [s, x, y, z]$, $s^2 + x^2 + y^2 + z^2 = 1$), one can easily determine the corresponding rotation matrix by rotating the three unit vectors that correspond to the principal axes. The rotated vectors are the columns of the equivalent rotation matrix (Eq. B.59).

$$M_q = \begin{bmatrix} 1 - 2y^2 - 2z^2 & 2xy - 2sz & 2xz + 2sy \\ 2xy + 2sz & 1 - 2x^2 - 2z^2 & 2yz - 2sx \\ 2xy - 2sy & 2yz + 2sx & 1 - 2x^2 - 2y^2 \end{bmatrix} \tag{B.59}$$

Given a rotation matrix, one can use the definitions for the terms of the matrix in Equation B.59 to solve for the elements of the equivalent unit quaternion. The fact that the unit quaternion has a magnitude of one ($s^2 + x^2 + y^2 + z^2 = 1$) makes it easy to see that the diagonal elements sum to $4 \cdot s^2 - 1$. Summing the diagonal elements of the matrix in Equation B.60 results in Equation B.61. The diagonal elements can also be used to solve for the remaining terms (Eq. B.62). The square roots of these last equations can be avoided if the off-diagonal elements are used to solve for x, y, and z at the expense of testing for a divide by an s that is equal to zero (in which case Eq. B.62 can be used).

$$\begin{bmatrix} m_{0,0} & m_{0,1} & m_{0,2} \\ m_{1,0} & m_{1,1} & m_{1,2} \\ m_{2,0} & m_{2,1} & m_{2,2} \end{bmatrix} \tag{B.60}$$

$$s = \frac{\sqrt{m_{0,0} + m_{1,1} + m_{2,2} + 1}}{2} \tag{B.61}$$

$$\begin{aligned}
m_{0,0} &= 1 - 2y^2 - 2z^2 \\
&= 1 - 2(y^2 + z^2) \\
&= 1 - 2(1 - x^2 - s^2) \\
&= -1 + 2x^2 + 2s
\end{aligned} \tag{B.62}$$

$$x = \frac{\sqrt{m_{0,0} + 1 - 2s^2}}{2}$$

$$y = \frac{\sqrt{m_{1,1} + 1 - 2s^2}}{2}$$ (B.62a)

$$z = \frac{\sqrt{m_{2,2} + 1 - 2s^2}}{2}$$

B.4 Denevit and Hartenberg representation for linked appendages
B.4.1 Denavit-Hartenberg notation

The Denavit-Hartenberg (DH) notation is a particular way of describing the relationship of a parent coordinate frame to a child coordinate frame. This convention is commonly used in robotics and often adopted for use in computer animation. Each frame is described relative to an adjacent frame by four parameters that describe the position and orientation of a child frame in relation to its parent's frame.

For revolute joints, the z-axis of the joint's frame corresponds to the axis of rotation (prismatic joints are discussed in the following paragraph). The link associated with the joint extends down the x-axis of the frame. First, consider a simple configuration in which the joints and the axes of rotation are coplanar. The distance down the x-axis from one joint to the next is the *link length*, a_i. The *joint angle*, θ_{i+1}, is specified by the rotation of the $i + 1$ joint's x-axis, x_{i+1}, about its z-axis relative to the ith frame's x-axis direction, x_i (see Figure B.27).

Nonplanar configurations can be represented by including the two other DH parameters. For this general case, the x-axis of the ith joint is defined as the line segment perpendicular to the z-axes of the ith and $i + 1$ frames. The *link twist* parameter, α_i, describes the rotation of the $i + 1$ frame's z-axis about this perpendicular relative to the z-axis of the ith frame. The *link offset* parameter, d_{i+1}, specifies the distance along the z-axis (rotated by α_i) of the $i + 1$ frame from the ith x-axis to the $i + 1$ x-axis (see Figure B.28).

Notice that the parameters associated with the ith joint do not all relate the ith frame to the $i + 1$ frame. The link length and link twist relate the ith and $i + 1$ frames; the link offset and joint rotation relate the $i - 1$ and ith frames (see Table B.1).

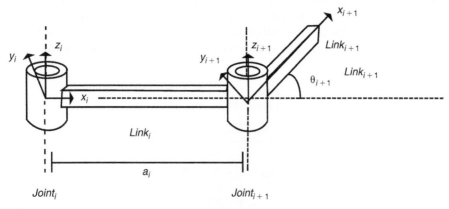

FIGURE B.27

DH parameters for planar joints.

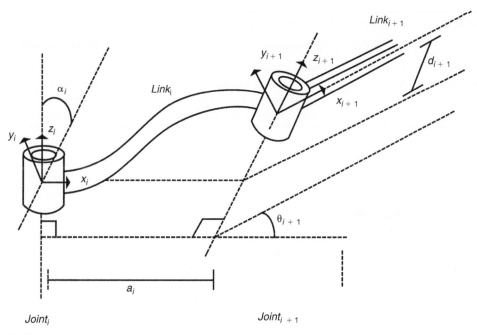

FIGURE B.28

DH parameters.

Table B.1 DH Joint Parameters for Joint i

Name	Symbol	Description
Link offset	d_i	Distance from x_{i-1} to x_i along z_i
Joint angle	θ_i	Angle between x_{i-1} and x_i about z_i
Link length	a_i	Distance from z_i to z_{i+1} along x_i
Link twist	α_i	Angle between z_i and z_{i+1} about x_i

Stated another way, the parameters that describe the relationship of the $i + 1$ frame to the ith frame are a combination of ith joint parameters and $i + 1$ joint parameters. The parameters can be paired off to define two *screw transformations*, each of which consists of a translation and rotation relative to a single axis. The offset (d_{i+1}) and angle ($\theta_i + 1$) are the translation and rotation of the $i + 1$ joint relative to the ith joint with respect to the ith joint's z-axis. The length (a_i) and twist (α_i) are the translation and rotation of the $i + 1$ joint with respect to the ith joint's x-axis (see Table B.2). The transformation of the $i + 1$ joint's frame from the ith frame can be constructed from a series of transformations, each of which corresponds to one of the DH parameters. As an example, consider a point, V_{i+1}, whose coordinates are given in the coordinate system of joint $i + 1$. To determine the point's coordinates in terms of the coordinate system of joint i, the transformation shown in Equation B.63 is applied.

In Equation B.63, T and R represent translation and rotation transformation matrices respectively; the parameter specifies the amount of rotation or translation, and the subscript specifies the axis involved. The matrix M maps a point defined in the $i + 1$ frame into a point in the ith frame. By forming the M matrix and its inverse associated with each pair of joints, one can convert points from one frame to another, up and down the hierarchy.

Table B.2 Parameters That Relate the ith Frame and the $i + 1$ Frame

Name	Symbol	Description	Screw Transformation
Link offset	d_{i+1}	Distance from x_i to x_{i+1} along z_{i+1}	Relative to z_{i+1}
Joint angle	θ_{i+1}	Angle between x_i and x_{i+1} about z_{i+1}	Relative to z_{i+1}
Link length	a_i	Distance from z_i to z_{i+1} along x_i	Relative to x_i
Link twist	α_i	Angle between z_i and z_{i+1} about x_i	Relative to x_i

B.4.2 A simple example

Consider the simple three-joint manipulator of Figure B.29. The DH parameters are given in Table B.3. The linkage is planar, so there are no displacement parameters and no twist parameters. Each successive frame is described by the joint angle and the length of the link.

$$V_i = T_x(\alpha_i)R_X(\alpha_i)T_Z(d_{i+1})R_Z(\theta_i+1)V_{i+1}$$

$$R_Z(\theta_i+1) = \begin{bmatrix} \cos(\theta_i+1) & (-\sin(\theta_i+1)) & 0 & 0 \\ \sin(\theta_i+1) & \cos(\theta_i+1) & 0 & 0 \\ 0 & 0 & 1 & 0 \\ 0 & 0 & 0 & 1 \end{bmatrix}$$

$$T_Z(d_{i+1}) = \begin{bmatrix} 1 & 0 & 0 & 0 \\ 0 & 1 & 0 & 0 \\ 0 & 0 & 1 & d_{i+1} \\ 0 & 0 & 0 & 1 \end{bmatrix}$$

$$R_X(\alpha_i) = \begin{bmatrix} 1 & 0 & 0 & 0 \\ 0 & \cos(\alpha_i) & -\sin(\alpha_i) & 0 \\ 0 & \sin(\alpha_i) & \cos(\alpha_i) & 0 \\ 0 & 0 & 0 & 1 \end{bmatrix} \qquad \text{(B.63)}$$

$$T_X(ai) = \begin{bmatrix} 1 & 0 & 0 & a_i \\ 0 & 1 & 0 & 0 \\ 0 & 0 & 1 & 0 \\ 0 & 0 & 0 & 1 \end{bmatrix}$$

$$V_i = M_i^{i+1}V_i + 1$$

$$M_i^{i+1} = \begin{bmatrix} \cos(\theta_{i+1}) & -\sin(\theta_{i+1}) & 0 & a_i \\ \cos(\alpha_i)\sin(\theta_{i+1}) & \cos(\alpha_i)\cos(\theta_{i+1}) & -\sin(\alpha_i) & (-d_{i+1})\sin(\alpha_i) \\ \sin(\alpha_i)\sin(\theta_{i+1}) & \sin(\alpha_i)\cos(\theta_{i+1}) & \cos(\alpha_i) & d_{i+1}\cos(\alpha_i) \\ 0 & 0 & 0 & 1 \end{bmatrix}$$

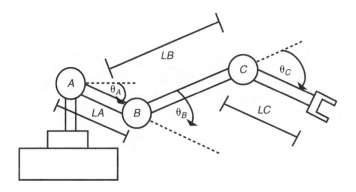

FIGURE B.29

Simple manipulator using three revolute joints.

Table B.3 Parameters for Three-revolute Joint Armature

Joint/Parameter	Link Displacement	Joint Angle	Link Length	Link Twist
A	0	θ_A	0	0
B	0	θ_B	LA	0
C	0	θ_C	LB	0

B.4.3 Including a ball-and-socket joint

Some human joints are conveniently modeled using a ball-and-socket joint. Consider an armature with a hinge joint, followed by a ball-and-socket joint followed by another hinge joint, as shown in Figure B.30.

The DH notation can represent the ball-and-socket joint by three single degree of freedom (DOF) joints with zero-length links between them (see Figure B.31).

Notice that in a default configuration with joint angles set to zero, the DH model of the ball-and-socket joint is in a gimbal lock position (incrementally changing two of the parameters results in rotation about the same axis). The first and third DOFs of that joint are aligned. The z-axes of these joints are colinear because the links between them are zero length and the two link twist parameters relating them are 90 degrees. This results in a total of 180 degrees and thus aligns the axes. As a consequence, the representation of the ball-and-socket joint is usually initialized with the middle of the three joint angles set to 90 degrees (see Table B.4).

B.4.4 Constructing the frame description

Because each frame's displacement and joint angle are defined relative to the previous frame, a Frame$_0$ is defined so that the Frame$_1$ displacement and angle can be defined relative to it. Frame$_0$ is typically defined so that it coincides with Frame$_1$ with zero displacement and zero joint angle. Similarly, because the link of the last frame does not connect to anything, the x-axis of the last frame is chosen so that it coincides with the x-axis of the previous frame when the joint angle is zero; the origin of the nth frame is

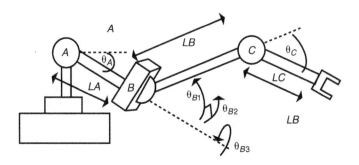

FIGURE B.30

Incorporating a ball-and-socket joint.

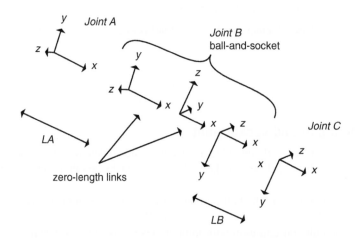

FIGURE B.31

Coordinate axes induced by the DH representation of a ball-and-socket joint.

Table B.4 Joint Parameters for Ball-and-socket Joint

Joint/Parameter	Link Displacement	Joint Angle	Link Length	Link Twist
A	0	θ_A	0	0
B1	0	θ_{B1}	LA	90
B2	0	$90 + \theta_{B2}$	0	90
B3	0	θ_{B3}	0	0
C	0	θ_C	LB	0

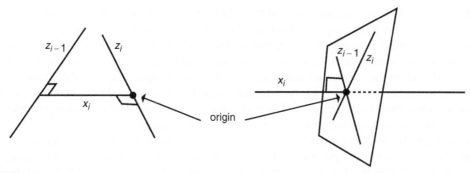

FIGURE B.32

Determining the origin and x-axis of the ith frame.

chosen as the intersection of the x-axis of the previous frame and the joint axis when the displacement is zero.

The following procedure can be used to construct the frames for intermediate joints.

1. For each joint, identify the axis of rotation for revolute joints and the axis of displacement for prismatic joints. Refer to this axis as the z-axis of the joint's frame.
2. For each adjacent pair of joints, the ith -1 and ith for i from 1 to n, construct the common perpendicular between the z-axes or, if they intersect, the perpendicular to the plane that contains them. Refer to the intersection of the perpendicular and the ith frame's z-axis (or the point of intersection of the two axes) as the origin of the ith frame. Refer to the perpendicular as the x-axis of the ith frame (see Figure B.32).
3. Construct the y-axis of each frame to be consistent with the right-hand rule (assuming right-hand space).

B.5 Interpolating and approximating curves

This section covers many of the basic terms and concepts needed to interpolate values in computer animation. It is not a complete treatise of curves but an overview of the important ones. While many of the terms and concepts discussed are applicable to functions in general, they are presented as they relate to functions having to do with the practical interpolation of points in Euclidean space as typically used in computer animation applications. For more complete discussions of the topics contained here, see, for example, Mortenson [14], Rogers and Adams [18], Farin [4], and Bartels, Beatty, and Barsky [1].

B.5.1 Equations: some basic terms

For present purposes, there are three types of equations: *explicit*, *implicit*, and *parametric*. *Explicit equations* are of the form $y = f(x)$. The explicit form is good for generating points because it generates a value of y for any value of x put into the function. The drawback of the explicit form is that it is dependent on the choice of coordinate axes, and it is ambiguous if there is more than one y for a given

x (such as $y = \sqrt{x}$, in which an input value of 4 would generate values of 2 or -2). *Implicit equations* are of the form $f(x, y) = 0$. The implicit form is good for testing to see if a point is on a curve because the coordinates of the point can easily be put into the equation for the curve and checked to see if the equation is satisfied. The drawback of the implicit form is that generating a series of points along a curve is often desired, and implicit forms are not generative. *Parametric equations* are of the form $x = f(t)$, $y = g(t)$. For any given value of t, a point (x, y) is generated. This form is good for generating a sequence of points as ordered values of t are given. The parametric form is also useful because it can be used for multivalued functions of x, which are problematic for explicit equations.

Functions can be classified according to the terms contained in them. Functions that contain only variables raised to a power are *polynomial* functions. If the highest power is one, then the function is *linear*. If the highest power is two, then the function is *quadratic*. If the highest power is three, then it is *cubic*. The highest power of a polynomial function of a single variable is referred to as the *degree* of the polynomial. If the function is not a simple polynomial but rather contains sines, cosines, log, or a variety of other functions, then it is called *transcendental*. In computer graphics, the most commonly encountered type of function is the cubic polynomial.

Continuity refers to how well behaved the curve is in a mathematical sense. For a value arbitrarily close to an x_0 if the function is arbitrarily close to $f(x_0)$, then it has *positional*, or *zeroth-order*, continuity (C^0) at that point. If the slope of the curve (or the first derivative of the function) is continuous, then the function has *tangential*, or *first-order*, continuity (C^1). This is extended to all of the function's derivatives, although for purposes of computer animation the concern is with first-order continuity or, possibly, *second-order*, or *curvature*, continuity (C^2). Polynomials are infinitely continuous.

If a curve is pieced together from individual curve segments, one can speak of *piecewise properties*—the properties of the individual pieces. For example, a sequence of straight line segments, sometimes called a *polyline* or a *wire*, is piecewise linear. A major concern regarding piecewise curves is the continuity conditions at the junctions of the curve segments. A junction has C^0 *continuity*, or is C^0 *continuous*, if one curve segment begins where the previous segment ends. This is referred to as *zeroth-order*, or *positional*, continuity at the junction. If the beginning tangent of one curve segment is the same as the ending tangent of the previous curve segment, then there is *first-order*, or *tangential*, continuity at the junction. The junction is C^1 continuous if it has tangential and positional continuity. If the beginning curvature of one curve segment is the same as the ending curvature of the previous curve segment, then there is *second-order*, or *curvature*, continuity at the junction. The junction is C^2 continuous if it has curvature, tangential, and positional continuity. Typically, computer animation is not concerned with continuity beyond second order.

Sometimes in discussions of the continuity at segment junctions, a distinction is made between *parametric continuity* and *geometric continuity* (e.g., [14]). So far the discussion has concerned parametric continuity. Geometric continuity is less restrictive. First-order parametric continuity, for example, requires that the ending tangent vector of the first segment be the same as the beginning tangent vector of the second. First-order geometric continuity, on the other hand, requires that only the direction of the tangents be the same, and it allows the magnitudes of the tangents to be different. Similar definitions exist for higher order geometric continuity. One distinction worth mentioning is that parametric continuity is sensitive to the rate at which the parameter varies relative to the length of the curve traced out. Geometric continuity is not sensitive to this rate.

When a curve is constructed from a set of points and the curve passes through the points, it is said to *interpolate* the points. However, if the points are used to control the general shape of the curve, with the curve not necessarily passing through them, then the curve is said to *approximate the points. Interpolation* is also used generally to refer to all approaches for constructing a curve from a set of points. For a given interpolation technique, if the resulting curve is guaranteed to lie within the convex hull of the set of points, then it is said to have the *convex hull property*.

B.5.2 Simple linear interpolation: geometric and algebraic forms

Simple linear interpolation is given by Equation B.64 and shown in Figure B.33. Notice that the interpolants, $1 - u$ and u, sum to one. This property ensures that the interpolating curve (in this case a straight line) falls within the convex hull of the geometric entities being interpolated (in this simple case the convex hull is the straight line itself).

$$P(u) = (1 - u)P_0 + uP_1 \tag{B.64}$$

Using more general notation, one can rewrite Equation 64 as in Equation B.65. Here, F_0 and F_1 are called blending functions. This is referred to as the geometric form because the geometric information, in this case P_0 and P_1, is explicit in the equation.

$$P(u) = F_0(u)P_0 + F_1(u)P_1 \tag{B.65}$$

The linear interpolation equation can also be rewritten as in Equation B.66. This form is typical of polynomial equations in which the terms are collected according to coefficients of the variable raised to a power. It is more generally written as Equation B.67. In this case there are only linear terms. This way of expressing the equation is referred to as the *algebraic* form.

$$P(u) = (P_1 - P_0)u + P_0 \tag{B.66}$$

$$P(u) = a_1u + a_0 \tag{B.67}$$

Alternatively, both of these forms can be put in a *matrix representation*. The geometric form becomes Equation B.68 and the algebraic form becomes Equation B.69. The geometric form is useful in situations in which the geometric information (the points defining the curve) needs to be frequently updated or replaced. The algebraic form is useful for repeated evaluation of a single curve for different values of the parameter. The fully expanded form is shown in Equation B.70. The curves discussed next can all be written in this form. Of course, depending on the actual curve type, the U (variable), M (coefficient), and B (geometric information) matrices will contain different values.

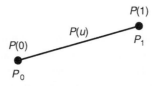

FIGURE B.33

Linear interpolation.

$$P(u) = \begin{bmatrix} F_0(u) \\ F_1(u) \end{bmatrix} [P_0 P_1] = FB^T \tag{B.68}$$

$$P(u) = [u \quad 1] \begin{bmatrix} a_1 \\ a_0 \end{bmatrix} = U^T A \tag{B.69}$$

$$P(u) = [u \quad 1] \begin{bmatrix} -1 & 1 \\ 1 & 0 \end{bmatrix} \begin{bmatrix} P_0 \\ P_1 \end{bmatrix} = U^T MB = FB = U^T A \tag{B.70}$$

B.5.3 Parameterization by arc length

It should be noted that in general there is not a linear relationship between changes in the parameter u and the distance traveled along a curve (its *arc length*). It happens to be true in the previous example concerning a straight line and the parameter u. However, as Mortenson [14] points out, there are other equations that trace out a straight line in space that are fairly convoluted in their relationship between changes in the parameter and distance traveled. For example, consider Equation B.71, which is linear in P_0 and P_1. That is, it traces out a straight line in space between P_0 and P_1. However, it is nonlinear in u. As a result, the curve is not traced out in a nice monotonic, constant-velocity manner. The nonlinear relationship is evident in most parameterized curves unless special care is taken to ensure constant velocity.

$$P(u) = P_0 + ((1-u)u + u)(P_1 - P_0) \tag{B.71}$$

B.5.4 Computing derivatives

One of the matrix forms for parametric curves, as shown in Equation B.70 for linear interpolation, is $U^T MB$. Parametric curves of any polynomial order can be put into this matrix form. Often, it is useful to compute the derivatives of a parametric curve. This can be done easily by taking the derivative of the U vector. For example, the first two derivatives of a cubic curve, shown in Equation B.72, are easily evaluated for any value of u.

$$\begin{aligned} P(u) &= U^T MB = |u^3 \ u^2 \ u \ 1 | MB \\ P'(u) &= U'^T MB = |3u^3 \ 2u \ 1 \ 0 | MB \\ P''(u) &= U''^T MB = |6u \ 2 \ 0 \ 0 | MB \end{aligned} \tag{B.72}$$

B.5.5 Hermite interpolation

Hermite interpolation generates a cubic polynomial from one point to another. In addition to specifying the beginning and ending points (P_i, P_{i+1}), the user needs to supply beginning and ending tangent vectors (P'_i, P'_{i+1}) as well (Figure B.34). The general matrix form for a curve is repeated in Equation B.73, and the Hermite matrices are given in Equation B.74.

$$P(u) = U^T MB \tag{B.73}$$

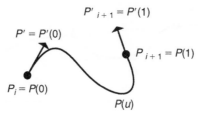

FIGURE B.34

Hermite interpolation.

$$U^T = \begin{bmatrix} u^3 & u^2 & u & 1 \end{bmatrix}$$

$$M = \begin{bmatrix} 2 & -2 & 1 & 1 \\ -3 & 3 & -2 & -1 \\ 0 & 0 & 1 & 0 \\ 1 & 0 & 0 & 0 \end{bmatrix}$$

$$B = \begin{bmatrix} P_i \\ P_{i+1} \\ P_{i+1} \\ P'_i \\ P'_{i+1} \end{bmatrix}$$

(B.74)

Continuity between beginning and ending tangent vectors of connected segments is ensured by merely using the ending tangent vector of one segment as the beginning tangent vector of the next. A composite Hermite curve (piecewise cubic with first-order continuity at the junctions) is shown in Figure B.35.

Trying to put a Hermite curve through a large number of points, which requires the user to specify all of the needed tangent vectors, can be a burden. There are several techniques to get around this. One is to enforce second-degree continuity. This requirement provides enough constraints so that the user does not have to provide interior tangent vectors; they can be calculated automatically. See Rogers and Adams [18] or Mortenson [14] for alternative formulations. A more common technique is the Catmull-Rom spline.

B.5.6 Catmull-Rom spline

The Catmull-Rom curve can be viewed as a Hermite curve in which the tangents at the interior control points are automatically generated according to a relatively simple geometric procedure (as opposed to the more involved numerical techniques referred to above). For each interior point, P_i, the tangent at

FIGURE B.35

Composite Hermite curve.

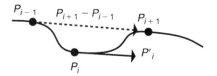

FIGURE B.36

Catmull-Rom spline.

that point, P'_i, is computed as one-half the vector from the previous control point, P_{i-1}, to the following control point, P_{i+1} (Eq. B.75), as shown in Figure B.36.[1] The matrices for the Catmull-Rom curve in general matrix form are given in Equation B.76. A Catmull-Rom spline is a specific type of cardinal spline.

$$P'_i = (1/2)(P_{i+1} - P_{i-1}) \tag{B.75}$$

$$U^T = \begin{bmatrix} u^3 & u^2 & u & 1 \end{bmatrix}$$

$$M = \frac{1}{2} \begin{bmatrix} -1 & 3 & -3 & 1 \\ 2 & -5 & 4 & -1 \\ -1 & 0 & 1 & 0 \\ 0 & 2 & 0 & 0 \end{bmatrix} \tag{B.76}$$

$$B = \begin{bmatrix} P_{i-1} \\ P_i \\ P_{i+1} \\ P_{i+2} \end{bmatrix}$$

For the end conditions, the user can provide tangent vectors at the very beginning and at the very end of the cubic curve. Alternatively, various automatic techniques can be used. For example, the beginning tangent vector can be defined as follows. The vector from the second point (P_1) to the third point (P_2) is subtracted from the second point and used as a virtual point to which the initial tangent is directed. This tangent is computed by Equation B.77. Figure B.37 shows the formation of the initial tangent curve according to the equation, and Figure B.38 shows a curve that uses this technique.

$$P'(0.0) = \frac{1}{2}(P_1 - (P_2 - P_1) - P_0) = \frac{1}{2}(2P_1 - P_2 - P_0) \tag{B.77}$$

A drawback of the Catmull-Rom formulation is that an internal tangent vector is not dependent on the position of the internal point relative to its two neighbors. In Figure B.39, all three positions (Q_i, P_i, R_i) for the ith point would have the same tangent vector.

---

[1] Farin [4] describes the Catmull-Rom spline curve in terms of a cubic Bezier curve by defining interior control points. Placement of the interior control points is determined by use of an auxiliary knot vector. With a uniform distance between knot values, the control points are displaced from the point to be interpolated by one-sixth of the vector from the previous interpolated point to the following interpolated point. Tangent vectors are three times the vector from an interior control point to the interpolated point. This results in the Catmull-Rom tangent vector described here.

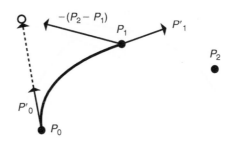

FIGURE B.37

Automatically forming the initial tangent of a Catmull-Rom spline.

FIGURE B.38

Catmull-Rom spline with end conditions using Equation B.77.

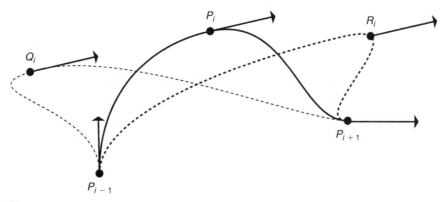

FIGURE B.39

Three curve segments, (P_{i-1}, P_i, P_{i+1}), (P_{i-1}, Q_i, P_{i+1}), and (P_{i-1}, R_i, P_{i+1}), using the standard Catmull-Rom form for computing the internal tangent.

An advantage of Catmull-Rom is that the calculation to compute the internal tangent vectors is extremely simple and fast. However, for each segment the tangent computation is a one-time-only cost. It is then used repeatedly in the computation for each new point in that segment. Therefore, it often makes sense to spend a little more time computing more appropriate internal tangent vectors to obtain a better set of points along the segment. One alternative is to use a vector perpendicular to the plane that bisects the angle made by $P_{i-1} - P_i$ and $P_{i+1} - P_i$ (Figure B.40). This can be computed easily by adding the normalized vector from P_{i-1} to P_i with the normalized vector from P_i to P_{i+1}.

Another modification, which can be used with the original Catmull-Rom tangent computation or with the previously mentioned bisector technique, is to use the relative position of the internal point (P_i)

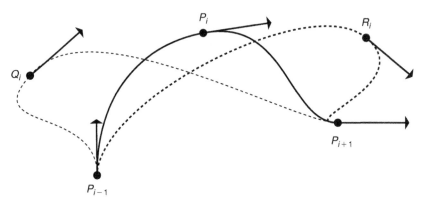

FIGURE B.40

Three curve segments, (P_{i-1}, P_i, P_{i+1}), (P_{i-1}, Q_i, P_{i+1}), and (P_{i-1}, R_i, P_{i+1}), using the perpendicular to the angle bisector for computing the internal tangent.

to independently determine the length of the tangent vector for each segment it is associated with. Thus, a point P_i has an ending tangent vector associated with it for the segment from P_{i-1} to P_i as well as a beginning tangent vector associated with it for the segment P_i to P_{i+1}. These tangents have the same direction but different lengths. This relaxes the C^1 continuity of the Catmull-Rom spline and uses G^1 continuity instead. For example, an initial tangent vector at an interior point is determined as the vector from P_{i-1} to P_{i+1}. The ending tangent vector for the segment P_{i-1} to P_i is computed by scaling this initial tangent vector by the ratio of the distance between the points P_i and P_{i-1} to the distance between points P_{i-1} and P_{i+1}. Referring to the segment between P_{i-1} and P_i as $P_{i-1}(u)$ results in Equation B.78. A similar calculation for the beginning tangent vector of the segment between P_i and P_{i+1} results in Equation B.79. These tangents can be seen in Figure B.41. The computational cost of this approach is only a little more than the standard Catmull-Rom spline and seems to give more intuitive results.

$$P'_{i-1}(1.0) = \frac{|P_i - P_{i-1}|}{|P_{i+1} - P_{i-1}|} (P_{i+1} - P_{i-1}) \tag{B.78}$$

$$P'_i(0.0) = \frac{|P_{i+1} - P_i|}{|P_{i+1} - P_{i-1}|} (P_{i+1} - P_{i-1}) \tag{B.79}$$

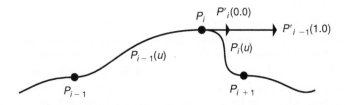

FIGURE B.41

Interior tangents based on relative segment lengths.

B.5.7 **Four-point form**

Fitting a cubic segment to four points (P_0, P_1, P_2, P_3) assigned to user-specified parametric values (u_0, u_1, u_2, u_3) can be accomplished by setting up the linear system of equations for the points (Eq. B.80) and solving for the unknown coefficient matrix. In the case of parametric values of 0, 1/3, 2/3, and 1, the matrix is given by Equation B.81. However, with this form it is difficult to join segments with C^1 continuity.

$$P(u) = \begin{bmatrix} u^3 & u^2 & u & 1 \end{bmatrix} \begin{bmatrix} m_{0,0} & m_{0,1} & m_{0,2} & m_{0,3} \\ m_{1,0} & m_{1,1} & m_{1,2} & m_{1,3} \\ m_{2,0} & m_{2,1} & m_{2,2} & m_{2,3} \\ m_{3,0} & m_{3,1} & m_{3,2} & m_{3,3} \end{bmatrix} \begin{bmatrix} P_0 \\ P_1 \\ P_2 \\ P_3 \end{bmatrix}$$

$$\begin{bmatrix} P_0 \\ P_1 \\ P_2 \\ P_3 \end{bmatrix} = \begin{bmatrix} u_0^3 & u_0^2 & u_0 & 1 \\ u_1^3 & u_1^2 & u_1 & 1 \\ u_2^3 & u_2^2 & u_2 & 1 \\ u_3^3 & u_3^2 & u_3 & 1 \end{bmatrix} \begin{bmatrix} m_{0,0} & m_{0,1} & m_{0,2} & m_{0,3} \\ m_{1,0} & m_{1,1} & m_{1,2} & m_{1,3} \\ m_{2,0} & m_{2,1} & m_{2,2} & m_{2,3} \\ m_{3,0} & m_{3,1} & m_{3,2} & m_{3,3} \end{bmatrix} \begin{bmatrix} P_0 \\ P_1 \\ P_2 \\ P_3 \end{bmatrix}$$

(B.80)

$$M = \frac{1}{2} \begin{bmatrix} -9 & 27 & -27 & - \\ 18 & -45 & 36 & -9 \\ -11 & 18 & -9 & 2 \\ 2 & 0 & 0 & 0 \end{bmatrix}$$

(B.81)

B.5.8 **Blended parabolas**

Blending overlapping parabolas to define a cubic segment is another approach to interpolating a curve through a set of points. In addition, the end conditions are handled by parabolic segments, which is consistent with how the interior segments are defined. Blending parabolas results in a formulation that is very similar to Catmull-Rom in that each segment is defined by four points, it is an interpolating curve, and local control is provided. Under the assumptions used here for Catmull-Rom and the blended parabolas, the interpolating matrices are identical.

For each overlapping triple of points, a parabolic curve is defined by the three points. A cubic curve segment is created by linearly interpolating between the two overlapping parabolic segments. More specifically, take the first three points, P_0, P_1, and P_2, and fit a parabola, $P(u)$, through them using the following constraints: $P(0.0) = P_0$, $P(0.5) = P_1$, $P(1.0) = P_2$. Take the next group of three points, P_1, P_2, P_3, which partially overlap the first set of three points, and fit a parabola, $R(u)$, through them using similar constraints: $R(0.0) = P_1$, $R(0.5) = P_2$, $R(1.0) = P_3$. Between points P_1 and P_2 the two parabolas overlap. Reparameterize this region into the range [0.0, 1.0] and linearly blend the two parabolic segments (Figure B.42). The result can be put in matrix form for a cubic curve using the four points as the geometric information together with the coefficient matrix shown in Equation B.82. To interpolate a list of points, this matrix is used by varying U as in previous examples.

$$M = \frac{1}{2} \begin{bmatrix} -1 & 3 & -3 & 1 \\ 2 & -5 & 4 & -1 \\ -1 & 0 & 1 & 0 \\ 0 & 2 & 0 & 0 \end{bmatrix}$$

(B.82)

FIGURE B.42

Parabolic blend segment.

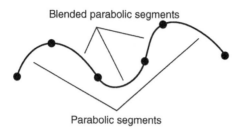

FIGURE B.43

Multiple parabolic blend segments.

End conditions can be handled by constructing parabolic arcs at the very beginning and very end (Figure B.43). For example, referring to the first three points as p_0, p_1, and p_2, Equation B.83 can be used to solve for the constants of the parabolic equation $P(u) = au^2 + bu + c$.

$$
\begin{aligned}
P(0) &= c = p_0 \\
P(0.5) &= a(0.5)^2 + b(0.5) + c = p_1 \\
P(1.0) &= a + b + c = p_2
\end{aligned}
\tag{B.83}
$$

This form assumes that all points are equally spaced in parametric space. Often it is the case that even spacing is not present. In such cases, relative chord length can be used to estimate parametric values. The derivation is a bit more involved [18], but the final result can still be formed into a 4 × 4 matrix and used to produce a cubic polynomial in the interior segments.

B.5.9 Bezier interpolation/approximation

A cubic Bezier curve is defined by the beginning point and the ending point, which are interpolated, and two interior points, which control the shape of the curve. The cubic Bezier curve is similar to the Hermite form. The Hermite form uses beginning and ending tangent vectors to control the shape of the curve; the Bezier form uses auxiliary control points to define tangent vectors. A cubic curve is defined by four points: P_0, P_1, P_2, and P_3. The beginning and ending points of the curve are P_0 and P_3, respectively. The interior control points used to control the shape of the curve and define the beginning and ending tangent vectors are P_1 and P_2 (see Figure B.44). The coefficient matrix for a single cubic Bezier curve is shown in Equation B.84. In the cubic case, $P'(0) = 3(P_1 - P_0)$ and $P'(1) = 3(P_3 - P_2)$.

$$
M = \begin{bmatrix}
-1 & 3 & -3 & 1 \\
3 & -6 & 3 & 0 \\
-3 & 3 & 0 & 0 \\
1 & 0 & 0 & 0
\end{bmatrix}
\tag{B.84}
$$

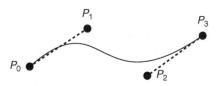

FIGURE B.44

Cubic Bezier curve segment.

FIGURE B.45

Composite cubic Bezier curve showing tangents and colinear control points.

Continuity between adjacent Bezier segments can be controlled by colinearity of the control points on either side of the shared beginning/ending point of the two curve segments where they join (Figure B.45). In addition, the Bezier curve form allows one to define a curve of arbitrary order. If three interior control points are used, then the resulting curve will be quartic; if four interior control points are used, then the resulting curve will be quintic. See Mortenson [14] for a more complete discussion.

B.5.10 De Casteljau construction of Bezier curves

The de Casteljau method is a way to geometrically construct a Bezier curve. Figure B.46 shows the construction of a point at $u = 1/3$. This method constructs a point u along the way between paired control points (identified by a "1" in Figure B.46). Then points are constructed u along the way between points just previously constructed. These new points are marked "2" in Figure B.46. In the cubic case, in which there were four initial points, there are two newly constructed points. The point on the curve is constructed by going u along the way between these two points. This can be done for any values of u and for any order of curve. Higher order Bezier curves require more iterations to produce the final point on the curve.

B.5.11 Tension, continuity, and bias control

Often an animator wants better control over the interpolation of key frames than the standard interpolating splines provide. For better control of the shape of an interpolating curve, Kochanek [11] suggests a parameterization of the internal tangent vectors based on the three values: tension, continuity, and bias. The three parameters are explained by decomposing each internal tangent vector into an incoming part and an outgoing part. These tangents are referred to as the left and right parts, respectively, and are notated by T_i^L and T_i^R for the tangents at P_i.

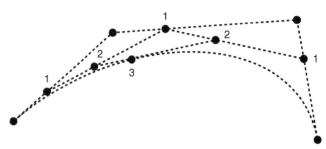

Interpolation steps

1. 1/3 of the way between paired points
2. 1/3 of the way between points of step 1
3. 1/3 of the way between points of step 2

FIGURE B.46

De Casteljau construction of a point on a cubic Bezier curve.

Tension controls the sharpness of the bend of the curve at P_i. It does this by means of a scale factor that changes the length of both the incoming and outgoing tangents at the control point (Eq. B.85). In the default case, $t = 0$ and the tangent vector is the average of the two adjacent chords or, equivalently, half of the chord between the two adjacent points, as in the Catmull-Rom spline. As the tension parameter, t, goes to one, the tangents become shorter until they reach zero. Shorter tangents at the control point mean that the curve is pulled closer to a straight line in the neighborhood of the control point (see Figure B.47).

$$T_i^L = T_i^R = (1 - t)\frac{1}{2}((P_{i+1} - P_i) + (P_i - P_{i-1})) \tag{B.85}$$

The continuity parameter, c, gives the user control over the continuity of the curve at the control point where the two curve segments join. The incoming (left) and outgoing (right) tangents at a control point are defined symmetrically with respect to the chords on either side of the control point. Assuming default tension, c blends the adjacent chords to form the two tangents, as shown in Equation B.86.

$$T_i^L = \frac{1 - c}{2}(P_i - P_{i-1}) + \frac{1 + c}{2}(P_{i+1} - P_i)$$

$$T_i^R = \frac{1 + c}{2}(P_i - P_{i-1}) + \frac{1 - c}{2}(P_{i+1} - P_i) \tag{B.86}$$

The default value for continuity is $c = 0$, which produces equal left and right tangent vectors, resulting in continuity at the joint. As c approaches -1, the left tangent approaches equality with the chord to the left of the control point and the right tangent approaches equality with the chord to the right of the control point. As c approaches $+1$, the definitions of the tangents reverse themselves, and the left tangent approaches the right chord and the right tangent approaches the left chord (see Figure B.48).

Bias, b, defines a common tangent vector, which is a blend between the chord left of the control point and the chord right of the control point (Eq. B.87). At the default value ($b = 0$), the tangent is an even blend of these two, resulting in a Catmull-Rom type of internal tangent vector. Values of b approaching -1 bias the tangent toward the chord to the left of the control point, while values of b approaching $+1$ bias the tangent toward the chord to the right (see Figure B.49).

$$T_i^R = T_i^L = \frac{1 + b}{2}(P_i - P_{i-1}) + \frac{1 - b}{2}(P_{i+1} - P_i) \tag{B.87}$$

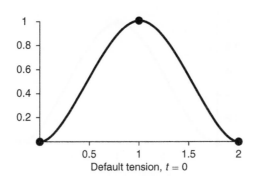

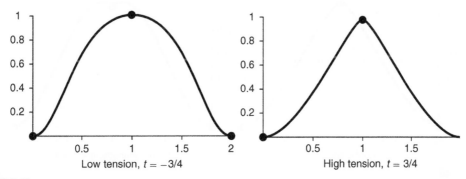

FIGURE B.47

The effect of varying the tension parameter.

The three parameters, tension, continuity, and bias, are combined in Equation B.88.

$$
T_i^R = \frac{((1-t)(1+c)(1+b))}{2}(P_i - P_{i-1}) + \frac{((1-t)(1-c)(1-b))}{2}(P_{i+1} - P_i)
$$

$$
T_i^L = \frac{((1-t)(1-c)(1+b))}{2}(P_i - P_{i-1}) + \frac{((1-t)(1+c)(1-b))}{2}(P_{i+1} - P_i)
$$

(B.88)

B.5.12 **B-splines**

B-splines are the most flexible and useful type of curve, but they are also more difficult to grasp intuitively. The formulation includes Bezier curves as a special case. The formulation for B-spline curves decouples the number of control points from the degree of the resulting polynomial. It accomplishes this with additional information contained in the *knot vector*. An example of a *uniform knot vector* is $[0, 1, 2, 3, 4, 5, 6, \ldots, n + k - 1]$, in which the knot values are uniformly spaced apart. In this knot vector, n is the number of control points and k is the degree of the B-spline curve. The parametric value

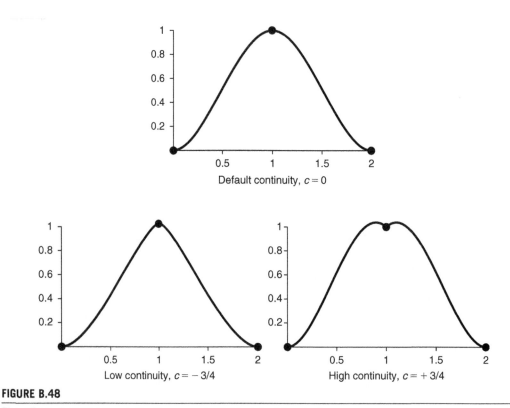

FIGURE B.48

The effect of varying the continuity parameter (with default tension).

varies between the first and last values of the knot vector. The knot vector establishes a relationship between the parametric value and the control points. With replication of values in the knot vector, the curve can be drawn closer to a particular control point up to the point where the curve actually passes through the control point.

A particularly simple, yet useful, type of B-spline curve is a uniform cubic B-spline curve. It is defined using four control points over the interval zero to one (Eq. B.89). A compound curve is generated from an arbitrary number of control points by constructing a curve segment from each four-tuple of adjacent control points: $(P_i, P_{i+1}, P_{i+2}, P_{i+3})$ for $i = 1, 2, \ldots, n-3$, where n is the total number of control points (Figure B.50). Each section of the curve is generated by multiplying the same 4×4 matrix by four adjacent control points with an interpolating parameter between zero and one. In this case, none of the control points is interpolated.

$$P(u) = \frac{1}{6} \begin{bmatrix} u^3 & u^2 & u & 1 \end{bmatrix} \begin{bmatrix} -1 & 3 & -3 & 1 \\ 3 & -6 & 3 & 0 \\ -3 & 0 & 3 & 0 \\ 1 & 4 & 1 & 0 \end{bmatrix} \begin{bmatrix} P_i \\ P_{i+1} \\ P_{i+2} \\ P_{i+3} \end{bmatrix} \tag{B.89}$$

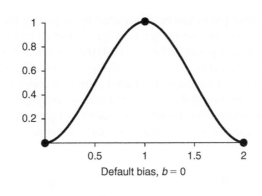

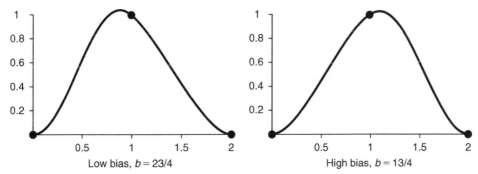

FIGURE B.49

The effect of varying the bias parameter (with default tension and continuity).

Segments of the curve defined by different sets of four points

FIGURE B.50

Compound cubic B-spline curve.

Nonuniform rational B-splines (NURBS), are even more flexible than basic B-splines. NURBS allow for exact representation of circular arcs, whereas Bezier and nonrational B-splines do not. This is often important in modeling, but for purposes of animation, the basic uniform cubic B-spline is usually sufficient.

B.5.13 Fitting curves to a given set of points

Sometimes it is desirable to interpolate a set of points using a Bezier formulation. The points to be interpolated can be designated as the endpoints of the Bezier curve segments, and the interior control points can be constructed by forming tangent vectors at the vertices, as with the Catmull-Rom formulation. The interior control points can be constructed by displacing the control points along the tangent lines just formed. For example, for the segment between given points b_i and b_{i11}, the first control point for the segment, c_i^1, can be positioned at $b_i + (1/3)\,(b_{i+1} - b_{i-1})$. The second control point for the segment, c_i^2, can be positioned at $b_{i+1} - (1/3) \cdot (b_{i+2} - b_i)$ (see Figure B.51).

Other methods exist. Farin [4] presents a more general method of constructing the Bezier curve and, from that, constructing the B-spline control points. Both Farin [4] and Rogers and Adams [18] present a method of constructing a composite Hermite curve through a set of points that automatically calculates internal tangent vectors by assuming second-order continuity at the segment joints.

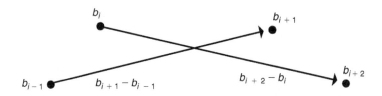

Four initial points with vectors drawn between pairs of
points adjacent to interior points (e.g., b_{i-1} and
b_{i+1} are adjacent to b_i)

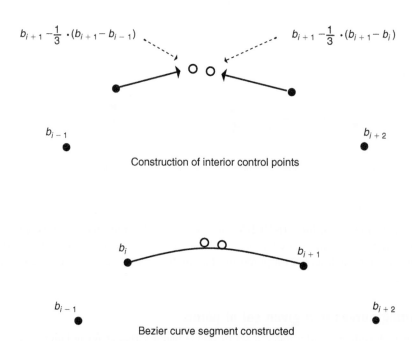

Construction of interior control points

Bezier curve segment constructed

FIGURE B.51

Constructing a Bezier segment that interpolates points.

B.6 Randomness

Introducing controlled randomness in both modeling and animation can often produce more interesting, realistic, natural-looking imagery. Noise and turbulence functions are often used in textures but also can be used in modeling natural phenomena such as smoke and clouds. The code for noise and turbulence that follows is from Peachey's chapter in Ebert [3]. Random perturbations are also useful in human-figure animation to make the motion less "robotic" looking. There are various algorithms proposed in the literature for generating random numbers; Gasch's [6] is presented at the end of this section.

B.6.1 Noise

The *noise* function uses a table of pseudorandom numbers between -1 and $+1$ to represent the integer lattice values. The table is created by *valueTableInit* the first time that *noise* is called. Lattice coordinates are used to index into a table of pseudorandom numbers. A simple function of the coordinates, such as their sum, is used to compute the index. However, this can result in unwanted patterns. To help avoid these artifacts, a table of random permutation values is used to modify the index before it is used. A four-point spline is used to interpolate among the lattice pseudorandom numbers (*FPspline*).

```
#define TABSIZE          256
#define TABMASK          (TABSIZE-1)
#define PERM(x)          perm[(x)&TABMASK]
#define INDEX(ix,iy,iz)  PERM((ix)+PERM((iy)+PERM(iz)))
#define FLOOR(x)         (int)(x)
/* PERMUTATION TABLE */
static unsigned char perm[TABSIZE] = {
225, 155, 210, 108, 175, 199, 221, 144, 203, 116, 70, 213, 69, 158, 33, 252, 5, 82, 173,
133, 222, 139, 174, 27, 9, 71, 90, 246, 75, 130, 91, 191, 169, 138, 2, 151, 194, 235, 81, 7,
25, 113, 228, 159, 205, 253, 134, 142, 248, 65, 224, 217, 22, 121, 229, 63, 89, 103, 96,
104, 156, 17, 201, 129, 36, 8, 165, 110, 237, 117, 231, 56, 132, 211, 152, 20, 181, 111,
239, 218, 170, 163, 51, 172, 157, 47, 80, 212, 176, 250, 87, 49, 99, 242, 136, 189, 162,
115, 44, 43, 124, 94, 150, 16, 141, 247, 32, 10, 198, 223, 255, 72, 53, 131, 84, 57, 220,
197, 58, 50, 208, 11, 241, 28, 3, 192, 62, 202, 18, 215, 153, 24, 76, 41, 15, 179, 39, 46,
55, 6, 128, 167, 23, 188, 106, 34, 187, 140, 164, 73, 112, 182, 244, 195, 227, 13, 35, 77,
196, 185, 26, 200, 226, 119, 31, 123, 168, 125, 249, 68, 183, 230, 177, 135, 160, 180, 12,
1, 243, 148, 102, 166, 38, 238, 251, 37, 240, 126, 64, 74, 161, 40, 184, 149, 171, 178, 101,
66, 29, 59, 146, 61, 254, 107, 42, 86, 154, 4, 236, 232, 120, 21, 233, 209, 45, 98, 193, 114,
78, 19, 206, 14, 118, 127, 48, 79, 147, 85, 30, 207, 219, 54, 88, 234, 190, 122, 95, 67, 143,
109, 137, 214, 145, 93, 92, 100, 245, 0, 216, 186, 60, 83, 105, 97, 204, 52
};
#define RANDNBR ww (((float)rand())/RAND_MASK)
float valueTab[TABSIZE];
/* ========================================================= */
/* VALUE TABLE INIT */
/* initialize the table of pseudorandom numbers */
void valueTableInit(int seed)
```

```
{
  float *table = valueTab;
  int i;
  srand(seed);
  for (i=0; i<TABSIZE; i++)
    *(table++) = 1.0 -2.0*RANDNBR;
}
/* ======================================================= */
/* LATTICE function */
/* returns a value corresponding to the lattice point */
float lattice(int ix, int iy, int iz)
{
  return valueTab[INDEX(ix,iy,iz)];
}
/* ======================================================= */
/* NOISE function */
float noise(float x, float y, float z)
{
  int ix,iy,iz;
  int i,j,k;
  float fx,fy,fz;
  float xknots[4],yknots[4],zknots[4];
  static int initialized = 0;
  if (!initialized) {
    valueTableInit(665);
    initialized = 1;
  }
  ix = FLOOR(x);
  fx = x - ix;
  iy = FLOOR(y);
  fy = y - iy;
  iz = FLOOR(z);
  fz = z-iz;
  for (k=-1; k<=2; k++) {
    for (j=-1; j<=2; j++) {
      for (i=-1; i<=2; i++) [i+1] = lattice(ix+i,iy+j,iz+k);
      yknots[j+1] = spline(fx,xknots);
    }
    zknots[k+1] = spline(fy,yknots);
  }
  return spline(fz,zknots);
}
#define FP00 -0.5
#define FP01 1.5
#define FP02 -1.5
#define FP03 0.5
#define FP10 1.0
```

```
#define FP11 -2.5
#define FP12 2.0
#define FP13 -0.5
#define FP20 -0.5
#define FP21 0.0
#define FP22 0.5
#define FP23 0.0
#define FP30 0.0
#define FP31 1.0
#define FP32 0.0
#define FP33 0.0
/* ========================================================== */
float spline(float u,float *knots)
{
  float c3,c2,c1,c0;
  c3 = FP00*knots[0] + FP01*knots[1] + FP02*knots[2] + FP03*knots[3];
  c2 = FP10*knots[0] + FP11*knots[1] + FP12*knots[2] + FP13*knots[3];
  c1 = FP20*knots[0] + FP21*knots[1] + FP22*knots[2] + FP23*knots[3];
  c0 = FP30*knots[0] + FP31*knots[1] + FP32*knots[2] + FP33*knots[3];
  return ((c3*u + c2)*u + c1)*u + c0;
}
```

B.6.2 Turbulence

Turbulence is a stochastic function with a "fractal" power spectrum [3]. The function is a sum of amplitude-varying frequencies. As frequency increases, the amplitude decreases.

```
/* TURBULENCE */
float turbulence (float x, float y, float z)
{
  float f;
  float value = 0;
  for (f = MINFREQ; f < MAXFREQ; f *= 2) value += fabs(noise(x*f, y*f, z*f))/f;
  return value;
}
```

B.6.3 Random number generator

This random number generator returns a random number in the range 0 to 999999999. An auxiliary routine would map this number into an arbitrary range of integers.

```
int r[100]; /* "global" pseudorandom table -- */
/* must be visible to rand and init_rand */
/* ========================================================== */
/* RAND */
/* return a random number in the range 0 to 999999999 */
int rand (void)
{
```

```
  int i = r[98];
  int j = r[99];
  int k;
  int t;
  if ((t = r[i] 2 r[j]) < 0) t += 1000000000 L;
  r[i] = t;
  r[98]--; r[99]--;
  if (r[98] == 0) r[98] = 55;
  if (r[99] == 0) r[99] = 55;
  k = r[100] % 42 + 56;
  r[100] = r[k];
  r[k] = t;
  return(r[100]);
}
/* ======================================================= */
/* INIT RAND */
/* seed the random number table */
void init_rand (char *seed)
{
  char buf[101];
  int i, j, k;
  if (strlen(seed) > 85) return(0);
  sprintf(buf, "aEbFcGdHeI%s", seed);
  while (strlen(buf) < 98) strcat(buf, "Q");
  for (i = 1; i < 98; i++) r[i] = buf[i] * 8171717 + i * 997;
  i = 97; j = 12;
  for (k = 1; k < 998; k++) {
    r[i] -= r[j];
    if (r[i] < 0) r[i] += 1000000000;
    i--; j--;
    if (i == 0) i=97;
    if (j == 0) j=97;
  }
  r[98] = 55;
  r[99] = 24;
  r[100] = 77;
}
/* ======================================================= */
/* RAND INT */
/* return a random int between a and b */
/* assumes init_rand already called. */
int rand_int(int a, int b)
{
  return (a + rand() % (b - a + 1));
}
```

B.7 Physics primer

Physically based motion is a limited simulation of physical reality. This can be as simple or as complex as the implementation requires. In the following sections are some of the equations of importance in simple physics simulation that can be found in any one of several standard texts. *The Mechanical Universe* [5] is used as the source for the brief discussion that follows.

B.7.1 Position, velocity, and acceleration

The fundamental equation relating position, distance, and speed is shown in Equation B.90. This can be used to control the positioning of an object for any particular frame of the animation because the frame number is tied directly to time (Eq. B.91). The average velocity of a body is the distance moved divided by the time it took to move, as stated by Equation B.92. Notice that the unit of velocity is distance per time, for example, feet/second.

$$\text{distance} = \text{speed} \times \text{time} \tag{B.90}$$

$$\text{time} = \text{frameNumber} \times \text{timePerFrame} \tag{B.91}$$

$$\text{average Velocity} = \text{distanceTraveled}/\text{time} \tag{B.92}$$

For this discussion, distance as a function of time is $s(t)$; the average velocity from time $t1$ to $t2$ is $(s(t2) - s(t1))/(t2 - t1)$. The instantaneous velocity is determined by moving $t2$ closer and closer to $t1$. In the limiting case this becomes the derivative of the distance function with respect to time. Similarly, the average acceleration of an object is the change in velocity divided by the time it took to effect the change. This is presented in Equation B.93, where $v(t)$ is a function that gives the velocity of the object at time t. Notice that the unit of acceleration is velocity per time or distance per time, for example, feet/second^2. In the same way, instantaneous acceleration is the derivative of $v(t)$ with respect to time (Eq. B.94). In the case of motion due to gravity, g is the acceleration due to gravity—a constant that has been measured to be 32 feet/second^2 or 9.8 meters/second^2 (Eq. B.95).

$$a_{\text{ave}} = (v(t_2) - v(t_1))/(t_2 - t_1) \tag{B.93}$$

$$a(t) = v'(t) = s''(t) \tag{B.94}$$

$$\begin{aligned} a(t) &= g \\ v(t) &= gt \\ s(t) &= (1/2)gt^2 \end{aligned} \tag{B.95}$$

B.7.2 Circular motion

Circular motion is important in physics and arises for a variety of phenomena, including the movement of planets and robotic armatures. Circular motion is easily specified by using polar coordinates. The position of a particle orbiting the origin at a distance r can be described using Equation B.96. Here, i and j are orthonormal unit vectors (at right angles to each other and unit length) and $p(t)$ is the positional vector of the particle. In a constant radius circular orbit, $\theta(t)$ varies as a function of time, and the distance r is constant. During uniform circular motion, $\theta(t)$ changes at a constant rate, and *angular velocity* is said to be constant. Angular velocity is referred to here as $\omega(t)$ (Eq. B.97). As for constant-velocity linear motion, in which the distance equals speed multiplied by time, for constant angular velocity the angle equals angular velocity multiplied by time (Eq. B.98). If $\theta(t)$ is measured in radians and time in seconds, then $\omega(t)$ is measured in radians per second. To simplify the following equations, the functional dependence on time will often be omitted when the dependence is obvious from the context.

$$p(t) = (r\cos(\theta(t)))i + (r\sin(\theta(t)))j \tag{B.96}$$

$$\frac{d}{dt}\theta(t) = \omega(t) \tag{B.97}$$

$$\theta(t) = \omega t \tag{B.98}$$

Taking the derivative of Equation B.96 with respect to time gives the instantaneous velocity (Eq. B.99). Notice that the velocity vector, $v(t)$, is perpendicular to the position vector, $p(t)$. This can be demonstrated by taking the dot product of the two vectors $v(t)$ and $p(t)$ and showing that it is identically zero.

$$v(t) = \frac{dp}{dt} = ((-r)\omega\sin(\omega t))i + (-r\omega\cos(\omega t))j \tag{B.99}$$

Computing the length of $v(t)$ shows that $|v| = r\,\omega$ and, therefore, that the velocity is independent of t (i.e., constant). Notice, however, that a constant circular motion still gives rise to an acceleration. Taking the derivative of Equation B.99 produces Equation B.100, which is called the centripetal acceleration. The centripetal acceleration resulting from uniform circular motion is directed radially inward and has constant magnitude. With the equation for the length of $v(t)$ from earlier, the magnitude of the acceleration can be written using Equation B.101.

$$a(t) = \frac{dv}{dt} = (-\omega^2)((r\cos(\omega t))i + (r\sin(\omega t))j) \tag{B.100}$$

$$= (-\omega^2)p(t)$$

$$a = v^2/r \tag{B.101}$$

For any particle in a rigid mass undergoing a rotation, that particle is undergoing the same rotation about its own center. In addition, if the particle is displaced from the center of rotation, then it is also undergoing an instantaneous positional translation as a result of its circular motion (Figure B.52).

B.7.3 Newton's laws of motion

It is useful to review Newton's laws of motion. The first law is the principle of inertia. The second law relates force to the acceleration of a mass (Eq. B.102). In another form, this law relates force to change in momentum (Eq. B.103), where momentum is mass times velocity ($m \bullet v$). The third law states that when an object pushes with a force on another object, the second object pushes back with an equal but opposite force. It is important to note that the force F used here is considered to be the sum of all external forces acting on an object. Force is a vector quantity, and these equations really represent three sets of equations, one for each coordinate (Eq. B.104). Newton's laws of motion are stated as follows:

First Law: If no force is acting on an object, then it will not accelerate. It will maintain a constant velocity.

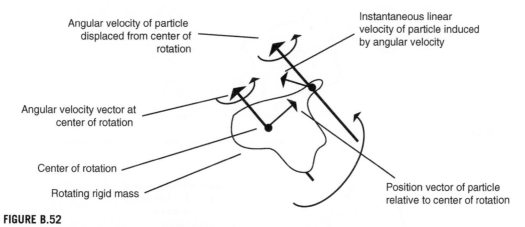

FIGURE B.52

Motion of particle in a rotating rigid mass.

Second Law: The change of motion of an object is proportional to the forces applied to it.
Third Law: To every action there is always opposed an equal and opposite reaction.

$$F = ma \tag{B.102}$$

$$F = \frac{d}{dt}(mv)$$

$$F_x = ma_x$$
$$F_x = ma_y \tag{B.103}$$

$$F_x = ma_z \tag{B.104}$$

B.7.4 Inertia and inertial reference frames

An *inertial frame* is a frame of reference (e.g., a local coordinate system of an object) in which the principle of inertia holds. Any frame that is not accelerating is an inertial frame. In an inertial frame one observes the laws of motion and has no way of determining whether one is at rest or moving in an "absolute" sense. (But then, what is "absolute"?)

B.7.5 Center of mass

The center of mass of an object is that point at which the object is balanced in all directions. If an external force is applied in line with the center of mass of an object, then the object moves as if all the mass were concentrated at the center ("*c*" in Figure B.53).

B.7.6 Torque

The tendency of a force to produce circular motion is called *torque*. Torque is produced by a force applied off-center from the center of mass of an object (Figure B.54) and is computed by Equation B.105.

$$\tau = r \times F \tag{B.105}$$

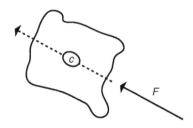

FIGURE B.53

Center of mass.

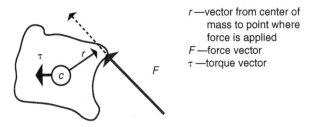

r—vector from center of
mass to point where
force is applied
F—force vector
τ —torque vector

FIGURE B.54

Applying a force off-center induces a torque.

It is important to note that *torque* refers to a specific axis of rotation and is perpendicular to both *r* and *F* in Figure B.54. The same force can exert different torques about different axes of rotation if it is applied at different locations on an object. A given torque vector for a rigid body is position independent. That is, for any particle in a rigid mass, the particle is undergoing that same torque.

B.7.7 Equilibrium: balancing forces

In the absence of acceleration, there is no change in the motion of an object (Newton's first law). In such cases, the net force and the torque acting on a body vanish (Eqs. B.106 and B.107). There may be forces present, but the vector sum is equal to zero.

$$\Sigma F = 0 \qquad\qquad (B.106)$$

$$\Sigma \tau = 0 \qquad\qquad (B.107)$$

B.7.8 Gravity

Equation B.108 is Newton's law of universal gravitation. It calculates the force of gravity between two masses (m_1 and m_2) whose centers of mass are a distance *r* apart. *G* is the universal gravitational constant equal to 6.67×10^{-11} Newton meter^2/kilogram^2. When two objects are not touching, each acts on the other as if all its mass were concentrated at its center of mass. When an object is on (or near) the earth's surface, the distance of the object to the center of mass of the earth, the mass of the earth, and the gravitational constant of Equation B.108 can all be combined to produce the object's acceleration (Eq. B.109). This acceleration is usually denoted as g.

$$F = G\frac{m_1 m_2}{r^2} \tag{B.108}$$

$$a = \frac{F}{m_{object}} = G\frac{m_{earth}}{(radius_{earth})^2} = 9.8\text{m/s}^2 = 32\text{f/s}^2 = g \tag{B.109}$$

B.7.9 Centripetal force

Centripetal force is any force that is directed inward, toward the center of an object's motion (centripetal means *center seeking*). In the case of a body in orbit, gravity is the centripetal force that holds the body in the orbit. Consider a body, such as the moon, with mass M_m and a distance r away from the earth. The unit vector from the earth to the moon is R. The earth has a mass M_e. Equation B.110 calculates the force vector that results from gravity. The acceleration induced by a circular orbit always points toward the center (centripetal), as in Equation B.111. As previously shown, the acceleration due to circular motion has magnitude v^2/r (Eq. B.100). This can be used to solve for velocity (Eq. B.112).

$$F = -G\frac{M_m M_e}{r^2}R \tag{B.110}$$

$$a = F/M_m$$
$$a = ((-GM_e)/r^2)R \tag{B.111}$$
$$a = (-v^2/r)R$$
$$a = (-GM_e/r^2)R$$
$$v = \sqrt{(GM_e)/r} \tag{B.112}$$

B.7.10 Contact forces

Gravity is one of the four fundamental forces (gravity, electromagnetic, strong, and weak) that act at a distance. Another important category is *contact forces*. The tension in a wire or rope is an example of a contact force, as is the compression in a rigid rod. These forces arise from the complex interactions of electric forces that tend to keep atoms a certain distance apart. The empirical law that governs such forces, and that is familiar when dealing with springs, is *Hooke's law* (Eq. B.113). Variable x is the change from the equilibrium length of the spring, k is the spring constant, and F is the restoring force of the spring to return to rest length. The constant k is a measure of the stiffness of the spring; the larger k is, the more sensitive the spring is to motion away from the rest position.

$$F = -kx \tag{B.113}$$

Another example of contact force, known as the *normal force*, is the result of repulsion between any two objects pressed against each other. It arises from the repulsion of the atoms of the two objects. It is always perpendicular to the surfaces, and its magnitude is proportional to how hard the two objects are pressed against each other. When objects in motion come in contact, an *impulse force* due to collision is produced. The impulse force is a short-duration force that is applied normal to the surface of contact on each of the two objects. Calculation of the impulse force due to collision is discussed in Chapter 7, Section 4.2. Other important examples of contact forces are friction and viscosity.

Friction

Friction arises from the interaction of surfaces in contact. It can be thought of as resulting from the multitude of collisions of molecules caused by the unevenness of the surfaces at the microscopic level.

The frictional force works against the relative motion of the two objects. Frictional forces from a surface do not exceed an amount proportional to the normal force exerted by the surface on the object. This is stated in Equation B.114, where s is the coefficient of static friction and f_N is the normal force (component of the force that is perpendicular to the surface). Variable s varies according to the two materials that are in contact.

$$f_s = sf_N \tag{B.114}$$

The frictional forces acting between surfaces at rest with respect to each other are called *forces of static friction*. For a force applied to an object sitting on another object that is parallel to the surfaces in contact, there will be a specific force at which the block starts to slip. As the force increases from zero up to that threshold force, the lateral force is counteracted by an equal force of friction in the opposite direction. Once the object begins to move, *kinetic friction* acts on the object and approximately obeys the empirical law of Equation B.115, where k is the coefficient of kinetic friction and f_N is the force normal to the surface. Kinetic friction is typically less than static friction. The force of kinetic friction is always opposite to the velocity of the object.

$$f_k = kf_N \tag{B.115}$$

Viscosity

The resistive force of an object moving in a medium is *viscosity*. It is another contact force and is extremely difficult to model accurately. When an object moves at low velocity through a fluid, the viscosity is approximately proportional to the velocity (Eq. B.116); K is the constant of proportionality, which depends on the size and the shape of the object, and n is the coefficient of viscosity, which depends on the properties of the fluid. The coefficient of viscosity, n, decreases with increasing temperature for liquids and increases with temperature for gases. Stokes's law for a sphere of radius R is given in Equation B.117.

$$F_{vis} = -Knv \tag{B.116}$$

$$K = 6\pi R \tag{B.117}$$

An object dropping through a liquid attains a constant speed, called the limiting or terminal velocity, at which gravity, acting downward, and the viscous force, acting upward, balance each other and there is no acceleration (e.g., Eq. B.118 for a sphere). Terminal velocity is given by Equation B.119. In a viscous medium, heavier bodies fall faster than lighter bodies. For spherical objects falling at a low velocity in a viscous medium, not necessarily at terminal velocity, change in momentum is given by Equation B.120.

$$mg = 6\pi Rnv \tag{B.118}$$

$$v = (mg)/(6\pi Rn) \tag{B.119}$$

$$m\frac{dv}{dt} = mg - 6\pi Rnv \tag{B.120}$$

B.7.11 Centrifugal force

Consider an object (a frame of reference) that rotates in uniform circular motion with respect to a post (an inertial frame) because it is held at a constant distance by a rope. Relative to the inertial frame, each point in the uniformly rotating frame has centripetal acceleration expressed by Equation B.121; r is the distance of the point from the axis of rotation, R is the unit vector from the inertial frame to the rotating frame, and v is the speed of the point. The tension in the rope supplies the force necessary to produce the centripetal acceleration. Relative to the rotating frame (not an inertial frame), the frame itself does not move and therefore the *centrifugal* force necessary to counteract the force supplied by the rope is calculated by using Equation B.122.

$$a = -\frac{v^2}{r}R \tag{B.121}$$

$$F_c = -(ma) = -\frac{mv^2R}{r} \tag{B.122}$$

B.7.12 Work and potential energy

For a constant force of magnitude F moving an object a distance h parallel to the force, the work W performed by the force is shown in Equation B.123. If a mass m is lifted up so that it does not accelerate, then the lifting force is equal to the weight (mass $\times$ gravitational acceleration) of the object. Since the weight is constant, the work done to raise the object up to a height h is presented in Equation B.124. Energy that a body has by virtue of its location is called *potential energy*. The work in this case is converted into potential energy.

$$W = Fh \tag{B.123}$$

$$W = mgh \tag{B.124}$$

B.7.13 Kinetic energy

Energy of motion is called *kinetic energy* and is shown in Equation B.125. The velocity of a falling body that started at a height h is calculated by Equation B.126. Its kinetic energy is therefore calculated by Equation B.127.

$$K = \frac{1}{2}mv^2 \tag{B.125}$$

$$v^2 = 2gh \tag{B.126}$$

$$K = \frac{1}{2}mv^2 = mgh \tag{B.127}$$

B.7.14 Conservation of energy

Potential plus kinetic energy of a closed system is conserved (Eq. B.128). This is useful, for example, when solving for an object's current height (current potential energy) when its initial height (initial potential energy), initial velocity (initial kinetic energy), and current velocity (current kinetic energy) are known.

$$U_a + K_a = U_b + K_b \tag{B.128}$$

B.7.15 Conservation of momentum

In a closed system, total momentum (mass times velocity) is conserved. This means that it does not change (Eq. B.129) and that the momenta of all objects in a closed system always sum to the same amount (Eq. B.130). This is useful, for example, when solving for velocities after a collision when the velocities before the collision are known.

$$\frac{d}{dt}((m_1v_1 + m_2v_2 + \ldots + m_nv_n)) = 0 \tag{B.129}$$

$$m_1v_1 + m_2v_2 + \ldots + m_nv_n = \text{constant} \tag{B.130}$$

B.7.16 Oscillatory motion

In some systems, the stability of an object is subject to a linear restoring force, F. The force is linearly proportional (using k as the constant of proportionality) to the distance x the object has been displaced from its equilibrium position (Eq. B.131). Associated with this force is the potential energy (Eq. B.132) that results from the object's displacement. These systems have the property that, if they are disturbed from equilibrium, the restoring force that acts on them tends to move them back into equilibrium. When disturbed from equilibrium, they tend to overshoot that point when they return, due to inertia. Then the restoring force acts in the opposite direction, trying to return the system to equilibrium. The result is that the system oscillates back and forth like a mass on the end of a spring or the weight at the end of a pendulum.

$$F = -kx \tag{B.131}$$

$$U = \frac{1}{2}kx^2 \tag{B.132}$$

Combining the basic equations of motion (force equals mass times acceleration) with Equation B.131, one can derive the differential equation for oscillatory motion. This equation is satisfied by the displacement function $x(t)$ (Eq. B.133).

$$\begin{aligned} F &= ma \\ F &= -kx \\ a &= \frac{d^2x}{dt} \\ m\frac{d^2x}{dt} &= -kx \\ \frac{d^2x}{dt} &= -kx/m \end{aligned} \tag{B.133}$$

B.7.17 **Damping**

The damping force can be modeled after Stokes's law, in which resistance is assumed to be linearly proportional to velocity (Eq. B.134). This is usually valid for oscillations of sufficiently small amplitude. The damping force opposes the motion. The constant, k_d, is called the *damping coefficient*. Adding the damping force to the spring force produces Equation B.135. Dividing through by m and collecting terms results in Equation B.136, where $b = k_d/m$ and $a^2 = k/m$.

$$F_d = -k_d \frac{dx}{dt} \tag{B.134}$$

$$m \frac{d^2 x}{dt} = -kx - k_d \frac{dx}{dt} \tag{B.135}$$

$$\frac{d^2 x}{dt} + b \frac{dx}{dt} + a^2 x = 0 \tag{B.136}$$

If there is a spring force but no damping, the general solution can be written as $x = C \cos(at + \theta_0)$. If there is damping but no spring force, the general solution turns out to be $x = C e^{-bt} + D$, with C and D constant. If both the spring force and the damping force are present, the solution takes the form shown in Equation B.137.

$$x = C e^{-bt} \cos(dt + \theta_0)$$
$$\text{where } d = \sqrt{a^2 - \frac{b^2}{4}} \tag{B.137}$$
$$\text{for } b < 2a$$

B.7.18 **Angular momentum**

Angular momentum is the rotational equivalent of linear momentum and can be computed by Equation B.138, where r is the vector from the center of rotation and p is momentum (mass $\times$ velocity). The temporal rate of change of angular momentum is equal to torque (Eq. B.139). Angular momentum, like linear momentum, is conserved in a closed system (Eq. B.140).

$$L = r \times p \tag{B.138}$$

$$\tau = \frac{dL}{dt} \tag{B.139}$$

$$\Sigma L_i = \text{constant} \tag{B.140}$$

B.7.19 **Inertia tensors**

An *inertia tensor*, or *angular mass*, describes the resistance of an object to a change in its angular momentum [5] [13]. It is represented as a matrix when the angular mass is related to the principal axes of the object (Eq. B.141). The terms of the matrix describe the distribution of the mass of the object relative to a local coordinate system (Eq. B.142). For objects that are symmetrical with respect to the local axes, the off-diagonal elements are zero (Eq. B.143). For a rectangular solid with mass M and dimensions $a, b,$ and c along its local axes, the inertia tensor at the center of mass is given by Equation B.144. For a sphere with radius R and mass M, the inertia tensor is given by Equation B.145. If the inertia tensor is known at one position by I, then the inertia tensor I' for parallel axes at a new position (X, Y, Z) relative to the original position is calculated by Equation B.146. If the inertia tensor for one set of axes is known by I, then the inertia tensor for a rotated frame is calculated by Equation B.147, where R is the rotation matrix describing the rotated frame relative to the original frame.

$$I = \begin{bmatrix} I_{xx} & -I_{xy} & -I_{xz} \\ -I_{yx} & I_{yy} & -I_{yz} \\ -I_{zx} & -I_{zy} & I_{zz} \end{bmatrix} \tag{B.141}$$

$$\begin{aligned} I_{xx} &= \int (y^2 + z^2) dm \\ I_{yy} &= \int (x^2 + z^2) dm \\ I_{zz} &= \int (x^2 + y^2) dm \\ I_{xy} &= \int xy\, dm \\ I_{xz} &= \int xz\, dm \\ I_{yz} &= \int yz\, dm \end{aligned} \tag{B.142}$$

$$I = \begin{bmatrix} I_{xx} & 0 & 0 \\ 0 & I_{yy} & 0 \\ 0 & 0 & I_{zz} \end{bmatrix} \tag{B.143}$$

$$I = \frac{1}{12} \begin{bmatrix} M(b^2 + c^2) & 0 & 0 \\ 0 & M(a^2 + c^2) & 0 \\ 0 & 0 & M(a^2 + b^2) \end{bmatrix} \tag{B.144}$$

$$I = \frac{2MR^2}{5} \begin{bmatrix} 1 & 0 & 0 \\ 0 & 1 & 0 \\ 0 & 0 & 1 \end{bmatrix} \tag{B.145}$$

$$I_{\text{translated}} = \begin{bmatrix} I_{xx} + M(Y^2 + Z^2) & -I_{xy} - MXY & -I_{xz} - MXZ \\ -I_{xy} - MXY & I_{yy} + M(X^2 + Z^2) & -I_{yz} - MXYZ \\ -I_{xz} - MXZ & -I_{yz} - MZYX & I_{zz} + MX^2 + Y^2 \end{bmatrix} \tag{B.146}$$

$$I_{\text{rotated}} = R I_{\text{object}} R^{-1} \tag{B.147}$$

B.8 Numerical integration techniques

Numerical integration is useful for finding the arc length of the curve, updating arbitrary function values using derivative information, and specifically updating the position of an object over time. A useful technique for arc length computation is Gaussian quadrature. With regard to general function value updating, Runge-Kutta, explicit Euler integration, and implicit Euler integration are discussed. Huen, Verlet, and Leapfrog integration are covered specifically in the context of position update using a known acceleration. As with many numerical techniques, Press et al. [16] is an extremely valuable reference.

B.8.1 **Function integration for arc length computation**

Given a function $f(x)$, Gaussian quadrature can be used to produce an arbitrarily close approximation to the integral of the function between two values, $\int_a^b f(x)$, if the function is sufficiently well behaved [16]. Gaussian quadrature approximates the integral as a sum of weighted evaluations of the function at specific values (abscissas). The number of evaluations controls the error in the approximation. In its general form, Gaussian quadrature incorporates a multiplicative function, $W(x)$, which can condition some functions for the approximation (Eq. B.148). Gauss-Legendre integration is a special case in which $W(x) = 1.0$, and it results in Equation B.149. The code in Figure B.55 for $n = 10$ duplicates that used in Chapter 3, Section 2.1 for computing arc length. Figure B.56 gives the code for computing Gauss-Legendre weights and abscissas for arbitrary n.

$$\int_a^b (W(x)f(x))dx \cong \sum_{i=1}^n w_i f(x_i) \tag{B.148}$$

$$\int_a^b f(x)dx \cong \sum_{i=1}^n wf(x_i) \tag{B.149}$$

```
/* ------------------------------------------------------------
INTEGRATE FUNCTION
use gaussian quadrature to integrate square root of given function in
the given interval */
double integrate_func(polynomial_td *func,interval_td *interval)
{
    double x[5]={.1488743389,.4333953941,.6794095682,.8650633666,.9739065285};
    double w[5]={.2966242247,.2692667193,.2190863625,.1494513491,.0666713443};
    double length, midu, dx, diff;
    int i;
    double evaluate_polynomial();
    double u1,u2;

    u1 = interval->u1;
    u2 = interval->u2;

    midu = (u1+u2)/2.0;
    diff = (u2-u1)/2.0;
    length = 0.0;
    for (i=0; i<5; i++) {
        dx = diff*x[i];
        length += w[i]*(sqrt(evaluate_polynomial(func,midu+dx)) +
        sqrt(evaluate_polynomial(func,midu-dx)));
    }
    length *= diff;
    return (length);
}
```

FIGURE B.55

Gauss-Legendre integration for $n = 10$.

```
/* GAUSS-LEGENDRE */

#define EPSILON 0.00000000001
/* calculate the weights and abscissas of the Gauss-Legendre n-point form */
void gaussWeights(float a, float b, float *x, float *w, int n)
{
  int i,j,m;
  float p1,p2,p3,p;
  float z,z1;
  float xave,xdiff;

  m = (n+1)/2;
  xave = (b+a)/2;
  xdiff = (b-a)/2;
  for (i = 0; i<m; i++) {
    z = cos(PI*((i+1)-0.25)/(n+0.5));
    do {
      p1 = 1.0;
      p2 = 0.0;
      for (j=0; j<n; j++) {
        p3 = p2;
        p2 = p1;
        p1 = ((2*(j+1) - 1.0)*z*p2-j*p3)/(j+1);
      }
      pp = n*(z*p1-p2)/(z*z-1);
      z1 = z;
      z = z1-p1/pp;
    } while (fabs(z-z1) > EPSILON);
    x[i] = xave - xdiff*z;
    x[n-1-i] = xave + xdiff*z;
    w[i] = 2.0*xdiff/((1.0-z*z)*pp*pp);
    w[n-1-i] = w[i];
  }
}
```

FIGURE B.56

Computing Gauss-Legendre weights and abscissas.

B.8.2 Updating function values

Integrating ordinary differential equations (ODEs) in computer animation typically means that the derivative function f' is available and that a numerical approximation to the function f is desired. For example, in a physically based simulation, the time-varying acceleration of an object is computed from the object's mass and the forces acting on the object in the environment. From the acceleration (the derivative function), the velocity (the function) is numerically calculated over time. Similarly, once the time-based velocity function is known (the derivative function), the time-varying position of the object (the function) can be calculated numerically.

The simple form of an ordinary differential equation involves a first-order derivative of a function of a single variable. In addition, it is usually the case that conditions at an initial point in time are known and that the numerical integration is used in a simulation of a system as time moves forward. Such problems are referred to as *initial value problems*.

The (explicit) Euler method

The *Euler method* is the most basic technique used for solving such simple ODE initial value problems. It is shown in Equation B.150, where h is the time step such that $x_{n+1} = h + x_n$. This method is not symmetrical in that it uses information at the beginning of the time step to advance to the end of the time step. The derivative at the beginning of the time step produces the vector, which is tangent to the curve representing the function at that point in time. The tangent at the beginning of the interval is used as a linear approximation to the behavior of the function over the entire interval (Figure B.57). The Euler method is neither stable nor very accurate. Other methods, such as Runge-Kutta, are more accurate for equivalent computational cost.

$$y_{n+1} = y_n + hf'(x_n, y_n) \tag{B.150}$$

Runge-Kutta

Runge-Kutta is a family of methods that is symmetrical with respect to the interval. The *second-order Runge-Kutta*, or *midpoint method*, is mentioned in Chapter 7, Section 4.1 in the discussion of physically based simulations. A half-step, using the explicit Euler method, is taken and the derivative is evaluated. This derivative is then used to update the original value (see Eq. B.151).

$$y_{n+1} = y_n + f'\left(x_n + \frac{h}{2}, y_n + \frac{h}{2}f'(x_n, y_n)\right) \tag{B.151}$$

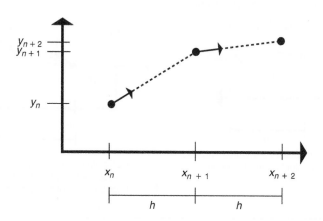

FIGURE B.57

The Euler method.

One of the most useful methods is *fourth-order Runge-Kutta* [16], shown in Equation B.152 and Figure B.58. It is referred to as a fourth-order method because its error term is on the order of the interval h to the fifth power. The advantage of using the method is that although each step requires more computation, larger step sizes can be used, resulting in an overall computational savings.

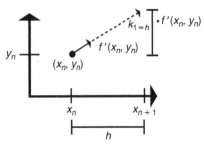

A   Compute the derivative at the beginning of the interval

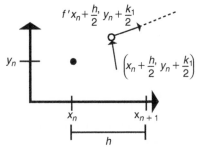

B   Step to midpoint (using derivative previously computed) and compute derivative

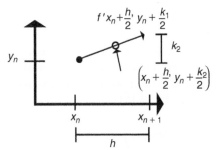

C   Step to new midpoint from initial point using midpoint's derivative just computed

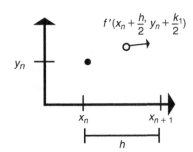

D   Compute the derivative at the new midpoint

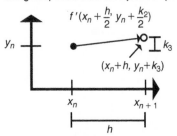

E   Use new midpoint's derivative and step from initial point to end of interval

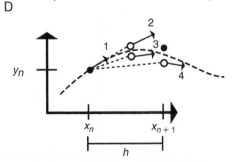

F   Compute derivative at end of interval and average with 3 previous derivatives to step from initial point to next function value

FIGURE B.58

The steps in computing fourth-order Runge-Kutta.

$$k_1 = hf'(x_n, y_n)$$

$$k_2 = hf'\left(x_n + \frac{h}{2}, y_n + \frac{k_1}{2}\right)$$

$$k_3 = hf'\left(x_n + \frac{h}{2}, y_n + \frac{k_2}{2}\right) \qquad \text{(B.152)}$$

$$k_4 = hf'(x_n + h, y_n + k_3)$$

$$y_{n+1} = y_n + \frac{k_1}{6} + \frac{k_2}{3} + \frac{k_3}{3} + \frac{k_4}{6} + O(h^5)$$

The implicit Euler method

The explicit Euler method updates the function value using the derivative at the current value. The danger with the implicit method is that if the time step is too large, the updated function values can be a poor approximation to the actual function. The implicit method, for comparison, finds a new position whose derivative can update the current value to the new value (Eq. B.153).

$$y_{n+1} = y_n + hf'(x_{n+1}, y_{n+1}) \qquad \text{(B.153)}$$

This requires solving an equation to find such a y_{n+1} using, for example, a Newton-Raphson method. It takes more time to take a step using the implicit Euler method but because this technique is more stable than explicit Euler, large time steps can be taken and, thus, an overall computational savings can often be achieved.

The semi-implicit Euler method

The *semi-implicit Euler method* is derived from the implicit Euler method by approximating the y_{n+1} term used in evaluating the derivative by an explicit Euler step, thus resulting in Equation B.154.

$$y_{n+1} = y_n + hf'(x_{n+1}, y_n + hf'(x_n, y_n)) \qquad \text{(B.154)}$$

The semi-implicit Euler method offers more stability than explicit Euler and is computationally much less expensive than implicit Euler.

B.8.3 Updating position

The next three methods are discussed specifically in the context of updating position from velocity and/or acceleration over a time period Δt.

The Heun method

The *Heun method* of integration, also called the *improved Euler method*, augments the basic Euler method. The Euler method takes the derivative at the current position to update to the next position. The Heun method takes the derivative at the position computed by the basic Euler method and averages it with the first-computed derivative and uses the average to update the position (Eq. B.155). This is a common integration method used in physics texts.

$$v(t_{i+1}) = v(t_i) + a(t_i)\Delta t$$
$$x(t_{i+1}) = x(t_i) + \frac{1}{2}(v(t_i) + v(t_{i+1}))\Delta t$$
$$x(t_{i+1}) = x(t_i) + v(t_i)\Delta t + \frac{1}{2}a(t_i)\Delta t^2$$

(B.155)

The Verlet method

Verlet integration is used to update the position directly from acceleration without explicitly calculating velocity. Essentially, the difference between the current position and the previous position is used as velocity. This vector is updated by acceleration and added to the current position to generate the next position (see Eq. B.156).

$$x(t_{i+1}) = 2x(t_i) - x(t_{i-1}) + a(t_i)\Delta t^2$$

(B.156)

The Leapfrog method

The *Leapfrog method* is unique in its temporal symmetry. It can be run forward in time or backward in time and it will trace out the same path. Position and acceleration are evaluated on the interval while velocity is evaluated on the half interval relative to the position (see Eq. B.157).

$$x(t_{i+1}) = x(t_i) + v(t_{i+\frac{1}{2}})\Delta t$$
$$v(t_{i+\frac{3}{2}}) = v(t_{i+\frac{1}{2}}) + a(t_{i+1})\Delta t$$

(B.157)

B.9 Optimization

As the field of computer graphics and computer animation matures, more complex and mathematically sophisticated techniques are being used. An increasingly important mathematical tool is *optimization*, also know as *mathematical programming*.[2] This discussion is intended for the reader who is well versed in computer graphics and familiar with the basics of calculus, but who has not had much, if any, exposure to mathematical techniques involving optimization and who might be a bit intimidated whenever discussion turns that direction. Optimization is a complex and expansive topic, and finds application in control theory, chemistry, meteorology, VLSI CAD, geophysics, power grid design, and many others. A complete treatment is beyond the scope of the current discussion and the interested reader should follow up with material from the graphics literature (e.g., [27] [28]) or the mass of material on optimization in general (e.g., [29]) or in the literature concerning a specific application of interest.

In computer graphics, optimization finds application in areas such as surface fitting, decimation, inverse kinematics, motion capture data editing, physically based motion, and path construction. These tasks have certain features in common. For example, problems in these areas usually have a large

---

[2] The term "programming" refers to the U.S. military's use of the term program as applied to logistics, the problem domain for the first early optimization techniques, referring to a well-defined procedure for doing something (i.e., linear programming), and has nothing to do with computer programming.

parameter space. Geometric problems that define vertex placement or selection can have tens of thousands of vertices and, of course, triple the number in coordinate values. Animation problems might have a relatively small set of parameters, such as velocity or torque, but require the values for these parameters over thousands of frames of animation. The resulting high-dimensional parameter space can be too complex for an animator to easily set the parameters and produce the desired motion. The other common feature is that the function that evaluates the parameters to find the "best" values is not an analytic function; rather, it is a rather computationally expensive procedure. Thus, techniques that require fewer function evaluations are preferred.

Optimization is finding the optimal, or best, value among a set of alternatives. For the current discussion, the minimum function value is sought over the space spanned by the function's parameters. For example, in Equation B.158, the minimum value is 1.

$$\min_{x \in \mathcal{R}^n} f(x) \tag{B.158}$$

Alternatively, optimization can also be expressed in terms of finding the values for a set of input parameters that minimize (or maximize) a function's single output value. For example, in Equation B.159, the input value that corresponds to the function's minimum is 0.

$$\arg\min_{x \in \mathcal{R}^n} f(x) \tag{B.159}$$

The number of parameters and their possible values define the search space. The input parameters may be *unconstrained*, as in "x in reals," or the optimization problem may be *constrainted*, as in "x greater than 0." The constraints can include equality as well as inequality constraints (Eq. B.160).

$$\begin{aligned} &\text{minimize } f(X) \\ &\text{Subject to } c_i(X) = 0 \quad i = 1, 2, \ldots, k_{eq} \\ &\text{Subject to } \hat{c}_j(X) = 0 \quad j = 1, 2, \ldots, k_{le} \end{aligned} \tag{B.160}$$

The function, f, can be anything that is able to be evaluated given values for its input parameters. As mentioned, in computer graphics applications, the function is often not analytically defined because the function is actually an arbitrarily complex procedure that produces a value such as the time it takes to get somewhere, or a motion that minimizes the maximum torque during some task.

This discussion of optimization will often consider minimizing a function, although it obviously does not matter whether we are trying to minimize or maximize something—a function to be maximized uses the same principles as a function to be minimized, or, if desired, a function to be maximized can be negated and the resulting function considered for minimization. Numerical optimization is often used in situations in which the function is not analytic and an exhaustive search over the parameter space is not feasible because the function is very expensive to evaluate, or because the parameter space is very large, or both.

Optimization techniques can be characterized by the type of function, the type of input parameters, the number and type of constraints, the type of the output value(s), and so forth. The function can be analytic, comprised of mathematical operations that can be evaluated given the input values. The function might, or might not, be continuous and differentiable. The input parameters and/or function value might be reals or could be restricted to integer values (*integer programming*) or might be a mixture of integers and reals (*mixed programming*). The function can be relatively simple to compute or it can be computationally expensive to evaluate. The derivative might have similar variations. There might be constraints on the values that the input parameters can have (*constrained optimization*), and the function might have multiple output values. Because optimization can be applied in a variety of situations under a variety of assumptions, strategies come in many forms. As a result, optimization does lend itself to a nice single taxonomy.

B.9.1 Analytic Solution

Many approaches to optimization are merely extensions of basic calculus taught in high school physics or calculus class. A basic analytic approach usually involves finding the optimal input x^* that minimizes (or maximizes) a function $f(x)$ by computing a solution to $f'(x^*) = 0$ where the second derivative at x indicates a minimum ($f''(x^*) > 0$) or a maximum ($f''(x^*) < 0$). In this case, optimization needs to find the roots of the first derivative.

If $f''(x) > 0$ for all x, then the function is considered *convex*. Convex means that the function's slope is always increasing. This is important because it means that, assuming $f'(x)$ is negative at negative infinity, it will increase until it reaches zero and that, once $f'(x) = 0$, it will continue to increase as x increases; that is, it will never go back down to zero. This means that $f(x^*)$ is a minimum ($f'(x^*) = 0$ and $f''(x^*) > 0$) and there is no other minimum because $f'(x)$ will never get back down to zero; that is, x^* is a *global* minimum of the function. Similarly, if $f''(x) < 0$ for all x, then the function is *concave* and x^* is a global maximum. It should be pointed out that if the derivative function is not easily available or computable, then approximations to the derivative by finite differencing can be used with a corresponding degree of inaccuracy. Functions that are not convex do not necessarily have a global minimum. In such cases, the initial guess is important as this will dictate which, if any, minimum is found.

B.9.2 Numerical methods

At the risk of oversimplifying the area, there are two basical approaches to numerical optimization. The first simply randomly samples values of the function and adaptively samples more often in areas of the parameter space where lower values are found. This approach is usually used for functions in a high dimensional space where there are many local optima, so the random sampling hopes to uncover the global optimum or at least an acceptable near-global optimum. This is guaranteed to succeed only in the limit (i.e., taking an infinite number of samples). The second approach to numerical optimization uses knowledge about local trends of the function value to search in a particular direction. The direction is based on derivatives of the function or finite difference approximations to the derivative. The derivative indicates which direction to follow in order to reduce (or increase) the function value. The local minimum (maximum) is found when the derivative is equal to zero and the second derivative is positive (negative).

Gradient information can be used to direct the search of parameter space to drive the function to a minimum or drive the derivative to zero. In cases where $f'(x) = 0$ cannot be solved analytically, numerical techniques can be used to find its zero values. The Taylor series expansion of a function $g(x + dx)$ is shown in Equation B.161 using the big O notation as a function of x^2 to represent the error term when using the first two terms as an approximation to $g(x + dx)$.

$$g(x + dx) = g(x) + g'(x)dx + O(dx^2) \tag{B.161}$$

If $x0$ is an initial guess, $x1 = x0 - a*g'(x0)$ can be computed if $g'(x)$ can be computed or is estimated. With the objective of driving the value of $g()$ to zero with successive guesses, the Taylor approximation (without the error term) is set to zero and rearranged to get Equation B.162. Thus, a succession of guesses are generated using Equation B.163.

$$0 = g(x) + g'(x)dx$$
$$dx = x_1 - x_0 = -\frac{g(x)}{g'(x)} \tag{B.162}$$

$$x_n = x_{n-1} - \frac{g(x)}{g'(x)} \tag{B.163}$$

In the current case, $f'(x) = g(x)$. If x is multidimensional, as it usually is in computer graphics applications, then $g(x)$ is the gradient of $f(x)$ (Eq. B.164) and $-g'(x)$ is the inverse of the Hessian (Eq. B.165), but the same strategy holds and becomes Equation B.166.

$$\nabla f(\mathbf{x}) = \left[\frac{\partial f}{\partial x_1}(\mathbf{x}), \frac{\partial f}{\partial x_2}(\mathbf{x}), \dots, \frac{\partial f}{\partial x_n}(\mathbf{x}) \right] \tag{B.164}$$

$$H(\mathbf{x}) = \left[\frac{\partial^2 f}{\partial x_i \partial x_j}(\mathbf{x}) \right] \tag{B.165}$$

$$x_n = x_{n-1} - [Hg(\mathbf{x}_{n-1})]^{-1} \nabla g(\mathbf{x}_{n-1}) \tag{B.166}$$

The negative of gradient is used because the gradient points in the direction of the greatest increase. This is called the *gradient descent method*.

It has been shown that always taking steps in the direction of the gradient can lead to situations that converge very slowly. Therefore, taking steps in directions orthogonal to previous steps helps the convergence. This is called the *conjugate gradient method*.

An alternative way to approach the minimization problem is to fit a quadratic approximation to the function f at x, and then solve the quadratic equation for its minimum point and use that in the next iteration. The Taylor series expansion of G is shown in Equation B.167.

$$f(x + dx) = f(x) + f'(x)(dx) + f''(x)\frac{(dx)^2}{2} + O((dx)^3) \tag{B.167}$$

The quadratic approximation can be easily solved for the minimum value. This is used as the next guess for the minimum and the procedure iterates. In this way, steps are taken toward the minimum of the function using the first and second derivative of the objective function.

So far, constraints on the parameter space have not been considered, and this is called unconstrained optimization. Often, there are constraints on valid input values, such as greater than zero.

Hard and soft constraints, equality constraints, and inequality constraints

The simplest constrained optimization is when the function is linear and the constrains are linear, which is referred to as linear programming. Think of a panel that is tilted representing the linear objective function, and a frame consisting of $2 \times 4s$, the constraints, bounding a region of the board. A well-known method of solving a linear programming problem is the simplex method. Essentially, this method says that the constrained minimum must occur at a corner point of the frame. The simplex method progresses from corner to corner, always going in a direction that reduces the objective function.

Nonlinear, constrained optimization is where things get interesting—and complicated. One strategy is to take a step in parameter space to reduce the objective function regardless of the constraints, followed by a step to restore constraint satisfaction.

A different common method for considering constraints is to incorporate constraints in the objective function. This adds a dimension to the parameter space. Assume $f(x)$ is the objective function and $g(x) - c = 0$ is a constraint. The new objective function is shown in Equation B.168, where λ is an unknown constant called the *Lagrange multiplier*. Differentiating produces Equation B.169. To deal with more complex constraint situations, *sequential quadratic programming* can be used.

$$\text{minimize} \, f(X) + \lambda(c(X) - d) \tag{B.168}$$

$$\frac{\partial F(X)}{\partial X} = \nabla f(X) + \lambda \nabla c(X) = 0$$

$$\frac{\partial F(X)}{\partial \lambda} = c(X) - d = 0 \tag{B.169}$$

If evaluation of the derivative of the function is expensive or unavailable, or if the parameter space is expansive with many local minima, the parameter space can be randomly sampled.

Completely random sampling is blended with increased sampling in areas where low values are found. Early sampling favors randomness while later sampling favors sampling in low value areas. Maintaining some randomness ensures that the global minimum will be found in the limit.

This is the basis for the approaches such as *genetic algorithms* and *genetic programming*. They are generally simple to program, but with finite resources do not guarantee convergence on the optimum. They are very useful when finding "good" values are sufficient.

B.10 Standards for moving pictures

When producing computer animation, one must decide what format to use for storing the sequence of images. Years ago the images were captured on film by taking pictures of refresh vector screens or by plotting the images directly onto the film. This was a long and expensive process requiring single-frame film cameras and the developing of nonreusable film stock. The advent of frame buffers and video encoders made recording on videotape a convenient alternative. With today's cheap disks, memory, and CPU cycles, all-digital desktop video production is a reality. While the entertainment industry is still based on film, most of the rest of computer animation is produced using digital images intended to be displayed on a raster-scan device in such forms as broadcast video, DVD video, animated Web banners, and streaming video. This section is intended to give the reader some idea of the various standards used for recording moving pictures. Standards related to the digital image are emphasized.

B.10.1 In the beginning, there was analog

Before the rise of digital video, the two common formats for moving pictures were film and (analog) video. Over the years there have been almost a hundred film formats. The formats differ in the size of the frame, in the placement, size, and number of perforations, and in the placement and type of audio tracks [15]. Silent film is played at 16 frames per second (fps). Some sound film is played at 18 fps, but 24 fps is more common. Film played at 24 fps is typically doubly projected; that is, each frame is displayed twice to reduce the effects of flicker.

Table B.5 Film Formats

Film Width	Notes
8 mm	An old format, introduced in 1932, 8 mm is used for inexpensive home movies. Cameras for regular 8 mm are no longer manufactured. Regular 8 mm uses 16 mm stock, which is recorded on both sides after flipping the film in the camera. This allows the 16 mm film stock to be split down the middle to produce two 8 mm reels. The frame is 0.192 × 0.145″. Super-8 was introduced in 1965 as an improvement over regular 8 mm. The perforations (the holes in the film stock used to advance and register the film) were made smaller and the frame size was increased to 0.224 × 0.163″. The film was placed into cassettes instead of on the reels of regular 8 mm film.
16 mm	16 mm is used for television and low-budget theatrical productions. It was introduced in 1923 and has a frame size of 0.404 × 0.295″.
35 mm	35 mm has been a standard film size since the turn of the twentieth century [19]. It first became popular because it could be derived from the original 70 mm film made by Kodak. It is the standard for theatrical work as well as television. The *standard academy frame*, the most popular of several 35 mm formats, is 0.864 × 0.630″.
65 mm	65 mm was the standard format for large-format cinematography. It is now gaining in popularity in special-venue and "ride" films.
70 mm	70 mm film is often a blow-up print of 35 mm film, produced for improved audio, better registration, and less grain of the release print. With better sound technology (e.g., digital) and the advent of multiplex theaters with smaller screen sizes, there is less demand for this type of 70 mm film. However, IMAX uses 70 mm film (69.6 × 48.5 mm) that is run at 24 fps.

Some of the film sizes (widths) are listed in Table B.5. Note that there are often several formats for each film size. Only the most popular film formats are listed here. See the Web pages of Norwood [15] and Rogge [19] for more information. With the rise of desktop video production, film is less of an issue for home-brew computer animation, although it remains useful for conventional animation and, of course, is still the standard medium for display of feature-length films in theaters, although even this is starting to change.

Broadcast video standard

In 1941, the National Television Standards Committee (NTSC) established 525-line, 60.00 Hz field rate, 2:1 interlaced monochrome television in the United States. In 1953, 525-line, 59.94 Hz field rate, 2:1 interlaced, composite color television signals were established as a standard. The image is displayed top to bottom, with each scanline displayed left to right. *2:1 interlaced* refers to the scanning pattern, with the information on the odd scanlines followed by the information on the even scanlines. Each set of scanlines is referred to as a *field*; there are two fields per *frame*. This standard is typically referred to by the initials of the committee—NTSC. Broadcast video in the United States must correspond to this standard. The standard sets a specific duration for a horizontal scanline, a frame time, the amplitude and duration of the various sync pulses, and so on. Home video recording units typically generate much sloppier signals and would not qualify for broadcast. There are encoders that can strip old sync signals off a video signal and re-encode it so that it conforms to broadcast quality standards.

There are a total of 525 scanline times per frame time in the NTSC format. The number of frames transmitted per second is 29.97. There is a 2:1 interlace of the scanlines in alternate fields. Of the 525 total scanline times, approximately 480 contain picture information. The remainder of the scanline times are occupied by the overhead involved in the scanning pattern: the time it takes the beam to

Table B.6 Video Format Comparison [10]

Standard	Lines	Scan Pattern	Field Rate (Hz)	Aspect Ratio
NTSC	525	2:1 interlaced	59.94	4:3
PAL	625	2:1 interlaced	50	4:3
SECAM	625	2:1 interlaced	50	4:3

go from the end of one scanline to the beginning of the next and the time it takes for the beam to go from the bottom of the image to the top. The aspect ratio of a 525-line television picture is 4:3, so equal vertical and horizontal resolutions are obtained at a horizontal resolution of 480 times 4/3, or 640 pixels per scanline. PAL and SECAM are the other two standards in use around the world (Table B.6). They differ from NTSC in specifics like the number of scanlines per frame, the field rate, and the frequency of the color subcarrier, but both are interlaced raster formats. One of the reasons that television technology uses interlaced scanning is that, when a camera is providing the image, the motion is updated every field, thus producing smoother motion.

Black-and-white signal
A black-and-white video signal is basically a single line that has the sync information and intensity signal (luminance) superimposed on one signal. The vertical and horizontal sync pulses are negative with respect to a reference level, with vertical sync a much longer pulse than horizontal sync. On either side of the sync pulses are reference levels called the front porch and the back porch. The active scanline interval is the period between horizontal sync pulses. During the active scanline interval, the intensity of the signal controls the intensity of the electron beam of the monitor as it scans out the image.

Incorporating color into the black-and-white signal
When color came on the scene in broadcast television, engineers were faced with incorporating the color information in such a way that black-and-white television sets could still display a color signal and color sets could still display black-and-white signals. The solution is to encode color into a high-frequency component that is superimposed on the intensity signal of the black-and-white video.

A reference signal for the color component, called the *color burst*, is added to the back porch of each horizontal sync pulse, with a frequency of 3.58 MHz. The color is encoded as an amplitude and phase shift with respect to this reference signal. A signal that has separate lines for the color signals is referred to as a *component* signal. A signal such as the color TV signal with all of the information superimposed on one line is referred to as a *composite* signal.

Because of the limited room for information in the color signal of the composite signal, the TV engineers optimized the color information for a particular hue they considered most important: Caucasian skin tone. Because of that, the RGB information has to be converted into a different color space: YUV. Refer to Figure B.59. Y is luminance and is essentially the intensity information found in the black-and-white signal. It is computed by Equation B.170.

$$Y = 0.299R + 0.587G + 0.114B \tag{B.170}$$

The U and V of television are scaled and filtered versions of $B-Y$ and $R-Y$, respectively. U and V are used to modulate the amplitude and phase shift of the 3.58 MHz color frequency reference signal. The phase of this chroma signal, C, conveys a quantity related to hue, and its amplitude conveys a quantity related to color saturation. In fact, the I and Q stand for "in phase" and "quadrature," respectively. The

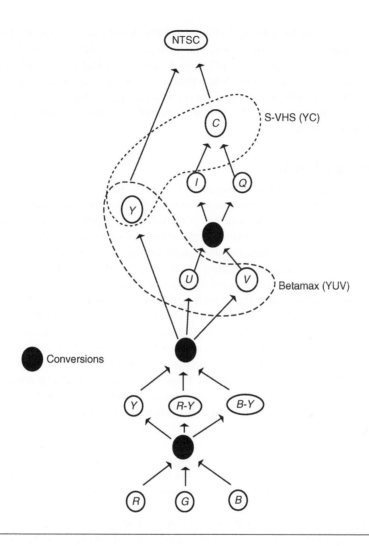

FIGURE B.59

Video signal.

NTSC system mixes Y and C together and conveys the result on one piece of wire. The result of this addition operation is not theoretically reversible; the process of separating luminance and color often confuses one for the other (e.g., the appearance of color patterns seen on TV shots of people wearing black-and-white seersucker suits).

Videotape formats

The size and speed of the tape and the encoding format contribute to the quality that can be supported by a particular video format. Common tape sizes are 1/2, 3/4, 1, and 2″. The first two sizes are of cassette tapes, and the other two are of open reel tapes. One-half inch supports consumer-grade formats (e.g., VHS, S-VHS, and the less popular Betamax), 3/4″ is industrial strength (U-Matic), and 1 and 2″ are professional broadcast-quality formats. One inch is the newer technology and has replaced some 2″ systems.

The common 1/2″ video formats are VHS and S-VHS. S-VHS is a format in which the *Y* and *C* signals are kept separate when played back, thus avoiding the problems created when the signals are superimposed. All video equipment actually records signals this way, but S-VHS allows the *Y* signal (luminance) to be recorded at a higher resolution than color. The color information is recorded with the same fidelity as on VHS. In addition, the sound is encoded differently than on regular VHS, also resulting in greater fidelity. The advantages of S-VHS are especially pronounced when it is played back on an S-VHS-compatible television.

B.10.2 In the digital world

Full-color (24 bits per pixel), video resolution (640 × 480), video rate (30 fps) animation requires approximately 1.6 gigabytes of information per minute of animation when stored as uncompressed RGB images. The problem is how to store and play back the animation. Various trade-offs must be considered when choosing a format, including the amount of compression, the time it takes for the compression/decompression, and the color and spatial resolution supported. The objective here is to provide an overview of the terminology, the important issues, and the most popular standards in recording animation. The discussion is at the level of consumer-grade technology and is not intended for professionals involved in the production of high-quality digital material.

Digital video sometimes refers specifically to digital versions of video to be broadcast for reception by television sets. In some literature this is also referred to as *digital TV* (DTV), which is the term used here. On the other hand, *digital video* (DV) is also used in the sense of computer-generated moving images intended to be played back on an RGB computer monitor. DV is used here to specifically denote material intended to be displayed by a computer. These two categories of digital representations, DTV and DV, have much in common. One of the most important common issues is the use of compression/decompression (codec) technology described in the following paragraphs.

Standards related to DTV have two additional concerns. First, the standards are concerned with images that will be encoded for broadcast. As a result, they rarely deal directly with RGB images. When destined for television, digital images are at least partially encoded in a format related to the broadcast video standard (NTSC) soon after they leave the camera (e.g., YUV). DTV has the advantage of encoding a better image (480p) in less bandwidth than its analog counterpart. As a result, DTV can broadcast additional information in the bandwidth given a channel.

Digital images synthesized for playback on a computer are typically generated as RGB images, compressed for storage and transmission, and then decompressed for playback on RGB monitors. Second, DTV standards are concerned with an associated recording format of the images and audio on tape; tape is the common storage medium for broadcast video studios. Because images for computer playback are typically stored digitally on a hard disk or removable disk, DV standards do not cover tape formats. However, there are DV standards for file formats and for digital movies. Digital movie formats organize the image and audio data into tracks with associated timing information.

Compression/decompression

The recent explosion in multimedia applications, especially as a result of the Web, has led to the development of a variety of compression/decompression schemes [25]. After the frames of an animation are computed, they are compressed and stored on a hard disk or CD-ROM. When playback of the animation is requested, the compressed animation is sent to the compute box, using disk I/O or over the Web, where the frames are decompressed and displayed. The decompression and playback can take place in real time as the data is transmitted. In this case, it is referred to as *streaming video*. If the decompression is not fast enough to support streaming video, the compressed animation is transmitted in its entirety to the compute box, decompressed into the complete animation, and then played back. In either case, the compression not only saves space on the storage device but also allows animation to be transferred to the computer much faster. Several of the codecs are proprietary and are used in workstation-based video products. The different schemes have various strengths and weaknesses and thus involve trade-offs, the most important of which are compression level, quality of video, and compression/decompression speed [21].

The amount of compression is usually traded off for image quality. With some codecs, the amount of compression can be set by a user-supplied parameter so that, with a trial-and-error process, the best compression level for the particular images at hand can be selected. Codecs with greater compression levels usually incorporate interframe compression as well as intraframe compression. *Intraframe compression* means that each frame is compressed individually. *Interframe compression* refers to the temporal compression possible when one processes a series of similar still images using techniques such as image differencing. However, when one edits a sequence, interframe compression means that more frames must be decompressed and recompressed to perform the edits.

The quality of the video after decompression is, of course, a big concern. The most fundamental feature of a compression scheme is whether it is *lossless* or *lossy*. With lossless compression, in which the final image is identical to the original, only nominal amounts of compression can be realized, usually in the range of 2:1. The codecs commonly used for video are lossy in order to attain the 500:1 compression levels necessary to realize the transmission speeds for pumping animations over the Web or from a CD-ROM. To get these levels of compression, the quality of the images must be compromised. Various compression schemes might do better than others with such image features as edges, large monochrome areas, or complex outdoor scenes. The amount of color resolution supported by the compression is also a concern. Some, such as animated GIF format, support only 8-bit color.

The compression and decompression speed is a concern for obvious reasons, but decompression speed is especially important in some applications, for example, streaming video. On the other hand, real-time compression is useful for applications that store compressed images as they are captured. If compression and decompression take the same amount of time, the codec is said to be *symmetric*. In some applications such as streaming video, it is acceptable to take a relatively long time for compression as long as the resulting decompression is very fast. In the case of unequal times for compression and decompression, the codec is said to be *asymmetric*. To attain acceptable decompression speeds on typical compute boxes, some codecs require hardware support.

A variety of compression techniques form the basis of the codec products. *Run-length encoding* is one of the oldest and most primitive schemes that have been applied to computer-generated images. Whenever a value repeats itself in the input stream, the string of repeating values is replaced by a single occurrence of the value along with a count of how many times it occurred. This was sufficient for early graphic images, which were simple and contained large areas of uniform color. This technique does not perform well with today's complex imagery. The Lempel-Ziv-Welch (LZW) technique was developed for compressing text. As the input is read in, a dictionary of strings and associated codes is generated and then used as the rest of the input is read in. *Vector quantization* simply refers to any scheme that uses a sample value to approximate a range of values. *YUV-9* is a technique in which the color is undersampled, so that, for example, a single color value is recorded for each 4×4 block of pixels. Discrete cosine transform (DCT) is a very popular technique that breaks a signal into a sum of cosine functions of various frequencies at specific amplitudes. The signal can be compressed by throwing away low-amplitude and/ or high-frequency components. DCT is an example of the more general *wavelet compression* in which the form of the base component function can be selected according to its properties. *Fractal compression* is based on the fact that many signals are self-similar under scale. A section of the signal can be composed of transformed copies of other sections of the signal. This is applied to rectangular blocks of the image. In fact, most of the compression schemes break the image into blocks and then compress the blocks.

Codecs employ one or more of the techniques mentioned earlier. The names of some of the more popular codecs are Video I, RLE, GIF, Motion JPEG, MPEG, Cinepak, Sorenson, and Indeo 3.2. Microsoft products are Video I and RLE. Video I employs DCT compression, and RLE uses run-length encoding. GIF is an 8-bit color image compression scheme based on the LZW compression. JPEG uses DCT compression.

Motion JPEG (MJPEG) is simply the application of JPEG for still images applied to a series of images. The compression/decompression is symmetric and is done in 1/30 of a second. JPEG compression can introduce some artifacts into some images with hard edges. The Moving Pictures Expert Group (MPEG) standards were designed specifically for digital video. MPEG uses the same algorithms as JPEG for intraframe compression of individual frames, called I-frames. MPEG then uses interframe compression to create B (bidirectional) and P (predicted) frames. MPEG allows quality settings to specify the amount of compression to use [24] and can be set individually for each frame type (I, P, B). MPEG-1 is designed for high-quality CD video, MPEG-2 is designed for high-quality DVD video, and MPEG-4 is designed for low-bandwidth Web applications [22].

Cinepak and Sorenson are products initially targeted for the MAC world, although Cinepak is now available for the PC. Cinepak uses block-oriented vector quantization. Sorenson uses YUV compression with 4×4 blocks and employs interframe compression [20]. Indeo 3.2 is an Intel product that also uses block-oriented vector quantization. Cinepak and Indeo are highly asymmetric, requiring on the order of 300 times longer for compression than for their efficient decompression. Indeo also incorporates color blending and run-length encoding into its scheme.

Digital video formats

The codec products (as opposed to the underlying codec techniques) used for DV have an associated file format. Some formats also include timing information, the ability to animate overlay images (sprites), and the ability to loop over a series of frames. MPEG and MJPEG are both common DV formats. GIF89a-based animation (animated GIF) is basically a number of GIF images stored in one file with interframe timing information but no interframe compression. Compressing continuous-tone images can result in color banding. The compression does well on line drawings but not on complex outdoor scenes. GIF animations can use delta frames, which, for example, overlay images on a previously transmitted background. This saves retransmitting static information for some animations. As used here, *movie format* refers to a format that is codec independent and that can handle audio as well as imagery. Both Quicktime (MOV) and Video for Windows (AVI) [2] are movie formats designed as open codec architectures. Any codec can be used with these standards as long as a compatible plug-in is available. Cinepak has been a standard codec used with Quicktime, but Quicktime can also accommodate other codecs such as Sorenson and JPEG [24]. Quicktime organizes data into tracks and includes timing information. Video for Windows from Microsoft uses Video I and RLE as standard codecs but can also accommodate others. Video for Windows allows interleaving of image and audio information.

Digital television formats

Most of the DTV formats are based on sampling scaled versions of the color difference signals (*B-Y*, *R-Y*). Luminance and the scaled color difference signals are referred to as YUV. Common formats are D1, D2, D3, D5, D6, Digital Betacam, Ampex DCT, Digital8, DV, DVCam, and DVCPRO. When digitally sampled, the YUV signals are referred to as YCrCb. A typical sampling scheme is 4:2:2, in which the color difference signals (Cr, Cb) are sampled at half the sampling rate of luminance (*Y*) in the horizontal direction. 4:1:1 sampling means that the color difference signals are also sampled at half the rate in the vertical direction, resulting in one-quarter of the luminance sampling.

The D1 standard was developed when the broadcast television industry thought it would make the composite-analog-to-component-digital transition in one fell swoop. But that did not happen because the cost was prohibitive. D1 uses YUV coding, so-called 4:2:2, which means that the *U* and *V* components are horizontally subsampled 2:1. Luminance is sampled at 13.5 MHz, resulting in 720 samples per picture width. There is no compression other than undersampling the chroma information. Aggregate data rate is roughly 27 megabytes per second (MB/s). The D2 standard was developed as a low-cost alternative to D1. D2 is a composite NTSC digital format (i.e., digitized NTSC). The composite signal is sampled at four-times-color-subcarrier, about 14.318 MHz at one byte per sample (aggregate data rate, of course, 14.318 MB/s). It has all the impairment of NTSC but the reliability and performance of digital. It uses the same 3/4″ cassette as D1. As with D1, D2 uses no compression other than undersampling the chroma information. Other uncompressed digital formats include D3, D5, and D6. D3 is a composite format that uses 1/2″ tape. D5 is a component format that uses 1/2″ tape. D6 is a component HDTV format.

Common compressed DTV formats include Digital Betacam, Ampex DCT, and Digital8 [9]. Digital Betacam uses 1/2″ tapes similar to the Betacam SP format with 2:1 compression based on DCT. Ampex DCT is a proprietary format; the *DCT* in its name stands for Digital Component Technology and not the compression scheme. The trio of DV, DVCam, and DVCPRO are similar formats using DCT compression. Depending on image content, the encoder decides whether to compress two fields separately or as a unit. Digital8 is a consumer-grade version of the DV format but uses cheaper Hi8 tapes. Newer formats include W-VHS, Digital S, Betacam SX, Sony HDD-1000, and D-VHS.

High-definition television and wide-screen format

In the United States the Grand Alliance was created by the FCC to develop the American HDTV specification. It is most commonly based on the MPEG-2 codec [8]. The basic idea behind high-definition television (HDTV) is to increase the percentage of the visual field occupied by the image [12]. HDTV is a type of DTV. Currently, there are two popular wide-screen HDTV formats: 720p and 1080i. The number refers to how many scanlines there are in an image. The "p" in 720p refers to progressive scan; the "i" in 1080i refers to interlaced scan. NTSC-format television has an aspect ratio (ratio of width to height) of 4:3, whereas the widescreen format more closely matches that found in movie theaters and is 16:9.

B.11 Camera calibration

For digitally capturing motion from a camera image, one must have a transformation from the image coordinate system to the global coordinate system. This requires knowledge of the camera's intrinsic parameters (focal length and aspect ratio) and extrinsic parameters (position and orientation in space). In the capture of an image, a point in global space is projected onto the camera's local coordinate system and then mapped to pixel coordinates. To establish the camera's parameters, one uses several points whose coordinates are known in the global coordinate system and whose corresponding pixel coordinates are also known. By setting up a system of equations that relates these coordinates through the camera's parameters, one can form a least-squares solution of the parameters [23].

Calibration is performed by imaging a set of points whose global coordinates are known and identifying the image coordinates of the points and the correspondence between the two sets of points. This results in a series of five-tuples, $(x_i, y_i, z_i, c_i, r_i)$ consisting of the three-dimensional global coordinates and two-dimensional image coordinates for each point. The two-dimensional image coordinates are a mapping of the local two-dimensional image plane of the camera located a distance f in front of the camera (Eq. B.171). The image plane is located relative to the three-dimensional local coordinate system (u, v, w) of the camera (Figure B.60). The imaging of a three-dimensional point is approximated using a pinhole camera model. The three-dimensional local coordinate system of the camera is related to the three-dimensional global coordinate system by a rotation and translation (Eq. B.172); the origin of the camera's local coordinate system is assumed to be at the focal point of the camera. The three-dimensional coordinates are related to the two-dimensional coordinates by the transformation to be determined. Equation B.173 expresses the relationship between a pixel's column and row number and the global coordinates of the point. These equations are rearranged and set equal to zero in Equation B.174. They can be put in the form of a system of linear equations (Eq. B.175) so that the unknowns are isolated (Eq. B.176) by using substitutions common in camera calibration (Eqs. B.176 and B.177). Temporarily dividing through by t_3 ensures that $t_3 \neq 0.0$ and therefore that the global origin is in front of the camera. This step results in Equation B.178, where A' is the first 11 columns of A; B' is the last column of A; and W' is the first 11 rows of W. Typically, enough points are captured to ensure an overdetermined system. Then a least-squares method, such as singular value decomposition, can be used to find the W' that satisfies Equation B.179. W' is related to W by Equation B.180, and the camera parameters can be recovered by undoing the substitutions made in Equation B.176 by Equation B.181. Because of numerical imprecision, the rotation matrix recovered may not be orthonormal, so it is best to reconstruct the rotation matrix first (Eq. B.182), massage it into orthonormality, and then use the new rotation matrix to generate the rest of the parameters (Eq. B.183).

$$
\begin{aligned}
c_i - c_0 &= s_u u_i \\
r - r_0 &= s_v v_i
\end{aligned}
\tag{B.171}
$$

$$
\begin{bmatrix} u_i \\ v_i \\ f \end{bmatrix} = R \begin{bmatrix} x_i \\ y_i \\ z_i \end{bmatrix} + T
$$

$$
R = \begin{bmatrix} R_0 \\ R_1 \\ R_2 \end{bmatrix} = \begin{bmatrix} r_{0,0} & r_{0,1} & r_{0,2} \\ r_{1,0} & r_{1,1} & r_{1,2} \\ r_{2,0} & r_{2,1} & r_{2,2} \end{bmatrix}
\tag{B.172}
$$

$$
T = \begin{bmatrix} t_0 \\ t_1 \\ t_2 \end{bmatrix}
$$

$$
\begin{aligned}
\frac{u_i}{f} &= \frac{c_i - c_0}{s_u f} = \frac{c_i - c_0}{f_u} = \frac{R_0 \cdot [x_i y_i z_i] + t_0}{R_2 \cdot [x_i y_i z_i] + t_2} \\
\frac{v_i}{f} &= \frac{r_i - r_0}{s_v f} = \frac{c_i - c_0}{f v} = \frac{R_1 \cdot [x_i y_i z_i] + t_1}{R_2 \cdot [x_i y_i z_i] + t_2}
\end{aligned}
\tag{B.173}
$$

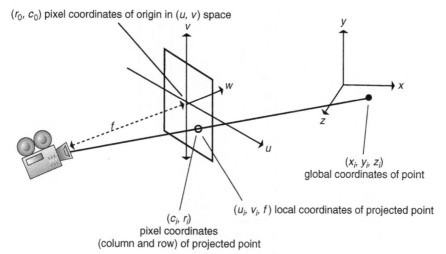

(r_0, c_0) pixel coordinates of origin in (u, v) space

y

v

w

x

f

z

u

(x_i, y_i, z_i)
global coordinates of point

(u_i, v_i, f) local coordinates of projected point

(c_i, r_i)
pixel coordinates
(column and row) of projected point

FIGURE B.60

Coordinate systems used in the projection of a global point to pixel coordinates.

$$(c_i - c_0)(R_2 \cdot [x_i y_i z_i] + t_2) - f_u(R_0 \cdot [x_i y_i z_i] + t_0) = 0$$
$$(r_i - r_0)(R_2 \cdot [x_i y_i z_i] + t_2) - f_v(R_1 \cdot [x_i y_i z_i] + t_1) = 0 \qquad (B.174)$$

$$AW = 0 \qquad (B.175)$$

$$\begin{bmatrix} -x_1 & -y_1 & -z_1 & 0 & 0 & 0 & r_1 x_1 & r_1 y_1 & r_1 z_1 & -1 & 0 & r_1 \\ 0 & 0 & 0 & -x_1 & -y_1 & -z_1 & c_1 x_1 & c_1 y_1 & c_1 z_1 & 0 & -1 & c_1 \\ \cdots & \cdots & \cdots & \cdots & \cdots & \cdots & \cdots & \cdots & \cdots & \cdots & \cdots & \cdots \\ -x_1 & -y_n & -z_n & 0 & 0 & 0 & r_n x_n & r_n y_n & r_n z_n & -1 & 0 & r_n \\ 0 & 0 & 0 & -x_n & -y_n & -z_n & c_n x_n & c_n y_n & c_n z_n & 0 & -1 & c_n \end{bmatrix}$$

$$\begin{aligned} W_0 &= f_u R_0 + c_0 R_2 = [w_0 w_1 w_2]^T \\ W_3 &= f_v R_1 + r_0 R_2 = [w_3 w_4 w_5]^T \\ W_6 &= R_2 = [w_6 w_7 w_8]^T \\ w_9 &= f_u t_0 + c_0 t_2 \\ w_{10} &= f_v f_1 + r_0 t_2 \\ w_{11} &= t_2 \end{aligned} \qquad (B.176)$$

$$W = [w_0\ w_1\ w_2\ w_3\ w_4\ w_5\ w_6\ w_7\ w_8\ w_9\ w_{10}\ w_{11}]^T \qquad (B.177)$$

$$A'W' + B' = 0 \qquad (B.178)$$

$$\min_w \|A'W' + B'\| \qquad (B.179)$$

$$W = \begin{bmatrix} W_0 \\ W_3 \\ W_6 \\ W_9 \\ W_{10} \\ W_{11} \end{bmatrix} = \frac{1}{\|W_6'\|} \times \begin{bmatrix} W_0' \\ W_3' \\ W_6' \\ W_9' \\ W_{10}' \\ 1 \end{bmatrix} \tag{B.180}$$

$$\begin{aligned}
c_0 &= W_0^T W_6 \\
r_0 &= W_1^T W_6 \\
f_u &= -\|W_0 - c_0 W_6\| \\
t_0 &= (w_9 - c_0)/f_u \\
t_1 &= (w_{10} - r_0)/f_v \\
t_2 &= w_{11} \\
R_0 &= (W_0 - c_0 W_6)/f_u \\
R_1 &= (W_3 - r_0 W_6)/f_v \\
R_2 &= W_6
\end{aligned} \tag{B.181}$$

$$\begin{aligned}
R_0 &= (W_0' - c_0 W_6')/f_u \\
R_1 &= (W_3' - r_0 W_6')/f_v \\
R_2 &= W_6'
\end{aligned} \tag{B.182}$$

$$\begin{aligned}
c_0 &= W_0^T R_2 \\
r_0 &= W_1^T R_2 \\
f_u &= -\|W_0 - c_0 R_2\| \\
f_v &= \|W_3 - r_0 R_2\| \\
t_0 &= (W_9 - c_0)/f_u \\
t_1 &= (W_{10} - r_0)/f_v \\
t_2 &= w_{11}
\end{aligned} \tag{B.183}$$

Given an approximation to a rotation matrix, $\tilde{R}$, the objective is to find the closest valid rotation matrix to the given matrix (Eq. B.184). This is of the form shown in Equation B.185, where the matrices C and D are notated as shown in Equation B.186. To solve this, define a matrix B as in Equation B.187. If $q = (q_0, q_1, q_2, q_3)^T$ is the eigenvector of B associated with the smallest eigenvalue, R is defined by Equation B.188 [26].

$$\min\|\tilde{R} - R\| \tag{B.184}$$

$$\|R \cdot C - D\| \tag{B.185}$$

$$\begin{aligned}
C &= [C_1, C_2, C_3] \\
D &= [D_1, D_2, D_3]
\end{aligned} \tag{B.186}$$

$$B = \sum_{i=1}^{3} B_i^T B_i$$

$$B_i = \begin{bmatrix} 0 & (C_i - D_i)^T \\ D_i - C_i & [D_i - C_i]_x \end{bmatrix}$$

$$[(x, y, z)]x^* = \begin{bmatrix} 0 & -z & y \\ -z & 0 & -x \\ -y & x & 0 \end{bmatrix}$$

(B.187)

$$R = \begin{bmatrix} q_0 - q_1 + q_2 - q_3 & 2(q_1 q_2 - q_0 q_3) & 2(q_1 q_3 + q_0 q_2) \\ 2(q_1 q_2 + q_0 q_3) & q_0 - q_1 + q_2 - q_3 & 2(q_3 q_2 - q_0 q_1) \\ 2(q_1 q_3 - q_0 q_2) & 2(q_3 q_1 + q_0 q_2) & q_0 - q_1 - q_2 + q_3 \end{bmatrix}$$

(B.188)

References

[1] Bartels R, Beatty J, Barsky B. An Introduction to Splines for Use in Computer Graphics and Geometric Modeling. San Francisco: Morgan-Kaufmann; 1987.

[2] Dixon D. AVI Formats, http://www.manifest-tech.com/pc_video/avi_formats/avi_formats. htm; January 2001.

[3] Ebert D, editor. Texturing and Modeling: A Procedural Approach. 2nd ed. Boston: AP Professional; 1998.

[4] Farin G. Curves and Surfaces for Computer Aided Design. Orlando, Fla.: Academic Press; 1990.

[5] Frautsch S, Olenick R, Apostol T, Goodstein D. The Mechanical Universe: Mechanics and Heat, the Advanced Edition. Cambridge: Cambridge University Press; 1986.

[6] Gasch S. 0.5.12.0 Source Code, http://wannabe.guru.org/alg/node136.html; January 2001.

[7] Gottschalk S, Lin M, Manocha D. OBB Tree: A Hierarchical Structure for Rapid Interference Detection. In: Rushmeier H, editor. Computer Graphics. Proceedings of SIGGRAPH 96, Annual Conference Series. New Orleans, La.: Addison-Wesley; August 1996. p. 171–80. ISBN 0-201-94800-1.

[8] Hromas D. Digital Television: The Television System of the Future . . ., http://www.cgg. cvut.cz/~xhromas/dtv/; January 2001.

[9] Iisakkila M. Video Recording Formats, http://www.hut.fi/u/iisakkil/videoformats.html; January 2001.

[10] King B. TV Systems: A Comparison, http://www.ee.surrey.ac.uk/Contrib/WorldTV/-compare.html; January 2001.

[11] Kochanek D. Interpolating Splines with Local Tension, Continuity and Bias Control. In: Computer Graphics. Proceedings of SIGGRAPH 84, vol. 18(3). Minneapolis, Minn.; July 1984. p. 33–41.

[12] Kuhn K. HDTV Television—an Introduction: EE 498, http://www.ee.washington.edu/conselec/CE/kuhn/hdtv/95x5.htm; January 2001.

[13] Madsen N. Three Dimensional Dynamics of Rigid Bodies, http://www.eng.auburn.edu/users/gflowers/me232/class_problems/angmom.html; January 2001.

[14] Mortenson M. Geometric Modeling. New York: John Wiley & Sons; 1985.

[15] Norwood S. rec.arts.movies.tech: Frequently Asked Questions (FAQ) Version 2.00, http://www.redballoon. net/~snorwood/faq2.html; January 2001.

[16] Press W, Flannery B, Teukolsky S, Vetterling W. Numerical Recipes: The Art of Scientific Computing. Cambridge: Cambridge University Press; 1986.

[17] Ritter J. An Efficient Bounding Sphere. In: Glassner A, editor. Graphics Gems. New York: Academic Press; 1990. p. 301–3.

[18] Rogers D, Adams J. Mathematical Elements for Computer Graphics. New York: McGraw-Hill; 1976.

[19] Rogge M. More Than One Hundred Years of Film Sizes, http://www.xs4all.nl/~wichm/filmsize.html; January 2001.

[20] Sorenson Media. Sorenson Video Tutorial: Sorenson Video Compression, http://www. sorenson.com/Sorenson-Video/tutorial/page03.html; January 2001.

[21] Speights II L. Video Compression Methods (Codecs), http://home.earthlink.net/~radse/Page9.html; January 2001.

[22] Terrain Interactive, Index of Codecs for Video, CD, DVD, and Audio, http://www. terran.com/CodecCentral/Codecs/index.html; January 2001.

[23] Tuceryan M, Greer G, Whitaker R, Breen D, Crampton C, Rose E, et al. Calibration Requirements and Procedures for a Monitor-Based Augmented Reality System. IEEE Trans Vis Comput Graph September 1995;1(3):255–73.

[24] Vahlenkamp H. GFDL Visualization Guide: Animation Formats, http://www.manifest-tech.com/pc_video/avi_formats/avi_formats.htm; January 2001.

[25] Waggoner B. Technology: A Web Video Guide to Codecs, http://www.dv.com/webvideo/ 2000/0900/waggoner0900.html; January 2001.

[26] Weng J, Cohen P, Herniou M. Camera Calibration with Distortion Models and Accuracy Evaluation. IEEE Trans Pattern Anal Mach Intell October 1992;14(10):965–80.

[27] Zordan V. Solving Computer Animation Problems with Numeric Optimization. GVU Technical Report GIT-GVU-02-13. Georgia Institute of Technology; 2002.

[28] Velho L, Carvalho P, Gomes J, de Figueiredo L. Mathematical Optimization in Computer Graphics and Vision. New York: Morgan Kaufmann; 2008.

[29] Bradley S, Hax A, Magnanti T. Applied mathematical Programming. New York: Addison Wesley; 1977.

Index

Note: Page numbers followed by *f* indicate figures and *t* indicate tables.

Printed and bound by CPI Group (UK) Ltd, Croydon, CR0 4YY

03/10/2024

01040326-0001